Lecture Notes in Mathematics

Edited by A. Dold and B. Eckmann

864

Complex Analysis and Spectral Theory

Seminar, Leningrad 1979/80

Edited by V. P. Havin and N. K. Nikol'skii

Springer-Verlag
Berlin Heidelberg New York 1981

Editors

Victor P. Havin
Nikolai K. Nikol'skii
Leningrad Branch, V.A. Steklov Institute of Mathematics
Academy of Sciences of the USSR
Fontanka 27, Leningrad, 191001, USSR

AMS Subject Classifications (1980): 30-XX, 47-XX

ISBN 3-540-10695-2 Springer-Verlag Berlin Heidelberg New York
ISBN 0-387-10695-2 Springer-Verlag New York Heidelberg Berlin

Printed in Germany

Printing and binding: Beltz Offsetdruck, Hemsbach/Bergstr.
2141/3140-543210

PREFACE

This book is a collection of works made by participants of the Seminar on Spectral Theory and Complex Analysis in 1979/80. This seminar consists mainly of mathematicians working in the Leningrad Branch of the Steklov Institute and in the Leningrad University and interested in problems arising in the Spectral Theory and in the Theory of Functions of a Complex Variable.

We are sure the interests of both directions essentially coincide and we hope this book corroborates our point of view. It may be considered as the third issue of selected works of the Seminar (the first and the second appeared in the series "Proceedings of the Steklov Institute of Mathematics", volumes 130 (1978), 155 (1980); their common title is "Spectral Theory of Functions and Operators"). Other works by members of the Seminar are systematically published in a special series "Investigations in Linear Operators and in Function Theory" edited by the Leningrad Branch of the Mathematical Institute (ten volumes of these "Investigations" have been published; they are partly translated into English by the publishing house "Plenum").

Only few of the articles below are to be definitely classified to represent exactly one of two directions mentioned above. E.g. the works by V.I.Vasyunin - N.G.Makarov and by S.V.Kisliakov represent the pure Operator Theory and those by S.A.Vinogradov - S.V.Hruščev and by E.M.Dyn'kin the classical Function Theory. As to the remaining articles they are a more or less regular blend of spectral and complex ideas, either their problems or their methods or their eventual results being easily included into the framework of both disciplines. A typical example is the treatise by A.B.Aleksandrov on the Hardy classes $H^p (0<p<1)$. It contains a new approach to the problem to characterize functions representable by a Cauchy potential but also a description of invariant subspaces of the shift operator and solutions of some spectral analysis-synthesis problems (not to mention many other things) all this giving an example of strong ties connecting the Spectral Theory with the Complex Analysis.

We are afraid readers have already noticed our English is far from being perfect. We beg everybody to be not too severe and hope our linguistic weaknesses at least won't prevent from the understanding of the mathematical contents of the book.

All results collected in this volume were reported to the Seminar in I979/I980.

We express our deep gratitude to L.N.Dovbysh and V.V.Peller for their assistance during the most fatiguing stage of the work at the text

V.Havin N.Nikol'skii

CONTENTS

Ленинградское Отделение Математического
Института им.В.А.Стеклова АН СССР

СЕМИНАР

ПО КОМПЛЕКСНОМУ И СПЕКТРАЛЬНОМУ АНАЛИЗУ

Редакторы

Н.К.НИКОЛЬСКИЙ В.П.ХАВИН

Ленинград, 1979/80

A.B.Aleksandrov

ESSAYS ON NON LOCALLY CONVEX HARDY CLASSES

Introduction

The principal objects of this article (continuing [14], [15], [16]) are the Hardy classes H^p $(0<p<1)$, their analogues and generalizations. The study of these classes is justified by the role they are playing now (after the paper of Fefferman and Stein) in analysis. The classical spaces $\Lambda^\alpha(\mathbb{R}^d)$ are dual to the classes $H^p(\mathbb{R}^d)$ considered as tempered distributions. Moreover, classes H^p emerge in connection with some problems of classical analysis whose statements don't even mention H^p. So for example using the H^p-techniques $(0<p<1)$ we obtain a very convenient new characterization of analytic functions representable in $\mathbb{C}\setminus\mathbb{T}$ ($\mathbb{T}$ being the unit circle) by the Cauchy potential $z\longmapsto\int\frac{d\mu(\zeta)}{1-\bar{\zeta}z}$ of a complex measure supported by $\mathbb{T}$ (Chapter 1). In Chapter 1 we complement the results of W.Rudin concerning Fourier coefficients of singular measures. Chapter 4 contains new results on partial sums of Fourier series of continuous functions (we describe sequences representable in the form $\{S_{n_k}(f)\}_{k\geq 0}$ where $\{n_k\}_{k\geq 0}$ is a very sparse family of integers, $S_m(f)\overset{def}{=}\sum_{|k|\leq m}\hat{f}(k)$, $\hat{f}(k)$ denoting the k-th Fourier coefficient of the continuous function f. We were led to these results by consideration of some spaces close to H^p, $p\in(0,1)$. The Appendix contains examples of application of the "H^p-ideology" to some generalizations of the F. and M. Riesz theorem on measures orthogonal to analytic functions (these generalizations were suggested by a paper of Joel H.Shapiro).

Now we are going to give a more detailed and systematic description of the article.

It consists of five almost independent chapters. In the first chapter all quasinormed symmetric spaces *) X of functions measurable on the unit circle $\mathbb{T}$ are described for which an analogue of the M.Riesz theorem on conjugate functions is valid: $\operatorname{Re} X=\operatorname{Re} X_A$, X_A being the correspondent space of analytic functions. This problem has been solved by D.W.Boyd for *normed* symmetric spaces in [21] (see also [6]). The proof shows that the result of Boyd remains valid for all quasinormed symmetric spaces $X\subset L^1=L^1(\mathbb{T})$.

---

*) Similar spaces of functions are called also rearrangement-invariant.

In the first chapter we consider the case of a quasinormed symmetric TVS X, $X \not\subset L^1$. Let $X_{\bar{A}}$ be the corresponding space of antianalytic functions (e.g. when $X = L^p$ $(0 < p < +\infty)$, $X_A = H^p$, $X_{\bar{A}} = H^p_- \overset{def}{=} \{f : \frac{1}{z} f(\frac{1}{z}) \in H^p\}$).

We discuss the following questions.

1) When is the linear span of the Cauchy kernels $\left\{\frac{1}{1-\bar{\zeta} z}\right\}$ $(\zeta \in \mathbb{T})$ dense in $X_A \cap X_{\bar{A}}$?

2) When $X_A \cap X_{\bar{A}} = \{0\}$?

3) When does $X_A \cap X_{\bar{A}}$ coincide with the set of all functions f representable by the Cauchy integral of a singular measure $\mu \in C^*(\mathbb{T})$: $f(z) = \int \frac{d\mu(\zeta)}{1-\bar{\zeta} z}$?

The first chapter contains the following criterion of the representability of an analytic function $f : \hat{\mathbb{C}} \setminus \mathbb{T} \to \mathbb{C}$ by the Cauchy potential of a measure supported by $\mathbb{T}$: it is representable by such a potential iff $f \mid \mathbb{D} \in H^p$, $\frac{1}{z} f(\frac{1}{z}) \mid \mathbb{D} \in H^p$ for all $p \in (0,1)$ ($\mathbb{D}$ being the open unit disc), $\| f \mid \mathbb{D} \|_{H^p} = O(\frac{1}{1-p})$ $(p \to 1-)$ and the function $\lim_{r \to 1-} (f(rz) - f(r^{-1} z))$ is summable on $\mathbb{T}$.

The problem of the characterization of functions representable by Cauchy integrals attracted attention in the past. At least two criteria are known: one due to V.P.Havin [37] and another to G.C. Tumarkin [57]. But there are situations when the above criterion is the most convenient one. We give two applications. This criterion admits multidimensional generalizations.

We prove in the first chapter that $\mathrm{Re} L^p(\mathbb{T}^d) = \mathrm{Re} H^p$ (here H^p denotes the Hardy class in the d-dimensional polydisc) and describe the closure in $L^p(\mathbb{T}^d)$ $(0 < p < 1)$ of the linear span of **the Cauchy Kernels** $\left\{\prod_{j=1}^d (1-\bar{\zeta}_j z_j)^{-1}\right\}_{(\zeta_j \in \mathbb{T})}$. We obtain also an analogue of the equality $L^p(\mathbb{T}^d) = H^p + \overline{H^p}$ $(0 < p < 1)$ for locally compact abelian groups and for abstract Hardy spaces. The first chapter contains also a strengthened variant of a result of W.Rudin concerning the so-called "modification sets".

In the second chapter we introduce a class of topological vector spaces (the locally holomorphic spaces). It contains all locally convex spaces and is adapted to the investigation of vectorvalued holomorphic functions.

We consider classes $L^p(X)$ and $H^p(X)$ of functions with values in a complete quasinormed space X and prove that $L^p(X) = H^p(X) + z^{-1} \overline{H^p(X)}$ $(p \in (0,1))$ and that the linear span of the X-valued Cauchy kernels is dense in $H^p(X)\cap z^{-1}\overline{H^p(X)}$ ($p\in(0,1)$).

Many natural properties of the scalar Hardy classes are destroyed by the passage to their vector analogues. So, for example, there is a non trivial holomorphic (in $\mathbb{D}$) L^p/H^p-valued function f $(0<p<1)$ continuous in the closed disc $\mathbb{D}$ and vanishing identically on $\mathbb{T}$.

We prove that holomorphic functions with values in a locally holomorphic space cannot have such pathologies; they have usual uniqueness properties and many other natural properties which however don't hold for general vector valued holomorphic functions.

Using the above example of the L^p/H^p-valued function we give a new proof of a result of N.J.Kalton [62] concerning the linear operators from L^p/H^p $(0<p<1)$ into the space $L^0(\mathbb{T})$ of all Lebesgue measurable functions.

Vectorvalued Hardy classes arise in a natural way in the investigations of scalar Hardy classes in polydiscs. So for instance the Hardy class H^p in a bidisc can be interpreted as the class $H^p(H^p)$ in $\mathbb{D}$ with values in the usual one-dimensional Hardy class H^p. This enables us to get another proof of some results of A.Frazier [63].

In the second chapter we construct one more example of a quasinormed space with a trivial dual space and with nontrivial compact operators. The first example of this kind (for an F-space) has been constructed by N.J.Kalton and J.H.Shapiro [40]. The second chapter contains an example of a rotation-invariant non finitely generated subspace of H^p $(0<p<1)$ in the bidisc.

In the third chapter we give the atomic decomposition for the space $\mathcal{H}^p_\alpha(\mathbb{R}^d)$ $(0<p\leq 1,\ \alpha>0)$ of all functions u harmonic in the upper half-space

$$\mathbb{R}^{d+1}_+ \overset{def}{=\!=} \{(x_0,x_1,\dots,x_d)\in\mathbb{R}^{d+1}: x_0>0\} \text{ with}$$

$$\int_{\mathbb{R}^{d+1}_+} |u(x_0,x_1,\dots,x_d)|^p x_0^{\alpha p-1}\,dx_0\,dx_1\dots dx_d<+\infty$$

(The case $p=1$ is not new). This space can be identified with a space of tempered distributions. The proof of our atomic theorem needs(in addition to the well-known R.Coifman - R.Latter theorem) the following result (which is, may be, of interest in itself): a distribution $f\in S'(\mathbb{R}^{d_1})$ $(d_1<d)$ belongs to $H^p(\mathbb{R}^d)$ (we mean the natural imbedding of $S'(\mathbb{R}^{d_1})$ into $S'(\mathbb{R}^d)$) iff $f\in \mathcal{H}^p_\alpha(\mathbb{R}^{d_1})$, $\alpha=(\frac{1}{p}-1)(d-d_1)$.

The fourth chapter is devoted essentially to the description of the space of all sequences $\{S_{2^n}(f)\}_{n\geq 0}$, $f\in C(\mathbb{T})$, $S_N(f)$ being the sum $\sum_{|k|\leq N}\hat{f}(k)$, $\hat{f}(k)$ the k-th Fourier coefficient of f. This description implies the following result:

$$\left\{\{S_{2^{2^n}}(f)\}_{n\geq 0}: f\in C(\mathbb{T})\right\}=\left\{\{a_n\}_{n\geq 0}: a_n=O(2^n)\ (n\to+\infty)\right\}.$$

The whole space $\left\{\{S_{2^n}(f)\}_{n\geq 0}: f\in C(\mathbb{T})\right\}$ is more complicated. It is isomorphic to the direct sum $\left(\sum_{n\geq 0}\oplus\, \ell^2_{2^n}\right)_{c_0}$ (of the type c_0) of finite-dimensional ℓ^2-spaces.

The above problem is intimately connected with the following question: what is the space $\left\{\{\hat{f}(2^k)\}_{k\geq 0}: f\in H^{1,\infty}\right\}$, where $H^{p,\infty}$ is the set of all $H^{p/2}$-functions f satisfying $m\{|f|>A\}=O(A^{-p})$ $(A\to+\infty)$ (m is the Lebesgue measure on $\mathbb{T}$). We consider also the spaces $\left\{\{\hat{f}(2^k)\}_{k\geq 0}: f\in H^{p,\infty}\right\}$, $\left\{\{\hat{f}(2^k)\}_{k\geq 0}: f\in H^p\right\}$ $(0<p<1)$ and the right-inverse of the operator mapping f into the sequence $\{\hat{f}(2^k)\}_{k\geq 0}$ (or into $\{S_{2^k}(f)\}_{k\geq 0}$).

The fifth (the last) chapter deals with functions $h\in H^\infty$ generating the Toeplitz operators $f\mapsto P(\bar{h}f)$ which act continuously in $H^p(0<p<1)$ (P denotes the Riesz projection). We consider also invariant subspaces of the backward shift operator $S^*: X_A\to X_A$ where X is a metric symmetric space. We obtain a complete description of these subspaces when $X=L^{1,\infty}_0 \stackrel{def}{=}$ $=\{f\in L^0: \lim_{t\to+\infty} t\,m\{|f|>t\}=0\}$.

Preliminary definitions and notations

A function $\|\ \|: X\mapsto[0,+\infty)$ defined on a vector space X will be called a quasinorm if

1) $\|x\|=0 \Rightarrow x=0$; 2) $\|\alpha x\|=|\alpha|\cdot\|x\|$ $(\alpha\in\mathbb{C}, x\in X)$;
3) there is a constant C such that $\|x+y\|\leq C(\|x\|+\|y\|)$ $(x,y\in X)$.

The collection of neighbourhoods of the origin $\{x\in X: \|x\|<\varepsilon\}$ $(\varepsilon>0)$ determines a linear Hausdorff locally bounded topology on X. Conversely, given any locally bounded

Hausdorff topological vector space (a TVS) (X,τ) there is a quasinorm on X generating τ.

A quasinorm is called a p-n o r m $(0<p\leqslant 1)$ if $\|x+y\|^p\leqslant \|x\|^p+\|y\|^p$ $(x,y\in X)$.

A space X endowed with such a quasinorm is said to be p-normed. The T.Aoki - S.Rolewicz theorem [8] asserts that every quasinormed space is isomorphic to a p-normed space $(0<p\leqslant 1)$.

Let $\mathcal{L}(X,Y)$ denote the set of all continuous linear operators mapping the TVS X into TVS Y. If X and Y are quasinormed, $\mathcal{L}(X,Y)$ has a natural quasinorm too: $\|A\|_{\mathcal{L}(X,Y)} \overset{def}{=\!=} \{\|Ax\|_Y : \|x\|\leqslant 1\}$. We shall write simply $\|A\|_X$ (instead of $\|A\|_{\mathcal{L}(X,Y)}$) when $X=Y$.

The space $\mathcal{L}(X,Y)$ (where X and Y are quasinormed TVS) is p-normed (complete) if Y is p-normed (resp.complete). In particular the dual space $X^* \overset{def}{=\!=} \mathcal{L}(X,\mathbb{C})$ is a Banach space for every quasinormed X.

A subset $\mathcal{U}$ of a vector space X is called p-c o n v e x if

$$\alpha,\beta\in\mathbb{C},\ |\alpha|^p+|\beta|^p\leqslant 1,\ x,y\in\mathcal{U}\Longrightarrow \alpha x+\beta y\in\mathcal{U}.$$

The intersection of all p-convex sets containing a set is called t h e p-c o n v e x **h u l l** o f the set.

Let $X\ (=(X,\tau))$ be a TVS . Consider the TVS $X_p=(X,\tau_p)$ whose base of neighbourhoods of the origin consists of all p-convex **hulls** of τ-neighbourhoods of the origin. The space X_p is not necessarily Hausdorff. Let $[X]_p$ denote the completion of the **corresponding** Hausdorff space (obtained from X_p after the suitable factorization). The space $[X]_p$ will be called the p-c o n v e x e n v e l o p e of X. It is easy to see that $[[X]_p]_q\simeq [X]_q$ whenever $0<p\leqslant q\leqslant 1$.

EXAMPLE 1. $[L^0(\mathbb{T})]_p\simeq\{0\}$ for every $p\in(0,1]$ (recall that $L^0(\mathbb{T})$ denotes the space of all Lebesgue measurable functions on $\mathbb{T}$).

Let X be a quasinormed space; the Minkowski functional of the p-convex **hull** of the unit ball $\{x\in X : \|x\|<1\}$ is defined by the formula

$$\|x\|_p \overset{def}{=\!=} \inf\Big\{\Big(\sum_{k=1}^{m}\|x_k\|^p\Big)^{\frac{1}{p}} : x=\sum_{k=1}^{m}x_k,\ m\in\mathbb{N},\ x_k\in X\Big\}\ (0<p\leqslant 1).$$

Therefore the p-convex envelope $[X]_p$ is automatically endowed with a p-norm.

Let j be the canonical map of a quasinormed space X into the Banach space X^{**}. Then $\|x\|_1 = \|j(x)\|_{X^{**}}$. Hence the 1-convex envelope of X is canonically isometric with the closure of $j(X)$ in X^{**}. The 1-convex envelope of a quasinormed space is usually called the Banach envelope (see [53]).

EXAMPLE 2. $[\ell^q]_p \simeq \ell^p \quad (0<q\leq p\leq 1)$.

EXAMPLE 3. $[L^q(\mathbb{T})]_p \simeq \{0\} \quad (0<q<p\leq 1)$.

P.L.Duren, B.V.Romberg and A.L.Shields [31] have found $[H^p]_1 \quad (p<1)$. In the Chapter III we shall describe $[H^p]_q$ for all values $q\in(p,1)$.

Now we enlist some notational conventions.

$\mathbb{Z}\,(\mathbb{Z}_+,\mathbb{Z}_-)$ will denote the set of all (nonnegative, negative) integers, $\mathbb{N}$ the set of all positive integers, $\mathbb{D}$ the open unit disc $\{\zeta\in\mathbb{C}:|\zeta|<1\}$, $\mathbb{D}_- \stackrel{\text{def}}{=} \{\zeta\in\hat{\mathbb{C}}:|\zeta|>1\}$, $\mathbb{T}\stackrel{\text{def}}{=}\{\zeta\in\mathbb{C}:|\zeta|=1\}$.

The symbol z will denote the identity map of a subset A of $\mathbb{C}$ or $\mathbb{C}^d$. In the last case $z_1,z_2,\dots,z_d$ will denote the coordinate functions of the identity map, i.e. $z=(z_1,z_2,\dots,z_d)$.

The d-dimensional Lebesgue measure will be denoted by m_d, $m\stackrel{\text{def}}{=}m_1$. We shall often write $dx_1\,dx_2\dots dx_d$ or dx under the integral sign (instead of dm_d).

Further, $|x|\stackrel{\text{def}}{=}\left(\sum_{k=1}^{d}|x_k|^2\right)^{1/2}$, $|x|_1\stackrel{\text{def}}{=}\sum_{k=1}^{d}|x_k|$ (the last symbol will be used for multiindices only, i.e. when $x\in\mathbb{Z}^d$).

We write $X\approx Y$ when X is isomorphic to Y, and $X\simeq Y$ when X is canonically isomorphic to Y.

The space of all distributions on $\mathbb{T}$ will be denoted by $\mathcal{D}'(\mathbb{T})$, $\mathcal{S}'(\mathbb{R}^d)$ will denote the space of all tempered distributions on $\mathbb{R}^d$.

The symbol $f|A$ will denote the restriction of the map f to the set A.

The set of all functions $f\in L^1(\mathbb{T})$ satisfying

$$\|f\|_{BMO}\stackrel{\text{def}}{=}\|f\|_{L^1}+\sup_I \frac{1}{m(I)}\int_I |f-f_I|\,dm<+\infty$$

$$\left(f_I\stackrel{\text{def}}{=}\frac{1}{m(I)}\int_I f\,dm\right)$$, the supremum being taken for all

subarcs $I \subset \mathbb{T}$, will be denoted by BMO; further, $BMO_A = H^1 \cap BMO = P(BMO)$ (P is the Riesz projection).

Let $f \in C(\mathbb{T})$, $\alpha > 0$, and suppose there is a number $C(f)$ such that for every bounded interval $I \subset \mathbb{R}$ there is a polynomial P_I of degree $\leq \alpha$ satisfying $\sup_{t \in I} |f(e^{it}) - P_I(e^{it})| \leq C(f)(m(I))^\alpha$. Then we shall write $f \in \Lambda^\alpha(\mathbb{T}) = \Lambda^\alpha$. Let $\|f\|_{\Lambda_0^\alpha}$ be the least of numbers $C(f)$. Put

$$\|f\|_{\Lambda^\alpha} \overset{def}{=} \|f\|_{\Lambda_0^\alpha} + \|f\|_{C(\mathbb{T})} .$$

The set $\Lambda^\alpha \cap C_A = P(\Lambda^\alpha)$ will be denoted by Λ_+^α (C_A is the disc-algebra).

Chapter One. SOME GENERALIZATIONS OF THE EQUALITY $H_+^p + H_-^p = L^p$ $(p<1)$

§0. Introduction

The following assertion is proved in [14] (see also [15] where a simpler proof is given).

THEOREM 0.1 Let $0 < p < 1$. Then

a) $L^p = H^p + H_-^p$;

b) the linear span of the Cauchy kernels $\left\{ \frac{1}{1-\bar{\zeta} z} \right\}$ $(\zeta \in \mathbb{T})$ is dense in $H^p \cap H_-^p$.

Here $L^p = L^p(\mathbb{T})$, H^p is the Hardy class in $\mathbb{D}$; this class is identified with a subspace of L^p associating with every $f \in H^p$ the function

$$\zeta \mapsto \lim_{r \to 1-} f(r\zeta) \ (\zeta \in \mathbb{T}) ; \quad H_-^p \overset{def}{=} \left\{ \tfrac{1}{z} f\left(\tfrac{1}{z}\right) : f \in H^p \right\} .$$

This chapter answers the following questions.

1) For which quasinormed symmetric spaces (besides L^p) is valid the natural analogue of Theorem 0.1?

2) Does Theorem 0.1 admit multidimensional generalizations?

3) How thick a subset A of $\mathbb{Z}$ has to be to ensure the decomposition

$$\mathbb{L}^p = \mathbb{L}^p_{A\cap\mathbb{Z}_+} + \mathbb{L}^p_{A\cap\mathbb{Z}_-}, \tag{1}$$

where $\mathbb{L}^p_B$ stands for the closure (in $\mathbb{L}^p$) of the linear span of the family $\{z^n\}$ $(n\in B)$?

Let us discuss each of these questions in more detail.

The most part of the chapter is devoted to the first question.

In §3 below we describe all quasinormed symmetric spaces X satisfying $X = X_A + X_{\bar{A}}$, X_A and $X_{\bar{A}}$ being the correspondent spaces of analytic and antianalytic functions in X (exact definitions see in §1). An analogous problem for normed symmetric spaces has been solved by D.W. Boyd [21].

In §2 we obtain **an analogue of the assertion b) of Th.0.1 valid** for a class of quasinormed symmetric spaces. The equality $X = X_A + X_{\bar{A}}$ and the density of rational functions in $X_A \cap X_{\bar{A}}$ (in the sense of b)) will be proved for some nonquasinormed symmetric spaces (§2). As an example may serve the space

$$N = N(\mathbb{T}) \overset{def}{=\!=} \{f\in \mathbb{L}^0(\mathbb{T}) : \int_{\mathbb{T}} \log(1+|f|)\,dm < +\infty\},$$

where $\mathbb{L}^0(\mathbb{T})$ is the space of all Lebesgue measurable functions. The metric is introduced in X by the formula $\rho(f,g) = \int \log(1+|f-g|)\,dm$. Let N_A denote the V.I. Smirnov class; the space N_A can be identified with the closure of the set of all polynomials (in z) in N (see [61]). Put $N_{\bar{A}} = \{\frac{1}{z} f(\frac{1}{z}) : f\in N_A\}$. The results of this chapter show that $N = N_A + N_{\bar{A}}$ and the closure in N of the linear span of the Cauchy kernels $\left\{\frac{1}{1-\bar{\zeta}z}\right\}$ $(\zeta\in\mathbb{T})$ coincides with $N_A \cap N_{\bar{A}}$ (§2).

The symbol $M = M(\mathbb{T})$ will denote the set of all functions $f : \hat{\mathbb{C}} \setminus \mathbb{T} \to \mathbb{C}$ representable by the Cauchy-type integrals of a measure;

$$f = \int_{\mathbb{T}} \frac{d\mu(\zeta)}{1-\bar{\zeta}z},$$

$\mu\in C^*(\mathbb{T})$. If **the measure** μ **is singular** (with respect to the Lebesgue measure on $\mathbb{T}$) we shall write $f\in M_s = M_s(\mathbb{T})$, and if μ is absolutely continuous we shall write $f\in M_a = M_a(\mathbb{T})$. Put $M/H^1_- \overset{def}{=\!=} \{f\,|\,\mathbb{D} : f\in M\}$.

We shall prove in §4 that if $X_A \cap X_{\bar{A}}$ (where X is a quasinormed symmetric space) is locally convex then either $X_A \cap X_{\bar{A}} = M_s$ or $X_A \cap X_{\bar{A}} = \{0\}$.

In §5 we give a sufficient and necessary condition for a function to belong to M (theorem 5.3). In §7 multidimensional generalizations of this criterion are discussed.

We give two applications of Th.5.3 in §6 (these applications can be generalized in the multidimensional situation by means of Th.7.5).

Let us turn now to the second question. Theorem 0.1 has a generalization concerning the space $H^p(\mathbb{R}^d)$ of functions harmonic in the upper half-space $\mathbb{R}^{d+1}_+$ (see the definition in [34] or in our Chapter Three).

THEOREM 0.2. Let $p \in (0,1)$, $d \in \mathbb{N}$. Then

a) for every $f \in L^p(\mathbb{R}^d)$ there is a function $u \in H^p(\mathbb{R}^d)$ such that $f = \lim_{t \to 0+} u(t, \cdot)$ a.e. in $\mathbb{R}^d$;

b) the closure of the set of all functions $u \in H^p(\mathbb{R}^d)$ **representable as a** linear combination of Poisson integrals of δ-measures coincides with the set of all functions $v \in H^p(\mathbb{R}^d)$ satisfying $\lim_{t \to 0+} v(t, \cdot) = 0$ a.e. in $\mathbb{R}^d$.

This theorem was stated in [14] (without proof) for $p \in (\frac{d-1}{d}, 1)$. Nor shall we prove it here, the proof being essentially the same as in the one-dimensional situation (see [14]). Part a) can be deduced also from Theorem 7.1 and some results of L.Carleson [27].

In §7 we prove an analogue of theorem 0.1 for spaces H^p in polydiscs (theorems 7.1 and 7.3). It is interesting to note that when $p > 1$ and $d > 1$ the analogue of the assertion a) from Th.0.1 is false.

We state without proof a generalization of Th.7.1 to abstract Hardy spaces containing sufficiently many inner functions.

I am unaware of any analogue of Th.0.1 (a) for the multidimensional ball $B^d \subset \mathbb{C}^d$. This question is probably connected with the problem of the existence of nontrivial inner functions in B^d (see [52]).

The third question is treated in §8. We prove that (1) holds if A has the outer density one. But there are A's satisfying (1) with the density zero. Both facts follow from Th.8.2. Then we show (Th.8.9) that Th.8.2 is sharp. In §8 we discuss the connection of Rudin "modification sets" ([50], [51]) with those satisfying (1).

In §9 Th.8.2 is generalized to locally compact abelian

groups (Th.9.1). This result strengthens the main result of [50].

One more generalization of Th.0.1 (for vector-valued classes L^p and H^p) is given in the second chapter (Th.II.2.1).

§1. Principal definitions

Let X be a non-null complete metric topological vector space over $\mathbb{C}$ contained in L^0, L^0 being the space of all complex measurable functions (mod 0) on a probabilistic Lebesgue space (see [49]). Let ρ be an invariant metric on X. The space X will be called a m e t r i c s y m m e t r i c s p a c e if

1) the convergence in X implies the convergence in measure, i.e. the inclusion map $X \to L^0$ is continuous;

2) if $f \in X$, $g \in L^0$, $|f|$ and $|g|$ being equimeasurable then $g \in X$ and $\rho(0,f) = \rho(0,g)$;

3) $f \in X$, $g \in L^0$, $|g| \leq |f|$ a.e. $\Rightarrow g \in X$, $\rho(0,g) \leq \rho(0,f)$.

Suppose the Lebesgue space is $(\mathbb{T}, m)$, m being the normalized Lebesgue measure on $\mathbb{T}$. If X is a separable metric symmetric space $(\subset L^0(\mathbb{T}))$ we shall denote by X_A the closure (in X) of the set of all polynomials (i.e. of linear combinations of $1, z, z^2, \dots$); $X_{\bar{A}} \overset{\text{def}}{=\!=} \{ \frac{1}{z} f(\frac{1}{z}) : f \in X \}$.

Sometimes wishing to indicate the Lebesgue space explicitly we shall write $X(T, \mathfrak{A}, \mu)$ instead of X. If $(T', \mathfrak{A}', \mu')$ is another probability Lebesgue space we may consider the metric symmetric space $X(T', \mathfrak{A}', \mu')$ consisting of all functions $f \in L^0(T', \mathfrak{A}', \mu')$ such that there is $g \in X(T, \mathfrak{A}, \mu)$ equimeasurable with $|f|$. Having this possibility in mind we shall very often think of a given symmetric space X with the accuracy "up to the change of the Lebesgue space".

A non-null complex complete quasinormed space X will be called a q u a s i n o r m e d s y m m e t r i c s p a c e if 1), 2), 3) are fulfilled with $\|h\|_X$ instead of $\rho(0,h)$.

Is is not hard to deduce from the Aoki - Rolewicz theorem that the quasinorm in a quasinormed symmetric space can be replaced by an equivalent one so that it becomes p-normed (with a

$p\in(0,1)$) and still remains a quasinormed symmetric space. Such spaces will be called p-n o r m e d s y m m e t r i c s p a c e s.

This shows that given a quasinormed symmetric space there is an invariant metric (which agrees with the given topology) such that the space becomes a metric symmetric space: we may assume that our space is p-normed and put $\rho(f,g)\overset{def}{=\!=}\|f-g\|^p$.

Clearly every metric symmetric space contains L^∞. It is not hard to prove that every p-normed symmetric space is contained in **the weak** L^p**-space** $L^{p,\infty}\overset{def}{=\!=}\{f\in L^0:\|f\|_{L^{p,\infty}}\overset{def}{=\!=}\sup_{\lambda>0}\lambda(\mu\{|f|>\lambda\})^{1/p}<\infty\}$.

Let X be a quasinormed symmetric space, $q\in(0,+\infty)$. Put $X^q\overset{def}{=\!=}\{f\in L^0:|f|^q\in X\}$, $\|f\|_{X^q}\overset{def}{=\!=}\||f|^q\|_X^{1/q}$. Clearly, X^q is a quasinormed symmetric space too.

Associate now with every quasinormed symmetric space two numbers α_X and β_X (lower and upper Boyd indices). Namely, consider the operator $\sigma_t:L^0[0,1]\longrightarrow L^0[0,1]$ $(t\in(0,+\infty))$

$$(\sigma_t f)(s)\overset{def}{=\!=}\begin{cases} f(\frac{s}{t}), & s\in[0,t]\cap[0,1] \\ 0, & s\notin[0,t]\cap[0,1]\ . \end{cases}$$

It is easy to see that $\sigma_t(X[0,1])\subset X[0,1]$ for every metric symmetric space X and for all positive numbers t. Now put

$$\alpha_X\overset{def}{=\!=}\lim_{\tau\to 0+}\frac{\log\|\sigma_\tau\|_X}{\log\tau}=\sup_{0<\tau<1}\frac{\log\|\sigma_\tau\|_X}{\log\tau},$$

$$\beta_X\overset{def}{=\!=}\lim_{\tau\to+\infty}\frac{\log\|\sigma_\tau\|_X}{\log\tau}=\inf_{\tau>1}\frac{\log\|\sigma_\tau\|_X}{\log\tau}.$$

It is not very hard to see that $0\leqslant\alpha_X\leqslant\beta_X\leqslant 1/p$ for every p-normed symmetric space X. Both Boyd indices are invariant under the change of the quasinorm of X by an equivalent one. Let us denote by $M:L^1[0,1]\longrightarrow L^1[0,1]$ the maximal operator of Hardy - Littlewood:

$$(Mf)(t)\overset{def}{=\!=}\sup_{0<h<\frac{1}{2}}\frac{1}{2h}\int_{x-h}^{x+h}|f(t)|\,dt$$

(we assume that f is extended periodically with period 1 onto $\mathbb{R}$).

It is known that $M(X[0,1]) \subset X[0,1]$ (where X is a normed symmetric space) iff $\beta_X < 1$ (see [6]). The proof of this assertion shows that it is valid for all quasinormed symmetric spaces $X \subset L^1$.

Associate with every quasinormed symmetric space X the "analytic" space $X_A \overset{def}{=\!=} X \cap N_A$ and the space of "antianalytic" functions $X_{\bar{A}} \overset{def}{=\!=} X \cap N_{\bar{A}}$ (spaces N_A and $N_{\bar{A}}$ were defined in §0; recall that in accordance with the above remark concerning the change of the underlying Lebesgue space we may think of elements of X as of functions defined on $\mathbb{T}$). We shall prove in §2 that this definition agrees with the definition of X_A and $X_{\bar{A}}$ given earlier.

It is clear that both X_A and $X_{\bar{A}}$ are closed in X because the embedding $X \subset N$ is continuous (by the Closed Graph Theorem) and N_A, $N_{\bar{A}}$ are closed in N.

A known theorem of Boyd [21] (see also [6]) asserts that if X is a normed symmetric space, then the equality $\mathrm{Re}\, X = \mathrm{Re}\, X_A$ (or in other words $X = X_A + X_{\bar{A}}$) is equivalent to the following inequalities: $\alpha_X > 0$, $\beta_X < 1$. The proof shows that this result holds for all quasinormed symmetric spaces X.

In this chapter an analogous problem is solved for quasinormed symmetric spaces X n o t c o n t a i n e d in L^1. We shall see that for these spaces the equality $X = X_A + X_{\bar{A}}$ holds iff $\alpha_X > 1$ (§3).

In the paper [22] certain space $H(X)$ of "analytic functions" is associated with every symmetric normed space X . This $H(X)$ coincides with our X_A only if X is a maximal symmetric space (see [6]).

§2. Rational approximation in $X_A \cap X_{\bar{A}}$. Equality $X = X_A + X_{\bar{A}}$ for a class of separable metric symmetric spaces

At first we prove a theorem on the maximal function. Further we obtain from this theorem the equivalence of two definitions (cf. §1) of the space X_A for separable quasi-normed symmetric spaces.

Denote by Γ_ζ the interior of the convex hull of the point $\zeta \in \mathbb{T}$ and of the disc of radius $1/2$ centered at the origin. For any function H in $\mathbb{D}$ set $(\mu H)(\zeta) \overset{def}{=\!=} \sup\{|H(z)| : z \in \Gamma_\zeta\}$. Every function $f \in X_A$,

$X_A = X(\mathbb{T}) \cap N_A$, can be extended into $\mathbb{D}$ as a function of the class N_A. We shall denote this extended function by the same symbol f, so we have an operator

$$\mathcal{M} : X(\mathbb{T}) \cap N_A \longrightarrow L^0(\mathbb{T})\,;\, (\mathcal{M}f)(\zeta) = \sup_{\Gamma_\zeta} |f| .$$

THEOREM 2.1. Let X be a quasinormed symmetric space, $X_A \overset{def}{=\!=} X(\mathbb{T}) \cap N_A$. Then $\mathcal{M}(X_A) \subset X$ and so

$$\| \mathcal{M}f \|_X \leqslant C_X \| f \|_{X_A} .$$

PROOF. Obviously $\beta_{X^p} = \frac{1}{p}\beta_X$ for $0 < p < +\infty$. Choose a number p so that $1/p\,\beta_X < 1$. Then $X^p \subset L^1$ and $\| Mf \|_{X^p} \leqslant C \| f \|_{X^p}$, $f \in X^p$, where M is the Hardy - - Littlewood maximal operator. Let $f \in X(\mathbb{T}) \cap N_A$. We can suppose f to be outer. Then

$$\| \mathcal{M}f \|_X = \| \mathcal{M}f^{\frac{1}{p}} \|^p_{X^p} \leqslant C \| M |f|^{\frac{1}{p}} \|^p_{X^p} \leqslant C \| |f|^{\frac{1}{p}} \|^p_{X^p} = C \| f \|_X . \bullet$$

COROLLARY 2.2. If X is a separable quasinormed symmetric space then the set $X(\mathbb{T}) \cap N_A$ coincides with the closure in X of the set of all polynomials.

PROOF. By Th.2.1 $f = \lim\limits_{r \to 1-} f_r$ in X, where f_r is the Poisson average of f. $\bullet$

REMARK 2.3. For general metric symmetric spaces the theorem 2.1 does not hold. It does not hold e.g. for $X = N$. To see this it is sufficient to consider the function $e^{f + i\tilde{f}}$ where $f \geqslant 0$, $f \in L^1(\mathbb{T})$ but $Mf \notin L^1(\mathbb{T})$. Then $e^{f + i\tilde{f}} \in N_A$ but $\mathcal{M}(e^{f+i\tilde{f}}) \notin N$, where $\tilde{f}$ is the conjugate function of f. $\bullet$

Now we turn to the main subject of the **section**. Consider the operator $R : L^0[0,1] \longrightarrow L^0[0,1]^2$,

$$(Rf)(x,y) = \frac{1}{y} f(x), \qquad x, y \in [0,1] .$$

THEOREM 2.4. Let X be a separable metric symmetric space and suppose $R(X) \subset X$. Then $X = X_A + X_{\bar{A}}$ and the closure of the span of **the Cauchy kernels** $\left\{\frac{1}{1 - \bar{\zeta} z}\right\}$ $(\zeta \in \mathbb{T})$ coincides with $X_A \cap X_{\bar{A}}$.

We begin with a lemma.

LEMMA 2.5. Let X be a metric symmetric space, let ρ be the metric of X, suppose $R(X)\subset X$ and $f\in C(\mathbb{T})$. Then

$$\lim_{n\to\infty} \rho\left(0,\frac{f(z)}{1-z^n}\right)=\rho\left(0,\frac{f(z_1)}{1-z_2}\right),$$

where we consider $\frac{f(z)}{1-z^n}$ as a function on $\mathbb{T}$ and $\frac{f(z_1)}{1-z_2}$ as a function on $\mathbb{T}^2$.

PROOF. Set

$$f_n(e^{it})=f(e^{\frac{2\pi ik}{n}}) \text{ for } t\in\left[\frac{2k\pi}{n},\frac{2(k+1)\pi}{n}\right),$$

$$0\leqslant k<n,\quad n\in\mathbb{N}.$$

Obviously $\lim_{n\to\infty}\|f_n-f\|_{L^\infty}=0$. Note that the functions $\frac{f_n(z)}{1-z^n}$ and $\frac{f_n(z_1)}{1-z_2}$ are equimeasurable. Hence

$$\rho\left(0,\frac{f_n(z)}{1-z^n}\right)=\rho\left(0,\frac{f_n(z_1)}{1-z_2}\right).$$

But $\lim_{n\to\infty}\frac{f_n(z)-f(z)}{1-z^n}=0$ in $X(\mathbb{T})$ and $\lim_{n\to\infty}\frac{f_n(z_1)-f(z_1)}{1-z_2}=0$ in $X(\mathbb{T}^2)$ because of $(1-z)^{-1}\in X$. ●

PROOF OF THEOREM 2.4 is analogous to the proof of proposition 1 of [15]. At first we prove the equality $X=X_A+X_{\bar{A}}$. To do this we associate with every trigonometric polynomial P, $P=\sum_{k\in\mathbb{Z}}a_kz^k$, two functions $Q_1\in X_A$, $Q_2\in X_{\bar{A}}$ such that $P=Q_1+Q_2$ and $\rho(0,Q_1)\leqslant\omega(\rho(0,P))$ where ω is a positive function on $(0,+\infty)$, $\omega(0+)=0$. Namely, set $Q_1=z^n\frac{P}{z^n-1}$, $Q_2=z^{-n}\frac{P}{z^{-n}-1}$. Clearly $Q_1\in X_A$, $Q_2\in X_{\bar{A}}$ if n is large. Our estimate is a consequence of lemma 2.5 and of the condition $R(X)\subset X$ if n is sufficiently large.

Represent now any function $f\in X(\mathbb{T})$ in the form $f=\sum_{k\geqslant1}P_k$ so that $\sum_{k\geqslant1}\rho(0,P_k)<+\infty$ and $\sum_{k\geqslant1}\omega(\rho(0,P_k))<+\infty$, where P_k are trigonometric polynomials. We can represent every P_k as a sum $P_k=Q_1^{(k)}+Q_2^{(k)}$, $Q_1^{(k)}\in X_A$, $Q_2^{(k)}\in X_{\bar{A}}$, $\rho(0,Q_1^{(k)})\leqslant\omega(\rho(0,P_n))$. Set $g=\sum_{k\geqslant1}Q_1^{(k)}$. Hence $g\in X_A$ and $f-g\in X_{\bar{A}}$.

Now we prove the second part of the theorem. Let $f\in X_A\cap X_{\bar{A}}$, $\varepsilon>0$. There exist polynomials P_1, P_2 such that $\rho(f,P_1)<\varepsilon$ and $\rho(f,z^{-1}P_2(z^{-1}))<\varepsilon$. Set $P=P_1-z^{-1}P_2(z^{-1})$. It is easy to check that the function $r=P_1-z^n\frac{P}{z^n-1}$ is a linear combination of Cauchy kernels $\left\{\frac{1}{1-\bar{\zeta} z}\right\}$ $(\zeta\in\mathbb{T})$ for sufficiently large n. As in the first part of the proof we can take n so large that $\rho(\tau, P_1)\leq\omega(2\varepsilon)$. Then $\rho(\tau, f)\leq$ $\leq\rho(\tau, P_1)+\rho(P_1, f)\leq\varepsilon+\omega(2\varepsilon)$. ●

COROLLARY 2.6. $N=N_A+N_{\bar{A}}$, Moreover $N_A\cap N_{\bar{A}}$ is the closure in N of the span of **the Cauchy kernels** $\left\{\frac{1}{1-\bar{\zeta} z}\right\}$ $(\zeta\in\mathbb{T})$.

PROOF. It is sufficient to check that $R(N)\subset N$. But

$$\int_0^1\int_0^1 \log\left(1+\frac{|f(x)|}{y}\right)dx\,dy=\int_0^1\log(1+|f(x)|)dx+\int_0^1|f(x)|\log\left(1+\frac{1}{|f(x)|}\right)dx<+\infty$$

for $f\in N$. ●

In the next **section** we shall prove that the condition $R(X)\subset X$ is essential for the equality $X=X_A+X_{\bar{A}}$, X being a quasinormed symmetric space. We shall see that in this case the separability condition in the first part of theorem 2.3 can be dropped.

§3. Quasinormed symmetric spaces X, satisfying $X=X_A+X_{\bar{A}}$

We begin with the lemma, which helps to decide whether Rf belongs to a symmetric metric space X. Consider a nonlinear operator $V: L^0\longrightarrow L^0[0,1]$,

$$(Vf)(x)\overset{def}{=\!=}\frac{1}{x}\int_x^1 f^*(t)\,dt\ ,$$

where f^* is the continuous on the left decreasing function on equimeasurable with $|f|$.

LEMMA 3.1. Let X be a metric symmetric space. Then $Rf\in X$ iff $Vf\in X$.

PROOF. Easy computations show that

$$(Vf)^*\leq(Rf)^*\leq(\sigma_4 Vf)^*.\quad ●$$

THEOREM 3.2. Let X be a quasinormed symmetric space (may be nonseparable), $X\not\subset L^1$. Then three following assertions are equivalent:

1. $X=X_A+X_{\bar{A}}$.
2. $R(X)\subset X$.
3. $\alpha_X>1$.

To prove this theorem we need two lemmas.

LEMMA 3.3. Let X be a quasinormed symmetric space, $R(X)\subset X$, $f\in L^\infty(\mathbb{T})$, $\varepsilon>0$. Suppose f is supported by a measurable set $E\subset\mathbb{T}$. Then for sufficiently large n the function $\frac{f}{1-z^n}$ may be rewritten as $\varphi+\psi$, where φ and ψ are supported by E and $\varphi^*\leqslant(Rf)^*$, $\|\psi\|_X<\varepsilon$.

PROOF. Take a function $g\in C(\mathbb{T})$ such that $\|f-g\|_X<\varepsilon$. Set $g_n(e^{ix})=g(e^{\frac{2\pi ik}{n}})$ for $x\in\left[\frac{2\pi k}{n},\frac{2\pi(k+1)}{n}\right)$. For n sufficiently large $\|g_n-g\|_X<\varepsilon$. The function $\frac{g_n(z)}{1-z^n}$ is equimeasurable with the function

$$\frac{g_n(z_1)}{1-z_2}=\frac{g_n(z_1)-f(z_1)}{1-z_2}+\frac{f(z_1)}{1-z_2}.$$

Obviously $\left\|\frac{g_n(z_1)-f(z_1)}{1-z_2}\right\|_X\leqslant C\varepsilon\|R\|_X$ and $\left(\frac{f(z_1)}{1-z_2}\right)^*\leqslant(Rf)^*$. Hence $\frac{f}{1-z^n}=\varphi+\psi$, where $\varphi^*\leqslant(Rf)^*$ and $\|\psi\|_X<C\varepsilon$. Remark finally that $\frac{f}{1-z^n}=\varphi\chi_E+\psi\chi_E$. ●

LEMMA 3.4. Let X be a quasinormed symmetric space, $R(X)\subset X$, $f\in X(\mathbb{T})$, $g\in L^0(\mathbb{T})$. Suppose that the circle $\mathbb{T}$ is divided into a countable family of measurable sets $\{E_k\}_{k\geqslant1}$. Suppose $(g\chi_{E_k})^*\leqslant(R(f\chi_{E_k}))^*$, $k\in\mathbb{N}$, Then $g\in X$ and $\|g\|_X\leqslant C_X\|f\|_X$.

PROOF. Clearly, $g^*\leqslant(Rf)^*$. Hence, $g\in X$ and $\|g\|_X\leqslant$ $\leqslant\|Rf\|_X\leqslant\|R\|_X\|f\|_X$. ●

PROOF OF THEOREM 3.2. $2\Longrightarrow1$ [*]. Let $f\in X$. We represent the function f as a sum of pairwise disjoint bounded functions: $f=\sum_{k\geqslant1}f_k$ $(f_k\in L^\infty(\mathbb{T}))$. In view of Lemma 3.3 and Lemma 3.4, for any $\varepsilon>0$ there exists a sequence of positive integers $\{m_k\}_{k\geqslant1}$ such that for any sequence $\{n_k\}_{k\geqslant1}$, $n_k\in\mathbb{N}$, $n_k>m_k$, the inequality

$$\left\|\sum_{k\geqslant1}\frac{z^{n_k}f}{z^{n_k}-1}\right\|_X\leqslant C\|f\|_X+\varepsilon$$

is valid. Choose now trigonometrical polynomials P_k so that $\|P_k-f_k\|_X\leqslant\frac{\varepsilon}{2^k}$. Then

---

[*] For separable X's, this implication follows from Th. 2.4; a reader interested in the separable case only can pass at once to the implication $1\Longrightarrow2$.

$$\Big\| \sum_{k\geq 1} \frac{z^{n_k} P_k}{z^{n_k}-1} \Big\|_X \leq C\|f\|_X + C\varepsilon ,$$

$$\sum_{k\geq 1} \frac{z^{n_k} P_k}{z^{n_k}-1} \in X_A , \quad \sum_{k\geq 1} \frac{z^{-n_k} P_k}{z^{-n_k}-1} \in X_{\bar A}$$

provided $n_k > \max(\deg P_k, m_k)$.

Hence, we have proved that for any function $f \in X$ and for any $\varepsilon > 0$ there exist functions $g \in X_A$ and $h \in X_{\bar A}$ such that $\|g\|_X \leq C\|f\|_X$ and $\|f-g-h\|_X < \varepsilon$.

Denote by g_0 and h_0 the functions g and h constructed for $\varepsilon = \frac{1}{2}$. Applying the same reasoning to the function $f-g_0-h_0$ and to $\varepsilon = (1/2)^2$, we find functions $g_1 \in X_A$ and $h_1 \in X_{\bar A}$ such that $\|g_1\|_X \leq \frac{1}{2} C$ and $\|f-g_0-g_1-h_0-h_1\|_X < \frac{1}{4}$. Proceeding inductively, we construct sequences of functions $\{g_k\}_{k\geq 0}$ and $\{h_k\}_{k\geq 0}$ such that $g_n \in X_A$ $h_n \in X_{\bar A}$, $\|g_n\|_X \leq C/2^n$, $\|f-\sum_{k=0}^{n}(g_k+h_k)\|_X \leq (1/2)^{n+1}$ for all $n \in \mathbb{N}$. Now the function f can be expressed in the form $f = \sum_{k\geq 0} g_k + \sum_{k\geq 0} h_k$ with $\sum_{k\geq 0} g_k \in X_A$, $\sum_{k\geq 0} h_k \in X_{\bar A}$.

$1 \Rightarrow 2$. Let $f \in X$. By hypothesis, the function f can be expressed in the form $f = g+h$, where $g \in X_A$, $h \in X_{\bar A}$. Consider the function H, $H(z) \overset{def}{=\!=} g(z) + h(1/\bar z)$ harmonic in $\mathbb{D}$. By Th.2.1, $\mathcal{M}H \in X$. Consider the open set $G_A \overset{def}{=\!=} \{\zeta \in \mathbb{T} : (\mathcal{M}H)(\zeta) > A\}$ and the closed set $F_A \overset{def}{=\!=} \mathbb{T} \setminus G_A$. Let H_A denote the harmonic continuation into $\mathbb{D}$ (the Poisson integral) of the function $f\chi_{F_A}$.

Let $\Omega_{F_A} \overset{def}{=\!=} \bigcup_{\zeta\in F_A} \Gamma_\zeta$. Denote by u_{G_A} the Poisson integral of the characteristic function of the set G_A. Then the inequality

$$|H(z) - H_A(z)| \leq c_0 A\, u_{G_A}(z) \tag{2}$$

is valid for any $z \in \Omega_{F_A}$, where c_0 is an absolute constant (see, e.g., Lemma 2 or the proof of Theorem 1 in [16]). This inequality implies

$$\Big| H(0) - \int_{\mathcal{M}H\leq A} f\,dm \Big| \leq c_0 A\, m\{\mathcal{M}H > A\} \tag{3}$$

provided $F_A \neq \emptyset$. Let $f \geqslant 0$. Then (3) implies the inequality

$$\int\limits_{\mathcal{M}H \leqslant A} f\,dm \leqslant C_f + C_0 A\, m\{\mathcal{M}H > A\} \tag{4}$$

which holds for all $A \in (0, +\infty)$. Set $A = (\mathcal{M}H)^*(t)$ $(t \in (0,1])$ in this inequality. We have

$$\frac{1}{t}\int_t^1 f^*(s)\,ds \leqslant \frac{1}{t}\int\limits_{\mathcal{M}H \leqslant A} f\,dm \leqslant \frac{C_f}{t} + C_0\,\mathcal{M}H^*(t)\,. \tag{5}$$

Let $f \in X \setminus L^1$ (by hypotheses, $X \not\subset L^1$), $f \geqslant 0$. Then from the inequality (5) it follows that $\lim\limits_{t \to 0+} t\,\mathcal{M}H^*(t) = +\infty$. Consequently, $t^{-1} \in X[0,1]$. Hence, we have proved that if $f \in X$, then $Vf \in X[0,1]$. Applying Lemma 3.1, we get $RX \subset X$.

$2 \Rightarrow 3$. The inclusion $R(R(X)) \subset X$ implies that the function $f(x)\frac{1}{yz}$ $(x, y, z \in [0,1])$ belongs to $X[0,1]^3$ provided $f \in X[0,1]$. Observe that $\left(\frac{1}{yz}\right)^*(t) \geqslant c\,\frac{\log 2/t}{t}$ $(t \in (0,1])$. Hence, if $f \in X[0,1]$ then $f(x)\frac{\log 2/y}{y} \in X[0,1]^2$. Therefore by the closed graph theorem

$$\left\| f(x)\left(\log\tfrac{2}{y}\right)\tfrac{1}{y}\right\|_X \leqslant C_X \|f\|_X\,.$$

On the other hand,

$$\left\| f(x)\frac{\log 2/y}{y}\right\|_X \geqslant \left\| f(x)\frac{\log 2/y}{y}\chi_{(\tau,2\tau)}(y)\right\|_X \geqslant \frac{\log 1/\tau}{2\tau}\|\sigma_\tau f\|_X$$

for any $\tau \in (0, 1/2)$. Consequently, $\|\sigma_\tau\|_X \leqslant$ $\leqslant C\frac{\tau}{\log 1/\tau}$ $(\tau \in (0, 1/2))$. Hence $\alpha_X > 1$, since

$$\alpha_X = \sup_{0<\tau<1}\frac{\log\|\sigma_\tau\|_X}{\log\tau} \geqslant \sup_{0<\tau<\frac12}\left(1 + \frac{\log\log\frac{1}{\tau} - \log c}{\log\frac{1}{\tau}}\right) > 1\,.$$

$3 \Rightarrow 2$. We need to verify that $RX \subset X$. We have: $|Rf(x,y)| \leqslant$ $\leqslant \sum_{n \geqslant 1} 2^n\,|f(x)\,\chi_{(2^{-n},2^{-n+1})}(y)|$. It remains to note that

$$2^n \| f(x)\chi_{(2^{-n},2^{-n+1})}(y)\|_{X[0,1]^2} = 2^n \|\sigma_{2^{-n}} f\|_X \leq$$

$\leq c2^{-n\varepsilon}$ for some $\varepsilon > 0$ since $\alpha_X > 1$. ●

REMARK 3.5. It is seen from the proof of the theorem that if X and Y are quasinormed symmetric spaces, $X \not\subset L^1$, then the following assertions are equivalent:

1. $X \subset Y_A + Y_{\bar{A}}$.
2. $R(X) \subset Y$.

In addition, if Y is a quasi-normed symmetric space, $f \in Y_A + Y_{\bar{A}}$, $f \geq 0$, $f \notin L^1$, then $Rf \in Y$.

§4. When is the intersection $X_A \cap X_{\bar{A}}$ trivial?

THEOREM 4.1. Let X be a quasi-normed symmetric space. Then $X_A \cap X_{\bar{A}} = \{0\}$ iff $X[0,1] \not\ni 1/t$.

PROOF. If $1/t \in X[0,1]$, then $\frac{1}{1-z} \in X_A \cap X_{\bar{A}}$. To prove the converse, let $f \in X_A \cap X_{\bar{A}}$. We set $H(z) \overset{def}{=\!=} f(z) - f(\frac{1}{\bar{z}})$ $(z \in \mathbb{D})$. Then H is a harmonic function; its nontangential boundary values are zero a.e. on $\mathbb{T}$, and $\mathcal{M}H \in X$. We have $\lim_{A\to\infty} Am\{\mathcal{M}H > A\} = 0$, because otherwise the function t^{-1} would belong to $X[0,1]$. Hence, $H = 0$ in view of Remark 4 [16]. ●

THEOREM 4.2. Let X be a quasinormed symmetric space, $\lim_{\tau\to 0+} \|\chi_{(0,\tau)} \cdot \frac{1}{t}\|_{X[0,1]} = 0$. Then the space $X_A \cap X_{\bar{A}}$ is not locally convex.

PROOF. The vector-valued function $\zeta \overset{\Phi}{\longmapsto} \frac{1}{1-\bar{\zeta}z}$ $(\zeta \in \mathbb{T})$ with values in $X_A \cap X_{\bar{A}} \subset X$ is continuous, since $\lim_{\tau\to 0+} \|\chi_{(0,\tau)} \cdot \frac{1}{t}\|_{X[0,1]} = 0$. Assume that the space $X_A \cap X_{\bar{A}}$ is locally convex. Then there exists the Riemann integral

$$\int_{\mathbb{T}} \Phi \, dm \overset{def}{=\!=} f \in X_A \cap X_{\bar{A}}.$$

The functional $g \longmapsto g(z)$ is continuous on $X_A \cap X_{\bar{A}}$ for any $z \in \mathbb{C} \setminus \mathbb{T}$. Hence,

$$f(z) = \int_{\mathbb{T}} \frac{dm(\zeta)}{1-\bar{\zeta}z} = \begin{cases} 1, & |z| < 1 \\ 0, & |z| > 1. \end{cases}$$

But this contradicts the inclusion $f \in X_A \cap X_{\bar{A}}$. ●

THEOREM 4.3. Let X be a quasi-normed symmetric space. Then

$X_A \cap X_{\bar{A}} = M_s$ iff $1/t \in X[0,1]$ and $\lim_{\varepsilon\to 0+} \|\chi_{(0,\varepsilon)} 1/t\|_{X[0,1]} > 0$.

PROOF. If $\lim_{\varepsilon\to 0+} \|\chi_{(0,\varepsilon)} 1/t\|_{X[0,1]} = 0$, then by Th.4.2, the space $X_A \cap X_{\bar{A}}$ is not locally convex, and, therefore, $M_s \neq X_A \cap X_{\bar{A}}$. If $1/t \notin X[0,1]$ then $X_A \cap X_{\bar{A}} = \{0\}$ by Th.4.1.

Let $1/t \in X[0,1]$ and $\lim_{\varepsilon\to 0+} \|\chi_{(0,\varepsilon)} 1/t\|_X > 0$. The inclusion $M_s \subset X_A \cap X_{\bar{A}}$ follows from the A.N.Kolmogorov's theorem. Let us prove the converse inclusion. Let $f \in X_A \cap X_{\bar{A}}$. Set $H(z) \overset{def}{=\!=} f(z) - f(\frac{1}{\bar{z}})$ $(z \in \mathbb{D})$. Then the boundary values of the harmonic function H are zero a.e. on $\mathbb{T}$ and $\mathcal{M}H \in X$. Clearly, $\lim_{t\to 0+} t\mathcal{M}H^*(t) < +\infty$ since $\lim_{\varepsilon\to 0+} \|\chi_{(0,\varepsilon)} \frac{1}{t}\|_{X[0,1]} > 0$. Hence, $\lim_{A\to +\infty} Am\{(\mathcal{M}H)^* > A\} < +\infty$, and by Remark 4 of [16] the function H is the Poisson integral of a singular measure. Hence $f \in M_s$. ●

COROLLARY 4.4. Let X be a quasi-normed **symmetric space**. Suppose that the space $X_A \cap X_{\bar{A}}$ is locally convex. Then either $X_A \cap X_{\bar{A}} = \{0\}$ or $X_A \cap X_{\bar{A}} = M_s$. ●

I don't know whether there is a function $f \in N_A \cap N_{\bar{A}}$, $f \not\equiv 0$ such that $\lim_{t\to+\infty} tm\{|f| > t\} = 0$. Theorem 4.1 gives no answer (even for $f \in H^p \cap H^p_-$, $p < 1$).

§ 5. Functions representable by **the Cauchy integral**

Denote by $H^p(\mathbb{T})$ the set of all functions $f : \hat{\mathbb{C}} \setminus \mathbb{T} \to \mathbb{C}$ such that $f | \mathbb{D} \in H^p$ and $f | \mathbb{D}_- \in H^p_-$.

A well-known theorem of V.I.Smirnov asserts that if $f \in M$, i.e.

$$f(z) = \int_{\mathbb{T}} \frac{d\mu(\zeta)}{1 - \bar{\zeta} z}$$

$(\mu \in (C(\mathbb{T}))^*)$, then $f \in H^p(\mathbb{T})$ and $\|f\|_{H^p(\mathbb{T})} = O(\frac{1}{1-p})$. In addition, $Tf \in L^1_i(\mathbb{T})$, where

$$(Tf)(\zeta) \overset{def}{=\!=} \lim_{r\to 1-} (f(r\zeta) - f(r^{-1}\zeta)).$$

We shall show that this assertion can be reversed (Theorem 5.3). It follows easily from the results of [16]. We present also another proof which does not appeal to the results of [16].

PROPOSITION 5.1. Let $f \in \bigcap_{p<1} H^p$, $f(0) \in \mathbb{R}$, $\mathrm{Re} f \in L^1(\mathbb{T})$, and $\lim_{p\to 1-} \|f\|_{H^p}(1-p) < +\infty$. Then there exists a real measure $\mu\in(C(\mathbb{T}))^*$ such that $f(z)=$ $=\int_{\mathbb{T}}\frac{1+\bar{\zeta}z}{1-\bar{\zeta}z}\,d\mu(\zeta)$, and

$$\|\mu\|\le\|\operatorname{Re}f\|_{L^1}+\varliminf_{p\to 1-}\|f\|_{H^p}(1-p)\frac{\pi}{2}\ .$$

Moreover, if $\varliminf_{p\to 1-}\|f\|_{H^p}(1-p)=0$, then the measure μ is absolutely continuous $(\mu=(\operatorname{Re}f)\cdot m)$.

PROOF. Suppose first $\operatorname{Re}f=0$ a.e. on $\mathbb{T}$. We shall make use of an idea of S.K.Pichorides (see [48]). Consider the function $\Phi_p(z)\overset{\text{def}}{=}|r|^p\cos pt$, where $z=re^{it}$ $(r\in\mathbb{R},\ t\in[-\frac{\pi}{2},\frac{\pi}{2}])$. This function is subharmonic, provided $p\in(0,2]$ (see [48]). Hence, the function $\Phi_p\circ f$ is subharmonic for $p\in(0,1)$. Since for $q\in(1,\frac{1}{p})$

$$\sup_{r<1}\int_{\mathbb{T}}\Phi_p^q(f(r\zeta))\,dm(\zeta)\le\|f\|_{H^{pq}}^{pq}<+\infty,$$

we have

$$\Phi_p(f(a))\le\int_{\mathbb{T}}\frac{1-|a|^2}{|1-\bar{a}\zeta|^2}\,|f(\zeta)|^p\cos\frac{p\pi}{2}\,dm(\zeta).$$

Note that by hypotheses

$$\varliminf_{p\to 1-}\int_{\mathbb{T}}|f(\zeta)|^p\cos\frac{p\pi}{2}\,dm(\zeta)<+\infty\ .$$

Consequently, using the weak-star compactness of the unit ball in the space of measures $(C(\mathbb{T}))^*$, we find

$$|\operatorname{Re}f(a)|=\lim_{p\to 1-}\Phi_p(f(a))\le\int_{\mathbb{T}}\frac{1-|a|^2}{(1-a\bar{\zeta})^2}\,d\nu(\zeta),$$

where ν is a finite positive measure, and $\|\nu\|\le\varliminf_{p\to 1-}\|f\|_{H^p}(1-p)\,\pi/2$.

Hence, (cf. [4]), there exists a real measure $\mu\in(C(\mathbb{T}))^*$ such that

$$\operatorname{Re}f(a)=\int_{\mathbb{T}}\frac{1-|a|^2}{|1-a\bar{\zeta}|^2}\,d\mu(\zeta)\qquad(a\in\mathbb{D}).$$

Then

$$f(a)=\int_{\mathbb{T}}\frac{1+\bar{\zeta}a}{1-\bar{\zeta}a}\,d\mu(\zeta)$$

and $\|\mu\| \le \|\nu\|$. The general case can be reduced easily to this special one.

REMARK 5.2. Proposition 5.1 admits multidimensional extensions. However in this situation the proposition does not seem to reduce to the case $\operatorname{Re} f = 0$ a.e. Instead we can use the inequality $\Phi_p(x+iy) \le \Phi_p(x) + \Phi_p(iy)$, which is equivalent to the following one proved in [48]: $\cos pt \le \cos\frac{p\pi}{2}\sin^p t + \cos^p t \quad (t \in [0, \frac{\pi}{2}],\ p \in (0, 1))$.

THEOREM 5.3. Let f be a function analytic in $\mathbb{C}\setminus\mathbb{T}$. Then

a) $f \in M$ iff $f \in \bigcap_{p<1} H^p(\mathbb{T})$, $\varliminf_{p\to 1-} \|f\|_{H^p}(1-p) < +\infty$ and $\mathcal{T}f \in L^1(\mathbb{T})$,

b) $f \in M_a$ iff $f \in \bigcap_{p<1} H^p(\mathbb{T})$, $\varliminf_{p\to 1-} \|f\|_{H^p}(1-p) = 0$ and $\mathcal{T}f \in L^1(\mathbb{T})$,

c) $f \in M_s$ iff $f \in \bigcap_{p<1} H^p(\mathbb{T})$, $\varliminf_{p\to 1-} \|f\|_{H^p}(1-p) < +\infty$ and $\mathcal{T}f = 0$ a.e. on $\mathbb{T}$.

PROOF. In any of these three cases we need only to verify the sufficiency. Let us begin with a). Suppose $f \in \bigcap_{p<1} H^p(\mathbb{T})$, $\varliminf_{p\to 1-} \|f\|_{H^p}(1-p) < +\infty$ and $\mathcal{T}f \in L^1(\mathbb{T})$. We may assume $f(0) = 0$. Set

$$f_1(z) \overset{\text{def}}{=\!=} -i\left(f(z) + \overline{f(1/\bar z)}\right) \qquad (z \in \hat{\mathbb{C}}\setminus\mathbb{T}),$$

$$f_2(z) \overset{\text{def}}{=\!=} f(z) - \overline{f(1/\bar z)} \qquad (z \in \hat{\mathbb{C}}\setminus\mathbb{T}).$$

By Proposition 5.1 there exist real measures μ_1 , and μ_2 such that

$$f_j(z) = \int_{\mathbb{T}} \frac{1+\bar\zeta z}{1-\bar\zeta z}\, d\mu_j(\zeta) \qquad (z \in \hat{\mathbb{C}}\setminus\mathbb{T}).$$

Set $\mu \overset{\text{def}}{=\!=} i\mu_1 + \mu_2$. Then

$$f(z) = \frac{1}{2}\int_{\mathbb{T}} \frac{1+\bar\zeta z}{1-\bar\zeta z}\, d\mu(\zeta) = \int_{\mathbb{T}} \frac{d\mu(\zeta)}{1-\bar\zeta z} - \frac{1}{2}\int_{\mathbb{T}} d\mu = \int_{\mathbb{T}} \frac{d\mu(\zeta)}{1-\bar\zeta z}$$

since $f_1(0) = f_2(0) = 0$.

The assertion c) is easily reduced to a). The assertion b) is proved in a similar fashion. ●

Theorem 5.3 strengthens the analog of Theorem 4 in [16] in the case of the circle $\mathbb{T}$.

We give another proof of Th.5.3 using the results of [16].

PROOF. We shall check only c). Let $f\in\bigcap_{p<1}H^p(\mathbb{T})$, $\varliminf_{p\to 1-}\|f\|_{H^p}(1-p)<+\infty$, $\mathcal{T}f=0$ a.e. Consider the harmonic function H, $H(z)\overset{def}{=\!=}f(z)-f(1/\bar z)$ $(z\in\mathbb{D})$. Then $\varliminf_{p\to 1-}\|\mathcal{M}H\|_{L^p}(1-p)<+\infty$, since $\|\mathcal{M}H\|_{L^p}\le C\|f\|_{H^p}$ $(p\in(1/2,1)$. Hence, $\varliminf_{A\to+\infty}Am\{\mathcal{M}H>A\}<+\infty$ and nontangential boundary values of H are zero a.e. on $\mathbb{T}$. Then, by Remark 4 in [16], the function H is the Poisson integral of a singular measure, thus $f\in M_s$. ●

§6. Some applications of Theorem 5.3

THEOREM 6.1. Let $h:\mathbb{D}\to\hat{\mathbb{C}}$ be a meromorphic function with the bounded characteristic, i.e. h can be expressed in the form $h=h_1/h_2$, with $h_1,h_2\in H^\infty$, $h_2\not\equiv 0$. Suppose that there exists a function g bounded and analytic in $\mathbb{D}_-$ such that

$$\lim_{r\to 1-}h(r\zeta)=\lim_{r\to 1+}g(r\zeta)$$

a.e. on $\mathbb{T}$. If the function f can be represented in $\mathbb{D}$ as the Cauchy integral of a measure from $(C(\mathbb{T}))^*$ (i.e. $f\in M/H^1_-$) and if $fh\in N_A$, then $fh\in M/H^1_-$.

PROOF. Let

$$f(z)=\int_{\mathbb{T}}\frac{d\mu(\zeta)}{1-\bar\zeta z}\qquad(z\in\mathbb{D}).$$

Consider the function $F:\hat{\mathbb{C}}\setminus\mathbb{T}\to\mathbb{C}$,

$$F(z)\overset{def}{=\!=}\begin{cases} h(z)\displaystyle\int_{\mathbb{T}}\frac{d\mu(\zeta)}{1-\bar\zeta z}, & z\in\mathbb{D}\\[2ex] g(z)\displaystyle\int_{\mathbb{T}}\frac{d\mu(\zeta)}{1-\bar\zeta z}, & z\in\mathbb{D}_-\end{cases}.$$

Clearly, $F\in M$ by Th.5.3. Hence, $fh=F|\mathbb{D}\in M/H^1_-$. ●

It is easy to deduce from this theorem a result of S.A.Vinogradov concerning the division property of Cauchy integrals.

COROLLARY 6.2 (S.A.Vinogradov, [59]). Suppose

I is an inner function, $fI^{-1}\in N_A$. Then $fI^{-1}\in M/H^1_-$.

Let φ be an inner function. By $H^p\cap\varphi H^p_-$ we denote the set of all functions f meromorphic in $\hat{\mathbb{C}}\setminus\mathbb{T}$ such that $f\,|\,\mathbb{D}\in H^p$ and $f/\tilde\varphi\,|\,\mathbb{D}_-\in H^p$, where

$$\tilde\varphi(z)\overset{def}{=\!=}\begin{cases}\varphi(z), & z\in\mathbb{D},\\ \overline{\varphi(1/\bar z)}, & z\in\mathbb{D}_-.\end{cases}$$

THEOREM 6.3. Let φ be an inner function. Then

a) for any function $f\in H^\infty\cap\varphi H^\infty_-$ and for any $\zeta\in\mathbb{T}$ the function $f/(\tilde\varphi-\zeta)$ belongs to M_s ,

b) for any function $f\in H^1\cap\varphi H^1_-$, the function $f/(\tilde\varphi-\zeta)$ belongs to M_s for a.e. $\zeta\in\mathbb{T}$.

PROOF. a) We can consider $\zeta=1$. Since $\mathrm{Re}\,\frac{1}{\varphi-1}\le 0$, we have $\frac{1}{\varphi-1}\in H^p$ $(p<1)$, and $\left\|\frac{1}{\varphi-1}\right\|_{H^p}=O\left(\frac{1}{1-p}\right)$ $(p\to 1-)$. Hence we can apply Theorem 5.3 c) to the function $\frac{f}{\varphi-1}$.

b) Define

$$h_p(\zeta)\overset{def}{=\!=}\int_{\mathbb{T}}\left|\frac{f(\xi)}{\varphi(\xi)-\zeta}\right|^p dm(\xi)\,.$$

We have

$$\int_{\mathbb{T}} h_p(\zeta)\,dm(\zeta)=\left\|\frac{1}{1-z}\right\|^p_{H^p}\|f\|^p_{H^p}\le\frac{c}{1-p}\,.$$

Using the Fatou lemma, we get

$$\varliminf_{p\to 1-} h_p(\zeta)(1-p)<+\infty$$

for a.e. $\zeta\in\mathbb{T}$, and Th.5.3 c) finishes the proof. ●

REMARK. Theorem 6.1 and Theorem 6.3 a) are also easy consequences of Theorem 4 [16], a weakened version of Theorem 5.3.

§ 7. Several multidimensional and other generalizations

Denote by $\{-1,1\}^d$ the set of all multiindices $\alpha\in\mathbb{Z}^d$ such that $|\alpha_j|=1$ for all $j\in\{1,2,\dots,d\}$.

Let $\varepsilon \in \{-1,1\}^d$. Denote by $H^p_\varepsilon(\mathbb{T}^d)$ the set of all functions holomorphic in $\mathbb{D}_\varepsilon \overset{def}{=} \prod_{j=1}^d \mathbb{D}_{\varepsilon_j} \subset \hat{\mathbb{C}}^d$ $(\mathbb{D}_1 \overset{def}{=} \mathbb{D},\ \mathbb{D}_{-1} \overset{def}{=} \mathbb{D}_-)$ and admitting the representation

$$f(z) = \sum_{\alpha \in \mathbb{Z}^d_\varepsilon} c_\alpha z^\alpha \quad (z \in \mathbb{D}^d_\varepsilon),$$

and satisfying

$$\|f\|^p_{H^p_\varepsilon(\mathbb{T}^d)} \overset{def}{=} \sup_{r_d < 1} \int_{\mathbb{T}^d} |f(\zeta_1 r_1^{\varepsilon_1}, \dots, \zeta_d r_d^{\varepsilon_d})|^p \, dm_d(\zeta_1, \dots, \zeta_d) < +\infty,$$

where $\mathbb{Z}^d_\varepsilon \overset{def}{=} \prod_{j=1}^d \mathbb{Z}_{\varepsilon_j}$ $(\mathbb{Z}_1 \overset{def}{=} \mathbb{Z}_+,\ \mathbb{Z}_{-1} \overset{def}{=} \mathbb{Z}_-)$. Such functions f have radial boundary values a.e. on $\mathbb{T}^d$ (see [10]). Thus, the space $H^p_\varepsilon(\mathbb{T}^d)$ can be isometrically identified with $L^p_{\mathbb{Z}^d_\varepsilon}(\mathbb{T}^d)$, where $L^p_A(\mathbb{T}^d)$ is the closed span of the family $\{z^n : n \in A\}$ in L^p.

We introduce an order in the set $\{-1,1\}^d$; $\varepsilon \geq \tau$ will mean $\varepsilon_j \geq \tau_j$ for all $j \in \{1,2,\dots,d\}$. The maximal element in $\{-1,1\}^d$ will be denoted by $+$, the minimal one by $-$.

THEOREM 7.1. Let $d \in \mathbb{N}$, $p \in (0,1)$. Then

$$L^p(\mathbb{T}^d) = H^p_+(\mathbb{T}^d) + H^p_-(\mathbb{T}^d).$$

In particular, for any real-valued function $h \in L^p(\mathbb{T}^d)$ there exists a function $f \in H^p_+(\mathbb{T}^d)$ such that $\operatorname{Re} f = h$ a.e. on $\mathbb{T}^d$.

PROOF. Let Q be a trigonometrical polynomial of d variables. It suffices to prove that there exist functions $Q_1 \in H^p_+(\mathbb{T}^d)$ and $Q_2 \in H^p_-(\mathbb{T}^d)$ such that $Q = Q_1 + Q_2$ and $\|Q_1\|_{L^p} \leq C(p) \|Q\|_{L^p}$. Set $Q_1^{(\zeta)} \overset{def}{=} z_1^n \dots z_d^n \dfrac{Q}{z_1^n \dots z_d^n - \zeta}$, $Q_2^{(\zeta)} \overset{def}{=} z_1^{-n} \dots z_d^{-n} \dfrac{Q}{z_1^{-n} \dots z_d^{-n} - \bar{\zeta}}$, where $\zeta \in \mathbb{T}$. It is clear that $Q = Q_1^{(\zeta)} + Q_2^{(\zeta)}$, and $\zeta \in \mathbb{T}$ can be chosen so that $\|Q_1^{(\zeta)}\|_{L^p} \leq C(p) \|Q\|_{L^p}$ (see the proof of Prop.1 in [15]). ●

Note that the similar assertion is false for $p \in [1,+\infty]$ and $d > 1$ since $\int_{\mathbb{T}^d} f z_1^{-1} z_2 \, dm_d = 0$ if $f \in H^p_+(\mathbb{T}^d) + H^p_-(\mathbb{T}^d)$ $(p \geq 1)$.

Theorem 7.1 has also the following extension to the case of an abstract Hardy space.

THEOREM 7.2. Let $(\varphi, H^\infty(X, \Sigma, \mu))$ be an abstract

Hardy space (see [1]) with the multiplicative functional φ, $\varphi(f)=\int f\,d\mu$. Suppose that the set of all functions $f\in L^p(X,\Sigma,\mu)$ $(p<1)$ such that $fI\in H^p(X,\Sigma,\mu)$ and $I\bar{f}\in H^p(X,\Sigma,\mu)$ for some inner function I, is dense in $L^p(X,\Sigma,\mu)$ (this holds, e.g., if the linear span of the family $\{I_1 I_2^{-1}\}$, I_1, I_2 being inner functions from $H^\infty(X,\Sigma,\mu)$, is dense in L^p). Then

$$L^p(X,\Sigma,\mu)=H^p(X,\Sigma,\mu)+\overline{H^p(X,\Sigma,\mu)}.$$

Denote by $H^p(\mathbb{T}^d)$ the set of all functions $f:\bigcup_{\varepsilon\in\{-1,1\}^d}\mathbb{D}_\varepsilon\longrightarrow\mathbb{C}$ such that $f\,|\,\mathbb{D}_\varepsilon\in H^p_\varepsilon(\mathbb{T}^d)$. Let $H^p_s(\mathbb{T}^d)$ denote the set of all functions $f\in H^p(\mathbb{T}^d)$ for which the limit $\lim_{r\to 1-} f(r^{\varepsilon_1}\zeta_1,\dots,r^{\varepsilon_d}\zeta_d)$ does not depend on the choice of $\varepsilon\in\{-1,1\}^d$ for a.e. $\zeta\in\mathbb{T}^d$.

Clearly, $H^p_s(\mathbb{T}^d)=\{0\}$, provided $p\geqslant 1$.

Note that if $\zeta\in\mathbb{T}^d$ then $\prod_{j=1}^d(1-\bar{\zeta}_j z_j)^{-1}\in H^p_s(\mathbb{T}^d)$ for any $p<1$.

THEOREM 7.3. If $d\in\mathbb{N}$, $p\in(0,1)$ then the closed (in L^p) span of the Cauchy kernels $\prod_{j=1}^d(1-\bar{\zeta}_j z_j)^{-1}(\zeta\in\mathbb{T}^d)$ coincides with $H^p_s(\mathbb{T}^d)$.

PROOF. Let $f\in H^p_s(\mathbb{T}^d)$, $\sigma>0$. For any $\varepsilon\in\{-1,1\}^d$ one can find a trigonometrical polynomial f_ε (of d variables) such that $f_\varepsilon\in L^p_{\mathbb{Z}^d_\varepsilon}(\mathbb{T}^d)$ and

$$\Big\|f-\Big(\prod_{j=1}^d\varepsilon_j\Big)f_\varepsilon\Big\|_{L^p}<\sigma.$$

Set $P_\varepsilon\overset{def}{=\!=}\sum_{\tau\geqslant\varepsilon}f_\tau$. Note that if $\varepsilon\neq +$, then $\|P_\varepsilon\|_{L^p}\leqslant 2^{d/p}\sigma$. Define

$$g_n\overset{def}{=\!=}\sum_{\varepsilon\in\{-1,1\}^d}P_\varepsilon\prod_{\varepsilon_j=-1}\frac{z_j^n}{1-z_j^n}.$$

We have

$$\|g_n-f\|^p_{L^p}\leqslant\|g_n-f_+\|^p_{L^p}+\|f_+-f\|^p_{L^p}\leqslant\sum_{\substack{\varepsilon\in\{-1,1\}^d\\ \varepsilon\neq +}}\Big\|P_\varepsilon\prod_{\varepsilon_j=-1}\frac{z_j^n}{1-z_j^n}\Big\|^p_{L^p}+$$

$$+\sigma^p\leqslant C_d(p)\sigma^p$$

for sufficiently large n, since

$$\lim_{n\to+\infty}\Big\| P_\varepsilon \prod_{\varepsilon_j=-1}\frac{z_j^n}{1-z_j^n}\Big\|_{L^p}=\|P_\varepsilon\|_{L^p}\cdot\Big\|\prod_{\varepsilon_j=-1}\frac{z_j}{1-z_j}\Big\|_{L^p}.$$

It remains to check that the function g_n is the linear combination of Cauchy kernels for sufficiently large values of n. This fact follows from the identity

$$g_n=\sum_{\varepsilon\in\{-1,1\}^d}\frac{f_\varepsilon\prod_{\varepsilon_j=-1}z_j^n}{(1-z_1^n)\dots(1-z_d^n)}. \tag{6}$$

This equality shows that g_n is a linear combination of Cauchy kernels, provided $\sum_{\varepsilon\in\{-1,1\}^d} f_\varepsilon=\sum_{\alpha\in\mathbb{Z}^d,|\alpha_j|<n} c_\alpha z^\alpha$

Let us verify (6):

$$g_n=\sum_{\varepsilon\in\{-1,1\}^d}\Big(\sum_{\tau\geq\varepsilon}f_\tau\Big)\prod_{\varepsilon_j=-1}\frac{z_j^n}{1-z_j^n}=$$

$$=\sum_{\tau\in\{-1,1\}^d}f_\tau\sum_{\varepsilon\leq\tau}\prod_{\varepsilon_j=-1}\frac{z_j^n}{1-z_j^n}=$$

$$=\sum_{\tau\in\{-1,1\}^d}f_\tau\prod_{j=1}^{d}\frac{1}{1-z_j^n}\sum_{\varepsilon\leq\tau}\prod_{\varepsilon_j=-1}z_j^n\prod_{\varepsilon_k=1}(1-z_k^n)=$$

$$=\sum_{\tau\in\{-1,1\}^d}f_\tau\frac{\prod_{\tau_j=-1}z_j^n}{\prod_{j=1}^{d}(1-z_j^n)}\prod_{\tau_j=1}(z_j^n+(1-z_j^n))=$$

$$= \sum_{\tau\in\{-1,1\}^d} \frac{f_\tau \prod_{\varepsilon_j=-1} z_j^n}{(1-z_1^n)\dots(1-z_d^n)} \quad . \bullet$$

Now we briefly consider functions representable by the Cauchy integral of a measure supported by $\mathbb{T}^d$.

Denote by $M(\mathbb{T}^d)$ the set of all functions $f:(\hat{\mathbb{C}}\setminus\mathbb{T})^d \to \mathbb{C}$ representable in the form

$$f(z_1,z_2,\dots,z_d) = \int_{\mathbb{T}^d} \frac{d\mu(\zeta_1,\dots,\zeta_d)}{\prod_{j=1}^{d}(1-\bar{\zeta}_j z_j)} \quad , \tag{7}$$

where $\mu\in(C(\mathbb{T}^d))^*$. Denote by $M_a(\mathbb{T}^d)$ $(M_s(\mathbb{T}^d))$ the set of all functions $f\in M(\mathbb{T}^d)$ with corresponding μ absolutely continuous (resp.singular).

It is known that in general a function of the class $M(\mathbb{T}^d)$ has radial limits on no set of positive measure (see [35]) if $d>1$; hence, $M(\mathbb{T}^d) \not\subset H^p(\mathbb{T}^d)$ for any p, $d>1$.

Define the operator $T_d : H^p(\mathbb{T}^d) \to L^p(\mathbb{T}^d)$,
$(T_d f)(\zeta_1,\dots,\zeta_d) \stackrel{def}{=\!=} \lim_{r\to 1-} \sum_{\varepsilon\in\{-1,1\}^d} (-1)^{(d-\Sigma\varepsilon_j)1/2} f(r_1^{\varepsilon_1}\zeta_1,\dots,r_d^{\varepsilon_d}\zeta_d)$.
Using Remark 4 of [16] and results of [34], we can prove the following theorem.

THEOREM 7.4. Let $f\in\bigcap_{p<1} H^p(\mathbb{T}^d)$, $T_d f\in L^1(\mathbb{T}^d)$ and $\lim_{p\to 1-} \|f\|_{H^p}(1-p) < +\infty$. Then $f\in M(\mathbb{T}^d)$. In addition, if $T_d f \equiv 0 \pmod 0$, then $f\in M_s(\mathbb{T}^d)$, and if $\lim_{p\to1-}\|f\|_{H^p}(1-p) = 0$, then $f\in M_a(\mathbb{T}^d)$.

Let $\mathcal{A}$ be a class of functions holomorphic in $(\hat{\mathbb{C}}\setminus\mathbb{T})^d$. Denote by $R\mathcal{A}$ the set of all functions $f\in A$ supported by $\mathbb{D}^d\cup\mathbb{D}_-^d$, i.e. $f|((\hat{\mathbb{C}}\setminus\mathbb{T})^d\setminus(\mathbb{D}^d\cup\mathbb{D}_-^d)) \equiv 0$. It follows from the well-known properties of the Calderon - Zygmund singular integrals that $RM(\mathbb{T}^d)\subset RH^p(\mathbb{T}^d)$ for any $p<1$ and any $d\in\mathbb{N}$.

The next theorem follows easily from Th.7.4, but the easiest way to establish it is to reason as in the proof of Th.5.3 (see Remark 5.2).

THEOREM 7.5.

$$RM(\mathbb{T}^d) = \{ f\in\bigcap_{p<1} RH^p(\mathbb{T}^d) : T_d f\in L^1(\mathbb{T}^d),\ \lim_{p\to1-}\|f\|_{H^p}(1-p) < +\infty \},$$

$$RM_a(\mathbb{T}^d)=\{f\in\bigcap_{p<1}RH^p(\mathbb{T}^d): T_d f\in L^1(\mathbb{T}^d),\ \lim_{p\to 1-}\|f\|_{H^p}(1-p)=0\},$$

$$RM_s(\mathbb{T}^d)=\{f\in\bigcap_{p<1}RH^p(\mathbb{T}^d): T_d f\equiv 0\,(mod\,0),\ \lim_{p\to 1-}\|f\|_{H^p}(1-p)<\infty\}. \bullet$$

This theorem enables us to obtain the multidimensional versions of the results of Section 6.

§8. On the equality $L^p = L^p_{A_+} + L^p_{A_-}$ $(0<p<1)$

Let $A\subset\mathbb{Z}$. Denote by A_+ the intersection $A\cap\mathbb{Z}_+$ $(A_-\overset{def}{=\!=}A\cap\mathbb{Z}_-)$. L^p_A denotes the closed span in $L^p=L^p(\mathbb{T}^d)$ of the family $\{z^n\}_{n\in A}$.

Recall that $H^p_s(\mathbb{T})$ is the set of all functions $f\in H^p(\mathbb{T}^d)$ such that $Tf=0$ a.e. on $\mathbb{T}$, $(Tf)(\zeta)\overset{def}{=\!=}\lim_{r\to 1-}(f(r\zeta)-f(r^{-1}\zeta))$ $(\zeta\in\mathbb{T})$.

For a function $f\in H^p(\mathbb{T})$,

$$f(z)=\begin{cases}\sum\limits_{n\geqslant 0}a_n z^n, & z\in\mathbb{D},\\ -\sum\limits_{n<0}a_n z^n, & z\in\mathbb{D}_-,\end{cases}$$

we set $\hat f(n)\overset{def}{=\!=}a_n$.

PROPOSITION 8.1. Let $A\subset\mathbb{Z}$, $p\in(0,1)$. The following assertions are equivalent:

1) $L^p = L^p_{A_+} + L^p_{A_-}$

2) for any function $f\in H^p(\mathbb{T})$ there exists a function $g\in H^p_s(\mathbb{T})$ such that $\hat g(n)=\hat f(n)$ for any $n\notin A$.

PROOF. 1)$\Rightarrow$2). Let $f\in H^p(\mathbb{T})$. We have $Tf=f_1-f_2$ where $f_1\in L^p_{A_+}\subset H^p$, and $f_2\in L^p_{A_-}\subset H^p_-$. Define

$$g(z)\overset{def}{=\!=}\begin{cases}f(z)-f_1(z), & z\in\mathbb{D},\\ f(z)-f_2(z), & z\in\mathbb{D}_-.\end{cases}$$

Clearly, $g\in H^p_s(\mathbb{T}^d)$ and $\hat g(n)=\hat f(n)$, provided $n\notin A$.

2) $\Longrightarrow$ 1). Let $h \in \mathbb{L}^p$. Then there exists a function $f \in H^p(\mathbb{T})$ such that $\mathrm{T} f = h$. By hypothes s, there exists a function $g \in H^p_s(\mathbb{T})$ such that $\hat{g}(n) = \hat{f}(n)$ for any $n \notin A$. Set $f_1 \overset{def}{=} (f-g) \mid \mathbb{D}$, $f_2 \overset{def}{=} (f-g) \mid \mathbb{D}_-$. Then $f_1 \in \mathbb{L}^p_{A_+}$, $f_2 \in \mathbb{L}^p_{A_-}$ and $h = f_1 - f_2$. ●

A set A^+, $A \subset \mathbb{Z}$, is called a modification set if for any $\mu \in C^*(\mathbb{T})$ there exists a singular measure $\nu \in C^*(\mathbb{T})$ such that $\hat{\nu}(n) = \hat{\mu}(n)$ for any $n \notin A$ (see [51]). This condition corresponds (in some sense) to the condition 2) of Prop.8.1 for $p \to 1-$.

W.Rudin provided in [51] an example of a modification set of zero density. J.Shapiro (see [54]) proved the existence of sets $A \subset \mathbb{Z}$ of zero density satisfying $\mathbb{L}^p = \mathbb{L}^p_A$. He deduced this assertion from the above mentioned result of W.Rudin, combining it with the following result: $\mathbb{L}^p = \mathbb{L}^p_A$ for any $p \in (0,1)$, provided A is a modification set.

We shall prove the equality $\mathbb{L}^p = \mathbb{L}^p_{A_+} + \mathbb{L}^p_{A_-}$ for a wide class of sets $A \subset \mathbb{Z}$ including some sets of zero density. Certainly, the equality $\mathbb{L}^p = \mathbb{L}^p_{A_+} + \mathbb{L}^p_{A_-}$ implies $\mathbb{L}^p = \mathbb{L}^p_A$.

THEOREM 8.2. Let $A \subset \mathbb{Z}^+$, $0 < p < 1$. Suppose that for any $n \in \mathbb{Z}$ there exists $a \in \mathbb{N}$ such that $([a-n, a+n] \cup \cup [-a-n, -a+n]) \cap \mathbb{Z} \subset A$. Then for any function $f \in \mathbb{L}^p$ there exist functions $g \in \mathbb{L}^p_{A_+}$ and $h \in \mathbb{L}^p_{A_-}$ such that $f = g + h$ and $\|g\|_{\mathbb{L}^p} + \|h\|_{\mathbb{L}^p} \le C_p \|f\|_{\mathbb{L}^p}$ with C_p not depending on the set A.

We prove first

LEMMA 8.3. Let $p \in (0,1)$. Then

$$\left\| 1 + \tfrac{1}{2}(z + z^{-1}) \right\|_{\mathbb{L}^p(\mathbb{T})} < 1 .$$

PROOF. Note that $\|1 + \frac{1}{2}(z+z^{-1})\|_{\mathbb{L}^1} = \int (1 + \frac{1}{2}(z+z^{-1}))\, dm = 1$. Since $|1 + \frac{1}{2}(z+z^{-1})|$ is not a constant function, $\|1 + \frac{1}{2}(z+z^{-1})\|_{\mathbb{L}^p} < 1$ ●

PROOF OF THEOREM 8.2. At first we prove that for any trigonometrical polynomial f there exist functions $g \in \mathcal{L}(A_+)$, $h \in \mathcal{L}(A_-)$ ($\mathcal{L}(B)$ is the linear span of the family $\{z^k\}_{k \in B}$) such that $\|g\|_{\mathbb{L}^p}$, $\|h\|_{\mathbb{L}^p} \le \|f\|_{\mathbb{L}^p}$ and $\|f - g - h\|_{\mathbb{L}^p} \le \varepsilon_p \|f\|_{\mathbb{L}^p}$, where $\varepsilon_p \in (0,1)$.

To do this, we choose $n \in \mathbb{N}$ so that $z^n f \in \mathcal{L}(A_+)$ and $z^{-n} g \in \mathcal{L}(A_-)$. Define $g^{(\zeta)} \overset{def}{=} \frac{1}{2} \zeta z^n f$, $h^{(\zeta)} \overset{def}{=} \frac{1}{2} \zeta^{-1} z^{-n} f$. Then $\int \| f - g^{(\zeta)} - h^{(\zeta)} \|^p dm = \|f\|^p_{\mathbb{L}^p} \varepsilon^p_p$. Hence, we can choose $\zeta \in \mathbb{T}$ so that $\|f - g^{(\zeta)} - h^{(\zeta)}\|_{\mathbb{L}^p} \le \varepsilon_p \|f\|_{\mathbb{L}^p}$ with $\varepsilon_p < 1$ in view of Lemma 8.3.

Next we prove that any trigonometrical polynomial f can be expressed in the form $f=g+h$, where $g\in L^p_{A_+}$, $h\in L^p_{A_-}$, and $\|g\|_{L^p}\leq C_p\|f\|_{L^p}$. Using what has been proved, it is easy to find by induction three sequences of trigonometrical polynomials $\{f_n\}_{n\geq 0}, \{g_n\}_{n\geq 0}, \{h_n\}_{n\geq 0}$ such that $f_0=f$, $g_n\in\mathcal{L}(A_+)$, $h_n\in\mathcal{L}(A_-)$, $\|f_n\|_{L^p}\leq\varepsilon_p^n\|f_0\|_{L^p}$, $\|g_n\|_{L^p}\leq\|f_n\|_{L^p}$, $\|h_n\|_{L^p}\leq\|f_n\|_{L^p}$, $f_{n+1}=f_n-g_n-h_n$. Set $g \stackrel{def}{=} \sum_{n\geq 0} g_n$, $h\stackrel{def}{=}\sum_{n\geq 0} h_n$. It is clear that $g\in L^p_{A_+}$, $h\in L^p_{A_-}$, $f=g+h$, and $\|g\|_{L^p}\leq c_p\|f\|_{L^p}$. Now let f be an arbitrary function from L^p. Represent f as a series $f= =\sum_{k\geq 0} f_k$, $\sum_{k\geq 0}\|f_k\|^p_{L^p}<+\infty$, f_k being trigonometrical polynomials. Then $f_k=g_k+h_k$, where $g_k\in L^p_{A_+}$, $h_k\in L^p_{A_-}$ and $\|g_k\|_{L^p}\leq C_p\|f_k\|_{L^p}$. Set $g\stackrel{def}{=} =\sum_{k\geq 0} g_k$. Clearly, $g\in L^p_{A_+}$, $f-g\in L^p_{A_-}$. ●

COROLLARY 8.4. There exists a set $A\subset\mathbb{Z}$ of zero density for which the equality

$$L^p = L^p_{A_+} + L^p_{A_-} \qquad (0<p<1)$$

holds. ●

COROLLARY 8.5. Let $A\subset\mathbb{Z}$ and

$$\overline{\lim_{n\to+\infty}}\,\frac{\operatorname{card}(A\cap[-n,n])}{2n+1}=1\,.$$

Then $L^p = L^p_{A_+} + L^p_{A_-}$. ●

REMARK 8.6. In Section 9 we shall prove [x] that under hypotheses of Th.8.2 (and, consequently, of Cor.8.5) A is a modification set. This assertion is implicitly contained in [51].

Under hypotheses of Th.8.2, the intersection $A\cap(-A)$ is nonempty; moreover, it satisfies the conditions of Th.8.2 itself. Therefore it is naturally to ask whether the condition (8) can be replaced by the following: for any $n\in\mathbb{N}$ there exist an arbitrarily great $a\in\mathbb{Z}_+$ and an arbitrarily small $b\in\mathbb{Z}_-$ such that

$$([a-n,a+n]\cup[b-n,b+n])\cap\mathbb{Z}\subset A\,. \qquad (9)$$

---

[x] This proof is essentially simpler in the case of the group $\mathbb{T}$

We shall construct a set $A\subset\mathbb{Z}$ satisfying (9) and such that the space of functionals $(L^p_A)^*$ separates points in L^p_A. This implies that $L^p_A \neq L^p$, and, consequently, $L^p \neq L^p_{A_+} + L^p_{A_-}$. This implies, moreover, that any measure $\mu\in C^*(\mathbb{T})$ with $\operatorname{supp}\hat{\mu}\subset A$ is absolutely continuous (see [54]), hence A is not a modification set.

Let $A, B\subset\mathbb{Z}$; $A\subset B$. Denote by $\mathcal{P}^B_A$ the rotation invariant projection of L^p_B onto L^p_A (if it exists). Note that the projection exists if A is finite and B is bounded from below (or from above).

LEMMA 8.7. Let A be a finite subset of $\mathbb{Z}$. Then $\lim\limits_{n\to+\infty}\|\mathcal{P}_A^{A\cup(n+\mathbb{Z}_+)}\|_{L^p}=1$ $(0<p<+\infty)$. Moreover, if p is separated from zero, then the convergence is uniform.

PROOF. We may assume $A\subset\mathbb{Z}$. Denote by $P_r : H^p\to H^p$ the Poisson mean $(P_r f)(z)=f(rz)$, $H^p\simeq L^p_{\mathbb{Z}_+}$. It remains to note that for any $\varepsilon>0$ there exists a number $r\in(0,1)$ such that for $p\geqslant\varepsilon$

$$\overline{\lim_{n\to+\infty}}\,\|\mathcal{P}_A^{A\cup(n+\mathbb{Z}_+)}-P_r|L^p_{A\cup(n+\mathbb{Z}_+)}\|\leqslant\varepsilon . \bullet$$

REMARK 8.8. It is easy to see that if $A\neq\emptyset$ and $A\cap(n+\mathbb{Z}_+)=\emptyset$, then

$$\|\mathcal{P}_{n+\mathbb{Z}_+}^{A\cup(n+\mathbb{Z}_+)}\|_{L^p}\geqslant\|\mathcal{P}_{\mathbb{N}}^{\mathbb{Z}_+}\|_{L^p}>1$$

provided $p\neq 2$.

THEOREM 8.9. There exists a set $A\subset\mathbb{Z}$ satisfying the condition (9) and such that $(L^p_A)^*$ separates points of L^p_A for all $p\in(0,1)$.

PROOF. According to Lemma 8.7 it is possible to construct an increasing sequence of sets $\{A_n\}_{n\geqslant1}$, $A_n\subset\mathbb{Z}$, such that

$$A_{2n+1}\setminus A_{2n}=[a_n-n,\,a_n+n]\cap\mathbb{Z},\quad A_{2n}\setminus A_{2n-1}=[b_n-n,\,b_n+n]\cap\mathbb{Z}$$

where $a_n,(-b_n)\in\mathbb{Z}_+$ are so great that $a_n>n$, $b_n<-n$

and $C_p\overset{\mathrm{def}}{=\!=}\prod_{n\geqslant1}\|\mathcal{P}_{A_n}^{A_{n+1}}\|_{L^p}<+\infty$ for any $p\in(0,+\infty]$. From this we conclude that $\|\mathcal{P}^A_{A_n}\|_{L^p}\leqslant c(p)$ for any $n\in\mathbb{N}$.

Consequently, for any function $f\in L^p_A$ $(0<p<+\infty)$ the equality $f = \lim_{n\to+\infty} \mathcal{P}^A_{A_n} f$ (in L^p) holds. To finish the proof, it suffices to note that the projection $\mathcal{P}^A_{\{k\}}$ exists for any $k\in A$. ●

REMARK 8.10. It is not difficult to construct a set $A\subset\mathbb{Z}$ such that

$$L^p = L^p_{A_+} + L^p_{A_-} \ (0<p<1) \quad \text{and} \quad A\cap(-A)=\emptyset .$$

In the next Section we shall prove that Th.8.2 naturally extends to the case of a locally compact abelian group.

§9. On the equality $L^p(G) = L^p_A(G)+L^p_B(G)$ $(0<p<1)$

Let G be a locally compact abelian group, Γ is the dual of G. Denote by $M(G)$ the space of all finite Borel measures on G. By $\hat{\mu}$ we denote the Fourier transform of $\mu\in M(G)$. Denote by $L^p_A(G)$ $(A\subset\Gamma)$ the closure in $L^p(G)$ of the set of all $f\in L^1(G)\cap L^p(G)$ with $\operatorname{supp}\hat{f}\subset A$ $(0<p<+\infty)$

THEOREM 9.1. Let G be a locally compact abelian group, Γ be its dual; $A, B\subset\Gamma$. Suppose that for any compact set $K\subset\Gamma$ there exists $\lambda\subset\Gamma$ such that $\lambda+K\subset A$ and $-\lambda+K\subset B$.

Then $L^p(G) = L^p_A(G)+L^p_B(G)$ for all $p<1$, and $A\cup B$ is a modification set, i.e. for any $\mu\in M(G)$ there exists a singular measure $\nu\in M(G)$ such that $\hat{\mu}=\hat{\nu}$ outside $A\cup B$.

From this Theorem, Th.7.1 follows again as well as its version for the Hardy spaces H^p in $\mathbb{C}^d_+$, where $\mathbb{C}_+ \overset{def}{=} \{z\in\mathbb{C} : \operatorname{Im} z>0\}$.

The proof is preceded by a sequence of lemmas.

LEMMA 9.2. Let G be a metrizable locally compact abelian group. Then there exists a sequence $\{\mathcal{K}_n\}_{n\geq 0}$ of $L^1(G)$-functions such that the following conditions are satisfied:

a) $\sup_{n\geq 0} \|\mathcal{K}_n\|_{L^1} < +\infty$,

b) for any neighbourhood V of zero in G

$$\lim_{n\to+\infty} \|\mathcal{K}_n(1-\chi_V)\|_{L^1} = 0 ,$$

c) $\operatorname{supp}\hat{\mathcal{K}}_n$ is a compact set containing zero,

d) $\operatorname{supp} \hat{\mathcal{K}}_n \subset \operatorname{int} \{\hat{\mathcal{K}}_{n+1} = 1\}$,

e) $\bigcup_{n\geq 0} \{\hat{\mathcal{K}}_n = 1\} = \Gamma$, where Γ is the dual of G

PROOF. From Th.2.6.8 [9] we conclude that there exists a sequence $\{\mathcal{K}_n\}_{n\geq 0}$ of functions from $L^1(G)$ satisfying the conditions a), c), d), e) and such that $\lim_{n\to+\infty} \|\mathcal{K}_n\|_{L^1} = 1$. Then $\{\mathcal{K}_n\}_{n\geq 0}$ satisfies b) too. ●

LEMMA 9.3. Let $\{\mathcal{K}_n\}_{n\geq 0}$ be the sequence from the previous lemma. Then for any $\mu \in M(G)$ and for any compact set $K \subset G$

$$\lim_{n\to+\infty} (\mathcal{K}_n * \mu)|K = f|K \quad (\text{ in } L^0(K)),$$

where f is the derivative of the measure μ with respect to the Haar measure on G.

PROOF see in [54]. ●

LEMMA 9.4. Let G be a locally compact abelian group, $p \in (0,2]$. Then the set S_p of all functions $f \in L^p(G) \cap L^\infty(G)$ with compact $\operatorname{supp} \hat{f}$ is dense in $L^p(G)$.

PROOF. It is sufficient to prove that the closure of S_p in $L^p(G)$ contains $L^p(G) \cap L^\infty(G)$. We can assume G to be metrizable. If $p \geq 1$ then $f = \lim_{n\to+\infty} f * \mathcal{K}_n$ (in L^p), where $\mathcal{K}_n$ are from Lemma 9.2, and $\mathcal{K}_n * f \in S_p$ as soon as $f \in L^p(G) \cap L^\infty(G)$. If $p < 1$, it suffices to choose $n \in \mathbb{N}$ such that $np \in [1,2]$ and to note that $S_{np}^n \subset S_p$. ●

COROLLARY 9.5. Let G be a locally compact abelian group, $p \in (0,1]$. Then the set of all $f \in L^p(G) \cap L^1(G)$ such that $\operatorname{supp} \hat{f}$ is compact, is dense in $L^p(G)$. ●

COROLLARY 9.6. Let G be a locally compact abelian group. Then any $f \in L^1(G)$ can be represented in the form

$$f = \sum_{k\geq 0} a_k g_k \quad , a_k \in \mathbb{C},$$

$\sum_{k\geq 0} |a_k| < +\infty$, $g_k \geq 0$, $\operatorname{supp} \hat{g}_k$ is compact, $\|g_k\|_{L^1} \leq 1$, and $g_k \in L^{1/2}(G)$.

PROOF. Using dual arguments it suffices to prove that for any $h \in L^\infty(G)$ there exists a function f, $f \geq 0$, $\|f\|_{L^1} \leq 1$, $f \in L^{1/2}(G)$, with compact $\operatorname{supp} \hat{f}$ such that $\left|\int_G f \cdot h\right| \geq c \cdot \|h\|_{L^\infty}$. Obviously, for given

$h \in L^\infty(G)$, $\|h\|_{L^\infty} = 1$, we can find a non-negative function $f \in L^1(G)$ satisfying the inequality $\left|\int_G f\cdot h\right| > \frac{1}{2}$. It is not hard to construct (cf. [9] , Th.2.6.6) a net of non--negative functions $\{\mathcal{K}_\alpha\}$ with compact $\operatorname{supp}\hat{\mathcal{K}}_\alpha$ such that $\|\mathcal{K}_\alpha\|_{L^1} = 1$ for all α and $\lim_\alpha \mathcal{K}_\alpha * f = f$ (in $L^1(G)$) . Consequently there exists a α satisfying the inequality

$\left|\int_G (\mathcal{K}_\alpha * f)h\right| > \frac{1}{2}$.

We are going to find a non-negative function $g \in L^1(G)$ with compact $\operatorname{supp}\hat{g}$ such that $\|(\mathcal{K}_\alpha * f) - g(\mathcal{K}_\alpha * f)\|_{L^1}$ is small (remark that $(\mathcal{K}_\alpha * f)g \in L^{1/2}(G)$).

Let $\mathcal{U}$ be a symmetric compact neighbourhood of zero in Γ . Set $\varphi_{\mathcal{U}}(x) \stackrel{def}{=} \hat{\chi}^2_{\mathcal{U}}(m(\mathcal{U}))^{-2}$ (here m denotes the Haar measure on Γ). It is clear that $\lim_{\mathcal{U}} \varphi_{\mathcal{U}}(x) = 1$ uniformly on compact subsets of G , $\varphi_{\mathcal{U}} \in L^1(G)$ and $0 \leqslant \varphi_{\mathcal{U}} \leqslant 1$. Consequently there exists a neighbourhood $\mathcal{U}$ of zero in Γ such that $\left|\int_G (\mathcal{K}_\alpha * f)\varphi_{\mathcal{U}}\cdot h\right| > \frac{1}{2}$.

It remains to note that $(\mathcal{K}_\alpha * f)\varphi_{\mathcal{U}} \in L^{1/2}(G)$, $(\mathcal{K}_\alpha * f)\varphi_{\mathcal{U}} \geqslant 0$, $\|(\mathcal{K}_\alpha * f)\varphi_{\mathcal{U}}\|_{L^1} \leqslant 1$ and $\widehat{(\mathcal{K}_\alpha * f)\varphi_{\mathcal{U}}}$ is supported by a compact subset of Γ . ●

LEMMA 9.7. Let the conditions of Th.9.1 be satisfied, $A \cap B \neq \Gamma$; K and L be two compacts in G . Then there exists $\lambda \notin L$ such that $K + \lambda \subset A$ and $K - \lambda \subset B$.

PROOF. Let M be a compact set such that $0 \in M$, $M = -M$ $M \supset K, L$; $M \not\subset A \cap B$. By the hypotheses there exists $\lambda \in \Gamma$, such that $\lambda + M + M \subset A$ and $-\lambda + M + M \subset B$ Suppose $\lambda \in L$. Then $M \subset A \cap B$ and we have a contradiction. ●

PROOF OF THEOREM 9.1. The first part of the theorem can be proved using the same arguments as in the proof of Theorem 8.2, only instead of trigonometrical polynomials we need to consider functions $f \in L^p(G) \cap L^1(G)$ with compact $\operatorname{supp}\hat{f}$. The set of all such functions f is dense in $L^p(G)$ according to Cor.9.5.

Prove now that $A \cup B$ is a modification set. Suppose first that G is metrizable. Then we can consider $A \cup B \neq \Gamma$. Let $f \in L^1(G) \cap L^{1/2}(G)$, $f \geqslant 0$ and suppose $\operatorname{supp}\hat{f}$ to be compact.

Then $K_0 = \operatorname{supp}\hat{f} \subset \operatorname{int}\{\hat{\mathcal{K}}_{n_0} = 1\}$ for some $n_0 \in \mathbb{N}$. By Lemma 9.7 there exists $\lambda \in \Gamma$ such that $K_0 + \lambda \subset A$, $K_0 - \lambda \subset B$ and $(K_0 \pm \lambda) \cap \operatorname{supp}\hat{\mathcal{K}}_{n_0} = \emptyset$. Define $g_\zeta(x) \stackrel{def}{=} \frac{1}{2}(x,\lambda)\zeta f(x)$, $h_\zeta(x) \stackrel{def}{=} \frac{1}{2}\frac{1}{(x,\lambda)\zeta}\cdot f(x)$, where $x \in G$ and $\zeta \in \mathbb{T}$. Choose a number $\zeta \in \mathbb{T}$ such that

$$\| f - g_\zeta - h_\zeta \|_{L^{1/2}} \leqslant \varepsilon_{1/2} \| f \|_{L^{1/2}}$$ (see the proof of Th.8.2). Clearly, $f - g_\zeta - h_\zeta \geqslant 0$ and $\| f - g_\zeta - h_\zeta \|_{L^1} = \hat{f}(0) = \| f \|_{L^1}$, since $\lambda \notin K_0 \cup (-K_0)$.

Proceeding by induction, we get a sequence $\{f_n\}_{n \geqslant 0}$ of functions from $L^1(G) \cap L^{1/2}(G)$ such that $f_0 = f$,

$\| f_j \|_{L^{1/2}} \leqslant \varepsilon_{1/2}^{j} \| f \|_{L^{1/2}}$, $supp(\hat{f}_{j+1} - \hat{f}_j) \subset A \cup B$, $K_j = supp\,\hat{f}_j$ is compact, $K_j \subset int\,\{\hat{\mathcal{K}}_{n_j} = 1\}$.

Since the sequence $\{\hat{f}_j\}_{j \geqslant 0}$ converges uniformly on compact sets, $\lim_{j \to \infty} \hat{f}_j$ is continuous. Hence, by the Bochner theorem there exists a positive measure $\mu \in M(G)$ such that $\hat{\mu} = \lim_{j \to \infty} \hat{f}_j$. Clearly, $f_j = \mathcal{K}_{n_j} * \mu$. Since $\lim_{j \to \infty} \| f_j \|_{L^{1/2}} = 0$ it follows from Lemma 9.3 that μ is singular.

Now let $f \in M(G)$. Of course we may assume that $f \in L^1(G)$. It remains to apply Lemma 9.6 to the function f.

Reduction to the case of a metrizable group. Let $f \in L^1(G)$ By induction it is easy to construct a sequence $\{M_n\}_{n \geqslant 1}$ of symmetric compact neighbourhoods of zero in Γ such that $M_n + M_n \subset M_{n+1}$, $M_n \supset \{|\hat{f}| \geqslant \frac{1}{n}\}$ and there exists $\lambda_n \in \Gamma$ such that $\lambda_n + M_n \subset A \cap M_{n+1}$, $-\lambda_n + M_n \subset B \cap M_{n+1}$. It is clear that $\Lambda \overset{def}{=} \bigcup_{n \geqslant 1} M_n$ is an open subgroup of Γ. Let H be its annihilator. Then G/H is metrizable. Put $A_0 \overset{def}{=} A \cap \Lambda$, $B_0 \overset{def}{=} B \cap \Lambda$. There exists a function $\tilde{f} \in L^1(G/H)$ such that $f = \tilde{f} \circ \pi$. It remains to note that the quadruple $(G/H, \Lambda, A_0, B_0)$ satisfies all conditions of Th.9.1. ●

COROLLARY 9.8. Let G be a locally compact abelian group and let Γ denote its dual group. Suppose Γ is endowed with an order which agrees with the group structure.

Assume that each compact subset of Γ has an upper (and thus lower) bound. Let $a, b \in \Gamma$. Put $A \overset{def}{=} \{\alpha \in \Gamma : \alpha \geqslant a\}$, $B \overset{def}{=} \{\alpha \in \Gamma : \alpha \leqslant b\}$. Then $L^p(G) = L^p_A(G) + L^p_B(G)$ and $A \cup B$ is a modification set. ●

Chapter Two. LOCALLY HOLOMORPHIC SPACES AND ROTATION-INVARIANT SUBSPACES OF H^p

§0. Introduction

In §1 we introduce a definition of a locally holomorphic space. The class of all locally holomorphic spaces is a class of topological vector spaces (TVS) containing all locally convex spaces. This class is well fit for the investigation of vector-valued holomorphic functions. Note that there are several non-equivalent definitions of holomorphic functions with values in a non locally convex space.

For example, the function $z \longmapsto \frac{1}{t-z}$ mapping $\mathbb{C}$ into $L^p(\mathbb{D}, m_2)$ $(p<1)$ is a differentiable function of the complex variable, but it does not belong to the class C^∞. On the other hand, this function, considered as an $L^0(\mathbb{D}, m_2)$-valued function is a differentiable function of z and belongs to C^∞, but it cannot be developed in a power series in any neighbourhood of the origin.

Holomorphic functions we shall consider will be always representable as the sums of power series in the disc $\mathbb{D}$, the series converging uniformly on compact subsets of $\mathbb{D}$.

In §2 we show that these holomorphic functions with values in a locally holomorphic space possess many usual properties of scalar holomorphic function.

In § 3 we give a new proof of the A.Frazier theorem about linear functionals on the space H^p $(0<p<1)$ in the polydisc.

We prove in §4 that the space L^p/H^p $(0<p<1)$ is not a locally holomorphic space. Holomorphic functions with values in this space differ essentially from scalar holomorphic functions. For instance, there exists a holomorphic function $f: \mathbb{D} \to L^p/H^p$ $(0<p<1)$ such that $\lim_{|z|\to 1} F(z) = 0$ (in L^p/H^p) and $F(0) \neq 0$.

In §5 we examine rotation-invariant subspaces of $H^p(0<p<1)$. These subspaces yield supplementary examples of non locally holomorphic spaces. Moreover, we construct a new example of a quasi-normed space with the trivial dual space and with the non-trivial space of compact endomorphisms. The first example of such F-space has been constructed by N.J.Kalton and J.H.Shapiro (see [40]).

Our §6 contains an example of a rotation-invariant subspace of $H^p(0<p<1)$ in the bidisc which is not finitely generated.

In this chapter we consider vector spaces over the field $\mathbb{C}$ only.

§1. Locally holomorphic spaces

We say that an open subset Ω of a TVS X is p s e u d o c o n v e x [x)] (or a domain of holomorphy) iff the intersection $E \cap \Omega$ is pseudoconvex for every finite-dimensional subspace E of X. This definition corresponds to the usual notion of a pseudoconvex domain of the space $\mathbb{C}^d$ (see [5]).

We say that a TVS X is l o c a l l y h o l o m o r p h i c (or locally pseudoconvex)[*)] iff there exists a base of balanced pseudoconvex neighbourhoods of the origin in X.

It is clear that each locally convex space is locally holomorphic. The definitions imply immediately the following assertion.

Let T be a linear continuous operator from a TVS X into a TVS Y and let U be a pseudoconvex subset of Y. Then $T^{-1}(U)$ is a pseudoconvex subset of X.

In particular, each subspace of a locally holomorphic space is locally holomorphic. We shall see that the analogous assertion for quotient-spaces fails (§4).

Let Ω be an open subset of a TVS X and let $u: \Omega \to [-\infty, +\infty)$ be an upper semicontinuous function. The function u is said to be p l u r i s u b h a r m o n i c iff the restriction $u|(\ell \cap \Omega)$ is a subharmonic function for every complex line $\ell \subset X$.

Let U be an open balanced neighbourhood of the origin in TVS X. Denote by P_U the Minkowski functional of the set U, $P_U(x) \overset{def}{=} \inf \{\alpha > 0 : \alpha^{-1} x \in U\}$. It is clear that P_U is an upper semicontinuous function.

Denote by $\mathcal{P}_A(X)$ the set of all polynomials with coefficients in a TVS X: $h \in \mathcal{P}_A(X) \Longleftrightarrow h = \sum_{k=0}^{N} h_k z^k (h_k \in X)$.

THEOREM 1.1. Let X be a TVS and let U be an open balanced neighbourhood of the origin in X, $0 < q < +\infty$. Then the following assertions are equivalent:

1) the function $\log P_U$ is plurisubharmonic,

$2)_q$ the function P_U^q is plurisubharmonic,

3) the set U is pseudoconvex,

---

x) Sometimes (see e.g. [8]) the terms "pseudoconvex" and "locally pseudoconvex" are used in quite different senses.

4) for every $F \in \mathcal{P}_A(X)$ the following inequality holds:

$$P_U(F(0)) \leq \max_{\zeta \in \mathbb{T}} P_U(F(\zeta)).$$

PROOF. Without loss of generality we can suppose that $X = \mathbb{C}^d$. The implications 1)⇒2)⇒3)⇒4) are rather obviously. The proof of the implication 4) ⇒ 1) uses the following lemma.

LEMMA 1.2. Let h be a positive continuous function on the circle $\mathbb{T}$. Then there exists a sequence of polynomials $\{g_n\}_{n \geq 0}$, $g_n \in \mathcal{P}_A(\mathbb{C})$ such that $|g_n| \leq h$ on $\mathbb{T}$ and

$$\lim_{n \to +\infty} \log |g_n(0)| = \int_{\mathbb{T}} \log h \, dm.$$

PROOF OF LEMMA. Let f be an outer function such that $|f| = h$ almost everywhere on $\mathbb{T}$. Denote by $\{\sigma_n(f)\}$ the sequence of Cesàro means of the function f. It is clear that $|\sigma_n(f)| \leq \sigma_n(h)$.

Take a sequence of positive numbers $\{a_n\}_{n \geq 0}$ such that $\lim_{n \to +\infty} a_n = 1$ and $a_n \sigma_n(h) \leq h$ everywhere on $\mathbb{T}$. It remains to put $g_n = a_n \sigma_n(f)$. ●

Let us continue the proof of the Theorem 1.1. It remains to prove that 4) ⇒ 1). We need to check that the following inequality holds (for arbitrary vectors x_0, x_1 in X)

$$\log P_U(x_0) \leq \int_{\mathbb{T}} \log P_U(x_0 + \zeta x_1) \, dm(\zeta). \tag{1}$$

Consider an arbitrary positive continuous function h such that $h(\zeta) \leq (P_U(x_0 + x_1\zeta))^{-1}$ for all $\zeta \in \mathbb{T}$. Applying Lemma 1.2 we obtain a sequence of polynomials $\{g_n\}_{n \geq 0}$.

Applying the assertion 4) to the vector polynomial $(x_0 + x_1\zeta) \times g_n(\zeta)$ we get $|g_n(0)| P_U(x_0) \leq 1$. Therefore

$$\log P_U(x_0) \leq -\log |g_n(0)| \longrightarrow -\int_{\mathbb{T}} \log h \, dm$$

for $n \to +\infty$. Since the function $-\log h$ is an arbitrary continuous majorant of the upper semicontinuous function $\log P_U(x_0 + x_1\zeta)$ $(\zeta \in \mathbb{T})$ the inequality (1) is proved. ●

COROLLARY 1.3. Let U be an open balanced neighbourhood of the origin in a TVS X. Then U is pseudoconvex if and only if the intersection $E \cap U$ is pseudoconvex for every two-dimensional subspace E of X. ●

LEMMA 1.4. Let U be an open balanced neighbourhood of the

origin in TVS X. Then there exists the smallest pseudoconvex set $\widetilde{U}$ containing U. Moreover, $\widetilde{U}$ is a balanced neighbourhood and if the neighbourhood U is p-convex then the neighbourhood $\widetilde{U}$ is p-convex.

PROOF. We shall inductively construct a sequence $\{U_n\}_{n\geqslant 0}$ of open subsets of X. Put $U_0 \overset{def}{=\!=} U$. Assume that U_n has been constructed. Denote by U_{n+1} the set of all vectors x in X such that there exists a polynomial $F_x(z)=\sum\limits_{k\geqslant 0} a_k z^k$ in $\mathcal{P}_A(X)$ satisfying the following condition: $a_0=x$, $F_x(\zeta)\in U_n$ for every $\zeta\in\mathbb{T}$. It is clear that all U_n are open and balanced. Moreover, if U_0 is p-convex, all U_n are p-convex $(0<p\leqslant 1)$. It is easy to see that $\widetilde{U}=\bigcup\limits_{n\geqslant 0} U_n$. ●

This lemma implies that if the set of all pseudoconvex neighbourhoods of the origin in TVS X is a base of neighbourhoods of the origin, then X is locally holomorphic.

With each TVS (X,τ) we may associate its pseudoconvex envelope $(X,\widetilde{\tau})$, a base of $\widetilde{\tau}$-neighbourhoods of the origin in X being the set of all balanced pseudoconvex τ-neighbourhoods of the origin in X. We shall write $\widetilde{X}$ instead of $(X,\widetilde{\tau})$.

Recall that $\mathcal{L}(X,Y)$ denotes the set of all linear continuous operators from TVS X into TVS Y.

PROPOSITION 1.5. Let X be a TVS. Then the following assertions are equivalent.

1) each non-empty pseudoconvex subset of X is X;
2) the topology of the space $\widetilde{X}$ is trivial;
3) $\mathcal{L}(X,Y)=\{0\}$ for each Hausdorff locally holomorphic space Y. ●

A n e x a m p l e o f a p- c o n v e x s u b s e t o f $\mathbb{C}^2$ w h i c h i s n o t p s e u d o c o n v e x. Suppose that $0<A<B$, $(A+B)2^{-1/p}\leqslant A$. Then the set $U\overset{def}{=\!=}\{(z_1,z_2)\in\mathbb{C}^2 : \max(|z_1|,|z_2|)<B,\ \min(|z_1|,|z_2|)<A\}$ is a p-convex neighbourhood, but it is not pseudoconvex (see e.g. [5]).

PROPOSITION 1.6. The space $L^0(\mathbb{T})$ is not locally holomorphic. Moreover, the equality $\mathcal{L}(L^0(\mathbb{T}),Y)=\{0\}$ holds for each Hausdorff locally holomorphic space Y.

PROOF. We shall prove that each pseudoconvex balanced neighbourhood U of zero in $L^0(\mathbb{T})$ is $L^0(\mathbb{T})$. Let $f\in L^0(\mathbb{T})$. Consider a sequence of polynomials $\{P_n\}_{n\geqslant 1}$ in $\mathcal{P}_A(\mathbb{C})$ such that $\lim\limits_{n\to+\infty} P_n=0$ (in $L^0(\mathbb{T})$) and $P_n(0)=1$ for all $n\in\mathbb{N}$. It is clear that $P_n(z\xi)f(\xi)\in\mathcal{P}_A(L^0(\mathbb{T}))$. Moreover, $P_n(z\xi)f(\xi)\in U$ for all $z\in\mathbb{T}$, if n is a sufficiently large number. Therefore, $f\in U$. ●

PROPOSITION 1.7. Let μ be a positive measure. Then $L^p(\mu)$ is a locally holomorphic space for all $p\in(0,+\infty]$.

PROOF. We may suppose that $p<1$. We shall prove that the function $f\longmapsto \|f\|^p_{L^p}$ $(f\in L^p)$ is plurisubharmonic, i.e. the inequality

$$\|f\|^p_{L^p}\leqslant\int_{\mathbb{T}}\|f+\zeta g\|^p_{L^p}\,dm(\zeta)$$

holds for all $f,g\in L^p$. This inequality may be proved by the integration with respect to the measure μ of the following obvious inequality

$$|f(\omega)|^p\leqslant\int_{\mathbb{T}}|f(\omega)+\zeta g(\omega)|^p\,dm(\zeta)$$

holding for μ-almost all ω. ●

Analogously one can prove that a quasinormed symmetric space X is locally holomorphic, if for some $q\in(0,+\infty)$ the space X^q (see chapter I) has a norm topology.

Another example of a locally holomorphic space is the space $N\overset{def}{=}\{f\in L^0(\mathbb{T}):\int\log(1+|f|)\,dm<+\infty\}$. It is easy to check that the function $f\mapsto\int\log(1+|f|)dm$ $(f\in N)$ is plurisubharmonic. Therefore for every $\varepsilon>0$ the neighbourhood $\{f\in N:\int\log(1+|f|)\,dm<\varepsilon\}$ is pseudoconvex.

Note that the quasinorm $\|f\|_{L^p}$ is plurisubharmonic for every $p>0$. It follows from the proof of Prop.1.7 and from Th.1.1.

The following theorem shows that this is not by chance.

THEOREM 1.8. Let X be a p-normed $(0<p<1)$ locally holomorphic space. Then there exists a plurisubharmonic p-norm on X inducing the same topology.

PROOF. This theorem follows from Lemma 1.4. ●

In §4 we shall construct a non-locally holomorphic p-normed space containing trivial pseudoconvex sets only. We shall use the following

THEOREM 1.9. Let F be a continuous function from $\overline{\mathbb{D}}$ into a Hausdorff TVS X such that $F\,|\,\mathbb{T}\equiv 0$ and

$$F(z)=\sum_{n\geqslant0}x_n z^n\qquad(z\in\mathbb{D}),\ x_n\in X,\tag{2}$$

the series (2) converging uniformly on compact subsets of $\mathbb{D}$. If

$A\in\mathcal{L}(X,Y)$, Y being a Hausdorff locally holomorphic space, then $Ax_n=0$ for all $n\in\mathbb{Z}_+$.

PROOF. Suppose that $Ax_n\neq 0$ for any $n\in\mathbb{Z}_+$. Denote by k the smallest number n such that $Ax_n\neq 0$. Let U be a pseudoconvex balanced neighbourhood of the origin in Y. It is clear that there exist $r\in(0,1)$ and $N>k$ such that $\sum_{n=0}^{N} r^{n-k}x_n\zeta^n\in A^{-1}(U)$ for all $\zeta\in\mathbb{T}$. Therefore $\sum_{n=0}^{N-k} r^n\zeta^n Ax_{n+k}\in U$ for all $\zeta\in\mathbb{T}$. Hence $Ax_k\in U$. Thus $Ax_k=0$ in view of the fact that Y is a Hausdorff locally holomorphic space and U is an arbitrary neighbourhood of the origin in Y. Therefore we get a contradiction. ●

Note that if X is a locally bounded space then the uniform convergence of the series (2) on compact subsets of $\mathbb{D}$ follows from its pointwise convergence.

§ 2. Hardy spaces H^p with values in a quasinormed space X

Let X be a complete quasinormed space. Define a quasinorm on the set of all X-valued polynomials $\mathcal{P}_A(X)$:

$$\|f\|^p_{H^p(X)} \overset{def}{=} \int_{\mathbb{T}} \|f(\zeta)\|^p\,dm(\zeta) \qquad (0<p<+\infty).$$

Denote by $H^p(X)$ the completion of the space $(\mathcal{P}_A(X), \|\ \|_{H^p(X)})$. A function $f:\mathbb{T}\to X(\mathrm{mod}\,0)$ is called m e a s u r a b l e iff f is the limit of an almost everywhere convergent sequence of step functions. Denote by $L^p(X)$ the space of all measurable functions $f:\mathbb{T}\to X$ such that

$$\|f\|^p_{L^p(X)} = \int_{\mathbb{T}} \|f(\zeta)\|^p\,dm(\zeta)<+\infty \quad (0<p<+\infty).$$

It is clear that the space $H^p(X)$ is canonically isometric with a subspace of $L^p(X)$. Denote by $H^p_-(X)$ the set of all functions $f\in L^p(X)$ such that $z^{-1}f(z^{-1})\in H^p(X)$. Note that if $\zeta\in\mathbb{T}$ and $x\in X$, then $x\frac{1}{1-\bar\zeta z}\in H^p(X)\cap H^p_-(X)$ for $p\in(0,1)$.

The results of this section are mostly not proved here, their proofs not differing from the proofs of respective scalar results.

In this section X will denote a complete quasinormed space.

THEOREM 2.1 (see [15] or §1.2). Let X be a (generally non-locally holomorphic) quasinormed complete space, $p\in(0,1)$. Then

a) $H^p(X)+H^p_-(X)=L^p(X)$,

b) the intersection $H^p(X)\cap H^p_-(X)$ is the closure in $L^p(X)$ of the linear span of the family of functions $\{x\frac{1}{1-\bar\zeta z}\}$ $(x\in X,\ \zeta\in\mathbb{T})$. ●

It is interesting to compare Th.2.1 a) with results of A.V. Buhvalov [24]. In the work [24] it is proved that there exist Banach spaces X such that $L^p(X)\neq H^p(X)+H^p_-(X)$ for all $p\in[1,+\infty)$.

THEOREM 2.2. Let $p\in(0,+\infty)$, $a\in\mathbb{D}$ and let X be a locally holomorphic space. Then the operator $f\mapsto f(a)$, initially defined for $f\in\mathcal{P}_A(X)$ has a bounded extension to $H^p(X)$, which we shall denote by $\Phi_a: H^p(X)\to X$. Moreover, $\|\Phi_a\|\leq\leq\left(\frac{1+|a|}{1-|a|}\right)^{1/p}$ if the quasinorm in X is plurisubharmonic. ●

COROLLARY 2.3. Let X be a locally holomorphic space. Then $L^p(X)\neq H^p(X)$ for all $p\in(0,+\infty)$.

PROOF. It is clear that $zH^p(X)\subset \mathrm{Ker}\,\Phi_0$. Therefore $zH^p(X)\neq H^p(X)$. Consequently $H^p(X)\neq z^{-1}H^p(X)$. It remains to note that $H^p(X)\subset z^{-1}H^p(X)\subset L^p(X)$. ●

We shall write $f(a)$ instead of $\Phi_a(f)$ $(a\in\mathbb{D})$ if X is a locally holomorphic space.

THEOREM 2.4. Let X be a locally holomorphic space, $p\in(0,+\infty)$. Then to each function $f\in H^p(X)$ there corresponds a unique sequence $\{x_k\}_{k\geq 0}$ in X such that $f(a)=$ $=\sum_{k\geq 0}a^k x_k$ for all $a\in\mathbb{D}$. ●

Let $\zeta\in\mathbb{T}$. Recall that Γ_ζ denotes the interior of the convex hull of the point ζ and of the disc of the radius $\frac{1}{2}$ centered at the origin.

THEOREM 2.5. Let X be a locally holomorphic space, $p\in(0,+\infty)$. Put $(\mathcal{M}f)(\zeta)\overset{def}{=}\sup\{\|f(a)\|: a\in\Gamma_\zeta\}$ $(\zeta\in\mathbb{T})$. Then $\|\mathcal{M}f\|_{L^p}\leq c(p,X)\|f\|_{H^p(X)}$. ●

COROLLARY 2.6. Let X be a locally holomorphic space and let $f\in H^p(X)$, $p\in(0,+\infty)$. Then

$$\lim_{z\to\zeta,\ z\in\Gamma_\zeta} f(z)=f(\zeta)$$

for almost all $\zeta\in\mathbb{T}$. ●

COROLLARY 2.7 (see [2]). Let X be a locally holomorphic

space and let μ be a finite positive Borel measure on $\mathbb{D}$, $0<p\leq q<+\infty$. Suppose the inequality $\mu\{z\in\mathbb{D}:|z-\zeta|<r\}\leq \leq Ar^{q/p}$ holds for all $\zeta\in\mathbb{T}$ and all $r>0$. Then

$$\Big(\int_{\mathbb{D}}\|f\|^q d\mu\Big)^{1/q}\leq c(A,p,q,X)\,\|f\|_{H^p(X)}\,. \quad \bullet$$

THEOREM 2.8. Let X be a locally holomorphic space, $0<p<+\infty$. Then

$$\int_{\mathbb{T}}\log\|f(\zeta)\|\,dm(\zeta)>-\infty$$

for all $f\in H^p(X)$, $f\not\equiv 0$.

REMARK 2.9. It follows from results of §4 that Th.2.2, Cor.2.3 and Th.2.8 fail for non-locally holomorphic spaces.

§3. Functionals on the space $H^p(0<p<1)$ in polydiscs

In this section we give a new proof of the A.Frazier theorem [63] which is a multidimensional generalization of the Duren-Romberg-Shields theorem concerning functionals on the space $H^p(0<p<1)$.

Denote by $H^p_+(\mathbb{T}^d)$ the set of all holomorphic functions $f:\mathbb{D}^d\to\mathbb{C}$ satisfying

$$\|f\|^p_{H^p_+(\mathbb{T}^d)}\overset{\text{def}}{=\!=}\sup_{r<1}\int_{\mathbb{T}^d}|f(r\zeta)|^p dm_d(\zeta)<+\infty\,.$$

Define the operator $\mathcal{D}=\mathcal{D}_d$ on the space of all scalar functions holomorphic in $\mathbb{D}^d$:

$$\mathcal{D}f=\!=\frac{\partial^d(z_1z_2\dots z_d f)}{\partial z_1\partial z_2\dots\partial z_d}\,.$$

Denote by $(\Lambda^\alpha_+)^d_n$ $(\alpha>0,\ n>\alpha,\ n\in\mathbb{N})$ the set of all holomorphic functions $f:\mathbb{D}^d\to\mathbb{C}$ such that

$$\|f\|_{(\Lambda^\alpha_+)^d_n}\overset{\text{def}}{=\!=}\sup\Big\{|(\mathcal{D}^n f)(z)|\prod_{j=1}^d(1-|z_j|)^{n-\alpha}:z\in\mathbb{D}^d\Big\}<+\infty\,.$$

It will follow from Th.3.2 that $(\Lambda^\alpha_+)^d_n$ does not depend on n.

Let $B^p_+(\mathbb{T}^d)$ denote the set of all holomorphic functions $f:\mathbb{D}^d\to\mathbb{C}$ such that

$$\|f\|_{B^p_+(\mathbb{T}^d)} \overset{def}{=\!=} \int_{\mathbb{D}^d} |f(z)| \prod_{j=1}^{d} (1-|z_j|)^{\frac{1}{p}-2}\, dm_{2d} < +\infty .$$

Using Cor.2.7 and the induction in d, it is not hard to prove the following

THEOREM 3.1. Let $p\in(0,1)$, $d\in\mathbb{N}$. Then

$$H^p_+(\mathbb{T}^d) \subset B^p_+(\mathbb{T}^d) . \quad \bullet$$

THEOREM 3.2. Let Φ be a linear continuous functional on the space $H^p_+(\mathbb{T}^d)$ $(d\in\mathbb{N},\ 0<p<1)$, $n\in\mathbb{N}$, $n>\frac{1}{p}-1$. Then there exists a unique function $g\in(\Lambda^{1/p-1}_+)^d_n$ such that

$$\Phi(f) = \lim_{r\to1-} \int_{\mathbb{T}^d} f(r\zeta)\,\overline{g(r\zeta)}\, dm_d(\zeta) \tag{3}$$

for all $f\in H^p_+(\mathbb{T}^d)$. Conversely, each function $g\in(\Lambda^{1/p-1}_+)^d_n$ defines a linear continuous functional on $H^p_+(\mathbb{T}^d)$ by the formula (3).

PROOF. Let $g\in(\Lambda^{1/p-1}_+)^d_n$. Check that the formula (3) defines a linear continuous functional on $H^p_+(\mathbb{T}^d)$. Note that

$$\int_{\mathbb{T}^d} f(r\zeta)\overline{g(r\zeta)}dm_d(\zeta) = \left(\frac{2^{n-1}}{r^2(n-1)!\pi}\right)^d \int_{r\mathbb{D}^d} f\overline{\mathcal{D}^n g}\left(\prod_{j=1}^{d}\log\frac{r}{|z_j|}\right)^{n-1} dm_{2d} .$$

Hence

$$\left|\int_{\mathbb{T}^d} f(r\zeta)\overline{g(r\zeta)}\, dm_d(\zeta)\right| \leq \left(\frac{2^{n-1}}{r^2(n-1)!\pi}\right)^d \int_{\mathbb{D}^d} |f||\mathcal{D}^n g|\left(\prod_{j=1}^{d}\log\frac{1}{|z_j|}\right)^{n-1} dm_{2d} \leq$$

$$\leq \frac{c(p,d,n)}{r^{2d}} \|f\|_{B^p_+(\mathbb{T}^d)} \|g\|_{(\Lambda^{1/p-1}_+)^d_n} .$$

It remains to note that

$$\lim_{r\to 1-}\left(\frac{2^{n-1}}{r^2(n-1)!\pi}\right)^d\int_{r\mathbb{D}^d} f\overline{\mathcal{D}^n g}\left(\prod_{j=1}\log\frac{r}{|z_j|}\right)^{n-1}dm_{2d}=$$

$$=\left(\frac{2^{n-1}}{(n-1)!\pi}\right)^d\int_{\mathbb{D}^d} f\overline{\mathcal{D}^n g}\left(\prod_{j=1}^d\log\frac{1}{|z_j|}\right)^{n-1}dm_{2d}\ .$$

Let now $\Phi\in(H^p_+(\mathbb{T}^d))^*$. Set $F_{z_1,z_2,\dots,z_d}(\zeta_1,\zeta_2,\dots,\zeta_d)\overset{\text{def}}{=\!=}$ $=\prod_{j=1}^d(1-\bar z_j\zeta_j)^{-1}$. It is clear that $F_{z_1,z_2,\dots,z_d}\in H^p_+(\mathbb{T}^d)$ for all $z\in\mathbb{D}^d$. Put $g(z_1,z_2,\dots,z_d)=\overline{\Phi(F_{z_1,z_2,\dots,z_d})}$.

The inclusion $g\in(\Lambda_+^{1/p-1})_n^d$ follows from the estimate:

$$\left\|\prod_{j=1}^d\frac{1}{(1-\bar z_j\zeta_j)^{n+1}}\right\|_{H^p_+(\mathbb{T}^d)}\le c(p,d,n)\prod_{j=1}^d(1-|z_j|)^{1/p-1-n}\ .\ \bullet$$

REMARK 3.3. The proof of this theorem shows that the Banach envelope of the space $H^p_+(\mathbb{T}^d)$ is canonically isomorphic with the space $B^p_+(\mathbb{T}^d)$.

§4. An example of a quasinormed space which is not locally holomorphic

It is well known (see [20]) that the space H^p is isomorphic with the space L^p provided $p\in(1,+\infty)$. The proof of this fact uses the M.Riesz theorem about conjugate functions.

Analogously, using theorem 1.0.1(a) we can prove the following

PROPOSITION 4.1. Let $p\in(0,1)$. Then the following three spaces are isomorphic:

$$L^p/H^p_-\ ,\quad H^p/H^p\cap H^p_-\ ,\quad L^p/H^p\cap H^p_-\ .\ \bullet$$

PROPOSITION 4.2. There exists a continuous function $F:\overline{\mathbb{D}}\to L^p/H^p_-$, $0<p<1$ such that

$$F(z)=\sum_{n\geq 0} z^n f_n\ (z\in\mathbb{D}),\quad f_n\in L^p/H^p_-,\quad F|\mathbb{T}\equiv 0$$

and the linear span of the family $\{f_n\}_{n\geq 0}$ is dense in L^p/H^p_-.

PROOF. Consider the function $G:\overline{\mathbb{D}}\to L^p$, $(G(z))(\zeta)\overset{def}{=}\frac{1}{1-z\zeta}\ (z\in\overline{\mathbb{D}},\ \zeta\in\mathbb{T})$. We have $G(z)=\sum_{n\geq 0} z^n\zeta^n$ for all $z\in\mathbb{D}$, the series converging in L^p. Let π be the quotient-map $\pi: L^p\to L^p/H^p_-$. It remains to set $F=\pi\circ G$. ●

COROLLARY 4.3. Let X be a Hausdorff locally holomorphic TVS, $p\in(0,1)$. Then $\mathcal{L}(L^p/H^p_-,X)=\{0\}$.

PROOF. It is sufficient to use Prop.4.2 and Th.1.9. ●

COROLLARY 4.4 (N.J.Kalton, 62). Let μ be a finite positive measure. Then $\mathcal{L}(L^p/H^p_-,L^0(\mu))=\{0\}$.

PROOF. A result of E.M.Nikishin implies that each operator $T\in\mathcal{L}(L^p/H^p_-,\ L^0(\mu))$ is factorizable through $L^{1/2}(\mu)$ (see [45]). It remains to apply Cor.4.3 and Prop.1.7. ●

REMARK 4.5. We can prove analogously that $\mathcal{L}(L^p/H^p_-,L^0(\mu,X))=\{0\}$ $(0<p<1)$, $L^0(\mu,X)$ denoting the space of all measurable functions with values in a locally holomorphic quasinormed space X. It is sufficient to note that $L^p(X)$ is locally holomorphic if X is a locally holomorphic quasinormed space.

REMARK 4.6. It is easy to prove that if $X=L^p/H^p_-$ $(0<p<1)$ then $H^q(X)=L^q(X)$ for all $q\in(0,+\infty)$.

REMARK 4.7. The proof of Proposition 4.2 shows that if X is a separable metric symmetric space and $X[0,1]\ni 1/t$ then there exists a continuous function $F:\overline{\mathbb{D}}\to X_A/X_A\cap X_{\bar{A}}$ (resp. $X/X_{\bar{A}}$) such that $F(z)=\sum_{n\geq 0} f_n z^n (z\in\mathbb{D})$, $F|\mathbb{T}\equiv 0$, $f_n\in X_A/X_A\cap X_{\bar{A}}$ (resp. $X/X_{\bar{A}}$) the linear span of family $\{f_n\}_{n\geq 0}$ being dense in $X_A/X_A\cap X_{\bar{A}}$ (resp. $X/X_{\bar{A}}$) and the series converging uniformly on compact sets of the disc $\mathbb{D}$.

§ 5. On rotation-invariant subspaces of H^p, $0<p<1$.

Let G be a compact abelian group. Denote by Γ the dual group. Let A be a subset of Γ. Recall that $L^p_A(G)$

denotes the closure (in $L^p(G)$) of the linear span of the set A $(0<p<+\infty)$. It is clear that $L^p_A(G)$ is a closed rotation-invariant subspace of $L^p(G)$ $(0<p<+\infty)$. It is well-known that these subspaces are the only invariant [x)] subspaces of $L^p(G)$ when $p\in[1,+\infty)$. Moreover, if G is a metrizable group then each invariant subspace of $L^p(G)$ $(1\leq p<+\infty)$ is cyclic i.e. generated by a single function f , such an f being e.g. $x\mapsto\sum_{\alpha\in\Lambda}\varepsilon_\alpha\cdot\alpha(x)$, where $\{\varepsilon_\alpha\}_{\alpha\in A}$ is a summable family of non-zero numbers.

K. de Leeuw [42] noticed that the invariant subspace $L^p_{\mathbb{Z}_+}(\mathbb{T})\cap L^p_{\mathbb{Z}_-}(\mathbb{T})\simeq H^p\cap H^p_-\neq\{0\}$ $(0<p<1)$ does not contain any character of the group $\mathbb{T}$. It follows from Th.1.0.1(b) that this subspace is cyclic. It is generated by the function $\frac{1}{1-z}$. However, the invariant subspace $\mathbb{C}+H^p\cap H^p_-$ $(0<p<1)$ is not cyclic, see Cor.5.2 below.

Let X be an invariant subspace of $L^p(G)$. Denote by X^*_0 the set of all rotation-invariant functionals on X . It is clear that X^*_0 is a subspace of the Banach space X^* . Note that $\dim X^*_0\leq 1$ if $p\in[1,+\infty)$.

PROPOSITION 5.1. Let G be a compact abelian group and let X be an invariant subspace of $L^p(G)$. Suppose that $n<\dim X^*_0$. Then the invariant subspace, generated by a collection of functions $\{f_k\}_{k=1}^n$ in X does not coincide with X.

PROOF. Consider the operator $T: X^*_0\longrightarrow\mathbb{C}^n$, $T\Phi\overset{def}{=}(\Phi(f_1),\Phi(f_2),\dots,\Phi(f_n))$. Since $n<\dim X^*_0$, there exists a functional $\Phi\in\operatorname{Ker}T$, $\Phi\neq 0$. To finish the proof note that $\{f_k\}_{k=1}^n\subset\operatorname{Ker}\Phi\subsetneq X$ and $\operatorname{Ker}\Phi$ is an invariant subspace of $L^p(G)$. ●

COROLLARY 5.2. The invariant subspace $H^p\cap H^p_-+\mathbb{C}$ of $L^p(\mathbb{T})$ is not cyclic $(0<p<1)$.

PROOF. It is easy to check that $\dim X^*_0=2$, X denoting $H^p\cap H^p_-+\mathbb{C}$. The space X^*_0 is the linear span of two functionals: $f\longmapsto f(0)$, $f\longmapsto f(\infty)$. ●

I don't know whether there exists an invariant subspace of $L^p(\mathbb{T})$ (H^p) , $0<p<1$, which cannot be generated by two functions. In the following section we shall provide an example of an invariant subspace of $H^p_+(\mathbb{T}^2)\simeq L^p_{\mathbb{Z}^2_+}(\mathbb{T}^2)\subset L^p(\mathbb{T}^2)$, $0<p<1$ such that $\dim X^*_0=+\infty$. In particular, X is not finitely generated.

[x)] In this chapter an invariant subspace means a rotation-invariant closed subspace.

Let's prove a generalization of Prop.4.2. Let X be an invariant subspace of H^p. Denote by $\sigma(X)$ the set of all $n\in\mathbb{Z}_+$, such that $\hat{f}(n)\neq 0$ for some function $f\in X$. It is clear that $X\subset L^p_{\sigma(X)}=L^p_{\sigma(X)}(\mathbb{T})$. In general $X\neq$ $\neq L^p_{\sigma(X)}$ if $p<1$. For instance, if $X=H^p\cap H^p_-$ then $L^p_{\sigma(X)}=H^p\,(0<p<1)$.

PROPOSITION 5.3. Let X be an invariant subspace of H^p. Then there exists a continuous function $F:\overline{\mathbb{D}}\to L^p_{\sigma(X)}/X$ such that $F(z)=\sum_{n\geq 0} z^n f_n\,(z\in\mathbb{D})$, $F\,|\,\mathbb{T}\equiv 0$, $f_n\in L^p_{\sigma(X)}/X$, the linear span of the family $\{f_n\}_{n\geq 0}$ being dense in the space $L^p_{\sigma(X)}/X$.

PROOF. Note that for any $n\in\sigma(X)$ the set of all $f\in X$ such that $\hat{f}(n)\neq 0$, is a nowhere dense subspace of X. Therefore there exists a function $f\in X$ such that $\hat{f}(n)\neq 0$ for all $n\in\sigma(X)$. Put $F(z)=\sum_{n\geq 0} z^n\pi(\hat{f}(n)\zeta^n)$, $\pi: L^p_{\sigma(X)}\to L^p_{\sigma(X)}/X$ being the quotient-map. ●

It is easy to see that the analogs of Prop.4.3 and Prop.4.4 hold for the space $L^p_{\sigma(X)}/X$.

In the paper [40] an F-space is constructed with the trivial dual space and with the non-trivial space of compact endomorphisms. In conclusion of this section we shall construct another example of such space. Consider the operator of fractional integration

$$I_\alpha:\mathcal{D}'(\mathbb{T})\to\mathcal{D}'(\mathbb{T}),\ I_\alpha\Big(\sum_{k\in\mathbb{Z}} a_k z^k\Big)\stackrel{def}{=\!=}\sum_{k\in\mathbb{Z}}\frac{|k|!}{(\alpha+1)(\alpha+2)\dots(\alpha+|k|)}a_k z^k.$$

The well-known Hardy - Littlewood theorem asserts (see [36] or §3.0) that $I_\alpha(H^p)\subset H^q$ if $\frac{1}{p}-\frac{1}{q}=\alpha$ $(\alpha,p,q>0)$. However, this operator $I_\alpha:H^p\to H^q$ is not compact $(\frac{1}{p}-\frac{1}{q}=\alpha)$.

LEMMA 5.4. Let $\alpha,p,q>0$ and let $\frac{1}{p}-\frac{1}{q}<\alpha$. Then the operator $I_\alpha:H^p\to H^q$ is compact.

PROOF. We shall use Lemma 5.4 for $p,q<1$ only so we suppose that $p,q<1$. Denote by $\mathcal{H}^p_\alpha$ the set of all analytic functions $f:\mathbb{D}\to\mathbb{C}$ such that

$$\|f\|^p_{\mathcal{H}^p_\alpha}\stackrel{def}{=\!=}\int_{\mathbb{D}}|f|^p(1-|z|)^{\alpha p-1}dm_2<+\infty .$$

Without loss of generality suppose $q>p$. In [36] it is proved that $H^p\subset\mathcal{H}^q_{1/p-1/q}$. It is easy to check that the imbedding operator $\mathcal{H}^q_{1/p-1/q}\subset\mathcal{H}^q_\alpha$ is compact. It remains to use the inclusion $I_\alpha(\mathcal{H}^q_\alpha) \subset H^q$ $(q \leq 1)$ [36]. ●

EXAMPLE. Let $p, \alpha, q > 0$; $\frac{1}{p} - \frac{1}{q} < \alpha$; $(1+\alpha)p < 1$. Denote by X^p_α the invariant subspace of H^p generated by the function $(1-z)^{-1-\alpha}$ $(X^p_0 = H^p \cap H^p_-)$. Consider the compact operator $I_\alpha : H^p \longrightarrow H^q$. It is clear that $I_\alpha(X^p_\alpha) \subset \subset H^q \cap H^q_-$ since $I_\alpha\left(\frac{1}{(1-z)^{1+\alpha}}\right) = \frac{1}{1-z}$. Therefore we may consider the compact operator $\tilde{I}_\alpha : H^p/X^p_\alpha \longrightarrow H^q/X^q_0$. The equalities $(H^p/X^p_\alpha)^* = (H^q/X^q_0)^* = \{0\}$ follow from Prop.5.3. It remains to note that $\tilde{I}_\alpha \neq 0$ $(\tilde{I}_\alpha(1 + X^p_\alpha) = 1 + X^q_0 \neq X^q_0)$ and to consider $\beta \in \mathcal{L}(H^p/X^p_\alpha \times H^q/X^q_0)$, $\beta(f, g) \stackrel{def}{=} (0, \tilde{I}_\alpha f)$. ●

§6. An example of an invariant subspace of $L^p(\mathbb{T}^2)$ $(0<p<1)$ which is not finitely generated

Let α be an irrational number. Put $A_\alpha \stackrel{def}{=} \{(n_1, n_2) \in \mathbb{Z}^2 : n_1 - \alpha n_2 \geq 0\}$. The following proposition is proved in [54]. We give another proof.

PROPOSITION 6.1. Let $\Phi \in (L^p_{A_\alpha}(\mathbb{T}^2))^*$, $0<p<1$. Then $\Phi(z^{n_1}_1 z^{n_2}_2) = 0$ if $n_1 - \alpha n_2 > 0$.

PROOF. Let $n = (n_1, n_2) \in A_\alpha$ and let $n_1 > \alpha n_2$. It is clear that for each $m \in \mathbb{N}$ there exists $k \in \mathbb{Z}^2 \setminus \{(0,0)\}$ such that $n + jk \in A_\alpha$ for all $j \in [-m, m] \cap \mathbb{Z}$. Then Cor.4.2.3 implies the following inequality:

$$|\Phi(z^{n_1}_1 z^{n_2}_2)| \leq c_p m^{1-\frac{1}{p}}$$

for all $m \in \mathbb{N}$. ●

It follows from Prop.6.1 that $(L^p_{A_\alpha}(\mathbb{T}^2))^* = (L^p_{A_\alpha}(\mathbb{T}^2))^*_0$. It is proved in [3] that there exists a functional $\Phi_\alpha \in L^p_{A_\alpha}(\mathbb{T}^2)$ $(0<p<1)$ such that $\Phi_\alpha(1) = 1$, α being an irrational number. It is clear that $(L^p_{A_\alpha}(\mathbb{T}^2))^* = \mathbb{C} \cdot \Phi_\alpha$.

THEOREM 6.2. Let $p \in (0,1)$. Put

$$X \stackrel{def}{=} L^p_{\mathbb{Z}^2_+}(\mathbb{T}^2) \cap \bigcap_{\alpha \notin \mathbb{Q}, \alpha > 0} L^p_{A_\alpha}(\mathbb{T}^2).$$

Then $\dim X^*_0 = +\infty$.

PROOF. We shall prove that the space X^*_0 is not separable. Put $\Psi_\alpha \stackrel{def}{=} \Phi_\alpha | X$. Prove that $\|\Psi_\alpha - \Psi_\beta\| \geq c_p$ if $0 < \alpha < \beta$. To prove this construct a function $f \in X$ such that $\|f\|_{L^p} = c_p^{-1}$, $\Phi_\alpha(f) = 1$, $\Phi_\beta(f) = 0$.

Fix two positive integers n_1 and n_2 such that $n_1 < \beta n_2$ and $n_1 > \alpha n_2$. To finish the proof put $f = \frac{1}{1 - z_1^{n_1} z_2^{n_2}} = z_1^{-n_1} z_2^{-n_2} (z_1^{-n_1} z_2^{-n_2} - 1)^{-1}$. ●

COROLLARY 6.3. The invariant subspace $X = L^p_{\mathbb{Z}^2_+}(\mathbb{T}^2) \cap \bigcap_{\alpha \in \mathbb{Q}, \alpha > 0} L^p_{A_\alpha}(\mathbb{T}^2)$ is not finitely generated. ●

REMARK 6.4. Let $\mathfrak{A}$ be the set of all multiindices $(n_1, n_2) \in \mathbb{N}^2$ such that the numbers n_1 and n_2 are mutually prime. Let $B \subset \mathfrak{A}$. Denote by Y^p_B the invariant subspace of $L^p(\mathbb{T}^2)$ $(0 < p < 1)$ generated by the collection of functions $\{(1 - z_1^{n_1} z_2^{n_2})^{-1}\}_{(n_1, n_2) \in B}$. Then the space Y^p_B cannot be generated by any collection of functions $\{f_1, f_2, \dots, f_n\} \subset Y^p_B$ if $n < card(B)$.

Chapter Three. ATOMIC DECOMPOSITION OF $\mathcal{H}^p_\alpha(\mathbb{R}^d)$-FUNCTIONS

§0. Introduction

With a function u harmonic on the half-space $\mathbb{R}^{d+1}_+ \overset{def}{=} \{(x_0, x_1, \dots, x_d) \in \mathbb{R}^{d+1} : x_0 > 0\}$ we shall associate a maximal function $u^* : \mathbb{R}^d \to \mathbb{R}$, $u^*(x) \overset{def}{=} \sup\{|u(y,t)| : t > 0, y \in \mathbb{R}^d, |x - y| < t\}, x \in \mathbb{R}^d$. Denote by $H^p(\mathbb{R}^d)$ the class of all harmonic functions $u : \mathbb{R}^{d+1}_+ \to \mathbb{C}$ such that $u^* \in L^p(\mathbb{R}^d)$, $\|u\|_{H^p} \overset{def}{=} \|u^*\|_{L^p}$. The space $H^p(\mathbb{R}^d)$ can be naturally imbedded into the space of tempered distributions $S'(\mathbb{R}^d)$, $u \mapsto \lim_{t \to 0+} u(t, \cdot)$, the limit being taken in the space $S'(\mathbb{R}^d)$ (cf. [34]). We shall identify $u \in H^p(\mathbb{R}^d)$ with $\lim_{t \to 0+} u(t, \cdot) \in S'(\mathbb{R}^d)$ and so $H^p(\mathbb{R}^d) \subset S'(\mathbb{R}^d)$.

By the R.R.Coifman - R.Latter theorem ([28], [43]) each function $f \in H^p(\mathbb{R}^d)$ with $p \leq 1$ can be represented as $f = \sum_{k \geq 1} \alpha_k b_k$ (in $S'(\mathbb{R}^d)$), where $\alpha_k \in \mathbb{C}$, b_k is a p-atom and $\sum_{k \geq 1} |\alpha_k|^p < +\infty$.

Recall that a function $b \in L^\infty(\mathbb{R}^d)$ is called a d-*dimensional* p-*atom* if there exists a cube $Q \subset \mathbb{R}^d$ such that $supp\, b \subset Q$, $\|b\|_{L^\infty} \leq (m_d(Q))^{-1/p}$ and b is orthogonal to all polynomials of degree at most $\frac{d}{p} - d$.

In this chapter we describe the class of distributions f which admit a representation $f = \sum_{k \geq 1} \alpha_k b_k$ (in $S'(\mathbb{R}^d)$) where b_k's are p-atoms, $\alpha_k \in \mathbb{C}$ and $\sum |\alpha_k|^q < +\infty$ $(p < q \leq 1)$. The case $q = 1$ is well-known.

Denote by $\mathcal{H}^p_\alpha(\mathbb{R}^d)$ ($\alpha\in\mathbb{R}$, $p\in(0,+\infty)$) the space of all harmonic functions u on $\mathbb{R}^{d+1}_+$ such that

$$\|u\|^p_{\mathcal{H}^p_\alpha} \overset{def}{=\!=} \int_{\mathbb{R}^{d+1}_+} |u(x_0,x_1,\dots,x_d)|^p x_0^{\alpha p-1}\,dx < +\infty .$$

The spaces $\mathcal{H}^p_\alpha(\mathbb{R}^d)$ also will be identified with spaces of tempered distributions. Note that for $p\geqslant 1$ $\mathcal{H}^p_\alpha(\mathbb{R}^d)=\{0\}$ if and only if $\alpha\leqslant 0$ and for $p<1$ $\mathcal{H}^p_\alpha(\mathbb{R}^d)=\{0\}$ if and only if $\alpha+d(\frac{1}{p}-1)\leqslant 0$. All functions of the space $\mathcal{H}^p_\alpha(\mathbb{R}^d)$ have non-tangential boundary values almost everywhere on $\mathbb{R}^d$ if and only if $\alpha\leqslant 0$. If $\alpha\leqslant 0$ then $\mathcal{H}^p_\alpha(\mathbb{R}^d)\subset H^q(\mathbb{R}^d)$ for $q=\dfrac{pd}{p\alpha+d}$ and for any function $u\in\mathcal{H}^p_\alpha(\mathbb{R}^d)$ $\lim\limits_{t\to 0+} u(t,x)=0$ for almost all x in $\mathbb{R}^d$.

In this chapter we shall consider the classes $\mathcal{H}^p_\alpha(\mathbb{R}^d)$ with $\alpha>0$ and $0<p\leqslant 1$ only. It is known that $\mathcal{H}^p_\alpha(\mathbb{R}^d)\supset H^q(\mathbb{R}^d)$ if $\alpha=\frac{d}{q}-\frac{d}{p}>0$ (cf. G.H.Hardy, J.E. Littlewood [36] for the case $d=1$). Hence $\sum\limits_{k\geqslant 1}\alpha_k b_k\in\mathcal{H}^p_\alpha(\mathbb{R}^d)$ $(0<p\leqslant 1)$, if $\alpha_k\in\mathbb{C}$, $\sum\limits_{k\geqslant 1}|\alpha_k|^p<+\infty$ and b_k's are q-atoms.

The main result of this chapter (theorem 2.1) asserts that any function $f\in\mathcal{H}^p_\alpha(\mathbb{R}^d)$ $(0<p\leqslant 1)$ can be represented as such a sum of q-atoms. Recently R.R Coifman and R.Rochberg [56] obtained a similar result but instead of the atoms they considered other functions.

We deduce in §4 that the p-convex envelope $[H^q(\mathbb{R}^d)]_p$ of $H^q(\mathbb{R}^d)$ is naturally isomorphic to $\mathcal{H}^p_\alpha(\mathbb{R}^d)$ $(\alpha=\frac{d}{q}-\frac{d}{p}>0,\ 0<p\leqslant 1)$. An application of this result will be given in §4.

A measure $\mu\in C^*(\mathbb{R}^d)$ is called a g e n e r a l i z e d p-a t o m $(0<p<1)$ in $\mathbb{R}^d$ if it is orthogonal to all polynomials of degree at most $d/p-d$ and if there exists a cube Q such that $\operatorname{supp}\mu\subset Q$ and $\|\mu\|\leqslant(m_d(Q))^{1-1/p}$. It is easy to see that $\mu\in H^p(\mathbb{R}^d)$ and $\|\mu\|_{H^p}\leqslant c(p,d)$.

Using this remark and theorem 2.1 it is easy to see that the class of distributions $\sum\limits_{k\geqslant 1}\alpha_k b_k$, where $\alpha_k\in\mathbb{C}$, $\sum\limits_{k\geqslant 1}|\alpha_k|^p<+\infty$ and b_k's are generalized q-atoms $(\alpha=\frac{d}{q}-\frac{d}{p}>0,\ 0<p\leqslant 1)$, coincides with $\mathcal{H}^p_\alpha(\mathbb{R}^d)$. It turns out that the above assertion can be strengthened, namely it can be supposed that b_k's are standard q-atoms (Theorem 3.1).

Put

$$\mathcal{M}_q(a,b) \overset{def}{=\!=} N^{-\frac{d}{q}+d}|b|^{-\frac{d}{q}+d}2^{-N}\sum_{k=0}^{N}(-1)^k\binom{N}{k}\sigma_{a+kb} ,$$

$a, b \in \mathbb{R}^d$, $b \neq 0$, $N = [\frac{d}{q}] - d + 1$. The distribution $\mathcal{M}_q(a,b)$ is called a s t a n d a r d q - a t o m.

In §1 we construct an imbedding of $\mathcal{H}^p_\alpha(\mathbb{R}^{d_1})$ into $H^p(\mathbb{R}^d)$ $(d_1 < d)$. Using this imbedding, Coifman - Latter theorem and certain properties of the fractional integration operator we obtain the main result of the chapter.

In §3 we give its refinement (theorem 4.1). In §4 we describe the space $[H^p(\mathbb{R}^d)]_q$, $0 < p < q \leq 1$

Let us state some known assertions which will be used in this chapter.

THEOREM (L.Carleson, P.L.Duren). Let μ be a positive measure in $\mathbb{R}^{d+1}_+$. Suppose that for some $\alpha \geq 1$ $\mu(B(x,a)) \leq \leq c a^{\alpha d}$ for all $x \in \mathbb{R}^d$ and $a > 0$, where

$$B(x,a) \overset{def}{=\!=} \{(y_0, y_1, \dots, y_d) \in \mathbb{R}^{d+1}_+ : y_0^2 + \sum_{k=1}^{d}(x_k - y_k)^2 < a^2\}, x = (x_1, \dots, x_d).$$

Then $H^p(\mathbb{R}^d) \subset L^{\alpha p}(\mu)$.

This theorem is considered in [11] for $\alpha = 1$ and in [2] for $d = 1$ and $\alpha \geq 1$. The proof of this theorem in general case is just the same.

COROLLARY. Let $\alpha = \frac{d}{q} - \frac{d}{p} > 0$, $d \in \mathbb{N}$, $p, q \in (0, +\infty)$ Then $H^q(\mathbb{R}^d) \subset \mathcal{H}^p_\alpha(\mathbb{R}^d)$. ●

We need also some properties of the fractional integration operator I_α , $\alpha > 0$. Let u be a harmonic function on $\mathbb{R}^{d+1}_+$ sufficiently small at infinity. Put

$$(I_\alpha u)(t,x) \overset{def}{=\!=} \frac{1}{\Gamma(\alpha)} \int_t^{+\infty} (s-t)^{\alpha-1} u(s,x)\, ds .$$

It is easy to check that $I_\alpha u$ is a harmonic function.

THEOREM (G.H.Hardy, J.E.Littlewood, S.L.Sobolev). Let $\alpha = \frac{d}{p} - \frac{d}{q} > 0$, $d \in \mathbb{N}$, $p, q \in (0, +\infty)$. Then $I_\alpha(H^p(\mathbb{R}^d)) \subset \subset H^q(\mathbb{R}^d)$.

PROOF. Let $x, y \in \mathbb{R}^d$, $|x - y| < t$. Suppose that $\|u\|_{H^p} \leq 1$. To estimate $|(I_\alpha u)(t,y)|$ we use the following obvious inequalities:

$|u(s,y)| \leq c(p,d)\, s^{-\frac{d}{p}}$ for $s > 0$ and $y \in \mathbb{R}^d$

$|u(s,y)| \leq u^*(x)$ if $|x-y|<t$ and $s \geq t$

It follows that

$$\left|(I_\alpha u)(t,y)\right| \leq c(\alpha,p,d)\int_t^{+\infty}(s-t)^{\alpha-1}\min\left(u^*(x), s^{-\frac{d}{p}}\right)ds \leq$$

$$\leq c(\alpha,p,d)(u^*(x))^{\frac{p}{q}}$$

if $|x-y|<t$. Whence it follows that $((I_\alpha u)^*)^q \leq \leq c(\alpha,p,d)(u^*)^p$. ●

We need one more known result. The proof of the following assertion can be derived from [36], [31].

THEOREM 0.1. Let $p\in(0,+\infty)$, $d\in\mathbb{N}$, $\alpha>\beta>0$. Then $I_\beta(\mathcal{H}^p_\alpha(\mathbb{R}^d)) = \mathcal{H}^p_{\alpha-\beta}(\mathbb{R}^d)$.

Concluding the introduction we prove a lemma which will be used in §1.

LEMMA 0.2. Let u be a harmonic function on the set $(-a,a)\times\mathbb{R}^d$, $a>0$. Then

$$\int_{\mathbb{R}^d}\sup_{-\frac{a}{2}\leq t\leq\frac{a}{2}}|u(t,x)|^p dx \leq \frac{c(p,d)}{a}\int_{(-a,a)\times\mathbb{R}^d}|u(t,x)|^p dt\,dx .$$

PROOF. It follows from lemma 2 of [34] that

$$\sup_{-\frac{a}{2}\leq t\leq\frac{a}{2}}|u(t,x)|^p \leq \frac{c(p,d)}{a^{d+1}}\int_{|x-y|^2+t^2\leq a^2}|u(t,y)|^p dt\,dy .$$

It remains to integrate this inequality over $\mathbb{R}^d$ with respect to x. ●

§1. $H^p(\mathbb{R}^d)$-functions supported by $\mathbb{R}^{d_1}$ $(d_1<d)$.

Consider the natural embedding $\mathcal{E}^d_{d_1}: S'(\mathbb{R}^{d_1}) \longrightarrow S'(\mathbb{R}^d)$ $(\mathcal{E}^d_{d_1}f,\varphi) \overset{def}{=} (f, \varphi|\mathbb{R}^{d_1})$, where $\varphi\in S(\mathbb{R}^d)$.

THEOREM 1.1. $\mathcal{E}^d_{d_1}$ is a continuous operator from $\mathcal{H}^p_\alpha(\mathbb{R}^{d_1})$ into $H^p(\mathbb{R}^d)$ where $\alpha=(\frac{1}{p}-1)(d-d_1)$, $0<p<1$.

PROOF. Let $f\in\mathcal{H}^p_\alpha(\mathbb{R}^{d_1})$. Put $F=\mathcal{E}^d_{d_1}f$. It is sufficient to prove that $\|F\|_{H^p} \leq c(p,\alpha,d_1,d_2)\|f\|_{\mathcal{H}^p_\alpha}$

and we can suppose that $f\in L^{\infty}(\mathbb{R}^{d_1})$. To estimate $\|F\|_{H^p}$ we use the following C.Fefferman - E.M.Stein theorem: If $\Phi\in S(\mathbb{R}^d)$, $\int_{\mathbb{R}^d}\Phi\neq 0$, $\Phi_t(x)\overset{def}{=\!=}t^{-d}\Phi(\frac{x}{t})$, $F\in S'(\mathbb{R}^d)$ then $\|F\|_{H^p}\le c(p,d,\Phi)\,\|F^*_{\Phi}\|_{L^p}$, where $F^*_{\Phi}(x)\overset{def}{=\!=}\sup_{t>0}|(\Phi_t*f)(x)|$ (see [34]).

Let $\varphi\in C^{\infty}(\mathbb{R}^{d_2})$, $d_2=d-d_1$, $\varphi\ge 0$, $\operatorname{supp}\varphi\subset\{x\in\mathbb{R}^{d_2}:1<|x|<r\}$, $\max\varphi=1$. Put $\Phi^{(k)}(x,y)\overset{def}{=\!=}P_1(x)\varphi_{1/k}(y)$, where P_t is the d_1-dimensional Poisson kernel $\left(P_1(x)=\frac{\Gamma(\frac{d_1+1}{2})}{\pi^{\frac{d_1+1}{2}}}(1+|x|^2)^{-\frac{d_1+1}{2}}\right)$ and $\varphi_{1/k}(y)\overset{def}{=\!=}k^{d_2}\varphi(ky)$.

Let us estimate at first the L^p-norm of the maximal function $F^*_k\overset{def}{=\!=}\sup_{t>0}|\Phi^{(k)}_t*F|$. If $x\in\mathbb{R}^{d_1}$, $y\in\mathbb{R}^{d_2}$, then $(\Phi^{(k)}_t*F)(x,y)=u(t,x)\,k^{d_2}t^{-d_2}\varphi(\frac{ky}{t})$, where u is the Poisson integral of f . Therefore

$$F^*_k(x,y)\le 2^{d_2}\sup_{\frac{k|y|}{2}\le t\le k|y|}|u(t,x)|\cdot|y|^{-d_2}.$$

By Lemma 0.2 we have

$$\int_{\mathbb{R}^{d_1}}(F^*_k(x,y))^p\,dx\le c(p,d_1)\frac{1}{k|y|^{pd_2+1}}\int_{\frac{|y|k}{4}\le t\le 2k|y|}|u(t,x)|^p\,dt\,dx.$$

Integrating this inequality with respect to y we obtain that

$$\|F^*_k\|_{L^p}\le c(p,d_1,d_2)\,k^{d_2-\frac{d_2}{p}}\,\|f\|_{\mathcal{H}^p_\alpha}.$$

The function Φ we shall substitute in the above Fefferman - Stein inequality will be constructed as a series

$$\Phi=\sum_{k\ge 1}a_k\Phi^{(k)}_k=\sum_{k\ge 1}a_kP_k(x)\varphi(y),$$

where $\{a_k\}_{k\ge 1}$ is a rapidly decreasing sequence of real numbers. Note that $\Phi\in S(\mathbb{R}^d)$ if and only if $\mathcal{F}\Phi\in S(\mathbb{R}^d)$, where $\mathcal{F}$ is the Fourier transform.. We have

$$(\mathcal{F}\Phi)(x,y)=\sum_{k\ge 1}a_k\,e^{-2\pi k|x|}(\mathcal{F}\varphi)(y).$$

It is easy to see that $\mathcal{F}\Phi \in S(\mathbb{R}^d)$ if $\sum_{k\geq 1} k^n a_k = 0$ for all $n \in \mathbb{N}$. The condition $\int \Phi \neq 0$ is equivalent to $\sum_{k\geq 1} a_k \neq 0$. Put $g(z) \overset{def}{=} \sum_{k\geq 1} a_k z^k$. Then $g \in C^\infty(\overline{\mathbb{D}})$, $g(0) = 0$, $g^{(n)}(1) = 0$ for $n \in \mathbb{N}$ and $g(1) \neq 0$. For example we can take $g(z) = e^{-(\frac{1+z}{1-z})^{1/2}} - e^{-1}$. Now let us estimate $\| F_\Phi^* \|_{L^p}$. It is clear that $\| F_\Phi^* \|_{L^p}^p \leq \sum_{k\geq 1} |a_k|^p \| F_k^* \|_{L^p}^p \leq c(p, d_1, d_2) \| f \|_{\mathcal{H}_\alpha^p}^p$. ●

Now we intend to prove that under hypothesis if Th.1.1 the operator $\mathcal{E}_{d_1}^d : \mathcal{H}_\alpha^p(\mathbb{R}^{d_1}) \to H^p(\mathbb{R}^d)$ $(d_1 < d, 0<p<1)$ is left invertible. Consider the operator $\mathcal{J}_d^{d_1} : H^p(\mathbb{R}^d) \to S'(\mathbb{R}^{d_1})(d_1 < d, 0<p<1)$, $(\mathcal{J}_d^{d_1} f, \varphi) \overset{def}{=} (f, j_d^{d_1} \varphi)$, $\varphi \in S(\mathbb{R}^{d_1})$; $(j_d^{d_1}\varphi)(x,y) \overset{def}{=} \varphi(x)$ $(x \in \mathbb{R}^{d_1}, y \in \mathbb{R}^{d_2}, d_2 = d - d_1)$.

The operator $\mathcal{J}_d^{d_1}$ is well defined since $j_d^{d_1}(S(\mathbb{R}^{d_1})) \subset \subset (H^p(\mathbb{R}^d))^*$.

THEOREM 1.2. Let $d_1, d_2 \in \mathbb{N}$, $0<p<1, \alpha = (\frac{1}{p} - 1) d_2$, $d = d_1 + d_2$. Then $\mathcal{J}_d^{d_1}$ maps $H^p(\mathbb{R}^d)$ continuously into $\mathcal{H}_\alpha^p(\mathbb{R}^{d_1})$, $\mathcal{J}_d^{d_1} \mathcal{E}_{d_1}^d$ being the identity operator on $\mathcal{H}_\alpha^p(\mathbb{R}^{d_1})$.

PROOF. By R.R.Coifman - R.Latter theorem it is sufficient to prove that $\| \mathcal{J}_d^{d_1} b \|_{\mathcal{H}_\alpha^p} \leq c(p, d, d_1)$ for any p-atom b in $\mathbb{R}^d$. To prove this note that $\mathcal{J}_d^{d_1} b$ is a q-atom in $\mathbb{R}^{d_1}$, where $q = \frac{p d_1}{d - p d_2}$, since $(\mathcal{J}_d^{d_1} b)(x) = \int_{\mathbb{R}^{d_2}} b(x,y)\,dy$. Thus $\| \mathcal{J}_d^{d_1} b \|_{H^q} \leq c(p, d_1, d_2)$. The result follows from the inclusion $H^q(\mathbb{R}^{d_1}) \subset \mathcal{H}_\alpha^p(\mathbb{R}^d)$ (cf. §0). ●

Note that $\mathcal{J}_d^{d_1}$ can be defined by

$$(\mathcal{J}_d^{d_1} u)(t,x) = \int_{\mathbb{R}^{d_2}} u(t,x,y)\,dy \quad (t>0, x \in \mathbb{R}^{d_1}),$$

the integral absolutely converging if $u \in H^p(\mathbb{R}^d)$, $0 < p \leq 1$.

The following theorem which is probably of independent interest follows easily from the previous ones.

THEOREM 1.3. Let α, d_1, d, p be as in theorem 1.2, $f \in S'(\mathbb{R}^{d_1})$. Then $\mathcal{E}_{d_1}^d f \in H^p(\mathbb{R}^d)$ if and only if $f \in \mathcal{H}_\alpha^p(\mathbb{R}^{d_1})$, moreover

$$\| f \|_{\mathcal{H}_\alpha^p} \asymp \| \mathcal{E}_{d_1}^d f \|_{H^p}. \quad ●$$

It is interesting to compare Th.1.3 with a result of R.R.Coifman and R.Rochberg(see [56]). They have proved that each harmonic in $\mathbb{R}_+^{d+1}$ function $u \in \mathcal{H}_{1/p}^p(\mathbb{R}^d)$ may be considered as a distribution in $H^p(\mathbb{R}^{d+1})$.

§2. Atomic decomposition of $\mathcal{H}_\alpha^p(\mathbb{R}^d)$ -functions ($\alpha>0,\ 0<p\leq 1$).

THEOREM 2.1. Let $f \in \mathcal{H}_\alpha^p(\mathbb{R}^d)$, $0<p\leq 1$, $\alpha = \frac{d}{q} - \frac{d}{p} > 0$, $d \in \mathbb{N}$. Then f can be decomposed as follows:

$$f = \sum_{k \geq 1} \alpha_k b_k \qquad (\text{in } S'(\mathbb{R}^d)),$$

where b_k's are q-atoms, $\alpha_k \in \mathbb{C}$, and

$$A(p,q,d)\|f\|_{\mathcal{H}_\alpha^p} \leq \Big(\sum_{k\geq 1} |\alpha_k|^p\Big)^{1/p} \leq B(p,q,d)\|f\|_{\mathcal{H}_\alpha^p}.$$

PROOF. The case $p=1$ is known and can be obtained by duality arguments. Thus we suppose $p<1$. Note that the left inequality follows from $\|b\|_{H^q} \leq c(q,d)$ and $H^q(\mathbb{R}^d) \subset \mathcal{H}_\alpha^p(\mathbb{R}^d)$, and so it suffices to prove only the right inequality.

Suppose first that $\alpha = \frac{d_2}{p} - d_2$, $d_2 \in \mathbb{N}$. Then $\mathcal{E}_d^{d+d_2} f \in H^p(\mathbb{R}^{d+d_2})$ by theorem 1.1. By R.R.Coifman - R.Latter theorem

$$\mathcal{E}_d^{d+d_2} f = \sum_{k\geq 1} \alpha_k b_k ,$$

where b_k's are $(d+d_2)$-dimensional p-atoms and $\sum_{k\geq 1} |\alpha_k|^p \leq c(\alpha,p,d)\|f\|^p_{\mathcal{H}_\alpha^p}$. Applying theorem 1.2 we obtain

$$f = \sum_{k\geq 1} \alpha_k \mathcal{J}_{d+d_1}^d b_k .$$

The result follows from the fact that $\mathcal{J}_{d+d_2}^d b_k$ is a q-atom in $\mathbb{R}^d$, $k \in \mathbb{N}$.

Now let us proceed to the general case. Choose $\beta > 0$ such that $\frac{\alpha+\beta}{1/p-1} \in \mathbb{N}$. Let $f \in \mathcal{H}_\alpha^p(\mathbb{R}^d)$. Then $I_\beta^{-1} f \in \mathcal{H}_\alpha^p(\mathbb{R}^d)$ by Th.0.1. Applying the considered particular case we have

$$I_\beta^{-1} f = \sum_{k\geq 1} \alpha_k b_k$$

where b_k's are τ-atoms $(\alpha+\beta=\frac{d}{\tau}-\frac{d}{p})$ in $\mathbb{R}^d$ and $\sum_{k\geq 1}|\alpha_k|^p \leq c(p,\alpha,\beta,d)\|f\|^p_{\mathcal{H}^p_\alpha}$. Let us apply the fact that I_β is a continuous operator from $H^\tau(\mathbb{R}^d)$ into $H^q(\mathbb{R}^d)$. Therefore

$$f = \sum_{k\geq 1} \alpha_k I_\beta b_k .$$

Each function $I_\beta b_k$ by R.Coifman - R.Latter theorem can be represented as

$$I_\beta b_k = \sum_{j\geq 1} \gamma_{jk} b_{jk}$$

where b_{jk}'s are q-atoms in $\mathbb{R}^d$ and $\sum_{j\geq 1}|\gamma_{jk}|^q \leq$ $\leq c(p,\beta,d)$. Therefore

$$f = \sum_{k\geq 1}\sum_{j\geq 1} \alpha_k \gamma_{jk} b_{jk} .$$

The result follows from the inequalities

$$\sum_{k\geq 1}\sum_{j\geq 1} |\alpha_k|^p|\gamma_{jk}|^p \leq \sum_{k\geq 1}|\alpha_k|^p\Big(\sum_{j\geq 1}|\gamma_{jk}|^q\Big)^{p/q} \leq$$

$$\leq c(p,\beta,d)\sum_{k\geq 1}|\alpha_k|^p \leq c(p,\alpha,\beta,d)\|f\|^p_{\mathcal{H}^p_\alpha} . \quad \bullet$$

§3. A refinement of Theorem 2.1

Recall that a standard q-atom in $\mathbb{R}^d$ $(0<q<1)$ is a measure $\mathcal{M}_q(a,b)$,

$$\mathcal{M}_q(a,b) \stackrel{def}{=\!=} 2^{-N} N^{d-\frac{d}{q}} |b|^{d-\frac{d}{q}} \sum_{k=0}^{N} (-1)^k \binom{N}{k} \delta_{a+kb} ,$$

where $a, b\in\mathbb{R}^d$, $b\neq 0$, $N = N_{q,d} \stackrel{def}{=\!=} \left[\frac{d}{q}\right] - d+1$.

THEOREM 3.1. Let $f\in \mathcal{H}^p_\alpha(\mathbb{R}^d)$, $0<p\leq 1$, $\alpha = \frac{d}{q}-\frac{d}{p} > 0$. Then

$$f=\sum_{k\geq 1}\alpha_k b_k \qquad (\text{in } S'(\mathbb{R}^d)),$$

where b_k's are standard q-atoms, $\alpha_k\in\mathbb{C}$, and

$$A(p,q,d)\|f\|_{\mathcal{H}^p_\alpha}\leq\Big(\sum_{k\geq 1}|\alpha_k|^p\Big)^{1/p}\leq B(p,q,d)\|f\|_{\mathcal{H}^p_\alpha}.$$

LEMMA 3.2. Let $a=\{a_\alpha\}(\alpha\in\mathbb{Z}^d_+)$, $|\alpha|_1<N$ be a family of complex numbers. Then there exists a unique family $c=\{c_\alpha\}(\alpha\in\mathbb{Z}^d_+)$, $|\alpha|_1<N$ such that

$$\int_{\mathbb{R}^d} x^\alpha\, d\mu_c = a_\alpha$$

for all $\alpha\in\mathbb{Z}^d_+$, $|\alpha|_1<N$, where $\mu_c=\sum_{\alpha\in\mathbb{Z}^d_+,|\alpha|_1<N} c_\alpha\delta_\alpha$.

The proof can be easily obtained by induction in $d+N$. ●

LEMMA 3.3. Let $0<q<1$, $d\in\mathbb{N}$, $N=N_{q,d}$, Q be a closed cube in $\mathbb{R}^d$ divided into N^d equal cubes. Then there exists an operator $T_Q: C^*(Q)\to C^*(Q)\subset S'(\mathbb{R}^d)$ such that $\operatorname{supp}(f-T_Q f)$ is contained in the set of vertices of the small cubes for any $f\in C^*(Q)$ and

$$\|T_Q f\|_{H^q}\leq c(q,d)(m_d(Q))^{1/q-1}\|f\|_{C^*(Q)}.$$

PROOF. We can suppose that $Q=\{(x_1,x_2,\dots,x_d)\in\mathbb{R}^d: 0\leq x_j\leq N\}$. Let $f\in C^*(Q)$. By Lemma 3.2 there exists a unique measure μ supported by $\{\alpha\in\mathbb{Z}^d_+: |\alpha|_1<N\}$ such that $f-\mu$ is orthogonal to all polynomials of degree less than N. Let $T_Q f \overset{\text{def}}{=\!=} f-\mu$. It is clear that $\|T_Q f\|_{C^*(Q)}\leq$ $\leq c(q,d)\|f\|_{C^*(Q)}$. It suffices to note that $(c(d,q))^{-1}\|f\|^{-1}_{C^*(Q)}(m_d(Q))^{1-\frac{1}{q}}T_Q f$ is a generalized q-atom. ●

LEMMA 3.4. Let μ be a measure on $\mathbb{R}^d$ with a finite support in $\mathbb{Z}^d$, $0<q<1$. Suppose that μ is orthogonal to all polynomials of degree less than $N_{q,d}$. Then μ belongs to the linear span of standard q-atoms supported by $\mathbb{Z}^d$.

PROOF. Let us prove the following dual assertion. If $f:\mathbb{Z}^d\to\mathbb{C}$ is orthogonal to all standard q-atoms supported by $\mathbb{Z}^d$ then f is a polynomial of degree less than $N_{q,d}$. The case $d=1$ can be easily obtained by induction in N. The multidimensional case can be reduced to the one-dimensional one using the following simple assertion. If for any $a, b \in \mathbb{Z}^d$ the function $n \longmapsto f(an+b)$ is a polynomial of degree less than N then f is a polynomial and $\deg f < N$. ●

LEMMA 3.5. Let $0<q<1$, $d \in \mathbb{N}$, $Q \subset \mathbb{R}^d$, Q be a cube divided into $N_{q,d}^{2d}$ equal cubes, f be a measure supported by the vertices of the small cubes and let f be orthogonal to all polynomials of degree less than $N_{q,d}$. Then $f = \sum_{k \geq 1} \alpha_k b_k$, where

$$\Big(\sum_{k \geq 1} |\alpha_k|^q \Big)^{1/q} \leq c(q,d) \, \|f\|_{H^q}$$

and α_k's vanish except for a finite set of indices.

PROOF. Without loss of generality we can suppose that $Q = \{(x_1, x_2, \dots, x_d) \in \mathbb{R}^d : 0 \leq x_j \leq N_{q,d}^2\}$. The result follows from Lemma 3.4 and from the fact that the space of such f's is finite-dimensional. ●

PROOF OF THEOREM 3.1. The case $p = 1$ is known and follows from the duality arguments.

Let $p<1$. By theorem 2.1 it is sufficient to prove that every d-dimensional atom b can be decomposed as follows.

$$b = \sum_{k \geq 1} \alpha_k b_k \qquad (\text{ in } \ S'(\mathbb{R}^d)) ,$$

where b_k's are standard q-atoms, $\alpha_k \in \mathbb{R}$ and $\sum_{k \geq 1} |\alpha_k|^p \leq c(p,q,d)$ $(0<q<p\leq 1)$.

Let b be a q-atom, $\operatorname{supp} b \subset Q$ and $|b| \leq (m_d(Q))^{-1/q}$. Without loss of generality we can suppose that $Q = \{(x_1, \dots, x_d) \in \mathbb{R}^d : 0 \leq x_j \leq 1\}$. Let us divide Q into N^{nd} equal cubes $\{Q_\alpha^{(n)}\}_{1 \leq \alpha \leq N^{nd}}$, where $N = N_{q,d}$. Put

$$g_n \stackrel{def}{=} b - \sum_{\alpha=1}^{N^{nd}} T_{Q_\alpha^{(n)}} (b \chi_{Q_\alpha^{(n)}}) .$$

It is obvious that $g = 0$. Let us show that $\lim_{n \to \infty} g_n = b$ (in $S'(\mathbb{R}^d)$). Suppose $\varphi \in S(\mathbb{R}^d)$ and $x_\alpha^{(n)} \in Q_\alpha^{(n)}$. We have

$$|(b - g_n, \varphi)| = \Big| \sum_{\alpha=1}^{N^{nd}} (T_{Q_\alpha^{(n)}} (b \chi_{Q_\alpha^{(n)}}), \varphi) \Big| \leq$$

$$\leq \sum_{\alpha=1}^{N^{nd}} \big| (T_{Q_\alpha^{(n)}} (b \chi_{Q_\alpha^{(n)}}), \varphi - \varphi(x_\alpha^{(n)})) \big| \leq$$

$$\leq C(N)\|b\|_{L^1} \sup_{\substack{x\in Q_\alpha^{(n)} \\ 1\leq\alpha\leq N^{nd}}} |\varphi(x)-\varphi(x_\alpha^{(n)})|.$$

Since the last $\sup$ tends to zero for $n\to\infty$, the following equalities hold

$$b=\sum_{n\geq 0}(b_{n+1}-b_n)=\sum_{n\geq 0}\Big(\sum_{\alpha=1}^{N^{nd}}(T_{Q_\alpha^{(n)}}(b\chi_{Q_\alpha^{(n)}})-$$

$$-\sum_{I_\beta^{(n+1)}\subset I_\alpha^{(n)}} T_{Q_\beta^{(n+1)}}(b\chi_{Q_\beta^{(n+1)}}))\Big)\overset{def}{=\!=}\sum_{n\geq 0}\sum_{\alpha=1}^{N^{nd}} h_\alpha^{(n)}.$$

By Lemma 3.3, the following inequalities hold

$$\|h_\alpha^{(n)}\|_{H^q}\leq c(q,d)(m_d(Q))^{\frac{1}{q}-1}\|b\chi_{Q_\alpha^{(n)}}\|\leq C(q,d)N^{-\frac{nd}{q}}.$$

Whence it follows that $\sum_{n\geq 0}\sum_{k=1}^{N^{nd}}\|h_\alpha^{(n)}\|_{H^q}^p\leq c(p,q,d)$ if $p\in(q,1]$. The result follows from Lemma 3.5. ●

REMARK 3.6. It is easily seen from the proof that theorem 3.1 remains valid if in the definition of a standard q-atom $N_{q,d}$ will be replaced by any integer m , $m\geq N_{q,d}$. Moreover in this definition it can be supposed that all coordinates of b except one vanish.

The question of description of the set

$$X^q(\mathbb{R}^d)\overset{def}{=\!=}\{f=\sum_{k\geq 1}\alpha_k b_k : b_k\text{'s are standard } q\text{-atoms } \sum_{k\geq 1}|\alpha_k|^q<+\infty\},$$

$$0<q<1,$$

arises now in a natural way. It is clear that $X^q(\mathbb{R}^d)\subset$ $\subset H_s^q(\mathbb{R}^d)$, where $H_s^q(\mathbb{R}^d)$ is a subset of $H^q(\mathbb{R}^d)$

consisting of functions whose non-tangential boundary values vanish a.e. on $\mathbb{R}^d$. However $X^q(\mathbb{R}^d) \neq H_s^q(\mathbb{R}^d)$ because $X^q(\mathbb{R}^d) \subset \mathcal{H}_0^q(\mathbb{R}^d) \subsetneqq H_s^q(\mathbb{R}^d)$. I don't know whether $X^q(\mathbb{R}^d) = \mathcal{H}_0^q(\mathbb{R}^d)$ $(0 \le q < 1)$.

§4. A description of the p-convex envelope of $H^q(\mathbb{R}^d)$.

We need several lemmas. Let us fix $\varphi \in \mathcal{D}(\mathbb{R}^d)$ such that $\varphi(x) = 1$ if $|x| \le 1$. Let $V_N : \mathcal{S}'(\mathbb{R}^d) \to \mathcal{S}'(\mathbb{R}^d)$ be an operator defined by $V_N f \overset{def}{=\!=} \varphi f - \mu$, μ being a measure defined uniquely by the following conditions: $\operatorname{supp} \mu \subset \{\alpha \in \mathbb{Z}_+^d : |\alpha|_1 < N\}$, $\varphi f - \mu$ is orthogonal to all polynomials of degree less than N. The proof of the following lemma is like that of Lemma 9 from [15].

LEMMA 4.1. Let $0 < q < 1$, $d \in \mathbb{N}$, $N = N_{q,d} \overset{def}{=\!=} \left[\frac{d}{q}\right] - d + 1$. Then $V_N(H^q(\mathbb{R}^d)) \subset H^q(\mathbb{R}^d)$ and so $V_N(\mathcal{H}_\alpha^p(\mathbb{R}^d)) \subset \mathcal{H}_\alpha^p(\mathbb{R}^d)$ $\left(\frac{d}{q} - \frac{d}{p} = \alpha > 0,\ 0 < p \le 1\right)$. ●

Let P_σ be the Poisson convolution, i.e. $(P_\sigma u)(t, x) \overset{def}{=\!=} u(t + \sigma, x)$, where u is a harmonic function on $\mathbb{R}_+^{d+1}$.

LEMMA 4.2. Under hypotheses of Lemma 4.1 $P_\sigma V_N \mathcal{H}_\alpha^p(\mathbb{R}^d) \subset H^q(\mathbb{R}^d)$ for any $\sigma > 0$.

PROOF. Let $f \in \mathcal{H}_\alpha^p(\mathbb{R}^d)$. Then

$(P_\sigma V_N f)^* \in L^\infty(\mathbb{R}^d)$ and $(P_\sigma V_N f)^*(x) = O(|x|^{-N-d})$. ●

Denote by T_σ the dilation operator: $(T_\sigma f)(x) = f\left(\frac{x}{\sigma}\right)$.

Put $K_\sigma \overset{def}{=\!=} P_\sigma T_\sigma^{-1} V_N T_\sigma$, $\sigma > 0$.

LEMMA 4.3. Let $f \in H^q(\mathbb{R}^d)$, $0 < q < 1$, $d \in \mathbb{N}$, $N = N_{q,d}$. Then $\lim_{\sigma \to 0+} K_\sigma f = f$ (in H^q).

PROOF. It is clear that $\|K_\sigma\|_{H^q} \le C$ for $\sigma > 0$. Hence it is sufficient to prove the needed equality for a dense set of f's. Let $\operatorname{supp} f$ be compact. Then $T_\sigma^{-1} V_N T_\sigma f = f$ for sufficiently small σ. It remains to note that $\lim_{\sigma \to 0+} P_\sigma f = f$ if $f \in H^q(\mathbb{R}^d)$. ●

THEOREM 4.4. Let $0 < q < p \le 1$, $\alpha = \frac{d}{q} - \frac{d}{p}$, $d \in \mathbb{N}$. Then the p-convex envelope of $H^q(\mathbb{R}^d)$ is naturally isomorphic to $\mathcal{H}_\alpha^p(\mathbb{R}^d)$.

PROOF. The case $p = 1$ is well known and follows from the duality arguments.

We have to prove that the following two quasinorms on $H^q(\mathbb{R}^d)$ are equivalent:

$$\|f\|_{I} \stackrel{def}{=} \|f\|_{\mathcal{H}^p_\alpha} ,$$

$$\|f\|_{II} \stackrel{def}{=} \inf\left\{\left(\sum_{k=1}^{n} \|f_k\|^p_{H^q}\right)^{\frac{1}{p}} : n\in\mathbb{N},\ f=\sum_{k=1}^{n} f_k,\ f_k\in H^q(\mathbb{R}^d)\right\}.$$

Since $H^q(\mathbb{R}^d)$ is continuously imbedded into $\mathcal{H}^p_\alpha(\mathbb{R}^d)$, $\|f\|_I \leq c(\alpha,p,d)\|f\|_{II}$. Let us prove the inverse inequality. Let $f\in H^q(\mathbb{R}^d)$. By theorem 3.1 $f=\sum_{k\geq 1} f_k$ (in $\mathcal{H}^p_\alpha$) where $f_k\subset H^q(\mathbb{R}^d)$ and $(\sum_{k\geq 1}\|f_k\|^p_{H^q})^{1/p}\leq c(\alpha,p,d)\|f\|_I$. Let δ be a positive number so small that $\|f-K_\delta f\|_{H^q}\leq$ $\leq \|f\|_I$. We have $K_\delta f=\sum_{k\geq 1} K_\delta f_k$ and this series converges in $H^q(\mathbb{R}^d)$ by Lemma 4.2.

Take n so that $\|K_\delta f-\sum_{k=1}^{n} K_\delta f_k\|_{H^q}\leq \|f\|_I$. It follows that

$$\|f\|^p_{II} \leq \|f-K_\delta f\|^p_{H^q} + \|K_\delta f-\sum_{k=1}^{n} K_\delta f_k\|^p_{H^q} + \sum_{k=1}^{n}\|K_\delta f_k\|^p_{H^q} \leq$$

$$\leq 2\|f\|^p_I + c(\alpha,p,d)\sum_{k=1}^{n}\|f_k\|^p_{H^q} \leq c(\alpha,p,d)\|f\|^p_I . \quad \bullet$$

COROLLARY 4.5. Let $0<q<p\leq 1$, $0<\beta<\alpha<+\infty$ and $\alpha-\beta=\frac{d}{q}-\frac{d}{p}$. Then the p-convex envelope of $\mathcal{H}^q_\beta(\mathbb{R}^d)$ is canonically isomorphic to $\mathcal{H}^p_\alpha(\mathbb{R}^d)$.

PROOF. Put $r=\frac{d}{\alpha+d/p}$. Then $\alpha=\frac{d}{r}-\frac{d}{p}$, $\beta=\frac{d}{r}-\frac{d}{q}$. Therefore $[H^r(\mathbb{R}^d)]_p \simeq \mathcal{H}^p_\alpha(\mathbb{R}^d)$ and $[H^r(\mathbb{R}^d)]_q \simeq$ $\simeq \mathcal{H}^q_\beta(\mathbb{R}^d)$ by theorem 4.4. It follows that $[\mathcal{H}^q_\beta(\mathbb{R}^d)]_p \simeq$ $\simeq \mathcal{H}^p_\alpha(\mathbb{R}^d)$. $\bullet$

We present here an application of Th.4.4

THEOREM 4.6. Let $q<1$, $d\in\mathbb{N}$. If X is an infinite-dimensional complemented subspace of $H^q(\mathbb{R}^d)$ then $[X]_p\neq X$ for $q<p\leq 1$; in other words X is not isomorphic to any p-normed space, $q<p\leq 1$.

The following lemma is proved essentially in [43]. We only sketch the proof here.

LEMMA 4.7. Let $0<p<+\infty$, $d\in\mathbb{N}$, $\alpha\in\mathbb{R}$. Then $\mathcal{H}^p_\alpha(\mathbb{R}^d)$

is isomorphic to a subspace of ℓ^p.

PROOF. Using compactness arguments it follows that $\mathbb{R}_+^{d+1}$ can be divided into small disjoint subsets $\{E_k\}_{k\geq 1}$ such that for any function $f\in \mathcal{H}_\alpha^p(\mathbb{R}^d)$ the following inequalities hold

$$c_1(\alpha,p,d)\|f\|^p_{\mathcal{H}_\alpha^p} \leq \sum_{k\geq 1} |f(a_k)|^p \mu_k \leq c_2(\alpha,p,d)\|f\|^p_{\mathcal{H}_\alpha^p},$$

where $a_k \in E_k$ and $\mu_k = \int_{E_k} x_0^{\alpha p-1}\, dx_0\, dx_1 \dots dx_d$. ●

PROOF OF THEOREM 4.6. Let $H^q(\mathbb{R}^d) \approx X \oplus Y$ and $[X]_p \simeq X$ for some $p\in(q,1]$. Then $X \oplus [Y]_r \approx \mathcal{H}_{\frac{d}{q}-\frac{d}{r}}(\mathbb{R}^d)$ for all $r\in(q,p]$. From Lemma 4.7 it follows that X is isomorphic to a subspace of ℓ^r if $q<r\leq p$. Therefore $\dim X<+\infty$. ●

COROLLARY 4.8. No infinite-dimensional Banach space can be imbedded into $H^p(\mathbb{R}^d)$ as a complemented subspace $(0<p<1)$.

Similar but slightly more complicated considerations permit us to deduce from the results of §3 that

$$[H_0^q(\mathbb{R}^d)]_p \simeq \mathcal{H}_\alpha^p(\mathbb{R}^d),\quad \alpha=\frac{d}{q}-\frac{d}{p}>0,\quad 0<p\leq 1.$$

Recall that $H_0^q(\mathbb{R}^d) = \{u\in H^q(\mathbb{R}^d) : \lim_{t\to 0+} u(t,x)=0$ for almost all $x\in\mathbb{R}^d\}$.

Chapter Four. PARTIAL SUMS OF FOURIER SERIES OF CONTINUOUS FUNCTIONS AND FOURIER COEFFICIENTS OF H^p-FUNCTIONS $(0<p<1)$

§0. Introduction

Let M be an infinite subset of $\mathbb{Z}_+$. Associate with M the strictly increasing sequence $m=\{m_k\}_{k\geq 0}$ with range M.

Denote by R_M an operator $R_M f \overset{def}{=\!=} \{\hat f(m_k)\}_{k\geq 0}$ from the distribution space $\mathcal{D}'(\mathbb{T})$ into the space of all one-sided sequences of numbers (i.e. sequences indexed by $\mathbb{Z}_+$)

In this chapter we give a complete description of the range $R_M(H^p)$, $0<p<1$, for sets lacunary in the Hadamard's sense i.e. with $\inf\{m/n,\ m,n\in M,\ m>n\}>1$. This description is a simple consequence of a result of [33].

Furhter, in this chapter an analogous problem is solved for the spaces $H^{p,\infty}$, $0<p\leq 1$, where

$$H^{p,\infty} \overset{def}{=} \{f\in H^{p/2} : \|f\|_{H^{p,\infty}} \overset{def}{=} \sup_{A>0} A(m\{|f|>A\})^{1/p} < +\infty\} .$$

Let

$$H_0^{p,\infty} \overset{def}{=} \{ f\in H^{p,\infty} : \lim_{A\to+\infty} A^p m\{|f|>A\} = 0\} .$$

The case $p\in(1,+\infty)$ is of no interest in view of $H^\infty\subset H_0^{p,\infty}\subset$ $\subset H^{p,\infty}\subset H^1$, and it is also well known that $R_M(H^\infty) =$ $= R_M(H^1) = \ell^2$ if M is lacunary or even if M is a finite union of lacunary sets.

The case $p=1$ is peculiar and the most interesting. As we shall see in §3 for any lacunary $M\subset \mathbb{Z}$

$$R_M(H_0^{1,\infty}) = T_M(C(\mathbb{T})) , \quad R_M(H^{1,\infty}) = T_M(L^\infty(\mathbb{T})) \text{ , where}$$

$$T_M f \overset{def}{=} \{ \sum_{|j|\leq m_k} \hat{f}(j)\}_{k\geq 0} .$$

The behaviour of partial Fourier sums of bounded (continuous) functions (or what is essentially the same the behaviour of norms of linear combinations of the Dirichlet kernels) is the subject of a large number of papers. See e.g. [19,23,26,39,44, 46,47].

In §1 we obtain the estimates of Fourier coefficients of functions $f\in H^p$ (or $H^{p,\infty}$), $0<p\leq 1$.

In §2 a description of $R_M(H^p)$ and $R_M(H^{p,\infty})$ is given($0<p<1$).

In §3 the sets $R_M(H_0^{1,\infty}) = T_M(C(\mathbb{T}))$ and $R_M(H^{1,\infty}) =$ $= T_M(L^\infty(\mathbb{T}))$ are described for lacunary sets M.

In §4 we give analogous results for the space $\mathcal{H}_\alpha^p$ of all functions f analytic in $\mathbb{D}$ satisfying

$$\|f\|^p_{\mathcal{H}_\alpha^p} \overset{def}{=} \int_{\mathbb{D}} |f(z)|^p (1-|z|)^{\alpha p-1} dm_2(z) < +\infty .$$

§1. Fourier coefficients estimates for H^p and $H^{p,\infty}$-functions $(0<p\le 1)$

The following theorem is proved in [4]. Here we give a simpler proof.

THEOREM 1.1. Let $f\in H^p$, $0<p<1$. Then

$$|\hat{f}(n)|\le (n+1)^{\frac{1}{p}-1}\|f\|_{H^p}.$$

PROOF. The case $n=0$ is well known; it follows, for example, from the subharmonicity of $|f|^p$. We reduce the general case to this particular one.

Consider the function $h(z)=\frac{1}{n+1}\sum_{k=0}^{n} f\left(ze^{\frac{2\pi ik}{n+1}}\right)z^{-n}e^{-\frac{2\pi ikn}{n+1}}$.
It is clear that $h\in H^p$ and $\hat{f}(n)=\hat{h}(0)$. It remains to remark that $\|h\|_{H^p}\le (n+1)^{1/p-1}\|f\|_{H^p}$. ●

LEMMA 1.2. Let $f\in H^{p,\infty}$, $0<p<+\infty$. Then

$$|f(z)|\le c_p\|f\|_{H^{p,\infty}}(1-|z|)^{-1/p}.$$

PROOF. Associate with every point $z\in\mathbb{D}$ the arc $I_z\subset\mathbb{T}$, $I_z\overset{def}{=}\{\zeta\in\mathbb{T}: z\in\Gamma_\zeta\}$. Remind that Γ_ζ is the interior of the convex hull of the point ζ and of the disc of radius $1/2$, centered at the origin. It is easy to check that $m(I_z)\ge c(1-|z|)$. By Th.1.2.1

$$m\{\mathcal{M}f\ge A\}\le\frac{c_p\|f\|^p_{H^{p,\infty}}}{A^p},$$

where $(\mathcal{M}f)(\zeta)\overset{def}{=}\sup\{|f(z)|: z\in\Gamma_\zeta\}$. It remains to note that $m\{\mathcal{M}f\ge|f(z)|\}\ge m(I_z)\ge c(1-|z|)$. ●

LEMMA 1.3. Let $f\in H^{p,\infty}$, $0<p<1$. Then

$$\int_{\mathbb{T}}|f(r\zeta)|\,dm(\zeta)\le c_p\|f\|_{H^{p,\infty}}(1-r)^{1-\frac{1}{p}}.$$

If $f\in H^{1,\infty}$, then

$$\int_{\mathbb{T}}|f(r\zeta)|\,dm(\zeta)\le c\|f\|_{H^{1,\infty}}\log\frac{2}{1-r}.$$

PROOF. It is sufficient to use the estimates

$m\{\zeta\in\mathbb{T}: |f(r\zeta)|>A\}\leq \frac{c_p}{A^p}\|f\|^p_{H^{p,\infty}}$ and $|f(r\zeta)|\leq c_p\|f\|_{H^{p,\infty}}(1-r)^{-1/p}$, $\zeta\in\mathbb{T}$ (see Lemma 1.2). ●

THEOREM 1.4. Let $f\in H^{p,\infty}$, $0<p<1$. Then for all $n\in\mathbb{N}$

$$|\hat{f}(n)|\leq c_p\, n^{1/p-1}\,\|f\|_{H^{p,\infty}} .$$

Further if $f\in H_0^{p,\infty}$ then $\hat{f}(n)=o(n^{1/p-1})$ $(n\to+\infty)$.

PROOF. By Lemma 1.3.

$$|r^n\hat{f}(n)|\leq c_p\|f\|_{H^{p,\infty}}(1-r)^{1-\frac{1}{p}},\quad 0<r<1 .$$

Set now $r=(1-\frac{1}{n})$. ●

THEOREM 1.5. Let M, $M\subset\mathbb{Z}_+$, be a lacunary set and $f\in H^{1,\infty}$. Then for any $n\in\mathbb{N}$, $n>1$,

$$\sum_{k\subset M,\, k\leq n}|\hat{f}(k)|^2\leq C_M(\log^2 n)\|f\|^2_{H^{1,\infty}}$$

If $f\in H_0^{1,\infty}$ then $\sum\limits_{k\in M,\, k\leq n}|\hat{f}(k)|^2=o(\log^2 n)\,(n\to+\infty)$.

PROOF. By the Paley theorem (cf. [13]) and Lemma 1.3

$$\sum_{k\in M,\, k\leq n}|\hat{f}(k)|^2 r^{2k}\leq C_M\|f\|^2_{H^{1,\infty}}\log^2\frac{1}{1-r} .$$

Set now $r=1-\frac{1}{n}$. ●

§2. Fourier coefficients with sparse indices of H^p and $H^{p,\infty}$-functions, $0<p<1$

First prove a lemma.

LEMMA 2.1. Let $\varphi\in C^\infty(\mathbb{R})$, $\operatorname{supp}\varphi\subset[-1,1]$. Then

$$\sum_{k=-n}^{n}\varphi\left(\frac{k}{n}\right)e^{2\pi ikt}=n\hat{\varphi}(nt)+\psi_n(t),\quad t\in\left[-\frac{1}{2},\frac{1}{2}\right],$$

where $\hat{\varphi}(x)\overset{def}{=}\int_{-\infty}^{+\infty}\varphi(t)e^{-2\pi ixt}\,dt$ is the Fourier transform of φ and $\|\psi_n'\|_{L^\infty[-\frac{1}{2},\frac{1}{2}]}=o(n^{-N})$, $n\to+\infty$ for any $N\in\mathbb{N}$.

PROOF. It is sufficient to use the Poisson summation formula [12]. ●

COROLLARY 2.2. For any $n\in\mathbb{N}$ there exists a trigonometric polynomial $Q_n=\sum_{k=-n}^{n} a_k^{(n)}z^k$ such that $a_0^{(n)}=1$ and $\|Q_n\|_{L^p}\leq c_p n^{1-1/p}$. ●

COROLLARY 2.3. Denote by L^p_n the subspace of L^p consisting of all trigonometric polynomials of degree $\leq n$. Let Φ be a linear functional on L^p_n. Then

$$|\Phi(1)|\leq c_p n^{1-\frac{1}{p}}\|\Phi\| .$$

PROOF. Set $Q_n^{(\zeta)}(z)\overset{def}{=\!=}\sum_{k=-n}^{n} a_k^{(n)}\zeta^k z^k$. Then

$$\Phi(1)=\int_{\mathbb{T}}\Phi(Q_n^{(\zeta)})\,dm(\zeta) \quad \text{and so}$$

$$|\Phi(1)|\leq\int_{\mathbb{T}}\|\Phi\|\cdot\|Q_n^{(\zeta)}\|_{L^p}\,dm(\zeta)\leq c_p n^{1-\frac{1}{p}}\|\Phi\|. \quad ●$$

Of course, these corollaries are of interest only when $0<p<1$.

Let $w=\{w_k\}_{k\geq 0}$ be a positive sequence. Set $\ell^p(w)\overset{def}{=\!=}\{x=\{x_k\}_{k\geq 0}: xw\in\ell^p\}$, $0<p\leq+\infty$, $c_0(w)\overset{def}{=\!=}\{x=\{x_k\}_{k\geq 0}: xw\in c_0\}$, where $xw=\{x_k w_k\}_{k\geq 0}$.

THEOREM 2.4. Let $M, M\subset\mathbb{N}$, be lacunary and $0<p<1$. Then

$$R_M(H^p)=\ell^p(m^{1-\frac{1}{p}})$$

and the operator $R_M: H^p\longrightarrow\ell^p(m^{1-\frac{1}{p}})$ is right invertible.

PROOF. The inclusion $R_M(H^p)\subset\ell^p(m^{1-\frac{1}{p}})$ was proved in [33]. Prove the inverse one. By lacunarity there exists $\varepsilon>0$ such that $m_{k+1}/m_k>\frac{1+\varepsilon}{1-\varepsilon}$. By the Cor.2.2 there exists a sequence $\{Q_n\}_{n\geq 1}$ of trigonometrical polynomials such that $\int Q_n\,dm=1$, $\deg Q_n\leq\varepsilon n$ and $\|Q_n\|_{L^p}\leq c_p(\varepsilon)n^{1-\frac{1}{p}}$ for any $n\in\mathbb{N}$. Let $a\in\ell^p(m^{1-1/p})$.

Set $f=\sum_{k\geq 0} a_k Q_{m_k} z^{m_k}$.

Then $f\in H^p$ and $R_M(f)=a$. ●

REMARK 2.5. Let $0<p<1$. One can prove that $R_M(H^p)=\ell^p(m^{1-1/p})$ iff M is a finite union of lacunary sets.

Denote by $(1-z)^{-\alpha}H^\infty$ (resp. $(1-z)^{-\alpha}C_{A,0}$) the set of all functions f analytic in $\mathbb{D}$ satisfying $(1-z)^\alpha f\in H^\infty$ (resp. $(1-z)^\alpha f\in C_A$ and $\lim_{z\to 1} f(z)(1-z)^\alpha=0$). Obviously, $H^{p,\infty}\supset(1-z)^{-1/p}H^\infty$, $H^{p,\infty}_0\supset(1-z)^{-1/p}C_{A,0}$.

THEOREM 2.6. Let M be lacunary $M\subset\mathbb{N}$ and $0<p<1$. Then

$$R_M(H^{p,\infty})=R_M((1-z)^{-1/p}H^\infty)=\ell^\infty(m^{1-1/p}),$$

$$R_M(H_0^{p,\infty}) = R_M((1-z)^{-1/p}C_{A,0}) = c_0(m^{1-\frac{1}{p}})$$

and all these operators R_M are right invertible.

PROOF. We prove the first assertion the second being analogous. It is clear that $R_M((1-z)^{-1/p}H^\infty) \subset R_M(H^{p,\infty})$.

Furthermore $R_M(H^{p,\infty}) \subset \ell^p(m^{1-1/p})$ by Th.1.4. It remains to prove that $\ell^\infty(m^{1-1/p}) \subset R_M((1-z)^{-1/p}H^\infty)$. Choose again $\varepsilon > 0$ such that $m_{k+1}/m_k > \frac{1+\varepsilon}{1-\varepsilon}$. Take a function $\varphi \in C^\infty(\mathbb{R})$, $\varphi(0)=1$, $\operatorname{supp}\varphi \subset [-\varepsilon,\varepsilon]$. Set $Q_n(z) = \sum_{k=-n}^{n} \varphi(\frac{k}{n}) z^k$. Let $a \in \ell^\infty(m^{1-1/p})$. Take $f(z) = \sum_{k\geq 0} a_k \cdot Q_{m_k} \cdot z^{m_k}$. It is easy to check (using Lemma 2.1) that $f \in (1-z)^{-\frac{1}{p}}H^\infty$ and $R_M f = a$. ●

REMARK 2.7. Let $0<p<1$. One can prove that $R_M(H^{p,\infty}) = \ell^\infty(m^{1-1/p})$ ($R_M(H_0^{p,\infty}) = c_0(m^{1-1/p})$) iff M is a finite union of lacunary sets. On the other hand, the equality $R_M((1-z)^{-1/p}H^\infty) = \ell^\infty(m^{1-1/p})$ (or $R_M((1-z)^{-1/p}C_{A,0}) = c_0(m^{1-\frac{1}{p}})$) holds for lacunary sets only.

§3. Sparsely indexed Fourier sums of continuous functions

Set for $f \in \mathcal{D}'(\mathbb{T})$, $S_n(f) = \sum_{k=-n}^{n} \hat{f}(k)$, $n \in \mathbb{Z}_+$. Let M be an infinite subset of $\mathbb{Z}_+$. Denote by T_M the operator $T_M f \overset{def}{=\!=} \{S_{m_k} f\}_{k\geq 0}$ from $L^\infty(\mathbb{T})$ into the space of all numerical sequences,

THEOREM 3.1. Let $f \in L^\infty(\mathbb{T})$. Then

$$\sum_{n\geq 0} S_n(f) z^n \in H^{1,\infty}.$$

If $f \in C(\mathbb{T})$ and $f(1) = 0$ then

$$\sum_{n\geq 0} S_n(f) z^n \in H_0^{1,\infty}.$$

PROOF. Again we shall prove the first assertion only. We can suppose $\hat{f}(k) = \hat{f}(-k)$, $k \in \mathbb{N}$, and $\hat{f}(0) = 0$. Set $g(z) = \sum_{k\geq 0} \hat{f}(k) z^k$, then $g \in BMO_A$ and $g = \frac{1}{2}(f+h)$, where $h(z) = -\sum_{k<0} \hat{f}(k) z^k + \sum_{k\geq 0} \hat{f}(k) z^k$. Let I be an arc of $\mathbb{T}$, centered at 1. One of definitions of BMO implies

$$\int_I |h-h_I|^2 \leq cm(I)\,\|h\|^2_{BMO} \leq$$

$$\leq cm(I)\|f\|^2_{L^\infty}\,, \quad h_I \overset{def}{=\!=} \frac{1}{m(I)}\int_I h\,dm = 0\,,$$

so

$$\int_I |h|^2 dm \leq cm(I)\|f\|^2_{L^\infty} \quad \text{and}$$

$$\int_I |g|^2 dm \leq cm(I)\|f\|^2_{L^\infty}\ .$$

Note that $\sum_{n\geq 0} S_n(f)z^n = (1-z)^{-1}g(z)$. Now we can complete the proof by the following lemma.

LEMMA 3.2. Let $h\in L^1[0,1]$. Suppose that $\sup_{0<\varepsilon<1}\frac{1}{\varepsilon}\int_0^\varepsilon |h(t)|^2 dt < +\infty$. Then $m\{t\in[0,1]: |h(t)/t|>A\} = O(\frac{1}{A})\ (A\to+\infty)$.

PROOF OF LEMMA. Set $\Delta_n = (2^{-n}, 2^{-n+1})$, $n\in\mathbb{N}$. By the assumption $\int_{\Delta_n}|h|^2 \leq c2^{-n}$. So

$$m\{t\in\Delta_n : |h(t)/t|>A\} \leq m\{t\in\Delta_n : |h(t)|>A\cdot 2^{-n}\} \leq$$
$$\leq \frac{2^{2n}}{A^2}\int_{\Delta_n}|h|^2 \leq \frac{2^n c}{A^2}\ .$$

We have

$$m\{t\in[0,1]: |h(t)/t|>A\} \leq \sum_{2^n\geq A} 2^{-n} + c\sum_{2^n\leq A}\frac{2^n}{A^2} \leq \frac{c}{A}\ . \bullet\bullet$$

THEOREM 3.3. Let M be a lacunary set and $f\in L^\infty(\mathbb{T})$. Then

$$\sum_{k\in M,\, k\leq n} |S_k(f)|^2 \leq C_M \|f\|^2_{L^\infty} \log^2 n\,, \quad n\in\mathbb{N}\,, \quad n>1\,.$$

If $f \in C(\mathbb{T})$ then

$$\sum_{k\in M,\, k\le n} |S_k(f)|^2 = o(\log^2 n) \quad (n \to +\infty).$$

PROOF. Use Th.1.5 and Th.3.1. ●

Note that the paper [44] contains results essentially more general than Th.3.3. Denote by $\sigma_M(\sigma_M^\circ)$ the set of all sequences $\{a_k\}_{k\ge 0}$ such that

$$\sum_{j=0}^{k} |a_j|^2 = O(\log^2 m_k) \;(\text{ resp. } o(\log^2 m_k)),\; k \to +\infty .$$

Denote by S the shift operator in the space of all one-sided numerical sequences. Note that $R_M((1-z)^{-1}H^\infty) = T_M(H^\infty)$, because $R_M((1-z)^{-1}f) = T_M f$ for any $f \in H^\infty$.

THEOREM 3.4. Let M be lacunary. Then

$$T_M(L^\infty) = T_M(H^\infty) = R_M(H^{1,\infty}) = \sigma_M ,$$

$$T_M(C) = T_M(C_A) = R_M(H_0^{1,\infty}) = \sigma_M^\circ .$$

LEMMA 3.5. Let $a = \{a_k\}_{k\ge 0}$ be a complex sequence, M a lacunary set such that $m_k \le q^{k+1}$, $k \in \mathbb{Z}_+$, for some $q > 1$. Three following assertions are equivalent.

1) $a \in \sigma_M(\sigma_M^\circ)$

2) $\sup_{n\in\mathbb{Z}_+} 2^{-2n} \sum_{k=2^n}^{2^{n+1}-1} |a_k|^2 < +\infty \;(\lim_{n\to+\infty} 2^{-2n} \sum_{k=2^n}^{2^{n+1}-1} |a_k|^2 = 0)$,

3) there exists a sequence $\{x^{(n)}\}_{n\ge 0}$, $x^{(n)} \in \ell^2$, such that $\sup_{n\ge 0} \|x^{(n)}\|_{\ell^2} < +\infty$ $(\lim_{n\to+\infty} \|x^{(n)}\|_{\ell^2} = 0)$ and $a = \sum_{n\ge 0} S^n x^{(n)}$.

PROOF. The implication 1)$\Rightarrow$2) is obvious. Prove the implication 2)$\Rightarrow$3). Let $\sup_{n\ge 1} 2^{-2n} \sum_{k=2^n}^{2^{n+1}-1} |a_k|^2 < +\infty$. We can suppose $a_0 = 0$. Set

$$y_k^{(n)} \overset{def}{=\!=} \begin{cases} a_k, & 2^n \le k < 2^{n+1} \\ 0 & \text{for others } k . \end{cases}$$

It remains to note that $x = \sum_{n\ge 0} y^{(n)}$ and $y^{(n+1)} = \sum_{k=2^n}^{2^{n+1}-1} S^k (2^{-n} S^{*k} y^{(n+1)})$, where S^* is the backward shift operator $S^*(\{\xi_k\}) = \{\xi_{k+1}\}$ and furthermore $\|2^{-n}S^{*k}y^{(n+1)}\|_{\ell^2} \leq 2^{-n}\|y^{(n+1)}\|_{\ell^2} \leq c(a)$.

Prove now 3)$\Rightarrow$1):

$$\sum_{j=0}^{k} |a_j|^2 \leq \Big\|\sum_{j=0}^{k} S^j x^{(j)}\Big\|^2 \leq \Big(\sum_{j=0}^{k} \|x^{(j)}\|_{\ell^2}\Big)^2 \leq c(k+1)^2. \quad \bullet$$

PROOF OF Theorem 3.4. Again we shall prove the first assertion only. It is clear that $T_M(H^\infty) \subset L^\infty$. By Th.3.1 $T_M(L^\infty) \subset R_M(H^{1,\infty})$. Further, $R_M(H^{1,\infty}) \subset \sigma_M$ by Th.3.3. It remains to show that $\sigma_M \subset T_M(H^\infty)$. We can suppose for simplicity that $M = \{2^k : k \in \mathbb{Z}_+\}$. Let $a \in \sigma_M$. Then $a = \sum_{n \geq 0} S^n x^{(n)}$, where $x^{(n)} \in \ell^2$ and $\sup_{n\geq 0} \|x^{(n)}\|_{\ell^2} < +\infty$. It is well known that $R_M(C_A) = \ell^2$ (cf. [58]). So there exists a sequence $\{g_n\}_{n\geq 0} \subset C_A$ such that $\sup_{n \geq 0} \|g_n\|_{C_A} < +\infty$, $R_M g_n = x^{(n)}$ and $g_n(0) = 0$. Let $K_n(z) = \sum_{k=-n}^{n} (1 - \frac{|k|}{n}) z^k$ be the Fejer kernel and set $f_n(z) = g_n(z^{2n}) \cdot K_{2^n}(z)$. It is clear that $f_n \in C_A$ and $R_E f_n = S^n(x^{(n)})$. Set $f(z) = \sum_{n \geq 0} f_n(z)(1-z)$. Then $f \in H^\infty$ because

$$|K_n(z)| \leq \min\Big\{n, \frac{4}{n|z-1|^2}\Big\}, \quad z \in \mathbb{T}.$$

Obviously, $T_M f = a$. $\bullet$

COROLLARY 3.6. Let $E \subset \mathbb{Z}_+$ and $\log m_{k+1}/\log m_k > q$, $k \in \mathbb{Z}_+$, for some $q > 1$. Then

$$T_M(L^\infty) = T_M(H^\infty) = R_M(H^{1,\infty}) = \ell^\infty((\log m)^{-1}), \tag{1}$$

$$T_M(C) = T_M(C_A) = R_M(H_0^{1,\infty}) = c_0((\log m)^{-1}). \tag{2}$$

In particular $T_M(L^\infty) = \ell^\infty(2^{-n})$ and $T_M(C) = c_0(2^{-n})$ for $M = \{2^{2^n}\}_{n \geq 0}$. $\bullet$

We can obtain the following reinforcement of Cor.3.6.

THEOREM 3.7. Let $M, M \subset \mathbb{Z}_+$, be lacunary. Two following assertions are equivalent.

1) The equalities (1) (or (2)) holds.

2) The set M is a finite union of sets satisfying the condition of Cor.3.6.

REMARK 3.8. One can prove that if both assertions of Th.3.7

hold then all operators R_M and T_M are right invertible. On the other hand if the assertions of Th.3.7 don't hold then no operator T_M involved in the statement of Th.3.4 is right invertible.

To show this we note that the right inverbility of the operator $T_M | H^\infty(T_M|C_A)$ implies the right invertibility of the operator $T_M | L^\infty(T_M|C)$. Thus the space $\sigma_M \simeq L^\infty / \operatorname{Ker} T_M$ (or $\sigma_M^\circ \simeq C/\operatorname{Ker} T_M$) can be imbedded as a complemented subspace into L^∞ (or C). Hence the space $\sigma_M (\sigma_M^\circ)$ cannot contain the spaces $\{\ell^2_n\}_{n\geq 1}$ as a uniformly complemented family of subspaces. Consequently, the equivalent conditions of Th.3.7 hold.

§4. Sparsely indexed Fourier coefficients of $\mathcal{H}^p_\alpha$-functions

We can easy deduce the following theorem from the results of §2 and Ch.III.

THEOREM 4.1. Let $M \subset \mathbb{Z}_+$, $0<p\leq 1$, $\alpha>0$. Two following assertions are equivalent.

1) The set M is a finite union of lacunary sets;

2) $R_M(\mathcal{H}^p_\alpha) = \ell^p(m^{1-1/p-\alpha})$. ●

One can prove Th.4.1 not using the results of Ch.III. The case $p=1$ is not new: 1)$\Rightarrow$2) by [32] and 2)$\Rightarrow$1) follows from [60]. Moreover, there is an analog of Th.4.1 for $p\geq 1$ (in this case 2) has to be replaced by $R_M(\mathcal{H}^p_\alpha) = \ell^p(m^{-\alpha}))$. The implication 1)$\Rightarrow$ 2) was proved in [18] (true, for $\alpha = \frac{1}{p}$ only). The case $\alpha \neq 1/p$ can be considered analogously or can be reduced to $\alpha = 1/p$ by means of the fractional integration. The implication 2)$\Rightarrow$1) follows from the results of [60] as in the case $p=1$.

REMARK 4.2. If the equivalent conditions of Th.4.1 hold then the operator R_M is right invertible. Furthermore, in the case $p\geq 1$ (in contrast with the case $p<1$) we can take the operator $\{a_k\}_{k\geq 0} \longmapsto \sum_{k\geq 0} a_k z^{m_k}$ as a right inverse.

Chapter Five. SOME REMARKS ON THE BACKWARD SHIFT OPERATOR

§0. Introduction

One of well known descriptions of the invariant subspaces of the backward shift operator $S^*: H^2 \to H^2$, $S^* f \overset{def}{=} \frac{f - f(0)}{z}$, depends on the notion of the so called pseudocontinuation (see

[7]). To recall it let f be a function of the Smirnov class N_A. The function $\hat{f}:\mathbb{C}\setminus\mathbb{T}\to\mathbb{C}$ is said to be the pseudocontinuation of f if $\hat{f}|\mathbb{D}=f$, if the restriction $\hat{f}|\mathbb{D}_-$ is meromorphic with the bounded characteristic (in the sense of Nevanlinna) and if $\lim\limits_{r\to 1-} f(r\zeta)=\lim\limits_{r\to 1+} f(r\zeta)$ a.e. on $\mathbb{T}$. Because of the well known uniqueness theorem each function f, $f\in N_A$, cannot have more than one pseudocontinuation. Each inner function φ has the pseudocontinuation by the reflection principle: $\tilde{\varphi}(z)=\overline{\varphi(\bar{z}^{-1})}^{-1}$, $|z|>1$.

Let X be a subset of the class N_A. Denote by $X\cap\varphi N_{\bar{A}}$ the set of all functions $f, f\in X$, admitting a pseudocontinuation $\tilde{f}$ with $\tilde{f}/\tilde{\varphi}\in N_{\bar{A}}$. Here $N_{\bar{A}}\overset{def}{=\!=}\{f(\frac{1}{z})\frac{1}{z}: f\in N_A\}$.

It is easy to see that for every inner function φ the intersection $H^2\cap\varphi N_{\bar{A}}=H^2\cap\varphi H^2_-$ is a S^*-invariant subspace ($\neq H^2$) of the space H^2. Each invariant subspace [x] of the operator $S^*: H^2\to H^2$ has such form.

Let now Z be a space of functions analytic in the disc $\mathbb{D}$ and let $Z\subset N_A$. Let's suppose that $S^*Z\subset Z$. We shall say that the invariant subspaces of the operator $S^*: Z\to Z$ are standard iff the lattice of all S^*-invariant subspaces of Z (except Z itself) coincides with $\{Z\cap\varphi N_{\bar{A}}:$ φ is an inner function$\}$ and $Z\cap\varphi_1 N_{\bar{A}}=Z\cap\varphi_2 N_{\bar{A}}$ $\Leftrightarrow$ φ_1/φ_2 is a constant. Note that last equivalence holds if $Z\supset H^\infty$.

It is well-known (see [3]) that every nonzero invariant subspace of the shift operator $S: H^p\to H^p$, $0<p\leq+\infty$ (for $p=+\infty$ the space H^p is endowed with the weak-star topology) is representable as φH^p where φ is an inner function. This implies (by the duality arguments) that S^*-invariant subspaces of the spaces $H^p(1<p<+\infty)$, L^1/H^1_- and BMO_A are standard.

It has been noted in [15] that the invariant subspaces of the operator $S^*: H^1\to H^1$ are also standard. In the paper [17] we shall describe a whole class of locally convex spaces Z, $Z\subset N_A$, with standard S^*-invariant subspaces. In particular, this class includes the spaces $H^p(1\leq p\leq+\infty)$, L^1/H^1_-, BMO_A, $C_A\overset{def}{=\!=}C(\mathbb{T})\cap H^\infty$,

$$U_A^\infty\overset{def}{=\!=}\{f\in H^\infty: \sup_{n\geq 0}\|S^{*n}f\|_{H^\infty}<+\infty\},$$

---

[x] In this chapter "subspace" means a closed linear manifold.

$U_A \overset{def}{=} \{ f \in H^\infty : \lim_{n\to+\infty} \| S^{*n} f \|_{H^\infty} = 0 \}$, and many others. Therefore in this chapter as a rule we shall deal with the operator S^* on non locally convex spaces.

In §3 we shall prove that the invariant subspaces of the operator $S^*: H_0^{1,\infty} \to H_0^{1,\infty}$ are standard. Recall that

$$H_0^{1,+\infty} \overset{def}{=} \{ f \in N_A : m\{|f| > A\} = o(\tfrac{1}{A})(A \to +\infty)\}.$$

It is worth noting that S^*-invariant subspaces are not standard if $Z \cap N_{\bar{A}} \neq \{0\}$, in fact, the subspace $\{0\}$ cannot be represented as $Z \cap \varphi N_{\bar{A}}$ with some inner function φ . In particular this is the case if $\frac{1}{\zeta - z} \in Z$ for some $\zeta \in \mathbb{T}$.

In the paper [15] we describe the invariant subspaces of the operator $S^*: H^p \to H^p, 0<p<1$. The description of S^*-invariant subspaces of the space $H^p \cap H^p_-$ was of great importance for this.

In §2 (Th.2.1) for a class of separable symmetric [x)] metric function spaces X we describe the invariant subspaces of the operator $S^*: X_A \to X_A$ containing the subspace $X_A \cap X_{\bar{A}}$. From that we deduce a complete description of the invariant subspaces of operators $S^*: X_A \to X_A$ for the class $\{X : X = X_A \oplus X_{\bar{A}} , X \subset N\}$.

Moreover, §2 contains a description (in terms of pseudocontinuations) of noncyclic vectors of the operators $S^*: X_A \to X_A$ for all separable quasinormed symmetric spaces X such that $X \supset BMO$. A similar problem for the spaces $X = L^p$ $(1 \le p < +\infty)$ was solved in [30], and for the spaces $X = L^p$ $(0<p<1)$ in [15].

In §1 we define the maximal functional calculus for the operator $S^*: H^p \to H^p (0<p<1)$, in other words we describe the set of all functions $h, h \in H^\infty$, such that the Toeplitz operator

$$f \mapsto P(\bar{h} f), \quad f \in H^2,$$

can be continuously extended onto the whole space H^p as a continuous map[xx)]. The results of the §1 were announced in [15].

---

x) See for this subject Ch.I.

xx) Recall that by P we denote the Riesz projection.

A similar problem for $p=1$ was solved in [55], see also [38]. The case $p<1$ turned out to be more simple. When $1<p<+\infty$ it is well known that the operator $h(S^*)$ is continuous for every h, $h\in H^\infty$.

§1. Functional calculus for the operator $S^*: H^p \longrightarrow H^p\ (0<p<1)$.

It is clear that the spectrum of the operator $S^*: H^p \longrightarrow H^p$ $(0<p<1)$ coincides with the closed unit disc $\overline{\mathbb{D}}$: each point λ, $\lambda\in\overline{\mathbb{D}}$, is a simple eigenvalue of S^* $(S^*(1-\lambda z)^{-1}=\lambda(1-\lambda z)^{-1})$ and on the other hand for every λ, $\lambda\in\mathbb{D}$, the operator $I-\lambda S^*$ is invertible, and

$$(I-\lambda S^*)^{-1}f=\frac{zf-\lambda f(\lambda)}{z-\lambda}\ .$$

It follows that an operator [*] $\mathcal{U}: H^p \longrightarrow H^p$ commutes with the backward shift operator iff there exists a function ψ (automatically unique and holomorphic in $\mathbb{D}$) defined in $\overline{\mathbb{D}}$ such that

$$\mathcal{U}(1-\lambda z)^{-1}=\psi(\lambda)(1-\lambda z)^{-1},\quad \lambda\in\overline{\mathbb{D}}\ .$$

In this situation we shall write $\mathcal{U}=\psi(S^*)$. Conversely, if for a function ψ the operator $\mathcal{U}$ defined by the last formula admits a (linear and continuous) continuation as an $(H^p\longrightarrow H^p)$ -operator, we shall write $\psi(S^*)H^p\subset H^p$.

Before stating our results let us recall (see the list of symbols at the begining of the article) that by $\{\Lambda^\alpha_+\}_{\alpha>0}$ we denote the standard scale of functions analytic in the disc $\mathbb{D}$ and smooth up to the boundary.

THEOREM 1.1. Let $p\in(0,1)$ and let S^* be the backward shift operator in the space H^p . Then

$$\psi(S^*)H^p\subset H^p \Longleftrightarrow \psi\in\Lambda_+^{\frac{1}{p}-1}\ .$$

---

[*] In the present section all operators and functionals are supposed to be linear and continuous.

PROOF. Let $\psi(S^*)H^p \subset H^p$. Consider the functional $f \overset{\alpha}{\longmapsto} (\psi(S^*)f)(0)$, $f \in H^p$. It is clear that $\alpha((1-\lambda z)^{-1}) = \psi(\lambda)$. Hence $\psi \in \Lambda_+^{1/p-1}$ by the duality theorem of P.Duren, B.Romberg and A.Shields [31].

Let now $\psi \in \Lambda_+^{1/p-1}$. Show that $\psi(S^*)H^p \subset H^p$. By the above mentioned duality there exists a functional α, $\alpha \in (H^p)^*$, such that $\alpha((1-\lambda z)^{-1}) = \psi(\lambda)$, $\lambda \in \overline{\mathbb{D}}$. Define the operator $\mathcal{U}$ by the formula

$$(\mathcal{U}f)(w) = \alpha\left(\frac{zf - wf(w)}{z-w}\right), \quad f \in H^p, \quad w \in \mathbb{D}.$$

For checking $\mathcal{U}(H^p) \subset H^p$ it suffices to show that there exists a constant $C = C(p,\psi)$ such that $\|\mathcal{U}f\|_{H^p} \leq C\|f\|_{H^p}$ for every polynomials f. We have

$$\|\mathcal{U}f\|_{H^p}^p = \int_{\mathbb{T}} |(\mathcal{U}f)(w)|^p dm(w) \leq \|\alpha\|^p \int_{\mathbb{T}} \left\|\frac{zf - wf(w)}{z-w}\right\|_{H^p}^p dm(w) \leq$$

$$\leq 2\|\alpha\|^p \|f\|_{H^p} \left\|\frac{1}{1-z}\right\|_{H^p}^p .$$

It remains to show that $\mathcal{U} = \psi(S^*)$. ●

COROLLARY 1.2. Let $p \in (0,1)$, $\psi \in \Lambda_-^{1/p-1}$. Then there exists an operator $\mathcal{U}: H_-^p \longrightarrow H_-^p$ such that $\mathcal{U}((1-\lambda z)^{-1}) = \psi(\lambda)(1-\lambda z)^{-1}$ for all λ, $|\lambda| > 1$.

PROOF. The assertion can be reduced to the theorem by the isometry $f \longmapsto z^{-1}f(z^{-1})$ of the space H^p onto H_-^p. ●

Now we are going to consider an analogue of Th.1.1 for the operator $S_0^* \overset{def}{=\!=} S^*|(H^p \cap H_-^p)$. The spectrum of S_0^* coincides with the circle $\mathbb{T}$, $(I - \lambda S_0^*)^{-1}f = \frac{zf - \lambda f(\lambda)}{z-\lambda}$, $\lambda \in \mathbb{C} \setminus \mathbb{T}$. Let ψ be a function on the circle $\mathbb{T}$. If there exists an operator $\mathcal{U}: H^p \cap H_-^p \longrightarrow H^p \cap H_-^p$ such that $\mathcal{U}((1-\zeta z)^{-1}) = \psi(\zeta)(1-\zeta z)^{-1}$, $\zeta \in \mathbb{T}$, we write $\psi(S_0^*)(H^p \cap H_-^p) \subset H^p \cap H_-^p$. It should be noted that because of Th.1.0.1 (b).the operator $\mathcal{U}$ is uniquelly defined by the function ψ. Obviously, for a given operator $\mathcal{U}$ there exists a function ψ with $\mathcal{U} = \psi(S_0^*)$ if and only of $\mathcal{U}S_0^* = S_0^*\mathcal{U}$.

THEOREM 1.3. Let $p \in (0,1)$, $S_0^* \overset{def}{=\!=} S^*|(H^p \cap H_-^p)$

and let ψ be a function on the circle $\mathbb{T}$. Then
$$\psi(S_0^*)(H^p\cap H_-^p)\subset H^p\cap H_-^p \Longleftrightarrow \psi\in\Lambda^{1/p-1}.$$

PROOF. Let $\psi(S_0^*)(H^p\cap H_-^p)\subset H^p\cap H_-^p$. Consider the functional $\alpha\in(H^p\cap H_-^p)^*$, $\alpha(f)=(\psi(S_0^*)f)(0)$. Then $\psi\in\Lambda^{1/p-1}$ because of Th.3 of [14].

Let now $\psi\in\Lambda^{1/p-1}$. Consider first the case $\psi\in\Lambda_+^{1/p-1}$. If we define $\mathcal{U}_0 \overset{def}{=} \psi(S^*)|(H^p\cap H_-^p)$ then $\mathcal{U}_0(H^p\cap H_-^p)\subset \subset H^p\cap H_-^p$ and obviously $\mathcal{U}_0=\psi(S_0^*)$.

Let now $\psi\in\Lambda_-^{1/p-1}$. In view of Cor.1.2 there exists an operator $\mathcal{U}: H_-^p\longrightarrow H_-^p$ such that $\mathcal{U}((1-\lambda z)^{-1})= =\psi(\lambda)(1-\lambda z)^{-1}$ for every λ, $|\lambda|>1$. Set $\mathcal{U}_0= =\mathcal{U}|(H^p\cap H_-^p)$. Clearly, $\psi(S_0^*)(H^p\cap H_-^p)\subset H^p\cap H_-^p$.

To finish the proof it is enough to note that
$$\Lambda^{1/p-1}=\Lambda_+^{1/p-1}+\Lambda_-^{1/p-1}. \bullet$$

§2. Cyclic vectors of the operator $S^*: X_A\rightarrow X_A$.

Recall that $N\overset{def}{=}\{f\in L^0 : \log(1+|f|)\in L^1\}$.

THEOREM 2.1. Let X be a separable symmetric metric space, $X\subset N$, $X=X_A+X_{\bar A}$, and let E be an invariant subspace of the operator $S^*: X_A\rightarrow X_A$, $E\neq X_A$. If $X_A\cap X_{\bar A}\subset E$, then $E=X_A\cap\varphi N_{\bar A}$ for some inner function φ.

To prove Theorem 2.1 we need a description of invariant subcpaces of the shift operator $S: X(\mathbb{T})\longrightarrow X(\mathbb{T})$, $Sf\overset{def}{=} zf$, where X is a separable symmetric metric space. For the case $X=L^p$ $(0<p<+\infty)$ such a description can be found in [3]. The general case is quite analogous to the particular one.

THEOREM ([3]). Let X be a separable symmetric metric space, $X\subset N$, and let E be an invariant subspace of the operator $S: X\longrightarrow X$. Then there are following two possibilities:

1) $SE\neq E$, and then $E=\varphi X_A$, where $\varphi\in L^\infty$, $|\varphi|=1$ a.e.

2) $SE=E$, and then $E=\chi X$, where $\chi\in L^\infty$, $\chi^2=\chi$. $\bullet$

The following corollary follows immediately from the above theorem.

COROLLARY 2.2. Let X be a separable symmetric metric space, $X\subset N$. Set $S^*(f+X_{\bar A})\overset{def}{=}\bar z f+X_{\bar A}$. Then the family $\{\varphi X_{\bar A}/X_{\bar A}: \varphi$ is an inner function$\}$ is the family of all $(\neq X_A/X_{\bar A})$ S^*-invariant subspaces of the space $X/X_{\bar A}$. $\bullet$

PROOF OF THEOREM 2.1. It is enough to describe the invariant subspaces of the operator $S^*: X_A/X_A\cap X_{\bar A}\to X_A/X_A\cap X_{\bar A}$, $S^*(f+X_A\cap X_{\bar A})\overset{def}{=\!=}S^*f+X_A\cap X_{\bar A}$. To obtain such a description let us consider the canonical projection $X_A\xrightarrow{\mathcal{J}} X/X_{\bar A}$, $\mathcal{J}f\overset{def}{=\!=}f+X_{\bar A}$ and the mapping $j: X_A/X_A\cap X_{\bar A}\to X/X_{\bar A}$, $j(f+X_A\cap X_{\bar A})\overset{def}{=\!=}\mathcal{J}f$. The conditions of the theorem imply $\mathcal{J}(X_A)=X/X_{\bar A}$, $\operatorname{Ker}\mathcal{J}=X_A\cap X_{\bar A}$ and hence the mapping j is an isomorphism. To finish the proof it is enough to mention Cor.2.2 and the commutativity of the following diagram

$$\begin{array}{ccc} X/X_{\bar A} & \xrightarrow{S^*} & X/X_{\bar A} \\ \uparrow j & & \uparrow \mathcal{J} \\ X_A/X_A\cap X_{\bar A} & \xrightarrow{S^*} & X_A/X_A\cap X_{\bar A} \end{array} . \quad \bullet$$

COROLLARY 2.3. Let E be an invariant subspace of the operator $S^*: N_A\to N_A$, $E\neq N_A$, and let $E\supset N_A\cap N_{\bar A}$. Then $E=N_A\cap\varphi N_{\bar A}$ for some inner function φ. $\bullet$

It should be remarked that Th.2.1 contains a complete description of invariant subspaces of the operator $S^*: X_A\to X_A$ for all separable symmetric metric spaces X with the properties $X_A\cap X_{\bar A}=\{0\}$, $X\subset N$ and $X=X_A+X_{\bar A}$. However, if $X_A\cap X_{\bar A}\neq\{0\}$. Th.2.1 yields no description even for the collecton of the noncyclic vectors of the operator $S^*: X_A\to X_A$.

. In conclusion of this section we describe the above mentioned collection for one class of separable symmetric quasinormed spaces X (see Cor.2.6 below).

Let X be a separable quasinormed space and let $\log 2/t\in X[0,1]$. Hence $X\supset BMO$. With every S^*-invariant subspace E, $E\subsetneqq X$, we associate an inner function φ_E in the following way. Consider at first the intersection $E\cap\cap BMO_A\subsetneqq BMO_A$. Standard arguments show that this intersection is a S^*-invariant subspace closed in the star-weak topology. The well known description of S-invariant subspaces of the Hardy space H^1 implies that $E\cap BMO_A=BMO_A\cap\varphi N_{\bar A}$ for some inner function φ (defined by E uniquely up to a multiplicative constant). Set $\varphi_E = \varphi$.

THEOREM 2.4. Let X be a separable symmetric quasinormed space, $\log 2/t \in X[0,1]$, and let E be an invariant subspace of the operator $S^*: X_A \to X_A$, $E \neq X_A$.

Then $E \subset X_A \cap \varphi_E N_{\bar{A}}$.

Proof of the theorem is based on the following lemma, but in all other details it is completely analogous to the proof of Th.1 in [15].

LEMMA 2.5. Suppose the conditions of Th.2.4 are fulfilled and let $f \in E$, $h \in H^\infty$, $(\mathcal{M}f)\cdot h \in L^\infty$ (here $(\mathcal{M}f)(\zeta) \overset{def}{=} \sup_{0<r<1} |f(r\zeta)|, \zeta \in \mathbb{T}$). Then $P(\bar{h}f) \in X_A$.

PROOF OF LEMMA 2.5. repeats the proof of Lemma 1 in [15]. We have only to note that every sequence of functions bounded in the space BMO and convergent in meausre converges in fact in the space X. ●

COROLLARY 2.6. Suppose the conditions of Th.2.4 are fulfilled and let $f \in X_A$. The function f is noncyclic for S^* iff f possesses a pseudocontinuation. ●

§3. Invariant subspaces of the operator $S^*: H_0^{1,\infty} \to H_0^{1,\infty}$.

Denote by $L_0^{1,\infty}$ the set of all functions f, $f \in L^0$, such that $\|f\|_{L^{1,\infty}} \overset{def}{=} \sup_{A>0} A\, m\{|f|>A\} < +\infty$ and $m\{|f|>A\} = o(A^{-1})$ $(A \to +\infty)$. Recall that $H_0^{1,\infty} \overset{def}{=} (L_0^{1,\infty})_A$

THEOREM 3.1. Invariant subspaces of the operator $S^*: H_0^{1,\infty} \to H_0^{1,\infty}$ are standard.

PROOF. Let E be a S^*-invariant subspace $E \subsetneqq H_0^{1,\infty}$. In view of Th.2.4 we have

$$BMO_A \cap \varphi_E N_{\bar{A}} \subset E \subset H^{1,\infty} \cap \varphi_A N_{\bar{A}} .$$

It remains to show that $E \supset H^{1,\infty} \cap \varphi_A N_{\bar{A}}$. Let $f \in H_0^{1,\infty} \cap \varphi_E N_{\bar{A}}$. Let us consider a measure preserving transformation $h: \mathbb{T} \to (0,1)$ such that $\mathcal{M}f = (\mathcal{M}f)^* \circ h$. We recall that g^* denotes the decreasing continuous on the left permutation of the function $|g|$.

Set $\lambda(A) = m\{\mathcal{M}f > A\}$ and define the function $g_A: \mathbb{T} \to \mathbb{R}$ by the following equalities:

$$g_A(\zeta) = \begin{cases} 0, & h(\zeta) \geq \lambda(A) \\ \log \frac{h(\zeta)}{\lambda(A)}, & h(\zeta) \leq \lambda(A). \end{cases}$$

Continue the function g_A into the unit disc by the Poisson integral. Consider the outer function $F_A = e^{g_A + i\mathcal{H} g_A}$, g_A, $\mathcal{H} g_A$ being conjugate harmonic functions, $(\mathcal{H} g_A)(0) = 0$. By Lemma 2.5 $P(\bar{F}_A f) \in E$ for every $A > 0$. We shall prove that $f = \lim_{A\to+\infty} P(\bar{F}_A f)$ in the quasinorm of the space $H_0^{1,\infty}$. Remark that $|F_A| \leq 1$, and since $\|g_A\|^2_{L^2} = \int_0^{\lambda(A)} \log^2 \frac{t}{\lambda(A)} dt = 2\lambda(A) \to 0$ for $A \to +\infty$, we have $\lim_{A\to+\infty} F_A = 1$ (in L^0). It follows that $\lim_{A\to+\infty} F_A f = f$ (in $H_0^{1,\infty}$). Now it is enough to show that $\lim_{A\to+\infty} P(F_A f - \bar{F}_A f) = 0$. For this in view of Kolmogorov theorem it suffices to prove that $\lim_{A\to+\infty} \|(F_A - \bar{F}_A) f\|_{L^1} = 0$. We have

$$\int_{\mathbb{T}} |F_A - \bar{F}_A| |f| dm \leq 2 \int_{\mathcal{M}f \geq A} |F_A| \cdot |f| dm + \int_{\mathcal{M}f \leq A} |F_A - \bar{F}_A| \cdot |f| dm \leq$$

$$\leq 2 \frac{1}{\lambda(A)} \int_0^{\lambda(A)} t (\mathcal{M}f)^*(t) dt + \Big(\int_{\mathcal{M}f \leq A} |f|^2 dm\Big)^{1/2} \Big(\int_{\mathbb{T}} 4 |\mathcal{H} g_A|^2 dm\Big)^{1/2} \leq$$

$$\leq 2 \frac{1}{\lambda(A)} \int_0^{\lambda(A)} t (\mathcal{M}f)^*(t) dt + 2 \Big(\int_{\lambda(A)}^1 ((\mathcal{M}f)^*(t))^2 dt\Big)^{1/2} (2\lambda(A))^{1/2} \to 0$$

for $A \to +\infty$, because $(\mathcal{M}f)^*(t) = o\left(\frac{1}{t}\right)$ for $t \to 0+$. ●

We remark that the proof of the theorem above is very close to that of Th.6 in [16] (the last theorem deals with the Kolmogorov-Titchmarch notion of $\mathcal{A}$-integral).

Appendix

A GENERALIZATION OF THE BOCHNER THEOREM ON ANALYTIC MEASURES

Let G be a locally compact abelian group. Denote by Γ the group of all characters of G. Let $M_A(G)$, where

$A\subset\Gamma$, denote the set of all measures $\mu\in M(G)$ such that $supp\,\hat{\mu}\subset A$. We shall say that a set $A\subset\Gamma$ has property of F. and M. Riesz (or is an R-set) iff $M_A(G)\subset L^1(G)$.

J.H.Shapiro [54] has found a sufficient condition for a set to have the F. and M.Riesz property.

THEOREM (J.H.Shapiro [54]). Let G be a compact abelian group, $A\subset\Gamma$, $0<p<1$. Suppose that for each $K\in\Gamma$ the functional $f\longmapsto\hat{f}(K)\,(f\in L^1_{A\cup\{K\}}(G))$ has a continuous extension onto $L^p_{A\cup\{K\}}(G)$. Then A is an R-set.

In [54] it is shown also that this condition gives a new proof of the Bochner theorem. We prove here that a generalization of the Bochner theorem follows from this condition.

Moreover, we give a generalization of the Shapiro theorem for the case of a locally compact abelian group.

THEOREM 1. Let $A\subset\mathbb{Z}_+\times\mathbb{Z}$. Suppose that for every $n\in\mathbb{Z}_+$ the set $\{k\in\mathbb{Z}:(n,k)\in A\}$ is bounded (from above or from below).
Then A is an R-set.

PROOF. We use the result of J.H.Shapiro. Let $K=(K_1,K_2)$ and let f be a trigonometric polynomial on $\mathbb{T}$ with $\hat{f}$ concentrated on $A\cup\{K\}$. It is clear that for each $z_2\in\mathbb{T}$ the following inequality holds

$$\Big|\sum_{\ell\in\mathbb{Z}}\hat{f}(K_1,\ell)z_2^{\ell}\Big|^p\leq C(K,p)\int_{\mathbb{T}}|f(z_1,z_2)|^p\,dm(z_1).$$

Therefore,

$$|\hat{f}(K_1,K_2)|^p\leq C(K,p)\int_{\mathbb{T}}\Big|\sum_{\ell\in\mathbb{Z}}\hat{f}(K_1,\ell)z_2^{\ell}\Big|^p dm(z_2)\leq$$

$$\leq C(K,p)\|f\|^p_{L^p}\,.\quad\bullet$$

THEOREM 2. Let $\mu\in M(\mathbb{T}^2)$. Assume $supp\,\hat{\mu}\subset\mathbb{Z}_+\times\mathbb{Z}$. Denote by $\nu_n\,(n\in\mathbb{Z}_+)$ the measure on $\mathbb{T}$ with $\hat{\nu}_n=\hat{\mu}(n,\cdot)$. Then $\mu\in L^1(\mathbb{T}^2)$ if and only if $\nu_n\in L^1(\mathbb{T})$ for all $n\in\mathbb{Z}_+$.

PROOF. It is clear that if $\mu\in L^1(\mathbb{T}^2)$ then $\nu_n\in L^1(\mathbb{T})$

for all $n\in\mathbb{Z}_+$. Conversely, let $\nu_n\in L^1(\mathbb{T})$ for all $n\in\mathbb{Z}_+$. With each function ν_n we associate a trigonometric polynomial p_n such that $\|\nu_n-p_n\|_{L^1}<2^{-n}$. Consider the function $\lambda\overset{def}{=}\sum_{n\geq 0} z_1^n(\nu_n(z_2)-p_n(z_2))$. It is clear that $\lambda\in L^1(\mathbb{T}^2)$. It remains to note that $\mu-\lambda\in L^1(\mathbb{T}^2)$ by Th.1. ●

Note that Th.2 follows easily from a result of A.V.Buhvalov - A.A.Danilevich [25] about the L^1_1-valued Hardy spaces H^p.

Analogously one can prove the following

THEOREM 3. Let G, G_1 be locally compact abelian groups and let Γ, Γ_1 be the dual groups. Suppose that $A\subset\Gamma$ is a countable set satisfying the conditions of the Shapiro theorem. Let $\mu\in M(G\times G_1)$. Denote by ν_n the measure on G_1 with $\hat{\nu}_n=\hat{\mu}(n,\cdot)$. Then $\mu\in L^1(G\times G_1)$ if and only if $\nu_n\in L^1(G_1)$ for all $n\in A$. ●

THEOREM 4. Let $\mu\in M(\mathbb{R}^d)$. Assume that $\operatorname{supp}\hat{\mu}\subset$ $\subset\{(x_1,x_2,\dots,x_d)\in\mathbb{R}^d: x_1\geq 0\}$. Denote by ν_x $(x\geq 0)$ the measure in $M(\mathbb{R}^{d-1})$ with $\hat{\nu}_x=\hat{\mu}(x,\cdot)$. Then $\mu\in L^1(\mathbb{R}^d)$ if and only if $\nu_x\in L^1(\mathbb{R}^{d-1})$ for all $x\geq 0$.

PROOF. It is sufficient to prove that $\mu\in L^1(\mathbb{R}^d)$ if $\nu_x\in L^1(\mathbb{R}^{d-1})$ for all $x\geq 0$. Denote by μ_a $(a>0)$ the measure on $\mathbb{T}^d$ defined by the equality $(\mu_a,\varphi)=$ $=(\mu,\varphi(e^{2\pi i a x_1},\dots,e^{2\pi i a x_d}),\ \varphi\in C(\mathbb{T}^d)$. It is easy to see that if $\mu\notin L^1(\mathbb{R}^d)$ then $\mu_a\notin L^1(\mathbb{T}^d)$ for all sufficient large values of $a>0$. It remains to apply Th.3 with $G=\mathbb{T}$, $G_1=\mathbb{T}^{d-1}$, $A=\mathbb{Z}_+$. ●

Notice that Th.4 implies a result of R.R.Coifman and B.Dahlberg [29].

Now we give without proof a generalization of the Shapiro theorem for the case of locally compact abelian group.

THEOREM 5. Let G be locally compact abelian group. Denote by Γ the group of all characters on G . Let $A\subset\Gamma$, $0<p<1$. Suppose that for each $\kappa\in\Gamma$ there exists a neighbourhood of zero in Γ such that the functional $f\longrightarrow\hat{f}(\kappa)$ ($f\in L^1_{(A\cup\{\kappa\})+u}(G)\cap L^p(G)$) can be continued onto $L^p_{(A\cup\{\kappa\})+u}(G)$ as a continuous functional. Then A is R-set.

I would like to express my deep gratitude to E.M.Dyn'kin, V.P.Havin, N.G.Makarov, N.K.Nikolskii and V.V.Peller for translating into English this article and for a great number of helpful suggestions and corrections.

I am indebted espesially to V.P.Havin for reading the Russian and English versions.

Also I wish to thank S.V.Kisljakov and S.A.Vinogradov for helpful conversations.

References

1. B a r b e y K., K ö n i g H. Abstract analytic function theory and Hardy algebras. Lecture Notes, 593, 1977.
2. D u r e n P.L. Theory of H^p spaces. Academic Press, New York, 1970.
3. G a m e l i n T.W. Uniform algebras. Prentice-Hall, Inc.Englewood Cliffs, New Jersey, 1969.
4. П р и в а л о в И.И. Граничные свойства аналитических функций. Москва-Ленинград, ГИТТЛ, 1950.
5. H ö r m a n d e r L. An introduction to complex analysis in several variables. Princeton, New Jersey, Toronto, New York, London, 1966.
6. К р е й н С.Г., П е т у н и н Ю.И., С е м е н о в Е.М. Интерполяция линейных операторов. Москва,"Наука", 1978.
7. Н и к о л ь с к и й Н.К. Лекции об операторе сдвига. Москва, "Наука", 1980.
8. R o l e w i c z S. Metric linear spaces. Warszawa, 1972.
9. R u d i n W. Fourier analysis on groups, Interscience, New York, 1962.
10. R u d i n W. Function theory in polydiscs. W.A.Benjamin, Inc., New York, Amsterdam, 1969.
11. S t e i n E.M. Singular integrals and differentiability properties of functions. Princeton Univ.Press, Princeton, New Jersey, 1970.
12. S t e i n E.M., W e i s s G. Introduction to Fourier analysis on Euclidean spaces. Princeton Univ.Press, Princeton, New Jersey, 1971.
13. Z y g m u n d A. Trigonometric series, V.I.Cambridge Univ. Press, New York, 1959.
14. А л е к с а н д р о в А.Б. Аппроксимация рациональными функциями и аналог теоремы М.Рисса о сопряженных функциях для пространств L^p с $p\in(0,1)$. Мат.сб., 1978, 107(149), № 1(9), 3-19.
15. А л е к с а н д р о в А.Б. Инвариантные подпространства оператора обратного сдвига в пространстве $H^p(p\in(0,1))$. Зап.научн.семинаров ЛОМИ, 1979, 92, 7-29.

16. А л е к с а н д р о в А.Б. Об $\mathcal{A}$-интегрируемости граничных значений гармонических функций. Мат.заметки, to appear.

17. А л е к с а н д р о в А.Б. Инвариантные подпространства операторов сдвига. Аксиоматический подход, to appear.

18. А н д р и а н о в а Т.Н. Коэффициенты Тейлора с редкими номерами для функций, суммируемых по площади. Зап.научн.семинаров ЛОМИ, 1976, 65, 161–163.

19. Б а л а ш о в Л.А., Т е л я к о в с к и й С.А. Некоторые свойства лакунарных рядов и интегрирование тригонометрических рядов. Труды Мат. ин-та им. В.А. Стеклова, 1977, 43, 32–41.

20. B o a s R. Isomorphism between H^p and L^p. Amer.J. Math., 1965, 77, 655-656.

21. B o y d D.W. The Hilbert transform on rearrangement-invariant spaces. Can.J.Math., 1967, 19, N 3, 599-616.

22. Б р ы с к и н И.Б., С е д а е в А.А. О геометрических свойствах единичного шара в пространствах типа классов Харди. Зап.научн.семинаров ЛОМИ, 1974, 39, 7–16.

23. Б у г р о в Я.С. О регулярности линейных методов суммирования рядов Фурье. Докл.АН СССР, 1974, 217, № 3, 505–507.

24. Б у х в а л о в А.В. Пространства Харди векторнозначных функций. Зап.научн.семинаров ЛОМИ, 1976, 65, 5–16.

25. Б у х в а л о в А.В., Д а н и л е в и ч А.А. Граничные свойства векторнозначных аналитических функций. Мат.заметки, to appear.

26. B u s k o E. Fonctions continues et fonctions bornées non adhérentes dans $L^\infty(\mathbb{T})$ à la suite de leurs sommes partielles de Fourier. Studia Math., 1970, 34, 319-337.

27. C a r l e s o n L. Two remarks on H^1 and BMO. Adv. Math., 1976, 22, 269-277.

28. C o i f m a n R.R. A real variable characterization of H^p. Studia Math., 1974, 51, N 3, 269-274.

29. C o i f m a n R.R., D a h l b e r g B. Singular integral characterizations of nonisotropic H^p spaces and F. and M.Riesz theorem. Proceedings of Symposia in Pure Mathematics, 1979, 35, Part 1, 231-234.

30. D o u g l a s R.G., S h a p i r o H.S., S h i e l d s A.L. Cyclic vectors and invariant subspaces for the backward shift operator. Ann.Inst.Fourier, 1970, 20, N 1, 37-76.

31. D u r e n P.L., R o m b e r g B.W., S h i e l d s A.L. Linear functionals on H^p spaces with $0<p<1$. J. reine und angew Math., 1969, 238, 32-60.

32. D u r e n P.L., S h i e l d s A.L. Properties of H^p $(0<p<1)$ and its containing Banach space. Trans.Amer. Math. Soc., 1969, 141, 255-262.

33. D u r e n P.L., S h i e l d s A.L. Coefficient multipliers of H^p and B^p spaces. Pac.J.Math., 1970, 32, N 1, 69-78.

34. F e f f e r m a n C., S t e i n E.M. H^p spaces of several variables. Acta Math., 1972, 129, 137-193.

35. G i l b e r t J.E. Nikišin - Stein theory and factorization with applications. Proceedings of Symposia in Pure Mathematics, 1979, 35, Part 2, 233-267.

36. H a r d y G.H., L i t t l e w o o d J.E. Some properties of fractional integral II. Math.Z, 1931/32, 34, N 3, 403-439.

37. Х а в и н В.П. Об аналитических функциях, представимых интегралом Коши-Стилтьеса. Вестн.Ленингр.ун-та, 1958, I, 66-78.

38. J a n s o n S. On functions with conditions on the mean oscillation. Ark.Math., 1976, 14, 189-196.

39. K a h a n e J.-P., K a t z n e l s o n Y. Séries de Fourier des fonctions bornées . Prépublications, Université de Paris-Sud, 1978.

40. K a l t o n N.J., S h a p i r o J.H. An F-space with trivial dual and non-trivial compact endomorphisms. Isr.J. Math., 1975, 20, N 3-4, 281-292.

41. L a t t e r R.H. A characterization of $H^p(\mathbb{R}^n)$ in terms of atoms. Studia Math., 1978, 62, 93-101.

42. d e L e e u w K. The failure of spectral analysis in L^p for $0<p<1$. Bull.Amer.Math.Soc., 1976, 82, N 1, 111-114.

43. L i n d e n s t r a u s s J., P e ł c z y ń s k i A. Contributions to the theory of the classical Banach spaces. J. Funct.Anal., 1971, 8, N 2, 225-249.

44. L o n g J.-L. Sommes partielles de Fourier des fonctions bornées . C.R.Acad.Sci.Paris, 1979, 288, N 22, A1009-A1011.

45. M a u r e y B. Théorèmes de factorization pour les opérateurs linéaires à valueurs dans les espaces L^p . Astérisque, 1974, N 11.

46. О с к о л к о в К.И. Оценка приближения непрерывной функции подпоследовательностью сумм Фурье. Труды Мат.ин-та им.В.А. Стеклова, 1975, 134, 240-253.

47. О с к о л к о в К.И. Последовательности норм сумм Фурье ог-

раниченных функций. Труды Мат.ин-та им.В.А.Стеклова, I977, I43, I29-I42.

48. P i c h o r i d e s S.K. On the best values of the constants in the theorems of M.Riesz, Zygmund and Kolmogorov. Studia Math., 1972, 44, N 2, 165-179.

49. Р о х л и н В.А. Об основных понятиях теории меры. Мат. сб., I949, 25, № I, I07-I50.

50. R u d i n W. Modifications of Fourier transforms. Proc.Amer. Math.Soc., 1968, 19, 1069-1074.

51. R u d i n W. Modification sets of density zero. Bull.Amer. Math.Soc., 1968, 74, 526-528,

52. R u d i n W., The inner function problem in balls. Зап.научн.семинаров ЛОМИ, I978, 8I, 278-280.

53. S h a p i r o J.H. Remarks on F-spaces of analytic functions. Lecture Notes, 1977, 604, 107-124.

54. S h a p i r o J.H. Subspaces of $L^p(G)$ spanned by characters: $0<p<1$. Isr.J.Math., 1978, 29, N 2-3, 248-264.

55. S t e g e n g a D. Bounded Toeplitz operators on H^1 and applications of the duality between H^1 and the functions of bounded mean oscillation. Amer.J.Math., 1976, 98, 573-598.

56. C o i f m a n R.R., R o c h b e r g R. Representation theorems for holomorphic and harmonic functions. Astérisque, I980, N 77, I2-66.

57. Т у м а р к и н Г.И. Об интегралах типа Коши-Стилтьеса. Успехи мат.наук, I956, II, № 4, I63-I66.

58. В и н о г р а д о в С.А. Интерполяционные теоремы Банаха--Рудина-Карлесона и нормы операторов вложения для некоторых классов аналитических функций. Зап.научн.семинаров ЛОМИ, I970, I9, 6-54.

59. В и н о г р а д о в С.А. Свойства мультипликаторов интегралов типа Коши-Стилтьеса и некоторые задачи факторизации аналитических функций. Труды Седьмой Зимней Школы, Дрогобыч, I974, 5-39.

60. В и н о г р а д о в С.А. Базисы из показательных функций и свободная интерполяция в банаховых пространствах с L^p-нормой. Зап.научн.семинаров ЛОМИ, I976, 65, I7-68.

61. Y a n a g i h a r a N. Multipliers and linear functionals for the class N^+. Trans.Amer.Math.Soc., 1973, 180, 449-461.

62. K a l t o n N.J. Compact and stricly singular operators on Orlicz-spaces. Isr.J.Math., 1977, 26, 126-136.

63. F r a z i e r A.P. The dual space of H^p of the polydisc for $0<p<1$. Duke Math.J., 1972, 39, 369-379.

E.M. Dyn'kin

THE RATE OF POLYNOMIAL APPROXIMATION IN THE COMPLEX DOMAIN

Introduction

The classical subject of the approximation theory is the connection of the rate of polynomial approximation of a given function with its local smoothness. In the complex domain this connection depends on geometrical characteristics of the boundary of a region where the approximation problem is considered.

The aim of this paper is a description of the complicated interaction between these three factors.

Let f be a function analytic in a plane region G. Usually, the rate of its polynomial approximation is measured by the sequence $\{E_n(f)\}_1^\infty$ of the best polynomial approximations in some sense (uniform, weighted, in the mean etc). The moduli of smoothness $\omega(f,\delta)$, $\delta>0$, of different kinds (uniform, weighted, in the mean ...) are used to measure its local smoothness. Geometrical properties of the boundary ∂G are expressed in terms of the conformal mapping φ of the exterior of G onto the exterior of the unit disk.

The geometry of regions with smooth boundary is essentially the same as that of the disk $(0<c_1<|\varphi'|<c_2<\infty)$ and the standard form of Jackson-Bernstein theorem may be transferred to these regions without modifications (see for example S.Ya. Alper [1, 2]). If the region has corners, uniform approximations and homogeneous conditions of Lipschitz type become not adjusted one to another. The whole situation **becomes complicated** and there are new effects in the case of L^P-approximation on the boundary or in the region.

The origin of the theory, which is the subject of our paper, can be found in W.K.Dzyadyk's papers [3-5] , where he obtained a constructive characterization of the class of analytic functions satisfying a Lipschitz condition of given order in a piecewise smooth region G . Now the theory is complete at least for the regions with Lipschitz boundaries. This paper contains its systematic exposition.

We have no place to describe the history of **the question.** Some historical comments are contained in §§9 and 10. Here we only note the essential contribution into the developement of the theory by W.K.Dzyadyk with co-authors [3-7] , N.A.Lebedev, P.M. Tamrazov and N.A.Shirokov [8-14] , V.I.Belyi and V.M.Mikljukov [15, 16] , T.Kövari and Ch.Pommerenke [17, 18] and T.H.Ganelius [19] ; see also the author's papers [20-25] . For further historical information we send a reader to cited works and also to surveys [26, 12, 27] .

In §1 of the present paper the main facts concerning function spaces, classes of regions and conformal mappings are collected.

In §2 outer functions in the sense of Beurling [28] are studied, whose boundary moduli satisfy well-known Muckenhoupt condition [30, 31] . It appears, that on such a way one can give real-variable proofs of main properties of conformal mappings and their level curves. From the geometric function theory we use only distortion theorems of Koebe and Lavrentiev [32, 33] .

§3 is rather technical and contains some estimates of Cauchy potentials.

In §4 we introduce the main approximation characteristics of functions, best weighted approximations. The weight may be any function on the boundary with the logarithm of bounded mean oscillation. All standard weights satisfy this condition. Further, we connect best approximations with pseudoanalytic continuation of functions [34, 21, 25] , which gives an appropriate intermediate language.

In §5 we introduce moduli of smoothness for functions of

$L^p(\partial G)$. Their main difference from usual ones is the variable step, defined by conformal mapping φ . Further, we connect moduli of smoothness with pseudoanalytic continuation and get an "inverse" theorem of approximation theory which gives an upper estimate of the moduli of smoothness of a function by its best approximations.

For the proof of "direct" theorems, i.e. for the estimation of best approximations by moduli of smoothness, we need some constructions of polynomial approximants. At present one knows two such constructions. An "elementary" construction of approximants by means of Faber operators was proposed independently by T.Kövari [18] , T.M.Ganelius [19] and the author [20] for the uniform approximation. It is based on a deep result of Ch. Pommerenke and T.Kövari [17] . Another construction, more complicated but more powerful, including approximation of Cauchy kernel, was proposed by W.K.Dzyadyk [3, 4] and was developed further by N.A.Lebedev and N.A.Shirokov [9, 11] .

In §6 we expose the theory of Faber operators and completely describe the connection of best approximations with moduli of smoothness in the case of slowly varying weights, in particular for the uniform approximation and L^p -approximation without weight.

In §7 we expose the polynomial approximation of the Cauchy kernel and obtain the final result for general weights.

In §8 we study the area approximation. The results of §8 were announced by the author [23] but are published here with proofs for the first time.

In §9 we formulate the general theory of §§4-8 in particular forms for standard weights. We discuss a constructive characterization of Lipschitz classes, uniform approximation problem, approximation in the mean etc.

At last, in §10 we discuss some unsolved problems of the complex approximation theory.

In the present paper the recent papers by V.I.Belyi, V.V. Andrievsky and P.M.Tamrazov [16, 35-37] are briefly mentioned. They have extended an important part of the theory to regions with quasiconformal, may be nonrectifiable, boundaries. The author is not a specialist in this domain and sends the reader to papers [16, 35-37] and to V.I.Belyi's monograph [38] .

In conclusion, the author expresses his deep gratitude to N.A. Shirokov, L.I.Potepun and V.I.Belyi for stimulating discussions and to N.K.Nikolsky and V.P.Havin for their permanent support of this work.

List of symbols.

Every symbol is accompanied by the number of section containing its definition. The notation A.B means subsection B. in section A. Throughout the paper C and c denote various constants; ● denotes the end of proof.

Symbol	Section	Symbol	Section	Symbol	Section
$A_p(\Gamma)$	1.9	$I(z)$	5.2	w_I	9.1
$\widetilde{A}(\Gamma)$	1.9	$K(z,\tau)$	1.1	z^*	1.5
$BMO(\Gamma)$	1.9	$K_p^s(G,w)$	4.1	$\hat{z}$	1.5
$\mathbb{C}$	1.1	$K_{pq}^s(G,w)$	9.5	z_τ	1.6
$C_A(B)$	1.2	$L^p(\Gamma,w)$	1.2	Γ	1.4
$C_A^s(B)$	1.2	$Lip_A(p,s)$	1.3	Γ_τ	1.6
$\mathbb{D}$	1.1	M^*f	1.7	$\gamma(w)$	2.4
$\mathcal{D}_m(f,I)$	5.1	Mf	1.9	ν	8.1
$\mathcal{D}_m(f,I,w)_p$	5.1	M_p	2.1	$\Pi^s(B)$	10.1
$E^p(G)$	1.2	p'	1.1	$\rho(z,E)$	1.1
$E^p(G,w)$	1.2	$\mathcal{P}_I f$	5.1	$\sigma_p(f,\tau,w)$	4.3
$E_n(f,w)_p$	4.1	$Q(w)$	2.4	φ	1.5
$E_n^o(f,w)_p$	4.1	$q(G)$	2.2	ψ	1.5
$\mathcal{E}_n(f,w)_p$	8.1	$S_{p,w}$	6.1	$\omega_m(f,\delta,w)_p$	5.3
$e_n(f)_p$	1.3	$\mathbb{T}$	1.1	$\omega_m^o(f,\delta,w)_p$	5.3
$f_+,\ f_-$	1.8	$T_{p,w}$	6.1	$\omega_m(f,I)$	9.1
f_+^*	1.9	T_1, T_∞	6.1	$\lvert I \rvert$	1.1
G	1.4	V, v	2.2	$\bar{\partial} f$	1.1
$H_n(\zeta,\cdot)$	7.1	w_i, w_e	2.2	$\lVert f \rVert_X$	1.2
		w_τ	2.4		

§ 1. Preliminaries.

1.1 Some notations.

$\mathbb{C}$ is the plane of the complex variable $z = x + iy$ or $\zeta = \xi + i\eta$.

$\mathbb{D}=\{z, |z|<1\}$ is the open unit disk.

$\mathbb{T}=\{z, |z|=1\}$ is the unit circle.

$\rho(z,E)$ is the distance from a point z to a set E.

$K(z,r)=\{\zeta, |\zeta-z|<r\}$ is the disk with the centre z and the radius r.

$|I|$ is the length of the arc I.

$p'=p/p^{-1}$ is the conjugate exponent.

$\bar\partial f(z)=1/2(\partial f/\partial x+i\,\partial f/\partial y)$ is the Cauchy-Riemann derivative.

1.2. Function spaces.

$\|f\|_X$ is the norm of the function in the Banach space X.

Let Γ be a rectifiable Jordan curve, w a nonnegative function on Γ.

$L^p(\Gamma,w)=\{f, fw\in L^p(\Gamma)\}$ is a Banach space with the norm $\|f\|_{L^p(\Gamma,w)}=\|fw\|_{L^p(\Gamma)}$.

Let G be a plane region bounded by a simple rectifiable curve Γ.

$E^p(G)$, $0<p\leqslant\infty$, is the Smirnov class [28, 29] of analytic functions with norm $\|f\|_{E^p(G)}=\|f\|_{L^p(\Gamma)}$.

Let w be a nonnegative function on Γ with $w^{-1}\in L^{p'}(\Gamma)$. For $1\leqslant p\leqslant\infty$ we set

$E^p(G,w)=\{f\in E^1(G), fw\in L^p(\Gamma)\}$ with the norm $\|fw\|_{L^p(\Gamma)}$. If $G=\mathbb{D}$ then $E^p(G)=H^p$ is the standard Hardy class [28].

If B is a plane compact, then $C_A(B)$ is the space of all continious functions on B analytic in the interior of B with the supremum norm. Its Lipschitz subspace $C_A^s(B)$, $0<s<1$, is defined by

$$C_A^s(B)=\{f, f\in C_A(B), |f(z)-f(z')|\leqslant C|z-z'|^s, z,z'\in B\}.$$

1.3. Spaces of smooth functions.

Let $1\leqslant p\leqslant\infty$ and $f\in H^p$. L^p-modulus of smoothness of order m, $m=0,1,\ldots$ is defined by the formula

$$\omega_m(f,\delta)_p=\|\Delta_\delta^{m+1}f\|_{L^p(\mathbb{T})}, \quad \delta>0,$$

where $\Delta_\delta f(e^{it})\overset{def}{=\!=}f(e^{i(t+\delta)})-f(e^{it})$, $\Delta_\delta^{k+1}f\overset{def}{=\!=}\Delta_\delta(\Delta_\delta^k f)$.

Let $s>0$. We define the class $Lip_A(p,s)$ on $\mathbb{D}$ by

$$Lip_A(p,s)=\{f\in H^p,\ \omega_m(f,\delta)_p=O(\delta^s),\ m>s\}.$$

The functions of $Lip_A(p,s)$ satisfy the Lipschitz condition of order s in the mean.

Let us define the best approximations of a function f , $f\in H^p$:

$$e_n(f)_p=\inf\|f-P_n\|_{H^p}\ ,\quad n=1,2,\dots$$

the infimum is taken over all polynomials P_n of degree n . It is well-known that [39, 40]

$$f\in Lip_A(p,s)\quad\text{iff}\quad e_n(f)_p=O(n^{-s}),\ n\to\infty\ . \tag{1.1}$$

1.4. Classes of regions.

For this section see I.I.Daniljuk [41] .

Let G be a plane region bounded by a simple rectifiable curve Γ . G will be called a Lipschitz region (and Γ a Lipschitz curve) if in a neighbourhood of any point $\zeta\in\Gamma$ the curve Γ can be defined by the equation $y=\varphi(x)$, $x\in I$, I being an interval, where

$$|\varphi(x)-\varphi(x')|\leqslant c|x-x'|\ ,\qquad x,x'\in I\ . \tag{1.2}$$

Such a curve Γ has a tangent in almost every point $\zeta\in\Gamma$; we denote the slope of this tangent by $\gamma(\zeta)$.

G will be called a Radon region (Γ a Radon curve) if this function γ has bounded variation on Γ and moduli of all its jumps are less than π .

REMARK. If γ is of bounded variation but some of its jumps are of modulus π then G will be called a region of bounded rotation.

EXAMPLES. (*i*) All convex regions are Radon regions.

(*ii*) A region with a piecewise smooth boundary without edges is Radon region, but with an edge on the boundary it is of bounded rotation only.

(*iii*) All Radon regions are Lipschitz regions, but not vice versa.

1.5. Conformal mappings.

For a Lipschitz region G we shall denote by φ , $\varphi:\mathbb{C}\setminus G\to\mathbb{C}\setminus\mathbb{D}$, the conformal homeomorphism such that $\varphi(\infty)=\infty$, $\varphi'(\infty)>0$, and by ψ its inverse mapping. Γ is a quasiconformal curve [42] because the length of any arc of Γ is commensurable with its diameter. Then the mappings φ and ψ admit a continuation to quasiconformal automorphisms of the whole plane. By Lavrentiev theorem φ and ψ have the so-called $\mathbb{D}$-property [33] : the image of any circle lies in a ring with a bounded ratio of radii. The C^1-quasiconformal reflection [42] with respect to Γ will be denoted $z \longrightarrow z^*$. On the other hand, boundary correspondences of Γ and $\mathbb{T}$ by φ and ψ are absolutely continious and φ' and ψ' are outer functions in the sense of Beurling [28, 29] , in $\mathbb{C}\setminus G$ and $\mathbb{C}\setminus\mathbb{D}$.

Now we remind some consequences of the $\mathbb{D}$-property and Koebe distorsion theorem.

(i) For $z\in\mathbb{C}\setminus G$

$$\frac{1}{4}|\varphi'(z)|\le\rho(\varphi(z),\mathbb{T})/\rho(z,\Gamma)\le 4|\varphi'(z)|, \tag{1.3}$$

$$|\varphi''(z)/\varphi'(z)|\le 4/\rho(z,\Gamma).$$

(ii) For $z\in\mathbb{C}\setminus\overline{G}$ set $\hat{z}=\psi(\varphi(z)/|\varphi(z)|)$ and for $z\in G$ set $\hat{z}=(z^*)^\wedge$. The point $\hat{z}$ is not the nearest to z on Γ , but $|z-\hat{z}|\asymp\rho(z,\Gamma)$.

(iii) A point $z\in\mathbb{C}\setminus\Gamma$ will be called reciprocal to the arc $I\subset\Gamma$ (and I will be called reciprocal to z) if $\rho(z,I)<c_1|I|<c_2\,\rho(z,\Gamma)$ (c_1 and c_2 are constant in each discussion). Thus, if z is reciprocal to I then $\varphi(z)$ is reciprocal to $\varphi(I)$ and vice versa. Further, we have

$$|\varphi'(z)|\asymp|\varphi(I)|/|I|=\frac{1}{|I|}\int_I|\varphi'|. \tag{1.4}$$

1.6. Level curves.

Let $z\in\mathbb{C}\setminus G$, $\tau>0$. Set $z_\tau=\psi|(1+\tau)\varphi(z)|$. The level curve Γ_τ is defined by $\Gamma_\tau=\{z_\tau, z\in\Gamma\}=\{z,|\varphi(z)|=1+\tau\}$. By the $\mathbb{D}$-property, for $z\in\Gamma$ we have

$$\rho(z,\Gamma_\tau)\asymp|z-z_\tau|\asymp\rho(z_\tau,\Gamma)\asymp\tau/\varphi'(z_\tau). \tag{1.5}$$

Let F be a function on $\mathbb{C}\setminus G$. Then

$$\iint\limits_{\mathbb{C}\setminus G} F(z)\,dx\,dy = \int_0^\infty \frac{dr}{r} \int_{\Gamma_r} F(z)\rho(z,\Gamma)\,\varkappa(z)\,dz \tag{1.6}$$

where $|\varkappa(z)| \asymp 1$. This formula is yielded by a simple combination of the change of variable $z = \psi(\zeta)$, of the application of polar coordinates and of the inverse change of variable. By the quasiconformal reflection we get for any function F on G

$$\iint\limits_{G} F(z)\,dx\,dy = \int_0^\infty dr \int_{\Gamma^*} F(z)\,\varkappa(z)\frac{dz}{\varphi'(z^*)}. \tag{1.7}$$

1.7. Carleson embedding theorem.

For $f \in L^1(\Gamma)$, $z \in G$, set

$$M^* f(z) = \sup\Big\{\frac{1}{|I|}\int_I |f|,\ I \subset \Gamma,\ |I| > \rho(z,\Gamma),\ \rho(z,I) < 2\rho(z,\Gamma)\Big\}.$$

We need the following variant of Carleson embedding theorem (the proof is standard [43]).

THEOREM. Let μ be a measure in $\overline{G}$ and $1 < p < \infty$. The following two assertions are equivalent.

(i) $\mu(K(z,\delta)) \le c\delta$, $z \in \Gamma$, $\delta > 0$.

(ii) $\iint\limits_{\overline{G}} (M^* f)^p\,d\mu \le c\|f\|_{L^p(\Gamma)}$, $f \in L^p(\Gamma)$. ●

Such a measure will be called **a Carleson measure. The length** on some curve Λ is a Carleson measure with respect to any region, if the length of the portion of Λ in any disk of radius δ is less than $c\delta$.

1.8. Cauchy type integrals.

Let $f \in L^1(\Gamma)$, Γ be a Lipschitz curve. Two functions

$$f_\pm(z) = \frac{1}{2\pi i}\int_\Gamma f(\zeta)(\zeta - z)^{-1}\,d\zeta, \quad z \notin \Gamma, \tag{1.8}$$

are analytic in G and $\mathbb{C}\setminus G$ respectively. Further, if $f \in L^p(\Gamma)$, $1 \le p \le \infty$, then $f_+ \in E^{p_1}(G)$, $f_- \in E^{p_1}(\mathbb{C}\setminus G)$ for all $p_1 < p$ and $f_+ - f_- = f$ almost everywhere on Γ. It was proved for Radon regions [41] that for $1 < p < \infty$ and $f \in L^p(\Gamma)$ both f_+ and f_- be-

long to E^p and $\|f_+\|_{E^p(G)}+\|f_-\|_{E^p(\mathbb{C}\setminus G)}\leqslant c\|f\|_{L^p(\Gamma)}$.
Recently A.P.Calderón [44] has proved such an estimate for some Lipschitz regions, but for general Lipschitz regions this problem is open.

1.9. Muckenhoupt condition.

Let Γ be a Lipschitz curve, w be a function on Γ, $w\geqslant 0$. We shall say that w satisfies Muckenhoupt condition (A_p), $1<p<\infty$, or $w\in A_p(\Gamma)$, if for any arc $I\subset\Gamma$

$$\Big(\int_I w^p\Big)^{1/p}\Big(\int_I w^{-p'}\Big)^{1/p'}\leqslant c|I|.$$

The whole theory of [30, 31] can be transferred to the Lipschitz curves without changes. In particular, if $w\in A_p(\Gamma)$, then $w^\alpha\in A_{p_1}(\Gamma)$ for some $\alpha>1$, $p_1<p$, and for an arc $I\subset\Gamma$ and $z\in I$, $\rho(z,\Gamma\setminus I)>q|I|$,

$$\int_{\Gamma\setminus I} w(\zeta)^p\frac{|I|^p}{|\zeta-z|^p}|d\zeta|\leqslant c(q)\int_I w^p. \tag{1.9}$$

For $f\in L^1(\Gamma)$ set $Mf(z)=\sup\limits_{I\ni z}\frac{1}{|I|}\int_I|f|$, $z\in\Gamma$.

If $w\in A_p(\Gamma)$ then $\|Mf\|_{L^p(\Gamma,w)}\leqslant c\|f\|_{L^p(\Gamma,w)}$.

If G is a Radon region $f\in L^1(\Gamma)$ and $w\in A_p(\Gamma)$ then

$$\|f_+\|_{E^p(G,w)}\leqslant c\|f\|_{L^p(\Gamma,w)}$$

and further [31]

$$\|f_+^*\|_{L^p(\Gamma,w)}\leqslant c\|f\|_{L^p(\Gamma,w)}, \tag{1.10}$$

where

$$f_+^*(z)\overset{def}{=\!=}\sup_{\varepsilon>0}\Big|\int_{|\zeta-z|>\varepsilon}f(\zeta)(\zeta-z)^{-1}d\zeta\Big|,\quad z\in\Gamma.$$

Finally, we define a class $\tilde A(\Gamma)$ of regular weights
$\tilde A(\Gamma)=\{w,\ w^\alpha\in A_p(\Gamma)$ for some $\alpha>0,1<p<\infty\}$.
One can prove like in [30, 31] that $w\in\tilde A(\Gamma)$ iff $\log w\in BMO(\Gamma)$, where

$$BMO(\Gamma)=\Big\{f,\ \sup_I\frac{1}{|I|}\int_I\Big|f-\frac{1}{|I|}\int_I f\Big|<\infty\Big\}.$$

In particular, if w_1 , $w_2 \in \widetilde{A}(\Gamma)$, then $w_1 w_2 \in \widetilde{A}(\Gamma)$ and $w_1/w_2 \in \widetilde{A}(\Gamma)$.

§ 2. Muckenhoupt condition and outer functions.

This section is an account of some of results of the author's paper [45] .

2.1. Muckenhoupt mappings.

Let Γ_1 and Γ_2 be Lipschitz curves, and suppose $\varphi : \Gamma_1 \to \Gamma_2$ and $\psi = \varphi^{-1} : \Gamma_2 \to \Gamma_1$ are absolutely continious homeomorphisms; φ will be called a Muckenhoupt mapping or, more precisely, will be said to belong to the class M_p , $1<p<\infty$, if $|\varphi'|^{1/p} \in A_p(\Gamma_1)$, where

$$\varphi'(z) = \lim_{\zeta \to z} [\varphi(\zeta)-\varphi(z)](\zeta - z)^{-1} \text{ a.e. on } \Gamma_1 .$$

LEMMA I. If $\varphi \in M_{p_1}$, $w \in A_{p_2}(\Gamma_2)$ and $\widetilde{w} = (w \circ \varphi)^{1/p_1} |\varphi'|^{\frac{1}{p_1 p_2}}$ then $\widetilde{w} \in A_{p_1 p_2}(\Gamma_1)$.

PROOF. Let $p = p_1 p_2$, I be an arc $I \subset \Gamma_1$, $J = \varphi(I) \subset \Gamma_2$. By Hölder inequality with the exponent $(p-1)/(p_2-1)$

$$\Big(\int_I \widetilde{w}^p\Big)^{1/p}\Big(\int_I \widetilde{w}^{-p'}\Big)^{1/p'} \leqslant \Big(\int_J w^{p_2}\Big)^{1/p}\Big(\int_J w^{-p_2'}\Big)^{\frac{1}{p_1 p_2'}}\Big(\int_I |\varphi'|^{1-p_1'}\Big)^{1/p_1'} \leqslant$$

$$\leqslant c|J|^{1/p_1}\Big(\int_I |\varphi'|^{1-p_1'}\Big)^{1/p_1'} = c\Big(\int_I |\varphi'|\Big)^{1/p_1}\Big(\int_I |\varphi'|^{1-p_1'}\Big)^{1/p_1'} \leqslant c|I| . \quad \bullet$$

COROLLARY. If $w \in \widetilde{A}(\Gamma_2)$ then $w \circ \varphi \in \widetilde{A}(\Gamma_1)$. $\bullet$

LEMMA 2. If $\varphi \in M_p$, then $\psi = \varphi^{-1} \in M_q$ for some $q>1$.

PROOF. By [31] for some $\delta > 0$ and for any arc $I \subset \Gamma_1$

$$\Big(\frac{1}{|I|}\int_I |\varphi'|^{1+\delta}\Big)^{\frac{1}{1+\delta}} \leqslant c \frac{1}{|I|}\int_I |\varphi'| .$$

Thus, for $J = \varphi(I)$ we have $|J| = \int_I |\varphi'|$, $|I| = \int_J |\psi'|$ and $\int_I |\varphi'|^{1+\delta} = \int_J |\psi'|^{-\delta}$, i.e. $\psi \in M_q$, $q = 1 + 1/\delta$. $\bullet$

Now we can prove the main distortion theorem for Muckenhoupt mappings.

THEOREM I. Let Γ_1 and Γ_2 be Lipschitz curves, $\varphi : \Gamma_1 \to \Gamma_2$ and $\psi = \varphi^{-1}$ be Muckenhoupt mappings. If $\varphi \in M_p$, $\psi \in M_q$, $1 < p$, $q < \infty$, then for any arcs I, J, $I \subset J \subset \Gamma_1$,

$$c_1\left(\frac{|I|}{|J|}\right)^p \leqslant \frac{|\varphi(I)|}{|\varphi(J)|} \leqslant c_2\left(\frac{|I|}{|J|}\right)^{1/q}. \qquad (2.1)$$

PROOF. The right inequality is implied by the left one for ψ. We shall prove the left inequality. First, $|\varphi(J)| = |\varphi(I)| + \int_{J\setminus I}|\varphi'|$ and by (1.9) for $z \in I$, $\rho(z, \Gamma_1 \setminus I) > \varepsilon |I|$, we have

$$\left(\frac{|I|}{|J|}\right)^p \int_{J\setminus I} |\varphi'| \leqslant c \int_{J\setminus I} \frac{|I|^p}{|\zeta - z|^p} |\varphi'(\zeta)||d\zeta| \leqslant c \int_I |\varphi'| = c|\varphi(I)|. \quad \bullet$$

2.2. Estimates for conformal mappings.

The results of n°2.1 give simple real-variable proofs of properties of conformal mappings. Let again $\varphi : \mathbb{C}\setminus G \to \mathbb{C}\setminus \mathbb{D}$ be the conformal mapping, normalized as usual, of the exterior of Lipschitz region G onto the exterior of the unit disk, and $\psi = \varphi^{-1}$ be **the inverse mapping**.

LEMMA 3. $\psi \in M_2$ on $\mathbb{T}$.

PROOF (cf. [41]). Let $\theta(t) = \arg \psi'(e^{it})$. G is a Lipschitz region and there exists an $\varepsilon > 0$ such that $|\theta(t) - \theta(t')| \leqslant c < \pi/2$ for $|t - t'| < \varepsilon$. Consequently, there exists a function $\theta_0 \in C^\infty[-\pi, \pi]$ such that $\sup|\theta - \theta_0| < \pi/2$. But ψ' is an outer function and so $\log|\psi'|$ is the conjugate function for $\theta = \theta_0 + (\theta - \theta_0)$. Further, **the conjugate function for** θ_0 is bounded and $\sup|\theta - \theta_0| < \pi/2$. Then, $|\psi'|^{1/2} \in A_2(\mathbb{T})$ by Helson-Szegö theorem [31]. $\bullet$

COROLLARY I. $\varphi \in M_p$ for some $p > 1$. $\bullet$

COROLLARY 2. $w \in BMO(\mathbb{T})$ iff $w \circ \varphi \in BMO(\Gamma)$. $\bullet$

REMARK. One can prove, that if G is a convex region, then $\varphi \in M_p$ for all $p > 1$.

In the next section we shall see that if $|\varphi'|^{1/p} \in A_p(\Gamma)$, then $|\varphi'|^{1/p} \in A_p(\Gamma_r)$ for all $r > 0$ uniformly. Using this fact, theorem 1, $\mathcal{D}$-property and estimates (1.5), we obtain a number of estimates for distances to level curves Γ_r, $r > 0$.

LEMMA 4. Let G be a Lipschitz region. There are numbers v and V, $0 < v < V < 2$, such that

(i) $c_1(r/\delta)^{v} \leq \rho(z,\Gamma_r)/\rho(z,\Gamma_\delta) \leq c_2(r/\delta)^{V}, \quad r>\delta>0, \ z\in\Gamma;$ (2.2)

(ii) $c_1(r/\delta)^{v-1} \leq |\varphi'(z_\delta)|/|\varphi'(z_r)| \leq c_2(r/\delta)^{V-1}; \quad r>\delta>0, \ z\in\Gamma;$ (2.3)

(iii) $\rho(\zeta,\Gamma_r) \leq c\rho(z,\Gamma_r)\left(1+\frac{|\zeta-z|}{\rho(z,\Gamma_r)}\right)^q, \quad r>0, \ \zeta, z\in\Gamma;$ (2.4)

(iv) $\rho(\zeta,\Gamma) \leq c\rho(z,\Gamma)\left(1+\frac{|\zeta-z|}{\rho(z,\Gamma)}\right)^q, \quad \zeta, z\in\Gamma_r, \quad r>0;$ (2.5)

where $q = 1 - v/V$. ●

2.3 Estimates for outer functions.

LEMMA 5. Let G be a Lipschitz region, $m>0$ and $f\in E^1(G)$. If $z\in G$ and $I\subset\Gamma$ is an arc, reciprocal to z (cf. n° 1.5), then

$$|f(z)| \leq \frac{c_1}{|I|}\int_I |f| + c_2\int_{\Gamma\setminus I} |f(\zeta)|\frac{|I|^m}{|\zeta-z|^{m+1}}\,|d\zeta|. \tag{2.6}$$

PROOF. We have $(\zeta-z)^{-1} = \sum_0^{m-1}(z-z^*)^k(\zeta-z^*)^{-k-1} + (z-z^*)^m(\zeta-z^*)^{-m}(\zeta-z)^{-1}$. But $z^*\in G$ and so

$$|f(z)| \leq c\rho(z,\Gamma)^m \int_\Gamma |f(\zeta)|\,|\zeta-z^*|^{-m}|\zeta-z|^{-1}|d\zeta|.$$

Further, $|\zeta-z| \asymp |\zeta-z^*| \asymp \rho(z,\Gamma) \asymp |I|$ on I and $|\zeta-z^*| \asymp |\zeta-z|$ on $\Gamma\setminus I$. ●

COROLLARY. $|f| \leq cM^*f$ (cf. n°1.7). ●

Let now $w\in\tilde{A}(\Gamma)$. The function w defines two outer functions w_i and w_e with the boundary modulus w respectively in G and $\mathbb{C}\setminus G$.

LEMMA 6. Let G be a Lipschitz region, $w^\alpha\in A_p(\Gamma)$, $\alpha>0$, $1<p<\infty$. If $z\in G$ and I is an arc reciprocal to z, then

$$|w_i(z)| \asymp |w_e(z^*)| \asymp \left(\frac{1}{|I|}\int_I w^{\alpha p}\right)^{1/\alpha p}. \tag{2.7}$$

PROOF. By (2.6) and (1.9) for $w_i^{\alpha p}$ we have

$$|w_i(z)| \leq \left(\frac{c}{|I|}\int_I w^{\alpha p}\right)^{1/\alpha p}.$$

The inverse estimate follows from (2.6) and (1.9) for $w_i^{\alpha p}$ by (A_p) condition. Finally, the estimate of $w_e(z^*)$ follows from the reciprocity of I to z^* in $\mathbb{C}\setminus G$. ●

COROLLARY 1. If $w\in\widetilde{A}(\Gamma)$ then $|w_i'(z)/w_i(z)|\leq c/\rho(z,\Gamma)$. ●

COROLLARY 2. If $w\in\widetilde{A}(\Gamma)$ then $|w_i(z)|\asymp|w_e(z^*)|, z\in G$. ●

COROLLARY 3. If $|\varphi'|^{1/p}\in A_p(\Gamma)$ then $|\varphi'|^{1/p}\in A_p(\Gamma_\tau)$ uniformly for $\tau\to 0$. ●

2.4 Regularization.

Let G be a Lipschitz region and $w\in\widetilde{A}(\Gamma)$. Set, for $\tau>0$, $w_\tau(z)=|w_e(z_\tau)|$, $z\in\Gamma$. This function w_τ on Γ will be called regularization of w. The following assertions follow from lemma 6 and previous estimates.

(i) Let $I\subset\Gamma$ be an arc. If $|I|<\rho(I,\Gamma_\tau)$ then $w_\tau\asymp const$ uniformly on I. If $|I|>\rho(I,\Gamma_\tau)$ then

$$\int_I w_\tau^{\alpha p}\asymp\int_{I_\tau}|w_e|^{\alpha p}\asymp\int_I w^{\alpha p}$$

where $w^\alpha\in A_p(\Gamma)$. In particular, for $w\in A_p(\Gamma)$, $w_\tau\in A_p(\Gamma)$ uniformly for $\tau\longrightarrow 0$.

(ii) There exists a critical exponent $\gamma(w)$ such that on Γ

$$c_1(\tau/\sigma)^{-\gamma(w)}\leqslant w_\tau/w_\sigma\leqslant c_2(\tau/\sigma)^{\gamma(w)},\quad 0<\sigma<\tau. \tag{2.8}$$

(iii) There exists an exponent $Q(w)$, $0<Q(w)<\infty$, such that for $\zeta, z\in\Gamma$ and $\tau>0$

$$w_\tau(\zeta)\leqslant c\,w_\tau(z)\left(1+\frac{|\zeta-z|}{\rho(z,\Gamma_\tau)}\right)^{Q(w)}. \tag{2.9}$$

§ 3. Estimates of potentials.

3.1. Formulation of results.

Let G be a Lipschitz region. Let m, ℓ, σ, τ be positive numbers and $\tau<\sigma$. For $f\in L^1(\Gamma_\tau)$ set

$$F(z)=\int_{\Gamma_\tau} f(\zeta)(\zeta-z)^{-1}d\zeta\ ;$$

$$G(z)=\int_{\Gamma_\tau} |f(\zeta)|\frac{\rho(z,\Gamma_\tau)^m\rho(\zeta,\Gamma)^\ell}{|\zeta-z|^{m+\ell+1}}|d\zeta|\ ;$$

$$H(z)=\int_{\Gamma_\tau}|f(\zeta)|\min\Big\{1,\frac{\rho(z,\Gamma_\delta)}{|\zeta-z|}\Big\}^m\frac{|d\zeta|}{|\zeta-z|},\quad z\in\Gamma.$$

We need the exponents $q=q(G)$ (cf. n°2.2) and $\gamma(w)$, $Q(w)$ for $w\in\widetilde{A}(\Gamma)$ (cf. n° 2.4).

LEMMA 7. If G is a Radon region, $1<p<\infty$, and $w\in A_p(\Gamma)$ then

$$\|Fw\|_{L^p(\Gamma)}\leq c\,\|fw_e\|_{L^p(\Gamma_\tau)}\,.$$

LEMMA 8. If $1\leq p\leq\infty$, $w\in\widetilde{A}(\Gamma)$ and $m+\ell(1-q)>Q(w)$ then

$$\|Gw_\tau\|_{L^p(\Gamma)}\leq c\,\|fw_e\|_{L^p(\Gamma_\tau)}\,.$$

LEMMA 9. If $1\leq p\leq\infty$, $w\in\widetilde{A}(\Gamma)$ and $m(1-q)>Q(w)$ then

$$\|Hw_\delta\|_{L^p(\Gamma)}\leq c\Big(\frac{\delta}{\tau}\Big)^{\gamma(w)}\log\delta/\tau\,\|fw_e\|_{L^p(\Gamma_\tau)}\,.$$

REMARK. The Radon condition in lemma 7 is needed only to assure the L^p-estimates of the Cauchy integral.

3.2. Proof of lemma 7.

By (1.10) $\|F\|_{L^p(\Gamma_\tau,|w_e|)}\lesssim c\,\|f\|_{L^p(\Gamma_\tau,|w_e|)}$.
Consider an outer function $\tilde{w}$ in the interior G_τ of the level curve Γ_τ with the boundary modulus $|w_e|/\Gamma_\tau$. For analytic function $F\tilde{w}$ in G_τ we have

$$\|F\tilde{w}\|_{E^p(G_\tau)}\leq c\,\|fw_e\|_{L^p(\Gamma_\tau)}\,.$$

By (2.6) and Carleson embedding theorem it is enough to check that the measure μ on Γ,

$$\mu(E)=\int_E w^p|\tilde{w}|^{-p}, \quad E\subset\Gamma,$$

is a Carleson measure in G_τ. Consider a disk $K(u,t)$, $u\in\Gamma_\tau$, $t>0$. If $t<\rho(u,\Gamma)$ then $K(u,t)\cap\Gamma=\emptyset$ and $\mu(K(u,t))=0$. Let $t\geq\rho(u,\Gamma)$. By the $\mathcal{D}$-property there is a finite set of arcs $I_1,I_2,\dots,I_N\subset\Gamma$, such that $K(u,t)\cap\Gamma\subset\bigcup_1^N I_k$, $|\varphi(I_k)|=\tau$ and $\sum_1^N|I_k|\leq ct$. By (2.7) on I_k, $|\tilde{w}|^p\asymp|(I_k)_\tau|^{-1}\int_{(I_k)_\tau}|w_e|^p$, and on $(I_k)_\tau$, $|w_e|^p\asymp|I_k|^{-1}\int_{I_k} w^p$. But $|(I_k)_\tau|\asymp|I_k|$ and $\mu(K(u,t))\leq c\sum_1^N\int_{I_k}w^p\{|I_k|^{-1}\int_{(I_k)_\tau}|I_k|^{-1}\int_{I_k}w^p\}^{-1}\leq c\sum_1^N|I_k|<ct$. ●

3.3 Proof of lemma 8.

By (2.5) and (2.9) for $\zeta\in\Gamma_\tau$, $z\in\Gamma$, $\rho(\zeta,\Gamma)\leq c\rho(z,\Gamma_\tau)^{1-q}|\zeta-z|^q$, $w_\tau(z)\leq c|w_e(\zeta)||\zeta-z|^Q\rho(\zeta,\Gamma)^{-Q}$ and so $w_\tau(z)G(z)\leq c\int_{\Gamma_\tau}\frac{\rho(z,\Gamma_\tau)^M}{|\zeta-z|^{M+1}}|f(\zeta)w_e(\zeta)||d\zeta|$, where $M=(\ell-Q)(1-q)+m>0$.

Consider the operator $U:L^p(\Gamma_\tau)\longrightarrow L^p(\Gamma)$ defined by

$$Ug(z)=\int_{\Gamma_\tau}g(\zeta)\frac{\rho(z,\Gamma_\tau)^M}{|\zeta-z|^{M+1}}|d\zeta|, \quad z\in\Gamma.$$

It is enough to verify that its norm is bounded for all p, $1\leq p\leq\infty$. By Riesz-Thorin theorem [46] we must prove this assertion only for $p=1,\infty$. If $p=\infty$ then

$$\|U\|=\sup_z\int_{\Gamma_\tau}\frac{\rho(z,\Gamma_\tau)^M}{|\zeta-z|^{M+1}}|d\zeta|\leq C<\infty.$$

On the other hand, if $p=1$ then

$$\|U\|=\sup_\zeta\int_\Gamma\frac{\rho(z,\Gamma_\tau)^M}{|\zeta-z|^{M+1}}|dz|.$$

But by (2.4) $\rho(z,\Gamma_\tau)\leq c\rho(\zeta,\Gamma)^{1-q}|\zeta-z|^q$ and

$$\|U\|\leq c\sup_\zeta\int_\Gamma\frac{\rho(\zeta,\Gamma)^{M(1-q)}}{|\zeta-z|^{M(1-q)+1}}|dz|\leq C<\infty. \quad ●$$

3.4. Proof of lemma 9.

Let $E_1(z) = \{\zeta \in \Gamma_r \quad , |\zeta - z| < \rho(z, \Gamma_\delta)\}$, $E_2(z) = \Gamma_r \setminus E_1(z)$. Divide H into two parts H_1 and H_2 , namely the integrals over $E_1(z)$ and $E_2(z)$ respectively. At first, estimate H_1 . We have for $\zeta \in E_1(z)$

$$|w_e(\zeta)| \geq c\,|w_\delta(z)|\,(r/\delta)^{\gamma(w)}$$

and so

$$H_1(z)\, w_\delta(z) \leq c \int_{E_1(z)} |f(\zeta) w_e(\zeta)| \frac{|d\zeta|}{|\zeta - z|} (\delta/r)^{\gamma(w)} .$$

Now we can repeat the reasoning from the previous proof for the operator $U_1 \;:\; L^p(\Gamma_r) \to L^p(\Gamma)$,

$$U_1\, g(z) = \int_{E_1(z)} g(\zeta) \frac{|d\zeta|}{|\zeta - z|}$$

and get

$$\| H_1 w_\delta \|_{L^p(\Gamma)} \leq c (\delta/r)^{\gamma(w)} \log \delta/r \, \| f w_e \|_{L^p(\Gamma_r)} .$$

Now estimate H_2 . By (2.5), (2.8), (2.9) for $\zeta \in \Gamma_r$, $z \in \Gamma$ and $\zeta = \tau_r$, $\tau \in \Gamma$, we have

$$w_\delta(z) \leq c\, w_\delta(\tau)\, |\zeta - z|^{Q} \rho(\tau, \Gamma_\delta)^{-Q} ,$$

$$\rho(z, \Gamma_\delta) \leq c \rho(\tau, \Gamma_\delta)^{1-q}\, |\zeta - z|^{q} ,$$

$$w_\delta(z) \leq c\, w_\delta(\tau)\, |\zeta - z|^{Q/1-q} \rho(z, \Gamma_\delta)^{-Q/1-q}$$

and, finally,

$$H_2(z)\, w_\delta(z) \leq c(\delta/r)^{\gamma(w)} \int_{E_2(z)} |f(\zeta) w_e(\zeta)| \frac{\rho(z, \Gamma_\delta)^{M}}{|\zeta - z|^{M+1}} |d\zeta|$$

where $M = m - \dfrac{Q}{1-q}$. We can repeat our previous arguments again. ●

§4. Best approximations.

4.1. Weighted best approximations. K_p^s classes.

In this section we turn to the main subject of the article, the theory of polynomial approximation. Let G be a Lipschitz region. Analytic functions in G have many different approximation characteristics. For example, one can consider for a function $f \in E^p(G)$, $1 \leq p \leq \infty$, its best approximations by polynomials of degree n in the E^p-norm. If $p=\infty$ it is the classical uniform approximation. On the other hand, the problem of the constructive characterization of Lipschitz classes leads to the best approximations with weights $\rho(z, \Gamma_{1/n})^s$ which are not analogous to the uniform ones. As we shall see, there is a rather simple general construction containing all these cases and their L^p-analogues.

Let $w \in \tilde{A}(\Gamma)$. Remind that it means $\log w \in BMO(\Gamma)$. Let $f \in E^p(G)$, $1 \leq p \leq \infty$. For $n = 1, 2, \ldots$ define weighted best approximations of f by the formula

$$E_n(f, w)_p = \inf_{P_n} \|(f - P_n) w_{1/n}\|_{L^p(\Gamma)} \tag{4.1}$$

where infimum is taken over all polynomials P_n of degree n. The quantity (4.1) contains the regularization $w_{1/n}$ of the weight w. If G is a Radon region and $w \in A_p(\Gamma)$ (this condition is stronger than $w \in \tilde{A}(\Gamma)$), we define nonregularized best approximations

$$\overset{\circ}{E}_n(f, w)_p = \inf_{P_n} \|(f - P_n) w\|_{L^p(\Gamma)} . \tag{4.2}$$

Evidently, $\overset{\circ}{E}_{n+1} \leq \overset{\circ}{E}_n$ for all n. This is not the case for E_n, but for $n \leq k \leq 2n$ we have $c_1 E_{2n} \leq E_k \leq c_2 E_n$.

Let now $1 \leq p \leq \infty$, $s > 0$ and $\gamma(w)$, $w \in \tilde{A}(\Gamma)$, be the critical exponent from n°2.4. For $s > \gamma(w)$ set

$$K_p^s(G, w) = \{ f \in E^p(G),\ E_n(f, w)_p = O(n^{-s}) \} .$$

If G is a Radon region, $1 < p < \infty$, $w \in A_p(\Gamma)$ and $s > 0$ set $K_p^s(G, w) = \{ f \in E^p(G, w),\ \overset{\circ}{E}_n(f, w)_p = O(n^{-s}) \}$.

The following lemma removes possible contradictinns.

LEMMA 10. If G is a Radon region, $1 < p < \infty$ and $w \in A_p(\Gamma)$ then for any polynomial P_n of degree n

$$\| P_n w_{1/n} \|_{L^p(\Gamma)} \asymp \| P_n w \|_{L^p(\Gamma)} \asymp \| P_n w_e \|_{L^p(\Gamma_{1/n})}$$

uniformly with respect to $n = 1, 2, \ldots$.

PROOF. Both functions Φ_1 and Φ_2 , $\Phi_1 = P_n w_e \varphi^{-n-1}$, $\Phi_2 = P_n w_{1/n} \varphi^{-n-1}$ are analytic in $\mathbb{C} \setminus G$. Then $\| \Phi_1 \|_{E^p} = \| P_n w \|_{L^p(\Gamma)}$, $\| \Phi_2 \|_{E^p} = \| P_n w_{1/n} \|_{L^p(\Gamma)}$ and $\| \Phi_1 \|_{L^p(\Gamma_{1/n})} \asymp \| \Phi_2 \|_{L^p(\Gamma_{1/n})} \asymp \| P_n w_e \|_{L^p(\Gamma_{1/n})}$. By the maximum principle $\| P_n w_e \|_{L^p(\Gamma_{1/n})} \leqslant c \cdot \min \{ \| P_n w \|_{L^p(\Gamma)}, \| P_n w_{1/n} \|_{L^p(\Gamma)} \}$. On the other hand for $z \in \Gamma$

$$P_n(z) = \frac{1}{2\pi i} \int_{\Gamma_{1/n}} P_n(\zeta)(\zeta - z)^{-1} d\zeta$$

and inverse estimates follow from (2.6) and lemmas 7-9. ●

COROLLARY. Under conditions of lemma IO for any $f \in E^p(G)$ $E_n(f, w)_p = O(n^{-s})$ iff $E_n^{\circ}(f, w)_p = O(n^{-s})$. ●

4.2. Examples. (i) Let $w \equiv 1$. Both (4.1) and (4.2) define best approximations in the E^p norm (uniform for $p = \infty$). The classes $K_p^s(G \setminus 1)$ ware discussed in works [7, 18-22, 47-49].

(ii) Let $w = |\varphi'|^s$. By n°2.2 $w \in \tilde{A}(\Gamma)$ and $\gamma(w) < s$. The function φ' being outer, we have

$$E_n(f, w)_p \asymp n^{-s} \inf_{P_n} \left\| \frac{f(z) - P_n(z)}{\rho(z, \Gamma_{1/n})^s} \right\|_{L^p(\Gamma)} . \qquad (4.3)$$

The classes $K_p^s(G, |\varphi'|^s)$ are classical objects of the complex approximation theory. If $p = \infty$ then $K_\infty^s(G, |\varphi'|^s)$ is the Lipschitz class of order s (see [3-16, 26, 27]). For $1 \leqslant p < \infty$ see [24, 25, 50, 51] .

(iii) Let Ω be a Lipschitz region with the boundary Λ and level curves $\{\Lambda_r\}$ and $\Phi : \mathbb{C} \setminus G \to \mathbb{C} \setminus \Omega$ be the conformal mapping. Set $w = |\varphi'|^s |\Phi'|^{-s}$. It is easy to prove that $w \in \tilde{A}(\Gamma)$ and for $z \in \Gamma$

$w_r(z) \asymp r^s \rho[\Phi(z), \Lambda_r]^{-s}$, hence

$$E_n(f, w)_p \asymp n^{-s} \inf_{P_n} \left\| \frac{f(z) - P_n(z)}{\rho(\Phi(z), \Lambda_{1/n})^s} \right\|_{L^p(\Gamma)} . \qquad (4.4)$$

For $\Omega = \mathbb{D}$ these $E_n(f, w)_p$ are usual uniform approximations, for $\Omega = G$ (4.4) coincides with (4.3). In general case

(4.4) gives a continious transition from uniform approximations to (4.3) [52] .

(iv). If $w(z)=|z-z_0|^{\alpha}$, $z_0 \in \Gamma$, then $w \in \widetilde{A}(\Gamma)$ for any α and $w \in A_p(\Gamma)$ for $-1/p < \alpha < 1/p'$. In general, the weight $w(z)=\exp \int \log|\zeta - z|\, d\mu(\zeta)$ with a finite plane measure μ belongs to $\widetilde{A}(\Gamma)$. Finally, if $w_1, w_2 \in \widetilde{A}(\Gamma)$, then $w_1 w_2 \in \widetilde{A}(\Gamma)$, i.e. the definition (4.1) contains different cases of mixed approximation [53, 52, 49] .

4.3. Pseudoanalytic continuation.

Our main problem is the description of classes K_p^s in terms of the local smoothness of functions. We begin with the translation of the definition of K_p^s into the language of the pseudoanalytic continuation. Pseudoanalytic continuation theory [34,21,25] describes the smoothness of a function f (belonging, say, to $E^1(G)$) by means of its continuation onto the whole plane such that $\bar{\partial} f$ decreases with a prescribed rate near Γ . For example [34] , $f \in C_A^s(\bar{G})$ iff it admits such a continuation and $|\bar{\partial} f(z)| \leq C \rho(z, \Gamma)^{s-1}$, $z \in \mathbb{C} \setminus G$. There exist different constructions of continuation connected with different characteristics of f . So, pseudoanalytic continuation gives a general language for the comparison of the different definitions of function spaces.

LEMMA II. Let G be a Lipschitz region and $f \in E^1(G)$. For any $q > 1$ three following assertions are equivalent.

(i) There is a function $F_1 \in W_q^1(\mathbb{C})$ such that $F_1|_G = f$.

(ii) There is a function $F_2 \in W_q^1(\mathbb{C} \setminus G)$ such that $F_2|_\Gamma = f|_\Gamma$.

(iii) The function f admits a representation

$$f(z)=\iint_{\mathbb{C} \setminus G} \lambda(\zeta)(\zeta - z)^{-1} d\xi\, d\eta \ , \ \lambda \in L^q(\mathbb{C} \setminus G) . \qquad (4.5)$$

REMARK. In section (ii) of lemma II $F_2 /_\Gamma$ is defined in the sense of embedding theorems [40, 54, 55] and $f|_\Gamma$ in the sense of boundary values [28, 29] . In both cases they may be regarded as angular limits almost everywhere.

PROOF OF LEMMA. The implication $(i) \Longrightarrow (ii)$ is evident. If (iii) holds then the right side of (4.5) by Calderon-Zygmund theorem [42, 55] defines a function from $W_q^1(\mathbb{C})$ and so (i) holds.

Finally, suppose (ii) holds. For $z \notin \bar{G}$

$$0=\frac{1}{2\pi i} \int_\Gamma f(\zeta)(\zeta - z)^{-1} d\zeta = \lim_{r \to 0} \frac{1}{2\pi i} \int_{\Gamma_r} F_2(\zeta)(\zeta - z)^{-1} d\zeta =$$

$$= F_2(z) + \frac{1}{\pi} \iint_{\mathbb{C}\setminus G} \bar\partial F_2(\zeta)(\zeta - z)^{-1} d\xi\, d\eta .$$

So, taking angular limits, we get (4.5) almost everywhere on Γ with $\lambda = \frac{1}{\pi} \bar\partial F_2$. But both sides of (4.5) belong to $E^1(G)$ and so coincide in the whole G. ●

If the assertions of lemma II hold, we say that the function f admits pseudoanalytical continuation (from W_q^1). Such a continuation - i.e. functions F_1 or F_2 of lemma II - will be denoted by the same symbol f. It doesn't lead to contradiction but we remind that continuation is not unique and all estimates mean c e r t a i n pseudoanalytic continuation with the same estimates. By (1.6) we can transform (4.5) to

$$f(z) = \int_0^\infty \frac{dr}{r} \int_{\Gamma_r} \lambda_1(\zeta)(\zeta - z)^{-1} d\zeta , \qquad (4.6)$$

where $|\lambda_1(\zeta)| \le c\rho(\zeta, \Gamma)^{r} |\bar\partial f(\zeta)|$.

Finally, let $w \in \widetilde{A}(\Gamma)$. Introduce the following characteristic of the pseudoanalytic continuation:

$$\sigma_p(f, r, w) = \| \rho(\zeta, \Gamma) \bar\partial f(\zeta) w_e(\zeta) \|_{L^p(\Gamma_r)} , \quad r > 0 . \qquad (4.7)$$

In particular for $p = \infty$

$$\sigma_\infty(f, r, w) = \sup_{\Gamma_r} | \rho(\zeta, \Gamma) \bar\partial f(\zeta) w_e(\zeta) | .$$

REMARK. The representation (4.6) and the norm (4.7) involve the conformal mapping $\varphi : \mathbb{C}\setminus G \to \mathbb{C}\setminus \mathbb{D}$. We could formulate similar definitions for other systems of curves, and first of all for the level curves of the usual distance. But we shall see that the best approximations (4.1) and (4.2) lead to the estimates of pseudoanalytic continuation exactly in terms of (4.7).

4.4 Continuation by approximation.

Let $f \in E^1(G)$ and let $\{P_n\}_1^\infty$ be a sequence of polynomials of degree $n = 1, 2, \ldots$ converging to f in $E^1(G)$. For example, it may be the sequence of best approximation polynomials of f in the sense of (4.1) and (4.2). For $z \in \mathbb{C}\setminus G$ and $2^{-n} \le |\varphi(z)| - 1 \le 2^{-n+1}$ set

$$\lambda(z) = \rho(z, \Gamma)^{-1} | P_{2^{n+1}}(z) - P_{2^n}(z) | . \qquad (4.8)$$

THEOREM 2. Let G be a Lipschitz region and f, $\{P_n\}$ and λ be as above. If $\lambda \in L^q(\mathbb{C}\setminus G)$, $q>1$, then f admits a pseudoanalytic continuation such that

$$|\bar\partial f| \leq c\lambda .$$

PROOF. Let $\eta \in C^\infty[0,+\infty)$, $0\leq\eta\leq 1$, $\eta(t)=1$ for $0\leq t\leq 1$, $\eta(t)=0$ for $t\geq 2$. Define the desired continuation for $z\in\mathbb{C}\setminus G$, $2^{-n}\leq|\varphi(z)|-1\leq 2^{-n+1}$ by the formula

$$f(z)=P_{2^n}(z)+\eta\left[2^n(|\varphi(z)|-1)\right](P_{2^{n+1}}(z)-P_{2^n}(z)) . \tag{4.9}$$

It is clear that $f\in C^\infty(\mathbb{C}\setminus\bar G)$ and $|\bar\partial f|\leq c\lambda$. Introduce a function F_N defined by (4.9) for $|\varphi(z)|-1>2^{-N}$ and coinciding with P_{2^N} for $|\varphi(z)|-1\leq 2^{-N}$ or in G. Then $F_N\in C^\infty(\mathbb{C})$, $\bar\partial F_N=0$ in the interior of the level curve $\Gamma_{2^{-N}}$ and $|\bar\partial F_N|=|\bar\partial f|\leq c\lambda$ in its exterior. By the Green's formula for $z\in G$

$$P_{2^N}(z)=F_N(z)=-\frac{1}{\pi}\iint\limits_{\mathbb{C}\setminus G}\bar\partial F_N(\zeta)(\zeta-z)^{-1}d\xi\, d\eta .$$

We get a representation (4.5) for f taking $N\to\infty$ in the last equality. ●

REMARK. If $w\in A_p(\Gamma)$ then $L^p(\Gamma,w)\subset L^q(\Gamma)$ for sufficiently small $q>1$. Further, if $w\in\tilde A(\Gamma)$ then $|w_{1/n}|>c\cdot n^{-\gamma(w)}$ on Γ and $\|g\|_{L^p(\Gamma)}\leq cn^{\gamma(w)}\|g w_{1/n}\|_{L^p(\Gamma)}$. Thus, the polynomials of best approximation in the sense (4.2) converge to f in $E^q(G)$. The polynomials of best approximation in the sense (4.1) converge to f in $E^1(G)$ if

$$n^{\gamma(w)}E_n(f,w)_p = o(1) .$$

THEOREM 3. Let G be a Lipschitz region, $f\in E^1(G)$.

(i) If $w\in\tilde A(\Gamma)$ and $E_n(f,w)_p=O(n^{-\gamma(w)-\varepsilon})$, $\varepsilon>0$, then f admits a pseudoanalytic continuation such that

$$\sigma_p(f,r,w)\leq c\,E_{[2/r]}(f,w)_p\ ,\quad 1\leq p\leq\infty . \tag{4.10}$$

(ii) If $1<p<\infty$, $w\in A_p(\Gamma)$ and $E_n^0(f,w)_p=O(n^{-\varepsilon})$, $\varepsilon>0$, then f admits a pseudoanalytic continuation such that

$$\sigma_p(f,r,w) \leq c E^{\circ}_{[2/r]}(f,w)_p . \tag{4.11}$$

COROLLARY. Any function $f \in K^s_p(G,w)$ admits a pseudoanalytic continuation such that

$$\sigma_p(f,r,w) = O(r^s) . \quad \bullet \tag{4.12}$$

PROOF OF THEOREM (i) Let us consider a function

$$\Phi(z) = \left[P_{2^{n+1}}(z) - P_{2^n}(z)\right] w_e(z_{2^{-n}})\,\varphi(z)^{-2^{n+1}} .$$

Evidently, $\Phi \in E^1(\mathbb{C} \setminus G)$ and $\Phi(\infty) = 0$. Further, $\|\Phi\|_{E^p} \leq c E_{2^n}(f,w)_p$. On the other hand on Γ_r , $2^{-n} \leq r \leq 2^{-n+1}$, $|\varphi|^{2^{n+1}} \leq e^2$, $|w_e(z_{2^{-n}})| \times$ $\times\, |w_e(z)|$ and in terms of (4.8) we have $\|\rho\lambda w_e\|_{L^p(\Gamma_r)} \leq$ $\leq c\|\Phi\|_{L^p(\Gamma_r)}$. It is well-known [29] that $\|\Phi\|_{L^p(\Gamma_1)} \leq \|\Phi\|_{E^p}$ and the theorem 2 gives a continuation with

$$\sigma_p(f,r,w) \leq c E_{2^n}(f,w)_p \leq c E_{[2/r]}(f,w)_p .$$

To check the condition $\lambda \in L^q(\mathbb{C} \setminus G)$ we remark that $\rho(\Gamma, \Gamma_r) \geq$ $\geq c r^2$ and by (1.6) and (4.1)

$$\iint\limits_{\mathbb{C} \setminus G} |\lambda|^q \leq c \sum_{n=0}^{\infty} 4^{n(q-1)}\, 2^{n\gamma(w)}\, E_{2^n}(f,w)_p^q < +\infty$$

for sufficiently small $q > 1$.

(ii) This case admits just a similar proof, but Φ must be defined as $\Phi = [P_{2^{n+1}} - P_{2^n}]\, w_e\, \varphi^{-2^{n+1}}$. $\bullet$

We shall see that the condition (4.12) is exact: if (4.12) holds then $f \in K^s_p(G,w)$. In the next paragraph we shall connect (4.12) with local characteristics of f such as moduli of smoothness. This roundabout way will be simpler than the "direct" connection of approximation and smoothness.

§ 5. Moduli of smoothness.

The theorems of approximation theory connect the rate of the polynomial approximation with the local smoothness of functions. The best approximations E_n , E°_n from §4 measure the rate of approximation. Here we introduce the moduli of smoothness for measuring the local smoothness. There is a number of different definitions of these moduli. The definition by Ju.A.Brudnyi

[56, 57] is the most convenient for our purpose.

5.1. Local polynomial approximation.

Let Γ be a Lipschitz curve, $f \in L^1(\Gamma)$ and $I \subset \Gamma$ be an arc. Consider the best approximations of f by polynomials of degree m, $m = 0, 1, \ldots$ in $L^1(I)$:

$$\mathcal{D}_m(f, I) = \inf_{P_m} \| f - P_m \|_{L^1(I)} . \tag{5.1}$$

In the local approximation theory m is fixed and I is variable in contrast with the global approximation theory of §4. For a fixed m we can replace (4.1) by the following interpolation construction.

Let P_0 be such a polynomial of degree m that

$$\int_0^1 P_0(x)\, dx = 1, \quad \int_0^1 x^k P_0(x)\, dx = 0, \quad k = 1, 2, \ldots, m .$$

For a Lipschitz arc J with the ends a, b set $P_J(z) = \frac{1}{b-a} P_0\left(\frac{z-a}{b-a}\right)$. Then for any polynomial T of degree m

$$T(a) = \int_J P_J(z) T(z)\, dz .$$

Let $I \subset \Gamma$ be an arc; divide I into $(m+1)$ arcs $I_0, \ldots$ $\ldots, I_m$ of equal length. Let $a_0, \ldots, a_m$ be their origins. Finally, for $f \in L^1(I)$ denote by $\mathcal{P}_I f$ the polynomial of degree m with values $\int_{I_k} f(z) P_{I_k}(z)\, dz$ at the points a_k, $k = 0, \ldots, m$. It is clear, that $\mathcal{P}_I T = T$ for any polynomial T of degree m and for any $\beta > 0$ we have

$$\max_{\rho(z, I) < \beta |I|} | \mathcal{P}_I f(z) | \leq \frac{c}{|I|} \int_I |f| \tag{5.2}$$

(c depends on β and m). So

$$\mathcal{D}_m(f, I) \leq \| f - \mathcal{P}_I f \|_{L^1(I)} \leq c\, \mathcal{D}_m(f, I)$$

and we can replace the approximation definition (5.1) by an interpolatory one. As in §4 we consider a weight analogue of (5.1). Let G be a Lipschitz region, $1 \leq p \leq \infty$ and $w \in L^p(\Gamma)$. Set for $f \in E^p(G)$

$$\mathcal{D}_m(f,I,w)_p = \inf_{P_m} \|(f-P_m)w\|_{L^p(I)} .$$

There will be two cases of interest for us: if $w\in A_p(\Gamma)$ and if $\max_I w \asymp \min_I w$. In both cases

$$\mathcal{D}_m(f,I) \leqslant c\mathcal{D}_m(f,I,w)_p \cdot |I| \cdot \|w\|^{-1}_{L^p(I)} . \tag{5.3}$$

5.2. Construction of continuation.

Connect with any $z\in \mathbb{C}\setminus G$ an arc $I(z)\subset\Gamma$ with the centre $\hat{z}$ (cf. n°1.5) and the diameter $5\rho(z,\Gamma)$. Set for $f\in E^1(G)$

$$\lambda(z) = \rho(z,\Gamma)^{-2}\mathcal{D}_m(f,I(z)), \quad z\in\mathbb{C}\setminus G . \tag{5.4}$$

THEOREM 4. Let G be a Lipschitz region, $f\in E^1(G)$ and let λ be defined by (5.4). If $\lambda\in L^q(\mathbb{C}\setminus G)$, $q>1$, then $f\in E^q(G)$ and f admits such pseudoanalytic continuation that $|\bar\partial f|\leqslant c\lambda$ in $\mathbb{C}\setminus G$.

PROOF. (i) Show that $f\in L^q(\Gamma)$ (and so $f\in E^q(G)$). Set for an arc $I\subset\Gamma$

$$\Omega(I) = \{ z\in\mathbb{C}\setminus G,\ \rho(z,I)<|I|/10,\ \rho(z,\Gamma)>|I|/100 \}$$

and further

$$\Omega_n = \{ z\in\mathbb{C}\setminus G,\ 2^{-n-10}\leqslant \rho(z,\Gamma)/|\Gamma| \leqslant 2^{-n+3}\},\ n=0,1,\dots .$$

If $z\in\Omega(I)$ then $\mathcal{D}_m(f,I)\leqslant\mathcal{D}_m(f,I(z))$. Consider the sequence of partitions Γ into arcs $\Gamma=\bigcup_{k=1}^{2^n} I_k^n$, $n=0,1,\dots$ where $I_1^0=\Gamma$ and every following partition arises from halving all arcs of the previous one. Set $T^n = \mathcal{P}_{I_k^n} f$ on I_k^n . Almost everywhere on Γ $f=T^0+\sum_0^\infty (T^{n+1}-T^n)$. Evidently, on I_k^{n+1} $|T^{n+1}-T^n|\leqslant c\cdot 2^n\mathcal{D}_m(f,\mathring{I}(z))$ for any $z\in\Omega(2I_k^{n+1})$,

$$\int_{I_k^{n+1}} |T^{n+1}-T^n|^q \leqslant c2^{-n(q-1)} \iint_{\Omega(2I_k^{n+1})} \lambda^q ,$$

$$\|T^{n+1}-T^{n}\|^q_{L^q} \leq c\,2^{-n(q-1)} \iint_{\Omega_n} \lambda^q .$$

$$So, \ \Sigma\|T^{n+1}-T^{n}\|_{L^q} < +\infty, \ f \in L^q(\Gamma).$$

(ii) Consider the Whitney partition of unity $1=\sum_K \chi_K$ in $\mathbb{C}\setminus G$ [55]. Let u_K be the centre of the support of χ_K and let I_K be an arc on Γ with centre $\hat{u}_K$ and length $1/10\,\rho(u_K,\Gamma)$. In $\mathbb{C}\setminus G$ set

$$F = \sum_K \chi_K \mathcal{P}_{I_K} f .$$

For any polynomial T

$$F = T + \Sigma \chi_K \cdot [\mathcal{P}_{I_K} f - T], \ \bar{\partial} F = \Sigma \bar{\partial}\chi_K \cdot [\mathcal{P}_{I_K} f - T].$$

For $T = \mathcal{P}_{I(z)} f$, $z \in \mathbb{C}\setminus G$, the inclusion $I_K \subset I(z)$ for k with $\chi_k(z) \neq 0$ implies $|\bar{\partial} F| \leq c\lambda$. In particular, $\bar{\partial} F \in L^q(\mathbb{C}\setminus G)$. Further, $|F(z)| \leq c M f(\hat{z})$, but $Mf \in L^q(\Gamma)$ and we can pass to the limit when $r \to 0$ in the formula

$$F(z) = \frac{1}{2\pi i}\int_{\Gamma_r} F(\zeta)(\zeta - z)^{-1} d\zeta - \frac{1}{\pi}\iint_{\Gamma_r^0} \bar{\partial} F(\zeta)(\zeta - z)^{-1} d\xi\, d\eta, \ z \in \mathbb{C}\setminus G.$$

So we get the representation (4.5) for f because $F|_\Gamma = f|_\Gamma$. ●

5.3. Definition of moduli of smoothness.

By §4 the global polynomial approximation gives the estimates of $\bar{\partial} f$ in $L^p(\Gamma_r)$, $r > 0$. But we see by theorem 4 that $\bar{\partial} f$ can be estimated by local approximation of f on arc $I(z)$ reciprocal to z. When z runs over Γ_r, $I(z)$ runs over Γ in such a way that $|I(z)|$ is always equivalent to the local distance to Γ_r. So, a local characteristic, adequate to global polynomial approximation, must involve some integral estimates of $\mathcal{D}_m(f, I)$ on such a family of arcs.

Let G be a Lipschitz region, $w \in \widetilde{A}(\Gamma)$ and $f \in E^p(G)$, $1 \leq p \leq \infty$. For $\delta > 0$ consider all partitions of Γ, $\Gamma = \bigcup_1^N I_K$ into arcs $\{I_K\}$ such that $\delta/2 \leq |\varphi(I_K)| \leq \delta$. Set

$$\omega_m(f,\delta,w)_p = \sup\Big(\sum_{k=1}^{N} \mathcal{D}_m(f,I_k,w_\delta)_p^p\Big)^{1/p}, \qquad (5.5)$$

where supremum is taken over all such partitions. $\omega_m(f,\delta,w)_p$ will be called the conformal modulus of smoothness of the function f in $L^p(\Gamma)$ of order m with weight w. If $w\in A_p(\Gamma)$, $1<p<\infty$, we can consider an alternative definition

$$\omega_m^0(f,\delta,w)_p = \sup\Big(\sum_{k=1}^{N} \mathcal{D}_m(f,I_k,w)_p^p\Big)^{1/p} \qquad (5.6)$$

without regularization of the weight. As in § 4 we need not distinguish between these definitions.

REMARK (i) Let T_δ^* be a piecewise polynomial function (spline) which coincides with $\mathcal{P}_{I_k} f$ on I_k. Evidently,

$$\omega_m(f,\delta,w)_p = \sup\| f-T_\delta^*\|_{L^p(\Gamma,w_\delta)} .$$

Thus, we can define the conformal modulus of smoothness in terms of the spline approximation.

(ii) The definitions (5.5), (5.6) are of conformal nature, because the supremum is taken over the partitions of Γ into arcs with equal length of their conformal images. We can formulate a completely analoguous definition with the supremum taken over all partitions $\Gamma = \bigcup\limits_{1}^{N} I_k$ with $\delta/2 \le |I_k| \le \delta$. Then we get "metrical" moduli of smoothness. Following to Ju.A.Brudnyi [56, 57] one can show that these metrical moduli are equivalent to usual moduli of smoothness defined in other terms (cf. n°1.3). But they are never equivalent to conformal moduli if the boundary Γ of G has corner points. In the approximation theory we need conformal moduli of smoothness.

(iii) If $m=0$ then $\omega_0(f,\delta,w)_p$ is equivalent to usual (weighted) modulus of continuity of $f\circ\psi$ in $L^p(\mathbb{T})$. But for $m\geq 1$ such a connection between $\omega_m(f,\delta,w)_p$ and characteristics of $f\circ\varphi$ does not exist.

5.4. Estimates of continuation.

THEOREM 5. Let G be a Lipschitz region, $w\in \tilde{A}(\Gamma)$, $1\le p\le\infty$ and $f\in E^p(G)$.

(i) If $\omega_m(f,\delta,w)_p = O(\delta^{\gamma(\omega)+\varepsilon})$, $\varepsilon>0$, then f admits a pseudoanalytic continuation such that

$$\sigma_p(f,r,w)\le c\,\omega_m(f,50r,w)_p .$$

(ii) If $1<p<\infty$, $w\in A_p(\Gamma)$ and $\omega_m^{\circ}(f,\delta,w)_p = O(\delta^{\varepsilon})$, $\varepsilon>0$, then f admits a pseudoanalytical continuation such that

$$\sigma_p(f,r,w)\le c\,\omega_m^{\circ}(f,50r,w)_p .$$

PROOF. The proofs of (i) and (ii) are completely analoguous. For example, by theorem 4 we have on Γ_r

$$|\rho(z,\Gamma)\bar\partial f(z) w_e(z)|\le c\,|w_e(z)|\,|I(z)|^{-1}\mathcal{D}_m(f,I(z)).$$

By (5.3) we get

$$\sigma_p(f,r,w)^p\le c\int_{\Gamma_r}\mathcal{D}_m(f,I(z),w_r)_p^p\,|dz|\le c\,\omega_m(f,50r,w)_p^p .$$

We need the condition $\omega_m(f,\delta,w)_p = O(\delta^{\gamma(w)+\varepsilon})$ only to check the inclusion $\lambda\in L^q(\mathbb{C}\setminus G)$ in theorem 4. Thus we can repeat the corresponding argument from the proof of theorem 3. ●

THEOREM 6. Let G be a Lipschitz region, $w\in\tilde A(\Gamma)$, $1\le p\le\infty$ and suppose $f\in E^1(G)$ admits a pseudoanalytic continuation.

(i) For m sufficiently large

$$\omega_m(f,\delta,w)_p\le c\int_0^{\infty}(\delta/r)^{\gamma(w)}\left|\log\frac{\delta}{r}\right|\sigma_p(f,r,w)\frac{1}{1+(r/\delta)^M}\,\frac{dr}{r} . \tag{5.7}$$

(ii) If G is a Radon region, $1<p<\infty$ and $w\in A_p(\Gamma)$ then for sufficiently large m

$$\omega_m^{\circ}(f,\delta,w)_p\le c\int_0^{\infty}\sigma_p(f,r,w)\frac{1}{1+(r/\delta)^M}\,\frac{dr}{r} . \tag{5.8}$$

Here $M=(m+1)v$ (for $v=v(G)$ see n° 2.2).

PROOF. (i) For $I\subset\Gamma$ define a polynomial of degree m

$$P(z)=-\frac{1}{\pi}\iint_{|\zeta-z_0|>2|I|}\bar\partial f(\zeta)\sum_{k=0}^{m}(z-z_0)^k(\zeta-z_0)^{-k-1}\,d\xi\,d\eta$$

where z_0 is the centre of I . On I

$$|f(z)-P(z)|\le c_1\iint_{|\zeta-z_0|<2|I|}|\bar\partial f(\zeta)|\frac{1}{|\zeta-z|}+c_2\iint_{|\zeta-z_0|>2|I|}|\bar\partial f(\zeta)|\frac{|I|^{m+1}}{|\zeta-z|^{m+2}} .$$

By definition in (5.5) $|I|\asymp\rho(z,\Gamma_\delta)$ for $z\in I$ and so

$$\omega_m(f,\delta,w)_p\le c\int_0^{\infty}\|h_r\|_{L^p(\Gamma,w_\delta)}\frac{dr}{r}$$

where

$$h_\tau(z)=\int_{\Gamma_\tau} \min\left\{1,\frac{\rho(z,\Gamma_\sigma)}{|\zeta-z|}\right\}^{m+1}|\rho(\zeta,\Gamma)\bar\partial f(\zeta)|\frac{|d\zeta|}{|\zeta-z|} .$$

Now the estimate (5.7) follows by lemma 9 for $\tau<\sigma$ and by lemma 8 and (2.2) for $\tau\geqslant\sigma$.

(ii) Set $\Omega=\{\zeta,\ |\varphi(\zeta)|>1+\sigma\}$ and

$$P(z)=-\frac{1}{\pi}\iint_{\Omega}\bar\partial f(\zeta)\sum_0^m (z-z_0)^k(\zeta-z_0)^{-k-1}d\xi\, d\eta .$$

Further argument is just analoguous to the previous one with the use of lemma 7 instead of lemma 9. ●

COROLLARY. For any $s>0$

$$\sigma_p(f,w,\tau)=O(\tau^s)\Longleftrightarrow\omega_m(f,\sigma,w)_p=O(\sigma^s)\Longleftrightarrow\overset{\circ}{\omega}_m(f,\sigma,w)_p=O(\sigma^s).$$

In particular, if $f\in K_p^s(G,w)$ then $\omega_m(f,\sigma,w)_p=O(\sigma^s)$. ●

Now we have got an "inverse" theorem of approximation: the estimate $E_n(f)=O(n^{-s})$ implies $\omega_m(f,\sigma)=O(\sigma^s)$. The "direct" theorems of estimates of $E_n(f)$ by $\omega_m(f,\sigma)$ are the subject of two next sections.

§6. Faber operators.

In §4 we have got the lower estimates of the best approximations by means of the pseudoanalytic continuation. For the upper estimates we must have certain construction of polynomial approximants. Two such constructions are known - the elementary construction by means of Faber operators and a more complicated but more powerful construction of the polynomial approximation of the Cauchy kernel. We expose the first of them in this section and the second in the next one.

6.1. Definition of Faber operators.

Let G be a Lipschitz region, $w\in A_p(\Gamma)$, $1<p<\infty$. Define Faber operator $T_{p,w}$ for functions $f\in H^p$ by

$$T_{p,w}f=\{(f\circ\varphi)\varphi'^{1/p}w_e^{-1}\}_+ \quad \text{or}$$

$$T_{p,w}f(z)=\frac{1}{2\pi i}\int_\Gamma f[\varphi(\zeta)]\varphi'(\zeta)^{1/p}w_e(\zeta)^{-1}(\zeta-z)^{-1}d\zeta,\ z\in G .$$

REMARK. Evidently, $(f\circ\varphi)\varphi'^{1/p}w_e^{-1}\in L^q(\Gamma)$ for some $q>1$ and so $T_{p,w}f\in E^q(G)$.

LEMMA 12. If G is a Radon region, $1<p<\infty$ and $w\in A_p(\Gamma)$, then

$$\| T_{p,w} f \|_{E^p(G,w)} \leq c \| f \|_{H^p}, \quad f \in H^p. \tag{6.1}$$

PROOF. Evidently, $\|(f\circ\varphi)\varphi'^{1/p} w_e^{-1}\cdot w\|_{L^p(\Gamma)} = \|f\|_{H^p}$.
So, (6.1) follows from Muckenhoupt condition. ●

For $p = 1, \infty$ we cannot prove such lemma, because Cauchy integral is not a bounded operator in $L^p(\Gamma)$. However, if $w \equiv 1$ then we can formulate the following result. Denote $T_1 = T_{1,1}$ and $T_\infty = T_{\infty,1}$, i.e.

$$T_1 f = \{(f\circ\varphi)\varphi'\}_+, \quad T_\infty f = \{f\circ\varphi\}_+.$$

T_1 and T_∞ are classical Faber operators without weights.

LEMMA I3. If G is a region of bounded rotation, then

(i) $\| T_\infty f \|_{C_A(\overline{G})} \leq c \| f \|_{C_A(\overline{\mathbb{D}})}, \quad f \in C_A(\overline{\mathbb{D}})$;

(ii) $\| T_1 f \|_{E^1(G)} \leq c \| f \|_{H^1}, \quad f \in H^1$.

Proof of this lemma is due to Ch.Pommerenke [58,17] and one can read it in J.E.Andersson's work [49]. ●

REMARK. In contrast with lemma I2, the condition of bounded rotation in the last lemma is not connected with the boundedness of the Cauchy integral operator.

Let us construct now an operator $S_{p,w}$ as an inverse of $T_{p,w}$. For $f \in E^p(G,w)$ set

$$S_{p,w} f = \{(f\circ\psi)\cdot(w_e\circ\psi)\cdot\psi'^{1/p}\}_+ \qquad \text{or}$$

$$S_{p,w} f(z) = \frac{1}{2\pi i}\int_{\mathbb{T}} f[\psi(\zeta)]\, w_e[\psi(\zeta)]\, \psi'(\zeta)^{1/p} (\zeta - z)^{-1} d\zeta, \quad z \in \mathbb{D}.$$

LEMMA I4. (i) $\| S_{p,w} f \|_{H^p} \leq c \| f \|_{E^p(G,w)}, \quad 1 < p < \infty$;

(ii) $S_{p,w}$ is an inverse operator of $T_{p,w}$.

PROOF. (i) is evident, because Cauchy integral is a bounded operator in $L^p(\mathbb{T})$, $1 < p < \infty$.

(ii) If $g = T_{p,w} f$ then

$$g = (f\circ\varphi)\varphi'^{1/p} w_e^{-1} + h, \quad h \in E^q(\mathbb{C}\setminus G), \quad h(\infty) = 0;$$

$$(g\circ\psi)(w_e\circ\psi)\psi'^{1/p} = f + (h\circ\psi)(w_e\circ\psi)\psi'^{1/p}.$$

But the second term belongs to $E^p(\mathbb{C}\setminus\mathbb{D})$ and vanishes in ∞. So $S_{p,w}\,g=f$. ●

REMARK. Clearly, the assertion (ii) of this lemma holds for $p=1$, ∞ and $w\equiv 1$.

Thus, operators $T_{p,w}$ and $S_{p,w}$ realize an isomorphism of spaces H^p and $E^p(G,w)$. But, evidently, for any polynomial P its image $T_{p,w}P$ is again a polynomial of the same degree. Consequently, if G is a Radon region, $1<p<\infty$ and $w\in A_p(\Gamma)$ then

$$E_n^\circ(T_{p,w}f,w)_p \asymp e_n(f)_p\ ,\quad f\in H^p\ . \tag{6.2}$$

By lemma I3 if $p=1,\infty$ and $w\equiv 1$ then

$$E_n^\circ(T_p,f,1)_p\leqslant c\,e_n(f)_p\ ,\quad f\in H^p\ .$$

The inequality (6.2) resolves the problem of the upper estimates for the best polynomial approximations. Now we shall discuss the action of Faber operators on the pseudoanalytic continuation.

REMARKS. (i) By lemma IO (§4) an estimate of type (6.2) holds for $E_n(T_{p,w}f,w)_p$.

(ii) It is known that the summation of the Taylor series gives good approximants (of the best order) for functions in H^p. The operator $T_{p,w}$ transforms this process into the summation of the Faber series (with respect to generalized Faber polynomials $T_{p,w}(z^n)$). Thus, by (6.2) such a summation gives polynomial approximants in $E^p(G,w)$ which are of the best order.

6.2. Faber operators and smoothness.

If $f\in H^p$ admits a pseudoanalytic continuation then by the Green's formula in G

$$T_{p,w}f(z)=-\frac{1}{\pi}\iint\limits_{\mathbb{C}\setminus G}\bar\partial f\,[\varphi(\zeta)]\,\varphi'(\zeta)^{1/p}\,\overline{\varphi'(\zeta)}\,w_e(\zeta)^{-1}(\zeta-z)^{-1}d\xi\,d\eta\ .$$

Then $T_{p,w}f$ admits a pseudoanalytic continuation and

$$|\,\bar\partial T_{p,w}f\,|\leqslant|\varphi'|^{1+1/p}\,|w_e|^{-1}\,|(\bar\partial f)\circ\varphi\,|\ .$$

Consequently, $\sigma_p(T_{p,w}f,\tau,w)\leqslant c\,\sigma_p(f,\tau,1)$.

All arguments are applicable to the inverse operator $S_{p,w}$ and so

$$\sigma_p(T_{p,w}f,\tau,w)\asymp\sigma_p(f,\tau,1)\ ,\quad \tau>0\ . \tag{6.3}$$

REMARK. It is clear, that (6.3) does not depend on the condition $1<p<\infty$; it holds for $p=1,\infty$ and $w\equiv 1$.

LEMMA I5. Let $1\leq p\leq \infty$, $s>0$, $f\in H^p$. Three following assertions are equivalent

(i) $f\in \mathrm{Lip}_A(p,s)$.

(ii) $f\in K_p^s(\mathbb{D},1)$.

(iii) f admits a pseudoanalytic continuation such that

$$\sigma_p(f,r,1)=O(r^s).$$

PROOF. By n° 1.3 (i) is equivalent to (ii). By theorem 3 §4 (ii) implies (iii). Finally, let (iii) hold. Then for $m>s$

$$|f^{(m)}(z)|\leq \iint\limits_{|\zeta|>1} |\bar\partial f(\zeta)|\frac{d\xi\, d\eta}{|\zeta-z|^{m+1}},\quad |z|<1;$$

$$\Big(\int\limits_{|z|=1-r} |f^{(m)}(z)|^p|dz|\Big)^{1/p}\leq c r^{s-m},\quad r\to 0 .$$

The last inequality is a well-known characterization of the class $\mathrm{Lip}_A(p,s)$ [55] . ●

REMARK. Of course, the proof of this simple lemma may be done without reference to theorem 3.

Now we can state the first complete theorem about K_p^s .

THEOREM 7. Let G be a Radon region, $1\leq p\leq\infty$, $w\in A_p(\Gamma)$ for $1<p<\infty$ and $w\equiv 1$ for $p=1,\infty$. Let $s>0$ and $f\in E^p(G,w)$. The following assertions are equivalent.

(i) $f\in K_p^s(G,w)$, i.e. $E_n^\circ(f,w)_p=O(n^{-s})$.

(ii) $\omega_m^\circ(f,\delta,w)_p=O(\delta^s)$ for $m>s/v-1$.

(iii) $S_{p,w}f\in \mathrm{Lip}_A(p,s)$.

PROOF. By theorems 3, 5, 6 and (6.2), (6.3) all these assertions are equivalent to the existence of a pseudoanalytic continuation of f with the estimate (4.12). ●

REMARKS. (i) Condition $w\in A_p(\Gamma)$ means a slow variation of w on Γ . If $w=|\varphi'|^s$ then $w\in A_p(\Gamma)$ only for sufficiently small s .

(ii) For $1<p<\infty$ the Radon condition to G is connected with the boundedness of the Cauchy integral in $L^p(\Gamma)$ and is not necessary.

§7. Approximation of Cauchy kernel.

7.1. Construction of approximation.

It was shown in §6 that Faber series summation gives almost best polynomial approximations. Here we shall use this summation in order to give more precise pointwise estimate for the polynomial approximation of Cauchy kernel.

Let G be a Lipschitz region and suppose $m>0$ is an integer.

THEOREM 8. For any $\zeta \in \mathbb{C} \setminus G$ there exists a polynomial $H_n(\zeta, \cdot)$ of degree n such that for $z \in \overline{G}$

$$\text{(i)} \quad |H_n(\zeta, z)| \leq c\rho(z^1, \Gamma_{1/n})^{-1} ; \tag{7.1}$$

$$\text{(ii)} \quad |(\zeta - z)^{-1} - H_n(\zeta, z)| \leq c \frac{|\zeta - \zeta_{1/n}|^m}{|\zeta - z|^{m+1}} . \tag{7.2}$$

PROOF. Let $\tau = \varphi(\zeta)$. Assume at first that $|\tau| > 1 + 1/n$, i.e. $\zeta \in \Gamma_r$, $r > 1 + 1/n$. Let $K_\tau(z) = (\tau - z)^{-1}$ be the Cauchy kernel. Clearly, $T_1 K_\tau = K_\zeta$, where T_1 is the Faber operator from §6, i.e. $T_1 g = \{(g \circ \varphi)\varphi'\}_+$.

Consider the standard Jackson kernel [39]

$$J_{k,n}(t) = \gamma_{k,n} (\sin nt/\sin t)^{2k}, \quad t \in [-\pi, \pi],$$

where $\gamma_{k,n}$ is defined by the condition $\int_{-\pi}^{\pi} J_{k,n} = 1$.

It is known that $J_{kn} \leq c_k \cdot n$ and for $|t| > 1/n$ $J_{k,n}(t) \leq c_k n^{-2k+1} |t|^{-2k}$.

The function

$$P_1(u) = \int_{-\pi}^{\pi} J_{kn}(t)(\tau e^{it} - u)^{-1} dt, \quad |u| < 1 + 1/n,$$

is evidently a polynomial of degree $2kn$; it is a partial sum of Taylor series for $(\tau - u)^{-1}$ obtained by its summation by the Jackson's method of order k. Applying the operator T_1 we obtain a polynomial of the same degree

$$P_n(z) = \int_{-\pi}^{\pi} J_{kn}(t) [\psi(\tau e^{it}) - z]^{-1} dt, \quad z \in \overline{G} .$$

The polynomial P_n is a result of summation of the Faber series for the Cauchy kernel by the Jackson's method. Now the equality $\int_{-\pi}^{\pi} J_{kn} = 1$ implies

$$\left|(\zeta-z)^{-1}-P_n(z)\right| \leq \int_{-\pi}^{\pi} J_{kn}(t)\frac{|\xi(t)-\zeta|}{|\xi(t)-z||\zeta-z|}\,dt$$

where $\xi(t) = \psi(\tau e^{it})$; so $\xi(0) = \zeta$.

Set $E_1 = \{t, |\xi(t)-z| > \frac{1}{10}|\zeta-z|\}$, $E_2 = [-\pi,\pi]\setminus E_1$. For $t\in E_1$

$$\frac{|\xi(t)-\zeta|}{|\xi(t)-z||\zeta-z|} \leq \frac{10}{|\zeta-z|^2}\,|\xi(t)-\zeta| .$$

But according to theorem I §2 and n° 2.2

$$|\xi(t)-\zeta| \leq c|\zeta-\zeta_{1/n}|(1+n|t|)^Q, \quad 0<Q<\infty , \tag{7.3}$$

and thus for $2k>Q$

$$\int_{E_1} J_{kn}(t)\frac{|\xi(t)-\zeta|}{|\xi(t)-z||\zeta-z|}\,dt \leq c\,\frac{|\zeta-\zeta_{1/n}|}{|\zeta-z|^2} .$$

For $t\in E_2$ we have $|\xi(t)-\zeta| > \frac{9}{10}|\zeta-z|$ and by (7.3)

$$|t| > c_1\frac{1}{n}\left(\frac{|\zeta-z|}{|\zeta-\zeta_{1/n}|}\right)^{1/Q}.$$

Hence, for $2k>Q$

$$\int_{E_2} J_{kn}(t)\frac{|\xi(t)-\zeta|}{|\xi(t)-z||\zeta-z|}\,dt \leq \frac{c}{|\zeta-z|}\int_{|t|>c_1\frac{1}{n}\left(\frac{|\zeta-z|}{|\zeta-\zeta_{1/n}|}\right)^{1/Q}} J_{kn}(t)\,dt \leq c\,\frac{|\zeta-\zeta_{1/n}|}{|\zeta-z|^2} .$$

So for $2k>Q$ and $\zeta\in\Gamma_\tau$, $\tau>1/n$, there exists a polynomial P_n of degree $2kn$ such that

$$\left|(\zeta-z)^{-1}-P_n(z)\right| \leq c\,\frac{|\zeta-\zeta_{1/n}|}{|\zeta-z|^2}, \quad z\in\overline{G} . \tag{7.4}$$

The assumption $\zeta\in\Gamma_\tau$, $\tau>1/n$ implies $|\zeta-z|>\rho(\hat{z},\Gamma_{1/n})$ and

$$|P_n(z)| \leq c\rho(\hat{z},\Gamma_{1/n})^{-1}.$$

Let now $\zeta \in \mathbb{C} \setminus G$ but $|\varphi(\zeta)| \leq 1 + \frac{1}{n}$ Construct the polynomial approximant P_n for the Cauchy kernel $K_{\zeta_{1/n}}$. As earlier we have $|P_n(z)| \leq c\rho(\hat{z}, \Gamma_{1/n})^{-1}$ and by (7.4)

$$|K_\zeta(z) - P_n(z)| \leq |K_\zeta(z) - K_{\zeta_{1/n}}(z)| + |K_{\zeta_{1/n}}(z) - P_n(z)| \leq$$
$$\leq c\,\frac{|\zeta - \zeta_{1/n}|}{|\zeta - z|^2}\,.$$

At last denote

$$H_n(\zeta, z) = \frac{1}{\zeta - z}\left\{1 - \left[1 - (\zeta - z)P_{[n/2km]}(z)\right]^m\right\}. \tag{7.5}$$

It is easy to check that $H_n(\zeta, \cdot)$ is a polynomial of degree at most n satisfying (7.1) and (7.2). ●

REMARKS. (i) The polynomial (7.5) is no more the result of some kind of summation of the Faber series of the Cauchy kernel. May be such a summation and the elementary approach of §6 cannot provide so strong approximation.

(ii) It is clear, that for $z \in G$ we can obtain from (7.2) the estimate $|K_\zeta - H_n(\zeta, \cdot)| \leq c_N n^{-N}$ for any N. The approximation in the interior of a region is better than at the boundary. We shall use this circumstance in the next section

7.2. Direct theorems of approximation theory.

Let G be a Lipschitz region, $w \in \tilde{A}(\Gamma)$ and $1 \leq p \leq \infty$.

THEOREM 9. If $f \in E^p(G)$ admits a pseudoanalytic continuation, then for any $N > 0$

$$E_n(f, w)_p \leq c_N \int_0^\infty (nr)^{-\gamma(w)} |\log nr|\, \sigma_p(f, r, w) \frac{1}{1 + (nr)^N} \frac{dr}{r}.$$

PROOF. Apply theorem 8. Set

$$P(\cdot) = -\frac{1}{\pi} \iint_{\mathbb{C} \setminus G} H_n(\zeta, \cdot)\, \bar{\partial} f(\zeta)\, d\xi\, d\eta.$$

Clearly,

$$E_n(f, w)_p \leq c \| A_1 + A_2 + A_3 + A_4 \|_{L^p(\Gamma, w_{1/n})},$$

where

$$A_1(z) = \iint_{\Omega_0} |\bar{\partial} f(\zeta)|\, |\zeta - z|^{-1} d\xi\, d\eta,$$

$$A_2(z)=\iint_{\Omega_0} |\bar{\partial} f(\zeta)|\, \rho(z,\Gamma_{1/n})^{-1} d\xi\, d\eta \leq A_1(z)\,,$$

$$A_3(z)=\iint_{\Omega_1} |\bar{\partial} f(\zeta)| \frac{|\zeta-\zeta_{1/n}|^m}{|\zeta-z|^{m+1}}\, d\xi\, d\eta\,,$$

$$A_4(z) \quad \iint_{\Omega_2} |\bar{\partial} f(\zeta)| \frac{|\zeta-\zeta_{1/n}|^m}{|\zeta-z|^{m+1}}\, d\xi\, d\eta\,,$$

$$\Omega_0=\{\zeta\in\mathbb{C}\setminus G,\ |\zeta-z|<\rho(z,\Gamma_{1/n})\},$$

$$\Omega_1=\{\zeta\in\mathbb{C}\setminus G\setminus\Omega_0,\ \ |\varphi(\zeta)|-1<1/n\},$$

$$\Omega_2=\{\zeta\in\mathbb{C}\setminus G\setminus\Omega_0,\ \ |\varphi(\zeta)|-1>1/n\}.$$

But for $\zeta\in\Omega_1$ by (2.4) $|\zeta-\zeta_{1/n}|\leq c\rho(z,\Gamma_{1/n})^{1-q}|\zeta-z|^q$. Therefore

$$A_1(z)+A_3(z)\leq c\int_0^{1/n}\frac{d\tau}{\tau}\int_{\Gamma_\tau} \min\left\{1,\frac{\rho(z,\Gamma_{1/n})}{|\zeta-z|}\right\}^m \rho(\zeta,\Gamma)|\bar{\partial} f(\zeta)|\frac{|d\zeta|}{|\zeta-z|}$$

and by lemma 9 §3 and by Minkowski inequality

$$\|A_1+A_2+A_3\|_{L^p(\Gamma,w_{1/n})}\leq c\int_0^{1/n}(n\tau)^{-\gamma(w)}|\log n\tau|\,\sigma_p(f,\tau,w)\frac{d\tau}{\tau}\,.$$

For $\zeta\in\Omega_2$, i.e. $\zeta\in\Gamma_\tau$, $\tau>1/n$ we have $|\zeta-\zeta_{1/n}|\asymp \rho(\zeta,\Gamma)/n\tau$. For m sufficiently large by lemma 8 §3 and Minkowski inequality

$$\|A_4\|_{L^p(\Gamma,w_{1/n})}\leq c\int_{1/n}^{\infty}(n\tau)^{-N}\sigma_p(f,\tau,w)\frac{d\tau}{\tau}\,. \quad \bullet$$

REMARK. An analoguous estimate can be proved for $E_n^0(f,w)_p$ if $w\in A_p(\Gamma)$:

$$E_n^0(f,w)_p\leq c_N\int_0^{\infty}\sigma_p(f,\tau,w)\frac{1}{1+(n\tau)^N}\frac{d\tau}{\tau}$$

by using lemma 7 instead of lemma 9. We shall not discuss this subject because the estimates for $\overset{\circ}{E}_n(f,w)_p$ follow from lemma IO or §6.

Now after theorems 3, 5, 6 and 9 we can formulate the following final result.

THEOREM IO. Let G be a Lipschitz region, $w\in\tilde{A}(\Gamma)$, $1\leqslant p\leqslant\infty$ and $s>\gamma(w)$. For $f\in E^p(G)$ three following assertions are equivalent.

(i) $f\in K_p^s(G,w)$, i.e. $E_n(f,w)_p=O(n^{-s})$.

(ii) $\omega_m(f,\delta,w)_p=O(\delta^s)$ for sufficiently large m.

(iii) f admits a pseudoanalytic continuation such that

$$\sigma_p(f,r,w)=O(r^s). \quad \bullet$$

§8. Area approximation.

8.1. Formulation of results.

Till now we considered approximations in various boundary L^p-metrics. There are some questions of great interest concerning area L^p-approximation. However, as can be seen from the results of §§4-7, best approximations depend first of all on the behaviour of the conformal mapping $\varphi:\mathbb{C}\setminus G\to\mathbb{C}\setminus\mathbb{D}$. This φ is defined on the boundary of the region, but how can this mapping of the exterior of G be included into estimates of the interior approximation? Now we shall see the way of overcoming this obstacle.

Let G be a Radon region, $1<p<\infty$ and $w\in A_p(\Gamma)$. For a function $f\in E^1(G)$ consider following approximation characteristics:

$$\mathcal{E}_n(f,w)_p=\inf_{P_n}\Big(\iint\limits_G|(f(z)-P_n(z))w_i(z)|^p|\varphi'(z^*)|dxdy\Big)^{1/p} \tag{8.1}$$

where infimum is taken over all polynomials P_n of degree n.

The main feature of (8.1) consists in the presence of the measure $d\nu(z)=|\varphi'(z^*)|dxdy$ in the area L^p-norm. Using this definition we are going to prove an embedding theorem, which is an approximation analogue of classical embedding theorems of different dimensions [40,54,55].

THEOREM II. Let G be a Radon region, $1<p<\infty$, $w\in A_p(\Gamma)$ and $f\in E^1(G)$. Then for $n=1,2,\ldots$

(i) $\mathcal{E}_n(f,w)_p \leq c n^{-1/p} E_n^{\circ}(f,w)_p$;

(ii) $E_n^{\circ}(f,w)_p \leq c \sum\limits_{k=n}^{\infty} k^{-1/p'} \mathcal{E}_k(f,w)_p$.

COROLLARY. $f \in K_p^{s}(G,w)$ iff $\mathcal{E}_n(f,w)_p = O(n^{-s-1/p})$. ●

REMARKS. (i) Roughly speaking, theorem II asserts that $E_n^{\circ}(f,w)_p \asymp n^{1/p} \mathcal{E}_n(f,w)_p$, i.e. the boundary smoothness exponent, as it ought to be in embedding theorems, is diminished by $1/p$.

(ii) Approximation without weight, i.e. in the norm of $L^p(G)$, corresponds to $w = |\varphi'|^{-1/p}$ and comes under the action of theorem II for $|\varphi'|^{-1/p} \in A_p(\Gamma)$.

(iii) Theorem II gives rather complete solution of the area approximation problem because the structure of the class K_p^{s} was completely described in theorems 7 and IO.

8.2. Proof of assertion (i)

LEMMA I6. For any $g \in E^1(G)$

$$\mathcal{E}_n(g,w)_p \leq c n^{-1/p} \|g\|_{L^p(\Gamma,w)} .$$

(i) follows immediately from lemma 16 by the application of this lemma to the function $g = f - P_n$ where P_n is the best polynomial approximant of f in the sense of $E_n^{\circ}(f,w)_p$.

PROOF OF LEMMA I6. For $z \in \overline{G}$ denote

$$P(\cdot) = \frac{1}{2\pi i} \int_{\Gamma} g(\zeta) H_n(\zeta,\cdot)\, d\zeta$$

where the polynomial $H_n(\zeta,\cdot)$ is defined in theorem 8 §7. In view of (7.1) and (7.2) we have on Γ

$$|g(z) - P(z)| \leq c_1 \rho(z,\Gamma_{1/n})^{-1} \int\limits_{|\zeta - z| < \rho(z,\Gamma_{1/n})} |g(\zeta)|\,|d\zeta| +$$

$$+ c_2 \rho(z,\Gamma_{1/n})^{m} \int\limits_{|\zeta - z| > \rho(z,\Gamma_{1/n})} \frac{|g(\zeta)|}{|\zeta - z|^{m+1}} |d\zeta| + \left| \int\limits_{|\zeta - z| < \rho(z,\Gamma_{1/n})} \frac{g(\zeta)}{\zeta - z} d\zeta \right| .$$

Two first temrs don't exceed $Mg(z)$ and the third term can be estimated by (1.10). So we have

$$\|g - P\|_{L^p(\Gamma,w)} \leq c \|g\|_{L^p(\Gamma,w)} . \qquad (8.2)$$

Note now that by (1.7)

$$\mathcal{E}_n(f,w)_p^p \leq \iint_G |g-P|^p |w_i|^p d\nu \asymp c\int_0^\infty d\tau \int_{\Gamma_\tau^*} |(g-P)w_i|^p . \tag{8.3}$$

Consider first the region $\tau < 1/n$ in (8.3). But $(g-P)w_i \in \in E^1(G)$ and we have by Carleson embedding theorem and (8.2)

$$\int_{\Gamma_\tau^*} |(g-P)w_i|^p \leq c\int_\Gamma |(g-P)w_i|^p \leq c\|g\|^p_{L^p(\Gamma,w)} .$$

Therefore the part of (8.3) corresponding to the region $\tau < 1/n$ does not exceed $cn^{-1}\|g\|^p_{L^p(\Gamma,w)}$.

It remains to estimate the contribution of the region $1/n < \tau < \infty$. But for $z \in \Gamma_\tau^*$, $\tau > 1/n$ it is evident that

$$|g(z)-P(z)| \leq \frac{c}{(n\tau)^{mv}} \int_\Gamma \frac{\rho(z,\Gamma)^{mv}}{|\zeta - z|^{mv+1}} |g(\zeta)||d\zeta|$$

(for v see n° 2.2).

Divide now Γ into arcs, $\Gamma = \bigcup_1^N I_k$, such that $\tau/2 \leq \leq |\varphi(I_k)| \leq \tau$. Let $J_k = ((I_k)_\tau)^*$ be correspondent arcs on Γ_τ^* . Clear that for $z \in J_k$, $\tau \in I_k$, $|g(z)-P(z)| \leq \leq C(n\tau)^{-mv} Mg(\tau)$.

Thus for all $\tau \in I_k$

$$\int_{J_k} |(g-P)w_i|^p \leq \frac{c}{(n\tau)^{mvp}} Mg(\tau)^p \int_{J_k} |w_i|^p \asymp \frac{c}{(n\tau)^{mvp}} Mg(\tau)^p \int_{I_k} w^p,$$

$$\int_{\Gamma_\tau^*} |(g-P)w_i|^p = \sum \int_{J_k} \leq \frac{c}{(n\tau)^{mvp}} \int_\Gamma (Mg)^p w^p \leq \frac{c}{(n\tau)^{mvp}} \|g\|^p_{L^p(\Gamma,w)} .$$

At last the contribution of the region $\tau > 1/n$ to (8.3) does not exceed

$$c\|g\|^p_{L^p(\Gamma,w)} \int_{1/n}^\infty (n\tau)^{-mvp} d\tau \leq cn^{-1}\|g\|^p_{L^p(\Gamma,w)}, \quad m > \frac{1}{pv} . \quad \bullet$$

8.3. Proof of the assertion (ii)

We shall state the following polynomial inequality.

LEMMA 17. If P is a polynomial of degree n, then

$$\|P\|_{L^p(\Gamma,w)} \leq c\, n^{1/p} \|P\|_G ,$$

where

$$\|P\|_G \overset{def}{=\!=} \Big(\iint_G |P w_i|^p d\nu\Big)^{1/p} .$$

Now for completing the proof of the theorem it is sufficient to apply lemma 17 to every term of the series $f-P_n=\sum_{k=0}^{\infty}(P_{2^{k+1}n}-P_{2^k n})$, $\{P_n\}_1^\infty$ being the best polynomial approximants for f in the sense of (8.1). ●

PROOF OF LEMMA 17. Consider the functional

$$A(g)=\int_\Gamma P(z)g(z)\varphi(z)^{-n}dz,\quad g\in C(\Gamma).$$

It is sufficient to show that

$$|A(g)|\leq c\|P\|_G \|g\|_{L^{p'}(\Gamma,w^{-1})}\, n^{1/p} . \tag{8.4}$$

First of all if $g\in E^1(\mathbb{C}\setminus G)$ and $g(\infty)=0$ then $A(g)=0$. Therefore $A(g)=A(g_+)$. Consider the function, defined in G : $F(z)=P(z)g_+(z)\varphi(z^*)^{-n}$.

By the Green's formula

$$A(g)=-2i\iint_G \bar\partial F(z)\,dx\,dy .$$

However

$$\bar\partial F(z)=P(z)g_+(z)\frac{n}{\varphi(z^*)^{n+1}}\varphi'(z^*)\,\bar\partial z^* .$$

The quasiconformal reflection belonging to the class C^1, we have $|\bar\partial z^*|\asymp 1$ and by the Hölder inequality

$$|A(g)|\leq cn\|P\|_G\Big(\iint_G |g_+(z)w_i(z)^{-1}\varphi(z^*)^{-n-1}|^{p'}d\nu(z)\Big)^{1/p'}.$$

But

$$\iint_G |g_+ w_i^{-1}\varphi^{-n-1}|^{p'}d\nu\leq c\int_0^\infty \frac{dr}{(1+r)^{np'}}\int_{\Gamma_r^*}|g_+ w_i^{-1}|^{p'} .$$

By Carleson embedding theorem and in view of $g_+ w_i^{-1}\in E^1(G)$

$$\int_{\Gamma_r^*}|g_+ w_i^{-1}|^{p'}\leq c\int_\Gamma |g_+ w_i^{-1}|^{p'},$$

$$|A(g)| \leq c n^{1/p} \|P\|_G \|g_+ w^{-1}\|_{L^{p'}(\Gamma)} .$$

If $w \in A_p(\Gamma)$ then $w^{-1} \in A_{p'}(\Gamma)$ and (8.4) holds. ●

§9. Special cases and generalizations.

9.1. Approximation in $C_A(\bar{G})$.

Let G be a Lipschitz region. For a function $f \in C_A(\bar{G})$ and an arc $I \subset \Gamma$ define the oscillation $\omega_m(f, I)$ of f on I (of order m) as the best approximation of f in $C(I)$ by polynomials of degree m or, equivalently, as the distance in $C(I)$ from f to its interpolation polynomial of degree m with interpolation points $a_0, a_1, \ldots, a_m \in I$, $|a_i - a_j| \geq |I| / 2(m+1)$. In particular, $\omega_0(f, I) = 1/2 \sup_I |f(\zeta) - f(z)|$ is simply the ocsillation of f on I .

Define for $w \in \tilde{A}(\Gamma)$ its mean value w_I on I by the equality $w_I = |w_e(z_I)|$ where the point z_I is reciprocal to I in the sense of n° 1.5.

Now we can formulate theorem 10 for $p = \infty$ as follows.

THEOREM 10'. Let G be a Lipschitz region, $w \in \tilde{A}(\Gamma)$, $s > \gamma(w)$ and $f \in C_A(\bar{G})$. Following two assertions are equivalent.

(i) For any n there exists a polynomial P_n of degree n such that

$$|f(z) - P_n(z)|\, w_{1/n}(z) \leq c n^{-s}, \quad z \in \Gamma . \tag{9.1}$$

(ii) For any arc $I \subset \Gamma$ and every sufficiently large m

$$\omega_m(f, I)\, w_I \leq c |\varphi(I)|^s . \tag{9.2}$$

REMARKS. (i) In P.M.Tamrazov's book [12] a number of different constructions of moduli of smoothness for continious functions on curves are discussed. Here we use the simplest variant of $\omega_m(f, I)$. The "right" definition of moduli of smoothness for complicated sets is far from evidence, but now the situation is clear after the work of P.M.Tamrazov [12,59], Ju.A. Brudnyi [56] and the author [20,34,60] .

(ii) The estimate (9.1) concerns the boundary of G . We remind that there is a stronger estimate at interior points of G . For example, we can show by means of constructions of §7 that for $f \in C_A(\bar{G})$ and $z \in G$

$$|f(z)-P_n(z)| \leq c_M(z)\, n^{-M} \tag{9.3}$$

for any $M>0$. This observation is due to N.A.Shirokov [10,11] (see also [26,27]). We have used it in §8. It is well-known that in $\mathbb{D}$ we can replace n^{-M} in (9.3) by q^n, $0<q<1$. N.A.Shirokov has proved (unpublished) that if G has corner points then one cannot replace n^{-M} in (9.3) by e^{-n^β}, were β, $0<\beta<1$, depends on G . It is unknown whether it is possible to replace n^{-M} in (9.3) by e^{-n^β} for piecewise smooth regions. It is unknown whether it is possible to improve (9.3) in a general Lipschitz region.

(iii) The proof of theorem IO shows that (9.2) holds for $m>as-b$ where a, $b>0$ depend on W and G . In particular, we can set $m=0$ and remove moduli of smoothness of higher order for sufficiently small s only.

9.2. Lipschitz classes.

Set $W=|\varphi'|^s$. Then (9.1) becomes

$$|f(z)-P_n(z)| \leq c\rho(z,\Gamma_{1/n})^s, \quad z\in\Gamma_1 \tag{9.4}$$

and (9.2) becomes

$$\omega_m(f,I) \leq c|I|^s . \tag{9.5}$$

It is easy to prove that (9.5) is equivalent to the usual Lipschitz condition of order s for $0<s<1$ and defines respective Hölder-Zygmund class for $s\geqslant 1$. Thus, theorem IO' transforms into the classical W.K.Dzyadyk's theorem [3-5] about constructive characterization of the Lipschitz classes. This was a standard case for a long time and the main progress in the theory was understood as the extension of the class of regions for which this characterization holds. After the work of W.K.Dzyadyk [3-7] , N.A.Lebedev and N.A.Shirokov [9-11,13] , V.I.Belyi [15,16] and others the corresponding "direct" theorems were proved for very general regions. V.I.Belyi [16] has proved such a theorem for any quasiconformal region (may be with nonrectifiable boundary). On the other hand N.A.Lebedev and P.M.Tamrazov [8,12,59] have proved an "inverse" theorem, i.e. implication (9.4)$\Longrightarrow$(9.5) for any continuum. However, this generality was superfluous. N.A.Shirokov [14] has constructed an example of a region of bounded rotation (with a zero interior corner) such that (9.4) is stricly stronger than (9.5). We shall descuss it in the next paragraph.

9.3. Uniform approximation.

For $w \equiv 1$ (9.1) and (9.2) become

$$|f-P_n| \leq cn^{-s}, \tag{9.6}$$

$$\omega_m(f,I) \leq c\,|\varphi(I)|^s . \tag{9.7}$$

The problem of description of the function class (9.6) has attracted the attention of specialists for a long time (see [7,18--21,26,27]). For $m=0$ (9.7) means that $f\circ\psi\in C^s(\mathbb{T})$. W.K.Dzyadyk in 1962 has conjectured that the condition $f\circ\psi\in C^s(\mathbb{T})$ is equivalent to (9.6). We see by (9.7) that this condition is sufficient for (9.6) because $\omega_m(f,I)\leq\omega_0(f,I)$, but it is necessary only for small s. For example, if $s>1$, one cannot set $m=0$ in (9.7). W.K.Dzyadyk and G.A.Alibekov [7] have proved sufficiency of $f\circ\psi\in C^s(\mathbb{T})$ for (9.6) in the case of piecewise smooth regions with some restrictions on s. Further this subject was discussed in [18,19,61] , where exceptional values of s depending on G were indicated. At the same time some attempts to find an alternative language for the description of the class have been made [53,20] . Finally, the author [21] has introduced the condition (9.7) which allowed to solve the problem completely.

If $w=|\varphi'|^s|\Phi'|^{-s}$ where $\Phi:\mathbb{C}\setminus G\to\mathbb{C}\setminus\Omega$, Ω being a Lipschitz region with level curves $\{\Lambda_r\}$ then, as we have seen in n° 4.4, (9.1) and (9.2) become [52]

$$|f(z)-P_n(z)|\leq c\rho(\Phi(z),\Lambda_{1/n})^s ,$$

$$\omega_m(f,I)\leq c\,|\Phi(I)|^s .$$

If $\Omega=G$ this is a Lipschitz class, if $\Omega=\mathbb{D}$ this is a uniform approximation problem. If $G=\mathbb{D}\neq\Omega$ we have a problem of non-uniform approximation in the disk.

9.4. Approximation in $E^p(G)$. Area approximation.

The case of $E^p(G)$ corresponds to $w\equiv 1$ in theorems 7 and IO. Here the difference between conformal moduli of smoothness and usual ones is essential (except for the modulus of order 0). Our description of $K^s_p(G,1)$ follows author's work [22] where Faber operators and conformal moduli of smoothness without weights were introduced. If $m=0$ then $\omega_0(f,\delta,1)_p$ is equivalent to the usual modulus of continuty in $L^p(\mathbb{T})$ of $f\circ\psi$ and so one can prove the sufficiency of the condition $f\circ\psi\in Lip_A(p,s)$ for $f\in K_p^s(G,1)$. This result has been independently proved by means of Faber operators by J.I. Mamedhanov and I.I.Ibragimov ([47] , see also [62]) and J.E. Andersson [48,49] . But the necessary and sufficient condition for large s cannot be found in such a way.

In the problem of area approximation the main difficulty is to include the conformal mapping φ of the exterior of G into the estimates of interior approximation. As we have seen, the exterior mapping controls polynomial approximation. By this cause the previous results of S.Ya.Alper [63] and V.M.Kokilashvili [64] , who have operated with interior conformal mapping $\psi:G\to\mathbb{D}$, are complete only for regions with smooth boundaries where ψ and φ have identical boundary behaviour.

Approximation in $L^p(G)$ corresponds to $W=|\varphi'|^{-1/p}$ in theorem II and we obtain a condition $|\varphi'|^{-1/p}\in A_p(\Gamma)$. There are two cases when this condition holds.

(i) G is a convex region and $p>2-\varepsilon$, $\varepsilon=\varepsilon(G)>0$.

(ii) G is a piecewise smooth region with exterior angles $\pi\gamma_1,\pi\gamma_2,\dots,\pi\gamma_N$, $0<\gamma_j<2$ in its vertices, $p>2-\min_j 1/\gamma_j$ and $\min_j\gamma_j>1/2$.

9.5. Besov classes.

Our classes $K_p^s(G,W)$ are particular cases of more general classes of Besov type

$$K_{pq}^s(G,W)=\Big\{f,\ \Big(\sum_{n=1}^{\infty}\frac{1}{n}(n^sE_n(f,W)_p)^q\Big)^{1/q}<+\infty\Big\},\ 1\le q\le\infty .$$

So, $K_p^s(G,W)=K_{p\,\infty}^s(G,W)$.

It is easy to show like in §§ 4-7 that $f\in K_{pq}^s(G,W)$ iff

$$\int_0^\infty\left[\delta^{-s}\omega_m(f,\delta,W)_p\right]^q\frac{d\delta}{\delta}<+\infty \tag{9.8}$$

and f admits a pseudoanalytic continuation such that

$$\int_0^\infty\left[r^{-s}\sigma_p(f,r,W)\right]^q\frac{dr}{r}<+\infty . \tag{9.9}$$

If $q \neq p$ then condition (9.9) is connected with the system of level curves $\{\Gamma_\tau\}$ and can be expressed in terms of conformal smoothness only. But if $q = p$ then one can rewrite (9.9) in a more invariant form

$$\iint_G |\bar{\partial} f(\zeta) \rho(\zeta,\Gamma) w_e(\zeta)(|\varphi(\zeta)|-1)^{-s}|^p \rho(\zeta,\Gamma)^{-1} d\xi\, d\eta < +\infty . \quad (9.10)$$

Condition (9.10) does not involve any choosen system of curves. On can rewrite it in the form of repeated integral with level curves of the usual distance and express (9.10) in terms of usual metrical moduli of smoothness. For example, in such a way we can get a constructive characterization of class $AB_p^s(G)$ of functions of $E^p(G)$ whose boundary values belong to Besov class [54] $B_{pp}^s(\Gamma)$. See author's work [25]; this characterization is the following: $AB_p^s(G) = K_{pp}^s(G, |\varphi'|^s)$. Sobolev classes $E_p^\ell(G) = \{f \in E^p(G), f^{(\ell)} \in E^p(G)\}$ also admit some constructive characterization, but of another form (analoguous to the theorem of dyadic decomposition in Fourier analysis) [24,25].

9.6. Approximation on a segment.

The segment $[-1,1]$ is not included into the domain of applicability of this article, but it is just formally. Usually this case is considered by pure real-variable methods [39,65]. But it is easy to check that all constructions and assertions of our article (except §8 of course) may be repeated for a segment. In such a way one can get the results on Lipschitz classes, uniform approximation, Besov classes and so on.

9.7. Mixed approximation.

Let $w_1 \in \tilde{A}(\Gamma)$ and $w_2 \in A_p(\Gamma)$, $1 < p < \infty$. Replace in all definitions of $E_n(f,w)_p$, $\omega_m(f,\delta,w)_p$ etc the regularization w_τ by the combination $(w_1)_\tau w_2$. This construction contains E_n and E_n^0 as particular cases (when $w_2 \equiv 1$ or $w_1 \equiv 1$). One can show that all results of §§4-8 hold for this "mixed" case. But the classe K_p^s do not depend on w_1 and w_2 but only on the whole weight $w = w_1 w_2$. Lemma 10 is really an example of such independence.

§ 10. Some unsolved problems.

10.1. N.A.Shirokov's counterexample.

Let $B \subset \mathbb{C}$ be a bounded continuum with the connected complement, $\varphi : \mathbb{C}\setminus B \to \mathbb{C}\setminus \mathbb{D}$, $\varphi(\infty)=\infty$, be a conformal mapping and $\Gamma_r = \{z\in\mathbb{C}\setminus B,\ |\varphi(z)|=1+r\}$ be its level curves. The classes $C_A(B)$ and $C_A^s(B)$, $0<s<1$, were defined in n° 1.2. Define now an approximation class $\Pi^s(B)$, $0<s<1$, consisting of all functions $f\in C_A(B)$ such that for any $n=1,2,\dots$ there exists a polynomial P_n of degree n for which

$$|f(z)-P_n(z)| \le c\rho(z,\Gamma_{1/n})^s,\quad z\in\partial B. \tag{10.1}$$

We have seen in §9 that $\Pi^s(B)=C_A^s(B)$ if $B=\overline{G}$ is the closure of a region G with the quasiconformal boundary or if $B=[-1,1]$. N.A.Lebedev and P.M.Tamrazov [8,12] have shown that $\Pi^s(B)\subset C_A^s(B)$ for any continuum B. But in 1977 N.A.Shirokov has observed the following fact.

THEOREM I2. Let $B=[0,1]\cup[0,i]$. Then $\Pi^s(B)\subset C_A^s(B)$ but $\Pi^s(B)\ne C_A^s(B)$, $0<s<1$.

PROOF. Set $B_0 = B\cap\{z,\ |z|\le 1/2\}$,

$$G_1=\{z\in\mathbb{C}\setminus B,\ |z|<1/2,\ 0<\arg z<\pi/2\},$$

$$G_2=\{z\in\mathbb{C}\setminus B,\ |z|<1/2,\ \tfrac{\pi}{2}<\arg z<2\pi\}.$$

Let $\Gamma_r^1=\Gamma_r\cap G_1$, $\Gamma_r^2=\Gamma_r\cap G_2$. It is easy to see that for $z\in B_0$

$$\rho(z,\Gamma_r^1)\asymp r^{1/2}\min(1, r^{1/2}|z|^{-1}),$$

$$\rho(z,\Gamma_r^2)\asymp r\max(r^{1/2},|z|^{1/3})$$

and hence $\rho(z,\Gamma_r)\asymp\rho(z,\Gamma_r^2)$.

Let $f\in\Pi^s(B)$. By theorem 2 §4 (B is not the closure of a region, but it does not influence the proof) f admits a pseudoanalytic continuation such that in $\mathbb{C}\setminus B$

$$|\bar\partial f(z)|\le c\rho(z,B)^{-1}|P_{2^{n+1}}(z)-P_{2^n}(z)|,\quad 2^{-n}\le|\varphi(z)|-1\le 2^{-n+1}.$$

Applying the estimate (2.6) to the function $(P_{2^{n+1}}-P_{2^n})\varphi^{-2^{n+1}}$ in $\mathbb{C}\setminus B$ we obtain

$$|\bar\partial f(z)|\le c\rho(z,B)^{s-1}\min(|z|,\rho(z,B))^{\frac{4}{3}s},\quad z\in G_1, \tag{10.2}$$

$$|\bar{\partial} f(z)| \leq c\rho(z,B)^{s-1}, \quad z \in G_2 . \qquad (10.3)$$

As we have noted above, the class $C_A^s(B)$ corresponds to the estimate (10.3) in the whole of $\mathbb{C} \setminus B$. But the estimate (10.2) is stricly stronger than (10.3) and so $\Pi^s(B) \neq C_A^s(B)$. The formal proof is as follows. Consider the Cauchy type integral Φ,

$$\Phi(z) = \int_B f(\zeta)(\zeta - z)^{-1} d\zeta , \quad z \in \mathbb{C} \setminus B .$$

It is easy to check from (10.2) by the Green's formula that

$$|\Phi'(-\delta e^{i\frac{\pi}{4}})| \leq c\delta^{\frac{4}{3}s-1} , \quad 0 < \delta < 1 . \qquad (10.4)$$

Now any function $f \in C_A^s(B)$ which does not satisfy (10.4) gives a counterexample. For instance we can take

$$f(z) = \begin{cases} |z|^s, & z \in [0, i], \\ |z|^s \exp \frac{2}{3}\pi s i, & z \in [0,1], \end{cases}$$

and get $|\Phi'(-\delta e^{i\frac{\pi}{4}})| \geq c\delta^{s-1}$, $0 < \delta < 1$, for this function. ●

Generalizing the idea, N.A.Shirokov [14] has constructed a region G such that

(i) the boundary of G has a tangent at all its points except one vertex, and a slope of the tangent has bounded variation;

(ii) there exist two one-sided tangents to ∂G at its vertex, making an angle π (zero interior angle at vertex);

(iii) $\Pi^s(\bar{G}) \neq C_A^s(\bar{G})$, $0 < s < 1$.

This counterexample disproves the conjecture, discussed for a long time, that (10.1) is an universal constructive characteristization of the class $C_A^s(B)$. The proof of this conjecture for general regions was the main subject of the development of the theory before 1977. The theorem I2, i.e. the failure of (10.1) for unclosed arcs, is rather elementary fact. The late understanding of this fact and its consequences was caused by neglecting the approximations on arcs.

It is not known, how the approximation problems must be posed for general sets. It is not known either whether there exists a constructive characterization of $C_A^s(B)$ on arcs by means of some modification of (10.1).

Note that in all proofs of §§4-9 one can limit the consideration by polynomial approximants of degrees $n=Q^k$, $Q>1$, $k=0,1,\ldots$, only. N.A.Shirokov [66] has shown that this is no longer true for $B=[0,1]\cup[0,i]$. All this means, probably, that any transformation of (10.1) into some estimate of type

$$|f(z)-P_n(z)|\leq c\gamma_n(z), \quad z\in\partial B, \tag{10.5}$$

is useless. The question of constructive characterization of $C_A^s(B)$ for general B is open.

10.2. Uniform approximation.

Let again B be a plane continuum with the connected complement. Denote by $K^s(B)$, $s>0$, the class of all functions $f\in C_A(B)$ such that

$$|f-P_n|\leq Cn^{-s}$$

on B for some polynomial P_n of degree $n=1,2,\ldots$. By §9 one can describe $K^s(B)$ for $B=\overline{G}$ (G is a Lipschitz region) or $B=[-1,1]$ in terms of local oscillations:

$$\omega_m(f,I)\leq c|\varphi(I)|^s. \tag{10.6}$$

The question of description of $K^s(B)$ for general B is open. This case has some psychological advantage over n°10.1 because the language for the description of K^s has been found quite recently even for simple regions and therefore this direction appears to be natural. Here we shall discuss only one difficulty of the problem. Let $B=\overline{G}$, region G having a piecewise smooth boundary with one corner point (a vertex) and suppose exterior angle at this vertex is $\pi\alpha$, $0<\alpha<2$. We see from (10.6) that all functions $f\in K^s(\overline{G})$ have the smoothness "of order s" on smooth parts of the boundary and the limit smoothness "of order s/α" at the vertex (for example, on the bisector of the interior angle f satisfies a Lipschitz condition of order s/α). Evidently, we must introduce moduli of smoothness of order $m>s/\alpha$ to describe such behaviour.

For $\alpha \to 2$ (zero interior angle) there are no difficulties, but for $\alpha \to 0$ (zero exterior angle) the condition (10.6) is not satisfactory for any m.

EXAMPLE. Let G be the interior of the cardioide Γ,

$$\Gamma = \{ z = re^{it}, \quad 0 \le t \le 2\pi, \quad r = 1 + \cos t \}.$$

G is not a Radon region because it has zero exterior angle at the origin but Γ is a piecewise smooth curve and so all proofs of §4 and §6 hold. Clearly, $f \in K^{s}(\overline{G})$ iff f admits a pseudoanalytic continuation such that

$$|\bar{\partial} f| \le c|\varphi'|(|\varphi|-1)^{s-1} \tag{10.7}$$

or iff

$$f = T_{\infty} g , \quad g \in C_A^s(\mathbb{D}), \tag{10.8}$$

where T_{∞} is Faber operator from n° 6.1.
But we cannot express conditions (10.7) and (10.8) in local terms. These conditions describe a class of functions whose smoothness changes very sharply along Γ : from Lips to C^{∞}. Really, the limit smoothness of $f \in K^{s}(\overline{G})$ at a vertex corresponds to some Gevrey class. It is unknown what local language is applicable to the description of so rapidly varying smoothness.

REMARK. Local and global approximation theories have the general concept of best approximation $E_n(f, I)$ to f on I by polynomials of degree n, but in the local theory n is fixed and I changes, and in the global theory $I = \Gamma$ is fixed and n increases. There is no theory for estimates of $E_n(f, I)$ uniformly with respect to both n and I. Such a method could yiels a solution of our problem, but now local and global approximation techniques differ too much.

10.3. Direct theorems without Fourier analysis.

In §6 and §7 we have constructed polynomial approximants by means of Faber operators and some artificial tricks. In the end, these approximants were connected with summation processes of Fourier series of a Faber-transformed function (or the Cauchy kernel). We propose now two problems.

(i) Is it possible to construct polynomial approximants in a region without any isomorphism of Faber-type with a function space in a disk?

(ii) Is it possible to prove Jackson's theorem on uniform approximation on a circle (or on a disk) without use of the Fourier analysis?

The approximation on disconnected sets (for example, on the union of two regions) gives an example of problem for which the usual approach is impossible and the solution of our problem (i) could be of interest. But now we don't see any way to solve these problems.

References

I. А л ь п е р С.Я. О равномерных приближениях функций комплексного переменного в замкнутой области. Изв.АН СССР, сер.матем., I955, I9, 423-444.

2. А л ь п е р С.Я. О приближении в среднем аналитических функций класса E_p. В сб. "Исследования по современным проблемам теории функций комплексного переменного". М., ФМ, I960, 273-286.

3. Д з я д ы к В.К. О проблеме С.М.Никольского в комплексной области. Изв.АН СССР, сер.матем., I959, 23, № 5, 697-736.

4. Д з я д ы к В.К. О проблеме С.М.Никольского, I. Изв.АН СССР, сер.матем., I962, 26, № 6, 796-824.

5. Д з я д ы к В.К. О проблеме С.М.Никольского; II. Изв.АН СССР, сер.матем., I963, 27, № 5, II35-II64.

6. Д з я д ы к В.К. Исследования по теории приближений аналитических функций, проводимые в институте математики АН УССР. Украинский матем.ж., I969, I9, № 5, 33-57.

7. Д з я д ы к В.К., А л и б е к о в Г.А. О равномерном приближении функций комплексного переменного на замкнутых множествах с углами. Матем.сборник, I968, 75, № 4, 502-557.

8. Л е б е д е в Н.А., Т а м р а з о в П.М. Обратные теоремы приближения на регулярных компактах комплексной плоскости. Изв.АН СССР, I970, 34, № 6, I340-I390.

9. Л е б е д е в Н.А., Ш и р о к о в Н.А. О равномерном приближении функций на замкнутых множествах, имеющих конечное число угловых точек с ненулевыми внешними углами. Изв. АН Арм.ССР, сер.матем., I97I, 6, № 4, 3II-34I.

I0. Ш и р о к о в Н.А. О равномерном приближении функций на замкнутых множествах, имеющих конечное число угловых точек с ненулевыми внешними углами. Докл.АН СССР, I972, 205, № 4,

798-800.

II. Ш и р о к о в Н.А. О равномерном приближении функций на замкнутых множествах с ненулевыми внешними углами. Изв. АН Арм.ССР, сер.матем., I974, 9, № I, 62-80.

I2. Т а м р а з о в П.М. Гладкости и полиномиальные приближения. Киев, "Наукова думка", I975.

I3. Ш и р о к о в Н.А. Приближение непрерывных аналитических функций в областях с ограниченным граничным вращением. Докл. АН СССР, I976, 228, № 4, 809-8I2.

I4. Ш и р о к о в Н.А. Аппроксимативная энтропия континуумов. Докл.АН СССР, I977, 235, № 3, 546-549.

I5. Б е л ы й В.И., М и к л ю к о в В.М. Некоторые свойства конформных и квазиконформных отображений и прямые теоремы конструктивной теории функций. Изв.АН СССР, сер.матем., I974, 38, № 6, I343-I36I.

I6. Б е л ы й В.И. Конформные отображения и приближение аналитических функций в областях с квазиконформной границей. Матем.сборник, I977, I02, № 3, 33I-36I.

I7. K ö v a r i T., P o m m e r e n k e Ch. On Faber polynomials and Faber expansions, Math.Zeit., 1967, 99, N 3, 193-206.

I8. K ö v a r i T. On the order of polynomial approximation for closed Jardan domains. J.Approximation theory. 1972, 5, N 4, 362-373.

I9. G a n e l i u s T.H. Degree of approximation by polynomials on compact plane sets. In "Approximation Theory", N.Y.-London, AP, 1973, 347-35I.

20. Д ы н ь к и н Е.М. О равномерном приближении функций многочленами в комплексной области. Зап.научн.семин.ЛОМИ АН СССР, I974, т.47, I60-I6I.

2I. Д ы н ь к и н Е.М. О равномерном приближении функций в жордановых областях. Сибирский матем.ж., I977, I8, № 4, 775-786.

22. Д ы н ь к и н Е.М. Скорость полиномиальной аппроксимации в $E^p(G)$. Докл. АН СССР, I976, 23I, № 3, 529-53I.

23. Д ы н ь к и н Е.М. Приближение многочленами в среднем по области. В сб. "Тезисы докладов Всесоюзного симпозиума по теории аппроксимации функций в комплексной области". Уфа, I976, 26-28.

24. Д ы н ь к и н Е.М. К конструктивной характеристике классов С.Л.Соболева и О.В.Бесова. Докл. АН СССР, I977, 233, № 5, 773-775.

25. Д ы н ь к и н Е.М. Конструктивная характеристика классов

С.Л.Соболева и О.В.Бесова. Труды Матем.ин-та АН СССР им.В.А. Стеклова, 1980, т.155.

26. Дзядык В.К. К теории приближения функций на замкнутых множествах комплексной плоскости. Труды Матем.ин-та АН СССР им.В.А.Стеклова, 1975, т.134, 63-114.
27. Дзядык В.К. Введение в теорию равномерного приближения функций полиномами. М., "Наука", 1977.
28. Duren P.L. Theory of H^p-spaces. N.Y.-London,AP,1970.
29. Привалов И.И. Граничные свойства аналитических функций. М.-Л., ГИТТЛ, 1950.
30. Hunt R.A., Muckenhoupt B., Wheeden R.L. Weighted norm inequalities for the conjugate function and Hilbert transform. Trans.Amer.Math.Soc., 1973, 176, 227--251.
31. Coifman R.R., Fefferman Ch. Weighted norm inequalities for maximal functions and singular integrals. Studia Math., 1974, 51, N 3, 241-250.
32. Голузин Г.М. Геометрическая теория функций комплексного переменного. М., "Наука", 1966.
33. Белинский П.П. Общие свойства квазиконформных отображений. М., "Наука", 1974.
34. Дынькин Е.М. Гладкие функции на плоских множествах. Докл. АН СССР,1973, 208, № 1, 25-27.
35. Андриевский В.В. О приближении функций частными суммами ряда по полиномам Фабера на континуумах с ненулевой локальной геометрической характеристикой. Украинский матем. ж., 1980, 32, № 1, 3-10.
36. Андриевский В.В. Прямые теоремы теории приближений на квазиконформных дугах. Изв.АН СССР, сер.матем., 1980, 44, № 2, 243-261.
37. Тамразов П.М., Белый В.И. Полиномиальные приближения и модули гладкости функций в областях с квазиконформной границей. Сибирский матем.ж., 1980, 21, № 3, 162-176.
38. Белый В.И. Конформные инварианты в теории приближения функций комплексного переменного. Киев, "Наукова Думка",1981.
39. Тиман А.Ф. Теория приближения функций действительного переменного. М., Физматгиз, 1960.
40. Никольский С.М. Приближение функций многих переменных и теоремы вложения. М., "Наука", 1977.
41. Данилюк И.И. Нерегулярные граничные задачи на плоскости. М., "Наука", 1975.
42. Ahlfors L.V. Lectures on quasiconformal mappings.

Toronto - N.Y.-London, Van Nostrand, 1966.

43. Виноградов С.А., Хавин В.П. Свободная интерполяция в H^{∞} и некоторых других классах функций. Зап. научн.семин.ЛОМИ АН СССР, 1974, т.47, 15-54.

44. Calderon A.P. Cauchy integrals on Lipschitz curves and related operators. Proc.Nat.Acad.Sci.USA, 1977, 74, 1324-1327.

45. Дынькин Е.М. Оценки аналитических функций в жордановых областях. Зап.научн.семин.ЛОМИ АН СССР, 1977, т.73, 70-90.

46. Dunford N., Schwartz J.T. Linear operators. N.Y.-London, Interscience, 1958.

47. Ибрагимов И.И., Мамедханов Дж.И. Конструктивная характеристика некоторого класса функций. Докл. АН СССР, 1975, 223, № 1, 35-37.

48. Andersson J.E. On the degree of polynomial approximation in $E^{p}(D)$. J. Approximation Theory, 1977, 19, N 1, 61-68.

49. Andersson J.E. On the degree of polynomial and rational approximation of holomorphic functions. Ph.D.Thesis, Univ. of Göteborg, 1975.

50. Андрашко М.И., Колесник Л.И. Приближение в среднем аналитических функций в областях с углами. В сб. "Метрические вопросы теории функций и отображений", вып. 2, Киев, 1970, 3-20.

51. Андрашко М.И., Колесник Л.И. Конструктивная характеристика некоторого подкласса функций класса E_p в областях с углами. В сб. "Метрические вопросы теории функций и отображений", вып. 6, Киев, 1975, 8-15.

52. Дынькин Е.М. К общей задаче приближения многочленами в жордановых областях. Зап.научн.семин.ЛОМИ АН СССР, 1976, т.65, 189-191.

53. Широков Н.А. О взвешенных приближениях на замкнутых множествах с углами. Докл. АН СССР, 1974, 214, № 2, 295-297.

54. Бесов О.В., Ильин В.П., Никольский С.М. Интегральные представления функций и теоремы вложения. М., "Наука", 1975.

55. Stein E.M. Singular integrals and differentiability properties of functions. Princeton Univ.Press, 1970.

56. Брудный Ю.А. Пространства, определяемые с помощью локальных приближений. Труды Моск.матем.об-ва, 1971, т.24, 69-132.

57. B r u d n y i Ju.A. Piecewise polynomial approximation, embedding theorems and rational approximation. Lecture notes math,, 1976, 556, 73-98.

58. P o m m e r e n k e Ch. Konforme Abbildung und Fekete-Punkte. Math. Zeit., 1965, 89, 422-438.

59. Т а м р а з о в П.М. Конечно-разностные гладкости и полиномиальные приближения. Препринт ИМ-75-I0, Ин-тматем.АН УССР, Киев, I975.

60. Д ы н ь к и н Е.М. Гладкость интегралов типа Коши. Зап. научн.семин.ЛОМИ АН СССР, I979, т.92, II5-I33.

6I. А н т о н ю к П.Е. К равномерному приближению функций,непрерывных на замкнутых множествах с углами. Докл. АН УССР, I97I, А, 2, 487-489.

62. И б р а г и м о в И.И., М а м е д х а н о в Дж.И. Прямые и обратные теоремы приближения в комплексной области. В сб. "Теория приближения функций", М., "Наука", I977, I90-I94.

63. А л ь п е р С.Я. О приближении аналитических функций в среднем по области. Докл. АН СССР, I96I, I36, № 2.

64. К о к и л а ш в и л и В.М. Об аппроксимации аналитических функций в среднем по области. Труды Тбилисского мат.ин-та АН Груз.ССР, I970, 38, 65-72.

65. П о т а п о в М.К. О структурных характеристиках классов функций с данным порядком наилучшего приближения. Труды Матем.ин-та АН СССР им.В.А.Стеклова, I975, I34, 260-277.

66. Ш и р о к о в Н.А. Аппроксимативные свойства одного континуума. Зап.научн.семин.ЛОМИ АН СССР, I979, 92, 24I-252.

V.P.Havin, B.Jöricke

ON A CLASS OF UNIQUENESS THEOREMS FOR CONVOLUTIONS

This article is an exposition of a seminar talk and gives a survey of results (mostly without proofs) published partly in [1] or prepared for publication (see [2]). Its theme is a phenomenon of quasianaliticity exhibited by many operators commuting with translations. This phenomenon can be roughly described in following terms: if a function and its image (under the operator) vanish both on a sufficiently large set then this function vanishes identically. Or in other words the knowledge of a function and of its image on a "big" set is sufficient to know the function as a whole. Here the words "a sufficiently large set" or "a big set" can be replaced very often by the word "open". What they really mean is the problem we intend to deal with.

1. The Hilbert transform. Let us begin with a well known example (which to be frank is as a rule - unfortunately - our single tool). The symbol $\mathcal{H}(f)$ will denote the Hilbert transform of a function $f \in L^2(\mathbb{R})$:

$$\mathcal{H}(f)(t) = \text{v.p.}\, \frac{i}{\pi} \int_{-\infty}^{+\infty} \frac{f(u)\,du}{t-u} \qquad (t \in \mathbb{R}).$$

Suppose f and $\mathcal{H}(f)$ vanish both on an i n t e r v a l E. Then the Cauchy transform $K(f): K(f)(z) = \frac{1}{2\pi i} \int_{-\infty}^{+\infty} \frac{f(t)\,dt}{t-z}$ $(z \in \mathbb{C} \setminus \mathbb{R})$ vanishes identically by the simplest uniqueness theorem for analytic functions. But $f(x) = \lim_{y \to +0} (K(f)(x+iy) - K(f)(x-iy))$ for almost every real x, whence $f = 0$.

When E is a set of positive Lebesgue measure and $f|E = \mathcal{H}(f)|E = 0$ the same conclusion holds (i.e. $f = 0$) as is seen from the same reasoning. We only have now to use a more subtle uniqueness theorem, namely that there is no nonzero H^2-function vanishing on E.

2. Riesz potentials. Another example is yielded by Riesz potentials $U_\alpha^\mu : U_\alpha^\mu(x) = \int_{-\infty}^{+\infty} \frac{d\mu(t)}{|x-t|^{1-\alpha}}$ $(x \in \mathbb{R})$. We suppose

$\alpha\in(0,1)$, μ is a (signed) measure sufficiently small at infinity (to ensure the convergence of the integral).

It is not hard to see that if E is an interval, $|\mu|(E)=0$ and $\mathcal{U}_\alpha^\mu(x)=0$ for all $x\in E$, then $\mu=0$. In [1] this assertion is given three different proofs. Here we choose the shortest.

PROPOSITION 1. Suppose $\varepsilon>0$, $\int_{-\infty}^{+\infty}\frac{d|\mu|(t)}{(1+|t|)^{1-\alpha}}<+\infty$, $|\mu|([-\varepsilon,\varepsilon])=0$, $\mathcal{U}_\alpha^\mu|(-\varepsilon,\varepsilon)=0$. Then $\mu=0$.

PROOF. Substituting $t=s^{-1}$ into the integral defining $\mathcal{U}_\alpha^\mu$ we obtain

$$0=|\sigma|^{\alpha-1}\mathcal{U}_\alpha^\mu(\sigma^{-1})=\int_{\mathbb{R}}\frac{|s|^{1-\alpha}dv(s)}{|\sigma-s|^{1-\alpha}}\qquad\left(|\sigma|>\frac{1}{\varepsilon}\right),$$

v being a measure concentrated on $[-\varepsilon^{-1},\varepsilon^{-1}]$. Put $\eta=|s|^{1-\alpha}v$. Then

$$0=\mathcal{U}_\alpha^\mu(\sigma)=\int_{-\varepsilon^{-1}}^{\varepsilon^{-1}}\frac{d\eta(s)}{|\sigma|^{1-\alpha}\left(1-\frac{s}{\sigma}\right)^{1-\alpha}}=|\sigma|^{\alpha-1}\sum_{j=0}^{\infty}\sigma^{-j}\binom{\alpha}{j}\int_{-\varepsilon^{-1}}^{\varepsilon^{-1}}s^j\,d\eta(s)$$

for all sufficiently large σ. All moments of η being zero (recall that $\binom{\alpha}{j}\neq 0$, $j=0,1,\dots$) we have $\eta=0$ whence $v=0$ and $\mu=0$ ●

Unlike the preceding example (concerning $\mathcal{H}$) it is not so clear how to replace here the interval $[-\varepsilon,\varepsilon]$ by a set without interior. We are able to do it under some thickness conditions imposed on the set (see section 12 below). We don't know whether the mere nonvanishing of its Lebesgue measure is sufficient for the uniqueness (but we suspect it is n o t).

Let us discuss briefly the values $\alpha\notin(0,1)$. Our proof works without changes for $\alpha\geqslant 1$, $\alpha\neq 1,2,\dots$ (the uniqueness fails for positive integer α's). When α becomes negative $\mathcal{U}_\alpha^\mu$ has no sense but it is still defined on a set E "free from μ" i.e. provided $\int_{-\infty}^{\infty}|\tau-t|^{\alpha-1}d|\mu|(t)<+\infty$ $(\tau\in E)$. The set of all measures μ satisfying the above condition and such that $|\mu|(E)=0$ will be denoted by $\mathcal{M}_\alpha^E$. If E contains an interval, $\alpha<0$, $\mu\in\mathcal{M}_\alpha^E$ and $\mathcal{U}_\alpha^\mu|E=0$, then $\mu=0$ (the same proof as in the proposition). The situation for E's without interior remains unclear - with one exception. If α is a negative odd integer we can associate with $\mathcal{U}_\alpha^\mu$ a function analytic in the upper half-plane, namely $z \to \int_{\mathbb{R}} (z-t)^{\alpha-1} d\mu(t)$. Suppose μ is finite, $\mu \in \mathcal{M}_\alpha^E$ and $\mathcal{U}_\alpha^\mu | E = 0$. Then this analytic function has zero angular limits on E and ([3], p.304) vanishes identically by the uniqueness theorem of Privalov provided $mes\, E > 0$, whence $\mu = 0$ (see [1], p.146 for the details). So we have the following

PROPOSITION 2. Suppose $\alpha = 1-2m$ $(m=1,2,\ldots)$, $E \subset \mathbb{R}$, $mes\, E > 0$, $\mu \in \mathcal{M}_\alpha^E$ a finite measure, $\mathcal{U}_\alpha^\mu | E = 0$. Then $\mu = 0$.

3. <u>Fractional integration</u>. Now we turn to the fractional integrals:

$$I_\alpha^\varphi(\tau) = \frac{1}{\Gamma(\alpha)} \int_{-\infty}^{\tau} \frac{\varphi(t)\, dt}{(\tau-t)^{1-\alpha}} \qquad (\tau \in \mathbb{R}).$$

Here we meet the analogous uniqueness phenomenon but in a slightly modified form. The vanishing of φ and I_α^φ on an interval, $(0,\varepsilon)$ say, does not imply $\varphi = 0$ but it is not hard to see that it does imply $\varphi | (-\infty, 0) = 0$. Indeed, $I_\alpha^\varphi(\tau) = \mathcal{U}_\alpha^{\varphi_1}(\tau)$ $(\tau \in (0,\varepsilon))$ where $\varphi_1 = \varphi | (-\infty, 0)$, and we may use proposition 1. Therefore the fractional integral I_α^φ "remembers the past of φ": knowing $\varphi | (0,\varepsilon)$ and $I_\alpha^\varphi | (0,\varepsilon)$ we authomatically know $\varphi | (-\infty, 0)$. The usual integral I_1^φ behaves differently.

We shall give another proof of the above property of I_α^φ $(0<\alpha<1)$. This proof was communicated to us by B.S.Rubin Suppose φ is continuous, small enough at infinity, $\varphi | (0,\varepsilon) = I_\alpha^\varphi | (0,\varepsilon) = 0$. Put

$$J_\alpha^\varphi(\tau) = \frac{1}{\Gamma(\alpha)} \int_0^\tau \frac{\varphi(t)\, dt}{(\tau-t)^{1-\alpha}} \qquad (\tau > 0).$$

A simple computation shows (see [4], p.378) that

$$I_\alpha^\varphi(\tau) = J_\alpha^\psi(\tau) \qquad (\tau > 0)$$

where $\psi(x) = \varphi(x) + \dfrac{\sin \alpha\pi}{\pi x^\alpha} \displaystyle\int_{-\infty}^{0} \frac{|t|^\alpha \varphi(t)}{x-t}\, dt \qquad (x > 0).$

The function J_α^ψ being zero on $(0,\varepsilon)$ we have $\psi | (0,\varepsilon) = 0$

(because of the Abel inversion formula: $\psi = J_{1-\alpha}(J_\alpha^\psi)$). But $\varphi|(0,\varepsilon)=0$, and therefore $\mathcal{H}(\Phi)|(0,\varepsilon)=0$, where $\Phi(t)=|t|^\alpha\varphi(t)$ $(t<0)$. So our uniqueness problem for I_α^φ is reduced to the analogous problem for the Hilbert transform $\mathcal{H}$ discussed in section 1.

4. Newton potentials. There are multidimensional situations where the same uniqueness phenomenon holds. A very interesting and intriguing one is connected with the Newton potential. Suppose $n\geq 2$ and let μ be a (signed) Borel measure in $\mathbb{R}^n$ concentrated outside an open set $G\subset\mathbb{R}^n$. If the potential $\mathcal{U}^\mu$:

$$\mathcal{U}^\mu(x) \overset{def}{=\!=} \int_{\mathbb{R}^n} \frac{d\mu(t)}{|t-x|^{n-1}}$$

vanishes on G, then $\mu=0$. This fact can be proved by a reasoning analogous to the proof of proposition 1. The same approach is applicable to the multidimensional Riesz potentials (i.e. to $\mathcal{U}^\mu$ with $(n-1)$ replaced by $(n-\alpha)$, $\alpha\neq 2m, m=1,2,\dots$). Another approach is the reduction to the Cauchy problem for the Laplace equation (this approach seems to be appropriate only to $\alpha=1$). Namely, $\mathcal{U}^\mu$ can be naturally defined as a function harmonic in the upper half-space $\mathbb{R}_+^{n+1}=\{(x_1,\dots,x_n,x_{n+1}):x_{n+1}>0\}$, $\mathbb{R}^n$ being its boundary. This function and its normal derivative $\frac{\partial \mathcal{U}^\mu}{\partial x_{n+1}}$ vanish on G. Therefore the odd and the even extensions of $\mathcal{U}^\mu$ from $\mathbb{R}_+^{n+1}$ into $\mathbb{R}_-^{n+1}$ are harmonic in $\mathbb{R}_+^{n+1}\cup\mathbb{R}_-^{n+1}\cup G$, whence $\mathcal{U}^\mu|\mathbb{R}_-^{n+1}=0$. The equality $\mathcal{U}^\mu|\mathbb{R}_+^{n+1}=0$ can be proved analogously. These equalities imply $\mu=0$.

But things are so easy only when G is open. It is not known for example whether $|\mu|(G)=0$, $\mathcal{U}^\mu|G=0$ imply $\mu=0$ in case $mes_n G>0$, mes_n being the n-dimensional Lebesgue measure. (Recall that the integral defining $\mathcal{U}^\mu$ is absolutely convergent almost everywhere on $\mathbb{R}^n$ with respect to mes_n, so that $\mathcal{U}^\mu$ can be restricted to an arbitrary G with $mes_n G>0$). The last remark is not too important because the problem does not become simpler even if we consider measures $\mu=f\cdot mes_n$ only, with f continuous and compactly supported. We are able to prove uniqueness only for G thick enough near a point (see [5] and [1], p.141).

5. Sets of uniqueness. The following definition sums up all preceding examples.

DEFINITION. Let K be a distribution in $\mathbb{R}^n$, X a class of distributions (in $\mathbb{R}^n$). Suppose the convolution $K*f$ has a sense for every $f\in X$. The set $E\subset\mathbb{R}^n$ will be called a (K,X)- set if

$$f\in X,\ (K*f)|E=0,\ \ f|E=0 \Longrightarrow f=0.$$

There are two obscure points in this definition. The first is the expression " $K*f$ has a sense" and the second is the restriction of f and $K*f$ to E . But in every particular situation these points become clear from the context.

Sometimes the distribution K (called also "the kernel" or even identified with the operator $f\longrightarrow K*f$) admits no interesting (K,X) -sets at all. So for example every differential operator (K a linear combination of δ and its derivatives) p r e s e r v e s open zero-sets: $(K*f)|E=0$ whenever $f|E=0$, E an open set. (K,X)-sets are met for essentially non-local operators K , i.e. such that $(K*f)|B$ is influenced by the whole of f and not by $f|B$ only as in the case with differential operators (B being an arbitrary ball). Usually it is not hard to see whether an operator is non-local in this sense. But the description of a l l (K,X) -sets (including those with empty interior) is much more interesting and difficult. Such description has a purely quantitative aspect discussed below in section 11.

6. Semirational symbols. The existence of (K,X)-sets depends on properties of the Fourier transform $\hat{K}$ (the symbol) of K . Suppose for example that K is a tempered distribution in $\mathbb{R}$, $\hat{K}\in L^\infty$, so that the convolution $K*f$ is defined (as an L^2 -function) whenever $f\in L^2$ $(\widehat{K*f}=\hat{K}\hat{f})$. If $\hat{K}$ is rational then no interval $(\neq\mathbb{R})$ is a (K,L^2)-set. Indeed suppose $\hat{K}=PQ^{-1}$, P and Q being polynomials. Take an arbitrary interval $I\neq\mathbb{R}$ and a compactly supported non-zero function $\varphi\in C^\infty$ vanishing on I ; let p,q be the differential operators with the respective symbols P,Q . Then

$$(K*q(\varphi))|I=p(\varphi)|I=0,\qquad q(\varphi)|I=0,\quad \varphi\neq 0.$$

But if $\hat{K}$ is composed by two d i f f e r e n t rational pieces (i.e. its restrictions onto rays $(-\infty,a)$, $(a,+\infty)$ are two d i f f e r e n t rational functions) the situation changes: such an operator has many (K,L^2) -sets.

DEFINITION. Let k be a Lebesque mesurable function on $\mathbb{R}$,

$b, c \in \mathbb{R}$, $b \leq c$. Suppose

1. $k|(c,+\infty) = r|(c,+\infty)$, r being a rational function;
2. $mes\{\xi \in (-\infty, b) : k(\xi) = r(\xi)\} = 0.$

Then we shall call k a s e m i r a t i o n a l function; r will be called its rational part.

The most important example is $k(\xi) = sgn\,\xi$ (the symbol of $\mathcal{H}$). Another example is $k(\xi) = |\xi|^{-1}$ (this semirational function can be interpreted as the symbol of the logarithmic potential: $f \longmapsto \int_{-\infty}^{\infty} f(t) \log|x-t|\,dt$).

Put $W_2^n = \{\varphi \in L^2 : \xi^n \hat{\varphi} \in L^2\}$ $(n=1,2,\ldots;\ \xi$ denotes the independent variable). The set W_2^n can be described as the class of all L^2-functions, $(n-1)$-times continuously differentiable with $\varphi^{(n-1)}$ absolutely continuous and $\varphi^{(n)} \in L^2$.

Every semirational function k determines an operator K , defined on the set $\mathcal{D}_K \overset{def}{=} \{\varphi \in L^2 : k\hat{\varphi} \in L^2\}$ and mapping this set into L^2 by the formula $\widehat{K(\varphi)} = k\hat{\varphi}$ (we shall write $K(\varphi) = K*\varphi$.).

THEOREM 1. Let k be a semirational function with the rational part pq^{-1} , p , q being polynomials without common zeros, $\deg p = m$, $|q(\xi)k(\xi)| \leq C|\xi|^m$ $(\xi \in \mathbb{R})$. Then every $E \subset \mathbb{R}$ with $mes\,E > 0$ is a $(K, W_2^m \cap \mathcal{D}_K)$ - set:

$$\varphi \in W_2^m \cap \mathcal{D}_K, \quad \varphi|E = (K*\varphi)|E = 0 \Longrightarrow \varphi = 0.$$

PROOF. Put $\psi = K*\varphi$ $(\varphi \in W_2^m \cap \mathcal{D}_K)$. Then $\psi \in L^2$, and

$$|\hat{\psi}(\xi)q(\xi)| = |q(\xi)||k(\xi)||\hat{\varphi}(\xi)| \leq C|\xi|^m|\hat{\varphi}(\xi)|,$$

whence $\psi \in W_2^n$. ($n \overset{def}{=} \deg q$) . Therefore $\hat{\psi}q = \widehat{Q(\psi)}$, Q being the differential operator with the symbol q . The function $Q(\psi)$ can be computed classicaly, all needed derivatives of ψ being well-defined. Suppose $(K*\varphi)|E = 0$ (i.e. $\psi|E = 0$). Then $Q(\psi)(t) = 0$ for almost every $t \in E$. The same reasoning shows that $P(\varphi)(t) = 0$ for almost every $t \in E$ (P being the differential operator with the symbol p). The function $h \overset{def}{=} Q(\psi) - P(\varphi)$ belongs to L^2 and $\hat{h}(\xi) = 0$ $(\xi > c)$ (the number c is taken from the definition of semirational functions). Therefore the function $t \longmapsto h(t)\exp(-ict)$ belongs to the Hardy class H^2_- , and the vanishing of h on E implies $h=0$. Thus $(qk-p)\hat{\varphi}=(=\hat{h})=0$. But $q(\xi)k(\xi)-p(\xi)=0$ almost nowhere on the ray $(-\infty,b)$ (see the definition of semirational functions), and $\hat{\varphi}|(-\infty,b)=0$. This means that the function $t\mapsto\varphi(t)\exp(-ibt)$ belongs to H^2_+, and now $\varphi|E=0$ implies $\varphi=0$ ●

EXAMPLES. 1. $k(\xi)=|\xi|^{-1}$. Put $\mathcal{U}^f(x)=\int_{-\infty}^{+\infty}f(u)\log|x-u|\,du$.

It is easy to see that in this case our theorem implies the following assertion:

$$\text{if}\quad f\in L^2,\ \int_{-\infty}^{+\infty}|f(t)||t|\,dt<+\infty,\ \int_{-\infty}^{+\infty}f(t)\,dt=0,\ E\subset\mathbb{R},$$

$\operatorname{mes}E>0$, then

$$f|E=\mathcal{U}^f|E=0\Longrightarrow f=0.$$

We shall return to the logarithmic potential later (see section 9 below).

2. Put $(If)(t)=(\text{v.p.})\int_{-\infty}^{\infty}f(u)\dfrac{\cos(t-u)}{t-u}\,du\quad(t\in\mathbb{R})$.

It is easy to see that I maps L^2 into itself, the symbol of I being semirational: $\hat{I}(\xi)=0\ (|\xi|<1)$, $\hat{I}(\xi)==\text{const}\cdot\operatorname{sgn}\xi\ (|\xi|>1)$. In this case $m=0$, and every set of positive Lebesque measure is a (I,L^2) -set.

3. Suppose the symbol k has the form $\operatorname{sgn}\xi\,\frac{p}{q}$, p and q being polynomials, and $k\in L^\infty$, so that $\mathcal{D}_K=L^2$. If $E\subset\mathbb{R}$, $\operatorname{mes}E>0$, $\deg p>0$, Theorem 1 does not permit to deduce the vanishing of $\varphi\in L^2$ from conditions $\varphi|E=0$, $(K*\varphi)|E=0$ only: we need some regularity condition to impose on φ $(\varphi\in W_2^m,\ m=\deg p)$. We suspect these regularity conditions cannot be dispensed with. Take for instance the symbol $\frac{\xi-i}{\xi-2i}\operatorname{sgn}\xi$. The corresponding operator represents a perturbation of the Hilbert transform (as does the operator in the preceding example). Theorem 1 guarantees every set E with $\operatorname{mes}E>0$ is a (K,W_2^1) -set. Is it a (K,L^2) -set? We don't know but the following theorem shows the answer is y e s provided E has a supplementary property.

7. <u>Semirational symbols and Carleson sets</u>. Let e be a compact set of real numbers, $\mathcal{L}(e)$ the set of all complementary intervals of e. Put

$$C(e) \overset{def}{=\!=} \sum_{\ell \in \mathcal{L}(e)} \lambda(|\ell|), \qquad \lambda(t) \overset{def}{=\!=} t \log^+ \frac{1}{t} \quad (t>0)$$

(in the last sum only ℓ's with $|\ell| < +\infty$ are present). The set e will be called a C a r l e s o n s e t if $C(e) < +\infty$.

THEOREM 2. Let k be a semirational function, K the corresponding operator. Then every Carleson set E with $mes\, E > 0$ is a $(K, \mathcal{D}_K)$ -set.

The proof of this theorem is much more complicated than the proof of Theorem 1. We shall only sketch it (for the complete proof see [1], p.p.156-167).

Suppose $\varphi \in L^2$, $\hat{\varphi} k \in L^2$, $\varphi | E = (K * \varphi) | E = 0$, $C(E) < +\infty$, $mes\, E > 0$. We have to prove $\varphi = 0$. Assume $c = 0$ (see the definition of semirationality), so that

$$\gamma(\xi) \overset{def}{=\!=} q(\xi)\hat{\psi}(\xi) - p(\xi)\hat{\varphi}(\xi) = 0 \quad (\xi > 0) \tag{1}$$

(here $\psi \overset{def}{=\!=} K * \varphi$, p, q are polynomials, pq^{-1} is the rational part of k : $k|(0,+\infty) = pq^{-1}|(0,+\infty)$.

We cannot assert now that $\gamma \in L^2$, but $\int_{-\infty}^{0} |\gamma(\xi)|^2 (1+|\xi|^2)^{-N} d\xi < +\infty$ for a positive N. This allows us to consider the Laplace transform

$$\Phi(\zeta) \overset{def}{=\!=} \int_{-\infty}^{0} \gamma(\xi) e^{i\xi\zeta} d\xi \qquad (\mathrm{Im}\, \zeta < 0).$$

It is sufficient to prove $\Phi = 0$, because then $\gamma = 0$ and the proof will be finished exactly as in Theorem 1.

To prove the vanishing of Φ we construct a sequence $\{\Phi_j\}$ of H^2-functions satisfying

(a) $\lim_{j\to\infty} \Phi_j(\zeta) = \Phi(\zeta) \quad (\mathrm{Im}\, \zeta < 0)$;

(b) there is a positive C such that $|\Phi_j(\zeta)| \le C |\mathrm{Im}\, \zeta|^{-C}$ $(-1 < \mathrm{Im}\, \zeta < 0,\ j = 1, 2, \dots)$;

(c) $\lim_{j\to\infty} \int_{\tilde{E}} |\Phi_j|^2 = 0$, $\tilde{E}$ being a Carleson subset of E with $mes\, \tilde{E} > 0$.

Conditions (a), (b), (c) imply $\Phi = 0$ by a theorem of S.V. Hruščev ([6], p.p.143-146). (By the way, S.V.Hruščev has proved that the existence of a part $\tilde{\tilde{E}}$ of $\tilde{E}$ satisfying the Carleson condition and with $mes\, \tilde{\tilde{E}} > 0$ is necessary for (a), (b), (c) to imply $\Phi = 0$).

To describe further steps of the proof let us suppose $\deg p = \deg q = 1$ (otherwise some details must be changed), so that our symbol $\frac{\xi - i}{\xi - 2i}\,\mathrm{sgn}\,\xi$ can be still considered. Replace the letter ξ (i.e. the symbol of differentiation) in $p(\xi)$ and $q(\xi)$ (see (1)) by $d_h(\xi) = ih^{-1}(1-e^{i\xi h})$, h being a small positive number ($d_h(\xi)$ is the symbol of the difference quotient with the step h). We obtain the function $\tilde{\gamma} = q(d_h)\hat{\psi} - p(d_h)\hat{\varphi}$ which is near (in a sense) to γ , which is the Fourier transform of a function vanishing on $E \cap (E-h)$ but needs not vanish on $(0,+\infty)$. So we pass to the function $\tilde{\gamma}\cdot\chi_{(-\infty,0)}$ ($\chi_{(-\infty,0)}$ is the characteristic function of the interval $(-\infty, 0)$) which is capable to serve as a Φ_j when h is small enough.

8. Another approach to the problem discussed in the preceding section is also based upon the cited Hruščev theorem. It is easy to deduce from this theorem the following

COROLLARY. Let $E \subset \mathbb{R}$ be a Carleson set with $\operatorname{mes} E > 0$ and let φ be a function belonging to H^2_+ and suppose that the derivative of $\varphi | E$ (computed along E) is zero almost everywhere on E , Then φ is constant.

P r o o f see in [1], p.167.

Let us consider now the operator K :

$$(Kf)(t) = (\mathcal{H}f)(t) + ce^{-\varepsilon t}\int_{-\infty}^{t} e^{\varepsilon\tau} f(\tau)\,d\tau \quad (f \in L^2, t \in \mathbb{R}).$$

Here ε is a positive number. The symbol of K is semirational (namely $\mathrm{sgn}\,\xi + \frac{\mathrm{const}}{\xi - i\varepsilon}$) and in accordance with Theorem 2 every carlesonian $E \subset \mathbb{R}$ with $\operatorname{mes} E > 0$ is a (K, L^2)-set. We are going to give a direct proof of this fact. Suppose $f \in L^2$, $f | E = K(f) | E = 0$. Introduce the Cauchy transform of f :

$$F(z) = \int_{\mathbb{R}} \frac{f(t)\,dt}{t - z} \qquad (\operatorname{Im} z > 0).$$

Clearly $F \in H^2_+$, $F | E = \mathrm{const}\,\mathcal{H}(f) | E$. Now the function Φ : $\Phi(z) \overset{def}{=} \exp(\varepsilon z) F(z)$ ($\operatorname{Im} z > 0$) coincides a.e. on E with the function $t \mapsto \mathrm{const} \int_{-\infty}^{t} e^{\varepsilon\tau} f(\tau)\,d\tau$ whose derivative vanishes almost everywhere on E . The corollary above implies the constancy of Φ. Thus $F(z) = const\, exp(-\varepsilon z)\ (\mathrm{Im}\, z>0)$ which is possible only when $const = 0$ (because $F \in H^2_+$). The proof is finished.

We conjecture the Carleson condition of Theorem 2 cannot be dropped.

9. Logarithmic potentials of compactly supported measures. Till now we have worked with convolutions defined on L^2 (or on some parts of L^2). This was not a mere matter of convenience; the widening of X is not always very easy when we are interested in the (K,X) -property of a set. We shall illustrate this by the example of the logarithmic potential.

Let μ be a complex Borel measure on $\mathbb{R}$ satisfying

$$\int_{|t|>1} \log|t|\, d|\mu|(t) < +\infty \qquad (2)$$

and $\mathcal{U}^\mu$ its potential:

$$\mathcal{U}^\mu(z) = \int_{\mathbb{R}} \log|z-t|\, d\mu(t) .$$

This integral is absolutely convergent for almost all real z. The set of all measures satisfying (2) will be denoted by $\mathcal{M}_{log}$.

THEOREM 3. Every set of reals with positive Lebesgue measure is a $(\log|x|, \mathcal{M}_{log})$-set. This means that

$$E \subset \mathbb{R},\ |\mu|(E) = 0,\ \mathcal{U}^\mu | E = 0,\ \mathrm{mes}\, E > 0 \Longrightarrow \mu = 0 .$$

The proof of this theorem goes along the same lines as the proof of Theorem 1 (see also the example 1 in section 6), but is much more delicate. The function $\mathcal{U}^\mu$ is in general nowhere differentiable and therefore we cannot reduce our uniqueness problem to the uniqueness of the Hilbert transform by means of the usual differentiation as was the case with the logarithmic potentials of L^2-functions. But fortunately we can use the so called approximate differentiation. Recall its definition.

Suppose a function f is defined on a set $E \subset \mathbb{R}$, $x_0 \in E$. A number L is called the approximate derivative of f at x_0 if there is a set $\mathcal{E} \subset E$ such that x_0 is a density point of $\mathcal{E}$ and

$$\lim_{x\in\mathcal{E},\, x\to x_0} \frac{f(x)-f(x_0)}{x-x_0} = L .$$

In this case we shall write $L = (\partial_{ap} f)(x_0)$.

Now we need the Hilbert transform of a finite measure μ , namely the function $\mathcal{H}(\mu)$ defined a.e. in $\mathbb{R}$ (with respact to the Lebesque measure) by the equality

$$\mathcal{H}(\mu)(t) = (\text{v.p.})\frac{1}{\pi}\int_{-\infty}^{\infty}\frac{d\mu(u)}{t-u} .$$

Our Theorem 3 is a corollary of the following formula valid for every $\mu\in\mathcal{M}_{log}$:

$$\partial_{ap} U^{\mu} = \pi\mathcal{H}(\mu) \qquad \text{a.e. in } \mathbb{R} \qquad (3)$$

This is a slight generalization of a theorem of Titchmarsh [7] concerning the approximate differentiation of conjugates of absolutely continuous functions. Its proof is rather long. We omit it (it will be published in [2]) and restrict ourselves by the deduction of Theorem 3 from (3).

If $U^{\mu}|E = 0$ then $(\partial_{ap} U^{\mu})(t) = 0$ for almost every $t\in E$. Thus $\mathcal{H}(\mu)(t) = 0$ for almost every $t\in E$. The Cauchy potential K^{μ} :

$$K^{\mu}(z) \stackrel{def}{=\!=} \frac{1}{\pi i}\int_{\mathbb{R}}\frac{d\mu(t)}{t-z} \qquad (z\in\mathbb{C}\setminus\mathbb{R})$$

belongs to H^p for every value of $p\in(0,1)$ in both upper and lower half-planes. Now

$$\lim_{y\to+0} K^{\mu}(x\pm iy) = \pm\frac{d\mu}{dm}(x) + i\mathcal{H}(\mu)(x)$$

for almost every real x (m denotes the Lebesgue measure). The equality $|\mu|(E)=0$ implies $\frac{d\mu}{dm}(x) = 0$ a.e. The uniqueness properties of H^p-functions imply $K^{\mu}(z)=0$ $(z\in\mathbb{C}\setminus\mathbb{R})$ and we have only to note that the Poisson transform of μ computed at a point $\zeta\ (\operatorname{Im}\zeta>0)$ is equal to $\frac{1}{2}(K^{\mu}(\zeta)-K^{\mu}(\bar{\zeta}))$.

10. Semirational symbols: free interpolation. Let X, Y be some spaces of distributions in $\mathbb{R}$, $X\subset Y$. Suppose a linear operator K maps X into Y. We shall call a set $E\subset\mathbb{R}$ interpolating (with respect to the triple (K,X,Y)) if for every pair φ,ψ of elements of Y there is $f\in X$ satisfying

$$f|E=\varphi|E,\quad K(f)|E=\psi|E \tag{4}$$

(the meaning of the restriction onto E will be made clear in every concrete situation). The solvability of equations (4) (with "the unknown" f) for all pairs $(\varphi,\psi)\in Y\times Y$ is analogous to the well known phenomenon of the free interpolation of analytic functions (see e.g. [8]).

The main object of this article is the homogeneous system (4) (i.e. with $\varphi=\psi=0$). It is natural to consider the corresponding non-homogeneous system as well. Interpolating sets are (in a sense) opposite to (K,X)-sets. So for example, no set $E\subset\mathbb{R}$ with $\operatorname{mes}E>0$ is interpolating with respect to $(\mathcal{H},L^2,L^2)$. Indeed let $E_1\subset E$, $\operatorname{mes}E_1>0$, $\operatorname{mes}(E\setminus E_1)>0$, $\psi\in L^2$, $\psi|E_1=0$, $\psi|(E\setminus E_1)\neq 0$; then no $f\in L^2$ can satisfy (4) with $\varphi=0$ and $K=\mathcal{H}$ because E_1 is a $(\mathcal{H},L^2)$-set.

In this section we shall be interested in interpolating sets for some perturbations of the Hilbert transform, namely for convolutions with semirational symbols $\hat{K}$ of the form

$$\hat{K}(\xi)=\operatorname{sgn}\xi\cdot R(\xi)\qquad(\xi\in\mathbb{R}), \tag{5}$$

R being a rational function bounded on $\mathbb{R}$ and bounded from zero on $\mathbb{R}$. We have met such symbols already. Recall that the question whether all sets with positive Lebesgue measure are (K,L^2)-sets remains open. Nevertheless our results concerning interpolating sets will be more satisfactory. Roughly speaking the possibility (or the impossibility) of the free interpolation is very weakly influenced by the perturbing factor R and the interpolatory properties of operators with symbols (5) coincide essentially with those of $\mathcal{H}$.

We intend to restrict our functions and their images under

K onto sets of zero Lebesgue measure. So we have to reduce our class of functions to be in a position to ascribe a value to f and $K(f)$ at every real point (not merely at a l m o s t every point).

Let C_r^0 denote the real Banach space of all real functions continuous in $\mathbb{R}$ and vanishing at infinity.

DEFINITION. We shall say that a function $f \in C_r^0$ belongs to the class $\mathcal{A}$ if the integral

$$\mathcal{H}(f)(x) \overset{def}{=\!=} \frac{1}{\pi} \lim_{\substack{\varepsilon \to +0 \\ \Delta \to +\infty}} \int_{\substack{\varepsilon < |t-x| \\ |t| > \Delta}} \frac{f(t)}{x-t}\, dt$$

exists for every $x \in \mathbb{R}$ and $\mathcal{H}(f) \in C_r^0$

Suppose a rational function R satisfyies

$$0 < \inf\{|R(\xi)| : \xi \in \mathbb{R}\}, \quad \sup\{|R(\xi)| : \xi \in \mathbb{R}\} < +\infty .$$

It gives rise to the operator A_R mapping the space S' of tempered distributions into itself by the formula

$$\widehat{(A_R f)} = R\hat{f} \qquad (f \in S').$$

It is not hard to see that $A_R | C^0$ is a one-to-one mapping of the space C^0 (of all functions continuous in $\mathbb{R}$ and vanishing at infinity) onto itself.

Suppose $R(\xi) = \overline{R(-\xi)}$ $(\xi \in \mathbb{R})$ and consider the operator

$$K_R : \mathcal{A} \longrightarrow C_r^0$$

defined by the equality $K_R(f) = A_R(\mathcal{H}(f))$ $(f \in \mathcal{A})$. Without loss of generality we may assume that

$$R(\xi) = 1 + \sum_{\gamma \in \Gamma} \frac{a_\gamma}{(\xi - \lambda_\gamma)^{m_\gamma}} \qquad (\xi \in \mathbb{R})$$

$\{a_j\}$, $\{m_j\}$, $\{\lambda_\gamma\}$ being finite families of complex, positive integer and complex non-real numbers. Then

$$A_R = I + \mathcal{J}, \tag{6}$$

where I is the identity mapping of C_r^0, $\mathcal{J}$ a convolution with a real summable and bounded function. Therefore $K_R(f) = \mathcal{H}(f) + \mathcal{J}(\mathcal{H}(f))$ $(f \in \mathcal{a})$. Noting that $(A_R)^{-1} = A_{R^{-1}}$ we have also

$$B \overset{def}{=\!=} (A_R^{-1})^* = I + \mathcal{J}^*, \tag{7}$$

I being the identity mapping of the space $\mathcal{M}_r$ of all real finite Borel measures (on $\mathbb{R}$) and $\mathcal{J}^*$ a convolution of the variable measure with a summable function.

DEFINITION. Suppose A is an invertible operator of C_r^0 onto itself satisfying (7) (with A instead of A_R). Then the operator $K : K(f) \overset{def}{=\!=} A(\mathcal{H}(f))$ $(f \in \mathcal{a})$ is called an a l m o s t H i l b e r t t r a n s f o r m.

THEOREM 4. Let K be an almost Hilbert transform, E a compact subset of $\mathbb{R}$. The following assertions are equivalent:
1. E is interpolating with respect to $(K, \mathcal{a}, C_r^0)$;
2. $mes\, E = 0$.

This theorem represents a generalization of the Rudin-Carleson interpolation theorem. The implication $2 \Longrightarrow 1$ can be proved by the usual argument (involving the F. and M.Riesz theorem on measures orthogonal to the disc-algebra). The inverse implication requires more efforts. In the classical situation (of the pure Hilbert transform, i.e. when $\mathcal{J} = 0$) one does not even mention it because of its triviality d u e h o w e v e r t o t h e u n i q u e n e s s p r o p e r t y o f a l l s e t s w i t h p o s i t i v e L e b e s g u e m e a s u r e (see the beginning of the section). But if we cannot use this property when $\mathcal{J} \neq 0$ (and we suspect it does not hold at all).

Using a standard duality argument we conclude that the assertion 1 is equivalent to the existence of a positive number γ satisfying

$$\inf\{var(\mu_1 - hm) + var(\mu_2 + B\mathcal{H}(h)m) : h \in H^1\} \geqslant \gamma(var\mu_1 + var\mu_2) \tag{7}$$

for every pair of real measures μ_1, μ_2 supported by E. Here

H^1 denotes the class of all real functions f summable on $\mathbb{R}$ with the summable Hilbert transform $\mathcal{H}(f)$, m denotes the Lebesgue measure, B is defined by (7) (with $A_{\mathbb{R}} = A$).

To prove $1 \Longrightarrow 2$ suppose $m(E)>0$, $h \in H^1$, $h|(\mathbb{R} \setminus E) = 0$ and put $\tilde{h} = \mathcal{H}(h)$, $\mu_1 = h\chi_E m$, $\mu_2 = -B(\tilde{h})\chi_E m$ (χ_E denotes the characteristic function of the set E). Then

$$var(\mu_1 - hm) + var(\mu_2 + B(\tilde{h})m) = \int_{CE} |B\tilde{h}|$$

and (7) implies

$$\int_{CE} |B\tilde{h}| \geq \gamma \left(\int_E |h| + \int_E |B\tilde{h}| \right) \geq \gamma \int_E |h|$$

for every $h \in H^1$ vanishing off E. We shall be done if we construct a family $\{h_\varepsilon\}$ $(\varepsilon > 0)$ of functions of the class H^1 vanishing off E and satisfying

$$(a) \lim_{\varepsilon \to +0} \frac{1}{\varepsilon} \int_E |h_\varepsilon| > 0 \qquad (b) \lim_{\varepsilon \to +0} \frac{1}{\varepsilon} \int_{CE} |B\tilde{h}_\varepsilon| = 0 .$$

To do this we remark that without loss of generality we may assume the origin to be a density point of E and E to be symmetric with respect to the origin (if not we shall consider $E \cap (-E)$ instead of E). Now put $E_\varepsilon = E \cap [-\varepsilon, \varepsilon]$, $E'_\varepsilon = [-\varepsilon, \varepsilon] \setminus E_\varepsilon$ and $h_\varepsilon(t) = \chi_{E_\varepsilon}(t)\, sgn(t)$ $(t \in \mathbb{R})$. The estimate (a) is true because 0 is a density point of E. The proof of (b) is somewhat more complicated and we omit it (it involves some standard estimates of singular integrals). The inclusion $h_\varepsilon \in H^1$ is almost obvious: $\mathcal{H}(h_\varepsilon)(x) = O(x^{-2})$ for great x because of the symmetry of E and $\mathcal{H}(h_\varepsilon) \in L^2_{loc}$. Now we are going to state an L^2-version of the second part of Theorem 4.

Suppose j is a function of the class $L^1(\mathbb{R}) \cap L^2(\mathbb{R})$, $c \in \mathbb{C}$. Define the operator K by the equality

$$\widehat{K(f)}(\xi) = sgn\, \xi (c + \hat{j}(\xi)) \hat{f}(\xi) \qquad (\xi \in \mathbb{R},\ f \in L^2). \qquad (8)$$

As an example we may take the operator (5) where the rational function R is bounded on $\mathbb{R}$ (but not necessarily bounded from zero). This enables us to compare the result stated below with theorems 1 and 2.

THEOREM 5. No set $E \subset \mathbb{R}$ with $mes\, E > 0$ can be interpolating with respect to the triple (K, L^2, L^2) (K is the operator (8)).

We shall only sketch the proof. As above we may assume E to be symmetric with respect to the origin which is a density point of E. Denote by $L^2(E)$ the set of all L^2-functions vanishing off E and put $R_E(F) = (F\chi_E, K(F)\chi_E)$ $(F \in L^2)$ Every pair $\Pi \overset{def}{=} (\varphi, \psi) \in L^2(E) \times L^2(E)$ gives rise to a family of numbers $\{(\Pi_\varepsilon)\}$ $(\varepsilon > 0)$:

$$(\Pi)_\varepsilon \overset{def}{=} -\int \varphi \cdot K^*(h_\varepsilon) + \int \psi \cdot h_\varepsilon ,$$

where $h_\varepsilon(x) = \chi_{E \cap [-\varepsilon, \varepsilon]}(x)\, sgn\, x$ $(x \in \mathbb{R}, \varepsilon > 0)$.

The proof consists of two parts. In the first we show that whenever $\Pi \in R_E(L^2)$

$$(\Pi)_\varepsilon = O(\alpha_E(\varepsilon)\sqrt{\varepsilon}), \quad \alpha_E(\varepsilon) = o(1) \quad (\varepsilon \to 0),$$

α_E depends on E only.

In the second we note that for every function $\beta : (0,1) \to (0, +\infty)$, tending to zero at the origin there is a $\psi \in L^2(E)$ satisfying

$$\overline{\lim_{\varepsilon \to 0}} \frac{(\Pi)_\varepsilon}{\beta(\varepsilon)\sqrt{\varepsilon}} = +\infty \tag{9}$$

where $\Pi = (0, \psi)$, so that this pair does not belong to $R_E(L^2)$ and cannot be interpolated on E in the sense of (4). To construct such a ψ we take a decreasing sequence of positive numbers $\{\varepsilon_k\}$ so that $\beta(\varepsilon_k) < k^{-3}$ $(k = 1, 2, \ldots)$ and put

$$\psi(x) = \sum_{k=1}^{\infty} \frac{\chi_{E \cap [-\varepsilon_k, \varepsilon_k]}(x)\, sgn\, x}{k^2 \sqrt{\varepsilon_k}} \quad (x \in \mathbb{R})$$

Then $$(\Pi)_{\varepsilon_k} = \int_E \psi h_{\varepsilon_k} \geqslant \int_E \frac{\chi_{E \cap [-\varepsilon_k, \varepsilon_k]} \cdot |h_{\varepsilon_k}|}{k^2 \sqrt{\varepsilon_k}} >$$

$$> \frac{mes(E \cap [-\varepsilon_k, \varepsilon_k])\, k \beta(\varepsilon_k)}{\sqrt{\varepsilon_k}} ,$$

and (9) follows.

11. (K,X) -property and Zweikonstantensatz. Suppose $E\subset\mathbb{R}$ is a (K,L^2)-set, $f\in L^2$, $f|E=0$ and $\|K(f)\|_{L^2(E)}$ is small. Then f as a whole must be small (the complete vanishing of $\|K(f)\|_{L^2(E)}$ implies $f=0$, so it is natural to look for a kind of the stability connected with the (K,X) -property). Simple examples show however that $\|f\|_{L^2}$ is not necessarily small when $f|E=0$ and $\|K(f)\|_{L^2(E)}$ is small. It is possible to guarantee the smallness of f in a weak sense.

DEFINITION. Let E be a Lebesgue measurable subset of $\mathbb{R}$, K a linear operator mapping L^2 into L^2, φ a linear functional defined on L^2. The function

$$R^{\varphi}_{K,E}:(0,+\infty)\longrightarrow(0,+\infty)$$

defined by the equality

$$R^{\varphi}_{K,E}(\varepsilon)=\sup\left\{|\varphi(f)|:f\in L^2(CE),\ \|f\|_{L^2}\leq 1,\ \|K(f)\|_{L^2(E)}\leq\varepsilon\right\}\ (\varepsilon>0)$$

will be called the φ-rigidity of K with respect to E. Recall that $L^2(e)$ stands always for $\{f\in L^2:f|(\mathbb{R}\setminus e)=0\}$.

We shall show that

$$\lim_{\varepsilon\to 0} R^{\varphi}_{K,E}(\varepsilon)=0 \tag{10}$$

whenever E is a (K,L^2)-set. So when $\|K(f)\|_{L^2(E)}$ is being pressed down to zero and f vanishes on E the quantity $|\varphi(f)|$ does not resist, does not exhibit any rigidity and goes to zero too.

We think the notion of the rigidity is very important for the investigation of uniqueness properties discussed here. If we can find good estimates of $R^{\varphi}_{K,E}$ be it only for a total set of functionals φ we have automatically the (K,L^2)-property of the set E. Moreover the following remark is obvious.

Suppose $\{\hat{K}_j\}$ is a sequence of L^∞-functions, $\hat{K}\in L^\infty$, satisfying

1) $\sup_j \|\hat{K}_j\|_\infty < +\infty$, $\lim_{j\to\infty} \hat{K}_j(t) = \hat{K}(t)$

a.e. in $\mathbb{R}$;

2) E is a (K_j, L^2)-set $(j = 1, 2, \dots)$;

3) $\lim_{\varepsilon\to 0} R^{\varphi}_{K_j E}(\varepsilon) = 0$ uniformly in j

for every $\varphi \in (L^2)^*$ (or at least for every $\varphi \in Y$, Y being a total set of functionals).

Then E is a (K, L^2)-set.

So if we only could find good estimates of rigidities of operators with semirational symbols (i.e. estimates taking into account-in an explicit form - geometric characteristics of the set E and parameters defining the symbol) we should be able to extend the uniqueness theorems of sections 6 and 7 onto much wider classes of symbols. Indeed the class of semirational symbols (or even of symbols (5)) is very rich and it is very often possible to construct sequences of such symbols $\{\hat{K}_j\}$ approximating a given $\hat{K}$ in the sense 1); the property 2) can be assured by Theorem 1 (or 2). The real difficulty arises when we try to obtain 3).

Our proof of (10) (provided E is a (K, L^2)-set) will be based upon considerations too abstract to serve a source of quantitative results needed in 3) above.

Consider two Banach spaces X and Y and a continuous linear operator A mapping X into Y . Put

$$B_\varepsilon \overset{def}{=} \{ x \in X : \|x\| \leq 1,\ \|A(x)\| \leq \varepsilon \} \qquad (\varepsilon > 0).$$

Let φ be a linear functional in X . Put $C_\varphi(\varepsilon) \overset{def}{=} \sup |\varphi(B_\varepsilon)|$

LEMMA. Suppose X is reflexive. The operator A is one-to-one iff

$$\lim_{\varepsilon\to 0} C_\varphi(\varepsilon) = 0$$

for every $\varphi \in X^*$.

The proof is easy (we have to use the weak compactness of the closed unit ball of X).

COROLLARY (the Zweikonstantensatz for (K, L^2)-sets). Let E be a Lebesgue measurable set of $\mathbb{R}$, $K \in S'(\mathbb{R})$, $\hat{K} \in L^\infty$. If E is a (K, L^2)-set and $\varphi \in (L^2)^*$ then

$$m, M > 0,\ x \in L^2(CE),\ \int_{-\infty}^{+\infty} |x|^2 \leq M^2;\ \int_E |K * x|^2 \leq m^2 \Longrightarrow$$
$$|\varphi(f)| \leq M R^{\varphi}_{K,E}(m),$$

where $\lim_{m \to 0} R^{\varphi}_{K,E}(m) = 0$.

The proof follows from the definition of the rigidity and from Lemma ($X = L^2(CE)$, $Y = L^2(E)$, $A(x) = \chi_E \cdot (K * x)$ $(x \in L^2(CE))$).

We conclude this section by two examples. It will be convenient here to replace the line $\mathbb{R}$ by the unit circle $\mathbb{T}$ (the definition of the (K, X)-property can be rephrased in a general group-theoretic context but we didn't do it because for the simplest groups our knowledge is too scanty). The open unit disc will be denoted by $\mathbb{D}$, the normalized Lebesgue measure on $\mathbb{T}$ by m; the same letter will denote the linear functional in $L^2(m)$: $m(x) = \int_{\mathbb{T}} x\, dm$ $(x \in L^2(m))$.

EXAMPLE 1. Put

$$h(x)(\zeta) = \frac{1}{2\pi i} \text{ v.p.} \int_{|\tau|=1} \frac{x(\tau)\, d\tau}{\tau - \zeta} \quad (x \in L^2(m),\ \zeta \in \mathbb{T}).$$

The following theorem yields a rough estimate of the m-rigidity of h with respect to a set $E \subset \mathbb{T}$.

THEOREM 6. Suppose $E \subset \mathbb{T}$, $m(E) > 0$. Then

$$\lim_{\varepsilon \to 0} \frac{\log R^m_{h,E}(\varepsilon)}{\log \varepsilon} = m(E).$$

This equality follows from the estimates

$$\left(1 + \varepsilon^2 \frac{m(CE)}{m(E)}\right)^{-1} (m(E))^{-\frac{m(E)}{2}} (m(CE))^{-\frac{m(CE)}{2}+1} \varepsilon^{m(E)} \leq$$

$$\leq R^m_{h,E}(\varepsilon) \leq (m(E))^{-\frac{m(E)}{2}} (m(CE))^{-\frac{m(CE)}{2}} \varepsilon^{m(E)}.$$

The right estimate is a simple consequence of the Jensen inequality, the left requires some work with outer functions and with the Ahlfors function of the set E .

EXAMPLE 2. Let ℓ denote the operator transforming every $x \in L^2(m)$ into $\ell(x)$:

$$\ell(x)(\zeta) \stackrel{def}{=} \int_{\mathbb{T}} x(\tau) \log|\tau-\zeta|\, dm(\tau) \qquad (\zeta \in \mathbb{T}).$$

So here we are going to deal with the m-rigidity of the logarithmic potential. The results of section 6 (modified from the $\mathbb{R}$-situation to the $\mathbb{T}$-situation) together with the abstract considerations exposed at the beginning of the section (see (10)) show that

$$E \subset \mathbb{T},\ m(E) > 0 \Longrightarrow \lim_{\varepsilon \to 0} R^{m}_{\ell,E}(\varepsilon) = 0.$$

It is however unclear which properties of E really influence the rate of decrease of $R^{m}_{\ell,E}(\varepsilon)$ when ε tends to zero.

Let us begin with some philosophy concerning the possible character of $R^{\varphi}_{K,E}$ provided every set $E \subset \mathbb{T}$ with $m(E) > 0$ is a (K, L^2)-set (here $L^2 = L^2(m)$, K a linear operator mapping L^2 into itself and commuting with rotations, $\varphi \in (L^2)^*$). It is very natural to conjecture in this situation that (given K and φ) the smaller is $m(E)$ the faster $R^{\varphi}_{K,E}$ is tending to zero. Speaking more precisely one could expect the following inequality:

$$R^{\varphi}_{K,E}(\varepsilon) \leq \Psi_{K,\varphi}(\varepsilon, m(E)) \qquad (\varepsilon > 0,\ E \subset \mathbb{T},\ m(E) > 0)$$

where $\Psi_{K,\varphi}$ is a positive function defined in $(0,+\infty) \times (0,+\infty)$ and such that

$$\lim_{\varepsilon \to 0} \sup\{\Psi_{K,\varphi}(\varepsilon, y) : y \geq \delta\} = 0 \qquad (\delta > 0). \tag{11}$$

This conjecture is confirmed by the preceding example $(K = h, \varphi = m)$. Namely we could take in that case

$$\Psi_{h,m}(\varepsilon, y) = \sqrt{2}\, \varepsilon^{y} . \tag{12}$$

But in general this conjecture is not true and $R^{\varphi}_{K,E}$ can tend to zero non-uniformly with respect to the class of all sets E with $m(E)$ fixed and positive. This possibility occurs when $K=\ell$, $\varphi=m$. We have no explicit estimate of $R^{m}_{\ell,E}(\varepsilon)$ but we can show this estimate cannot have "the Hadamard form" (i.e. depend on $m(E)$ o n l y as is described by (11) and (12) and as is the case with $R^{m}_{h,E}$).

Put $E_1=\mathbb{T}\cap\{\zeta\in\mathbb{C}:\ \Im m\,\zeta\geq 0\}$ and let E_n be the pre-image of E_1 under the mapping $\zeta\to\zeta^{2^{n-1}}$ $(n=1,2,\dots)$

THEOREM 7. The sequence $\{R^{m}_{\ell,E_n}(\frac{100}{2^n})\}_{n=1}^{\infty}$ is bounded away from zero (though $m(E_n)=\frac{1}{2}$, $n=1,2,\dots$).

PROOF. It is easy to check that

$$\ell(\mu)(\zeta)=\frac{1}{2}\ell(\lambda)(\zeta^2)\qquad(\zeta\in\mathbb{T})\tag{13}$$

where λ is an arbitrary L^1-function and $\mu(\zeta)\overset{def}{=}\lambda(\zeta^2)$ $(\zeta\in\mathbb{T})$. Indeed if $\zeta=e^{i\theta}$, $\theta\in\mathbb{R}$ we have

$$\ell(\mu)(\zeta)=\frac{1}{2\pi}\int_0^{2\pi}\lambda(e^{2i\tau})\log|e^{i\tau}-e^{i\theta}|\,d\tau=\frac{1}{4\pi}\int_0^{4\pi}\lambda(e^{i\tau})\log|e^{\frac{i\tau}{2}}-e^{i\theta}|\,d\tau=$$

$$=\frac{1}{4\pi}\left\{\int_0^{2\pi}+\int_{2\pi}^{4\pi}\right\}.$$

Substituting $\tau=\sigma+2\pi$ into the last integral we transform it to the integral $\int_0^{2\pi}\lambda(e^{i\tau})\log|e^{i\tau}-e^{i\theta}|\,d\tau$, and (13) follows.

Take now a function $f\in L^2(m)$ satisfying the following conditions: $\|f\|_{L^2(m)}=1$, $m(f)>0$, $f|(\mathbb{T}\setminus E_1)=0$. Put $F_1=f$, $F_{n+1}(\zeta)=F_n(\zeta^2)$ $(\zeta\in\mathbb{T},\ n=1,2,\dots)$. The function F_n is zero off E_n, $\|F_n\|_{L^2(m)}=\|f\|_{L^2(m)}=1$, $m(F_n)=m(f)$. Using (13) we obtain $\max_{\mathbb{T}}|\ell(F_{n+1})|=\frac{1}{2}\max_{\mathbb{T}}|\ell(F_n)|$ so that

$$\max_{\mathbb{T}}|\ell(F_n)|=\frac{1}{2^{n-1}}\max_{\mathbb{T}}|\ell(f)|\leq\frac{100}{2^n}\qquad(n=1,2,\dots)$$

because $|\ell(f)| \leq \|f\|_{L^2(m)} \left(\int_{\mathbb{T}} (\log|1-\zeta|)^2 dm(\zeta)\right)^{1/2} < 50$.
Therefore $\|\ell(F_n)\|_{L^2(E_n)} \leq \frac{100}{2^n}$ $(n=1,2,\dots)$ and the definition of the m-rigidity implies $m(f) = m(F_n) \leq R^m_{\ell,E_n}\left(\frac{100}{2^n}\right)$ $(n=1,2,\dots)$ ●

It is easy to show that the phenomenon described in the theorem (i.e. $\inf_n R^m_{K,E_n}(\varepsilon_n) > 0$ for a sequence $\{\varepsilon_n\}$ tending to zero) occurs for every operator $K: L^2(m) \longrightarrow L^2(m)$ defined by the equalities

$$[\widehat{K(x)}]_j = \kappa_j \hat{x}_j \qquad (x \in L^2(m))$$

provided $\hat{K}_0 = 0$ and $\lim_{|j|\to\infty} \hat{K}_j = 0$ ($\hat{x}_j$ denotes the j-th Fourier coefficient of x). When $K = \ell$ we have $\hat{K}_j = -\frac{1}{2}|j|^{-1}$ $(j \neq 0)$, $\hat{K}_0 = 0$.

We have no good upper estimate of $R^m_{\ell,E}$ but we shall give an upper estimate of its modification. This estimate suggests which characteristics of E may be involved into $R^m_{\ell,E}$.

Consider the subset W^1_2 of $L^2(m)$ consisting of all functions x absolutely continuous on $\mathbb{T}$ and with $x' \in L^2(m)$ $\left(x'(e^{i\theta}) \overset{def}{=\!=} \frac{d}{d\theta} x(e^{i\theta})\right)$. Put

$$\|x\|^2_{W^1_2} \overset{def}{=\!=} |\hat{x}_0|^2 + \sum_{n\in\mathbb{Z}} |n|^2 |\hat{x}_n|^2 \quad (x \in W^1_2).$$

Introduce now the "small rigidity" $r^m_{\ell,E}$ $(= r)$ not exceeding $R^m_{\ell,E}$: $r^m_{\ell,E}(\varepsilon) \overset{def}{=\!=} \sup\{|m(x)| : x \in W^1_2, \|x\|_{W^1_2} \leq 1, x|E = 0, \|\ell(x)\|_{L^2(E)} \leq \varepsilon\}$.

Now put $E_\sigma = e^{-i\sigma}E$ and

$$\omega_E(\sigma) = m(E \setminus E_\sigma) \qquad (\sigma > 0).$$

It is known that $\lim_{\sigma\to 0} \omega_E(\sigma) = 0$.

THEOREM 8. $\quad r^m_{\ell,E}(\varepsilon) \leq A\,\varepsilon^{\frac{1}{3}(m(E) - \omega_E(\varepsilon^{2/3}))}$,

A being an absolute constant.

The proof of this theorem is a quantitative variant of the argument used in Theorem 1. Roughly speaking we differentiate $\ell(x)$ and obtain the Hilbert transform $h(x)$ whose estimate was considered in the Example 1. To be a little more precise we not just differentiate $\ell(x)$ but take its difference quotient with a step σ depending on ε . In this very moment appears

E_δ. We see that the rate of decrease in the right-hand side is influenced not only by $m(E)$ but by ω_E also. The function ω_E characterizes "the smoothness" of E. It is interesting to note that the sets E_n of the theorem 7 are "not "uniformly smooth": $\omega_{E_n}\left(\frac{2\pi}{2^n}\right) = \frac{1}{2}\ (n=1,2,\dots)$. Probably this is the cause of the "non uniform rigidity" of ℓ (in n) with respect to E_n. We think ω_E must appear in the estimate of $R^m_{\ell,E}$ too.

And now we add that after Theorems 7 and 8 the proofs of Theorems 1 and 2 seem a bit less artificial: the apparition of ω_E (or of other possible characteristics of E connected with its s h i f t s) justifies the use of differentiation to reduce the uniqueness problem to the uniqueness of analytic functions.

12. MORE ABOUT M.RIESZ POTENTIALS. Put $K_\alpha(x) = |x|^{\alpha-1}$ $(0<\alpha<1,\ x\subset \mathbb{R})$, so that $K_\alpha * \mu = \mathcal{U}^\mu_\alpha$, the Riesz potential of the (signed) measure μ (recall that the integral $\int_{\mathbb{R}} |x-t|^{\alpha-1} d\mu(t)$ converges absolutely a.e. and $\mathcal{U}^\mu_\alpha \in L^1_{loc}$ whenever μ is finite as is easily seen from the Fubini theorem).

Suppose E is a Lebesgue measurable subset of $\mathbb{R}$ with a "strong" density point (the origin, say), i.e. $\delta^{-1}|(-\delta,\delta)\setminus E|$ tends to zero very rapidly when δ goes to zero (we shall write here $|e|$ instead of $m(e)$, the Lebesgue measure of e). Then following the reasonings utilized in [5] it is not hard to see that

$$\left.\begin{array}{l} f\in L^p(-A,A) \quad \text{for a } p>1 \text{ and an } A>0 \\ f|E = \mathcal{U}^f_\alpha|E = 0 \end{array}\right\} \Longrightarrow f=0$$

or that

$$E \text{ is a } (K_\alpha, L^1\cap L^p_{loc})\text{- set} \qquad (p>1). \tag{I4}$$

But we prefer to present here another version of the uniqueness theorem for K_α. This version shows that (I4) holds for some sets E whose a l l density points are arbitrarily"weak". We shall prove that E satisfies (I4) whenever there exist very small intervals "almost filled" with the points of E. But these intervals need not contain a fixed point. The weak point of the theorems we are going to prove is that they seem essentially one-dimensional, whereas the methods of [5] work in $\mathbb{R}^n$ as well.

Suppose $E \subset \mathbb{R}$ is Lebesgue measurable and put

$$h_E(\delta) = \inf\left\{\frac{|p \setminus E|}{|p|} : p \in \mathcal{P}_\delta\right\}, \quad (\delta > 0)$$

$\mathcal{P}_\delta$ being the set of all intervals p with $|p| = \delta$.

THEOREM 9. If

$$\lim_{\delta \to +0} \delta \log h_E(\delta) = -\infty \tag{I5}$$

then (14) holds.

To illustrate this theorem take a strictly decreasing sequence $\{x_k\}$ of positive numbers tending to zero and place a set E_k into each segment $I_k = [x_{2k-1}, x_{2k}]$ so that $|E_k|$ is very close to $|I_k|$ but all density points of E_k are arbitrarily "weak". Choosing $\{x_k\}$ so that $|I_k|$ tends to zero very rapidly we can make the density of $\bigcup_k I_k$ at the origin also arbitrarily "weak". So we obtain a set $E \overset{def}{=} \bigcup_k E_k$ satisfying (I5) but whose all density points are as "weak" as we please.

Theorem will be deduced from the following

THEOREM IO. Let μ be a finite (signed) Borel measure in $\mathbb{R}$. Suppose there are a positive number K and a sequence $\{p_j\}$ of intervals contained all in a bounded interval and such that

(a) $\lim_{j \to \infty} |p_j| = 0$; (b) every p_j contains a Lebesgue measurable subset E_j satisfying

$$\int_{E_j} |\mathcal{U}_\alpha^\mu| \, dm \le K |\mu| (p_j), \tag{I6}$$

and

$$\lim_{j \to \infty} |E_j| \log |\mu| (p_j) = -\infty . \tag{I7}$$

Then $\mu = 0$.

We begin with the deduction of Theorem 9 from Theorem IO. Put $\mu = fm$ (recall that m denotes the Lebesgue measure) where $f \in L^1 \cap L^p_{loc}$, $f | E = \mathcal{U}_\alpha^f | E = 0$ and prove that μ satisfies conditions of Theorem IO. Take a sequence $\{\delta_j\}$ of positive numbers tending to zero and such that $\lim_{j \to \infty} \delta_j \log h(\delta_j) = -\infty$ ($h \overset{def}{=} h_E$). To every j corresponds a $p_j \in \mathcal{P}(\sigma_j)$ satisfying

$$|p_j \setminus E| < (h(\sigma_j)+\varepsilon_j)\cdot\sigma_j , \tag{I8}$$

where $\varepsilon_j > 0$ is chosen so small that $\lim_{j\to\infty} \sigma_j \log(h(\sigma_j)+\varepsilon_j) = -\infty$. We may assume $h(\sigma_j)+\varepsilon_j < 1$ so that $E_j \stackrel{def}{=} p_j \cap E \neq \emptyset$ $(j=1,2,\dots)$, and the boundedness of E implies the boundedness of $\bigcup p_j$. The condition (I6) is trivially satisfied. Turn now to (I7). If R is large enough we have

$$|\mu|(p_j) = \int_{p_j\setminus E} |f|\,dm \le \Big(\int_{-R}^{R} |f|^p dm\Big)^{\frac{1}{p}} |p_j\setminus E|^{\frac{1}{q}} \le C\big[(h(\sigma_j)+\varepsilon_j)\sigma_j\big]^{\frac{1}{q}} ,$$

$$C = C(f,p,R) , \quad q = \frac{p}{p-1} .$$

Taking logarithms and noting that $\lim_{j\to\infty} \sigma_j^{-1}\cdot|E_j| = 1$ (see (I8)) we show that (I7) is satisfied, and $\mu = 0$ by Theorem IO.

PROOF OF THEOREM IO. We may assume both sequences of endpoints of p_j's tend to a point, say a. Introduce functions k_j defined on $\mathbb{R}$:

$$k_j(t) = \begin{cases} 0 & t \in p_j \\ 1 & t \notin p_j ,\ t \ge \max p_j \quad j=1,2,\dots \\ -e^{i\pi\alpha} & t \notin p_j ,\ t \le \min p_j \end{cases}$$

Clearly, $\lim_{j\to\infty} k_j(t) = -e^{i\pi\alpha} (t<a)$, $\lim_{j\to\infty} k_j(t) = 1$ $(t>a)$. If the set of j's satisfying $a \in p_j$ is infinite we may (and shall) assume $\lim_{j\to\infty} k_j(a) = 0$. In this case $|\mu|\{a\} = 0$, because (I7) implies $\lim_{j\to\infty} |\mu|(p_j) = 0$. If $a \notin p_j$, for all large values of j we may assume (taking subsequences if necessary) that $\lim_{j\to\infty} k_j(a)$ exists and is 1 or $-e^{+i\pi\alpha}$.

Let $\zeta \in \mathbb{C}$, $\operatorname{Im}\zeta \ge 0 : \zeta = |\zeta| e^{i\theta}$, $0 \le \theta \le \pi$. We shall write $\zeta^{\beta} \stackrel{def}{=} |\zeta| e^{i\beta\theta}$ $(\beta > 0)$. Put

$$K_j(\zeta) = \int_{\mathbb{R}} \frac{k_j(t)\,d\mu(t)}{(\zeta-t)^{1-\alpha}} \quad (\operatorname{Im}\zeta > 0,\ j=1,2,\dots) .$$

These functions are analytic in the upper half-plane $\mathbb{C}_+$, and $\lim_{y\to+0} K_j(x+iy) \big(\stackrel{def}{=} K_j(x)\big) = \mathcal{U}_{\alpha}^{\mu_j}(x)$ whenever x is an interior point of p_j, $\mu_j \stackrel{def}{=} \chi_{\mathbb{R}\setminus p_j}\mu$ (χ_e being the characteristic function of the set e).

Consider now a Jordan region $\mathcal{D} \subset \mathbb{C}_+$ with the smooth

(C^2 , say) boundary $\partial\mathcal{D}$ such that $\partial\mathcal{D}\cap\mathbb{R}=(-R,R)$, where R such that $\cup p_j=(-R,R)$.

It is not hard to see that K_j belongs to the Smirnov class $E_1(\mathcal{D})$ (see [3] , p.203), and $\int_{\partial\mathcal{D}}|K_j(\zeta)||d\zeta|\le L$, $L>1$ being a constant depending on μ , R and α only. If $x\in p_j$ we have $K_j(x)=\mathcal{U}_\alpha^{\mu_j}(x)=\mathcal{U}_\alpha^{\mu}(x)-\mathcal{U}_\alpha^{\mu|p_j}(x)$. But

$$\int_{E_j}|\mathcal{U}_\alpha^{\mu|p_j}(x)|\,dx\le\int_{p_j}d|\mu|(t)\int_{-R}^{R}\frac{dx}{|x-t|^{1-\alpha}}\le C|\mu|(p_j),\ C=C(\mathcal{D},\alpha)>1$$

and (I6) implies

$$\int_{E_j}|K_j(x)|\,dx\le(C+K)|\mu|(p_j)\qquad(j=1,2,\dots).$$

Let ω_z be the harmonic measure on $\partial\mathcal{D}$ (with respect to $\mathcal{D}$) computed at a point $z\in\mathcal{D}$. It follows from the smoothness of $\partial\mathcal{D}$ that ω_z is comparable with the Lebesgue measure (the arc-length) on $\partial\mathcal{D}$: $\gamma_z\cdot|E|\le\omega_z(E)\le\Gamma_z\cdot|E|$, E being an arbitrary Lebesgue measurable subset of the curve $\partial\mathcal{D}$, γ_z and Γ_z positive numbers not depending on E . Now

$$|K_j(z)|\le\exp\int_{\partial\mathcal{D}}\log|K_j|\,d\omega_z\le\left(\frac{1}{\omega_z(E_j)}\int_{E_j}|K_j|\,d\omega_z\right)^{\omega_z(E_j)}\left(\frac{1}{\omega_z(CE_j)}\int_{CE_j}|K_j|\,d\omega_z\right)^{\omega_z(CE_j)}\le$$

$$\le A\Gamma_z L\left(\int_{E_j}|K_j(x)|\,dx\right)^{\omega_z(E_j)}\le A\Gamma_z L\left[(C+K)|\mu|(p_j)\right]^{\omega_z(E_j)},$$

where $A=\max_{0\le x\le 1}\left[x^{-x}(1-x)^{-(1-x)}\right]$. This estimate, the inequality $\omega_z(E_j)\ge\gamma_z|E_j|$ and (I7) imply

$$\lim_{j\to\infty}K_j(z)=0\qquad(z\in\mathcal{D}).$$

But $\lim_{j\to\infty}K_j(z)=\int_{\mathbb{R}}\frac{k(t)\,d\mu(t)}{(z-t)^{1-\alpha}}$ $(\mathrm{Im}\,z>0)$, where $k(t)\overset{\mathrm{def}}{=}\lim_{j\to\infty}k_j(t)$. Hence $\int_{\mathbb{R}}\frac{k(t)\,d\mu(t)}{(z-t)^{1-\alpha}}=0$ $(\mathrm{Im}\,z>0)$.

It is not hard to deduce from the last identity that it holds with $\alpha=0$ too (see [1] , p.144) and therefore the measure $k\mu$ is m-absolutely continuous, $k\mu=fm$, f being a function from the Hardy class H_+^1 . But $\frac{1}{|p_j|}\int_{p_j}\log|f|\,dm\le$ $\le\log\frac{1}{|p_j|}\int_{p_j}|f|\,dm$, and

$$\int_{P_j} \log|f|\,dm \le |P_j|\log\frac{1}{|P_j|} + |P_j|\log\int_{P_j}|f|\,dm =$$
$$= |P_j|\log\frac{1}{|P_j|} + |P_j|\log|\mu|(P_j).$$

Clearly $|P_j|\log|\mu|(P_j) \le |E_j|\log|\mu|(P_j)$ for large j, and (I7) implies $\lim_{j\to\infty}\int_{P_j}\log|f| = -\infty$ which is impossible if $f\neq 0$. We are done, because $|k(t)|=1$ whenever $t\neq a$, and if $k(a)=0$, then $|\mu|\{a\}=0$.

The most essential feature of the Riesz kernel K_α used in this proof is that K_α is composed of two analytic pieces: $K_\alpha|(0,+\infty)$, $K_\alpha|(-\infty,0)$ coinciding both with boundary values of functions analytic in $\mathbb{C}_+$, (namely of $z^{1-\alpha}$ and of $(-e^{-i\pi\alpha})\cdot z^{1-\alpha}$). The article [2] contains uniqueness theorems applicable to many kernels of this kind.

References

I. Ё р и к к е Б., Х а в и н В.П. Принцип неопределенности для операторов, перестановочных со сдвигом.I. - Зап.науч. семин.ЛОМИ, I979, т.92, с.I34-I70.

2. Ё р и к к е Б., Х а в и н В.П. Принцип неопределенности для операторов, перестановочных со сдвигом.П. - will be published in the same "Записки".

3. П р и в а л о в И.И. Граничные свойства аналитических функций. М., I950, 336 с.

4. Р у б и н Б.С. О пространствах дробных интегралов на прямолинейном контуре. - Известия АН Армянской ССР, I972, т.7, № 5, 373-386.

5. М а з ь я В.Г., Х а в и н В.П. О решениях задачи Коши для уравнения Лапласа (единственность, нормальность, аппроксимация). - Тр.Московского мат.об-ва, I974, т.30, 6I-II4.

6. Х р у щ ё в С.В. Проблема одновременной аппроксимации и стирание особенностей интегралов типа Коши. - Тр.мат.ин-та АН СССР, I978, т.I30, I24-I95.

7. T i t c h m a r s h E.C. On conjugate functions. - Proc. London Math.Soc., 1928, vol.29, p.49-80.

8. В и н о г р а д о в С.А., Х а в и н В.П. Свободная интерполяция в H^∞ и в некоторых других классах функций I. - Зап. науч.семин.ЛОМИ,I974,47,I5-54; П - idem ,I976,56,I2-28.

S.V.Hruščëv, S.A.Vinogradov

FREE INTERPOLATION IN THE SPACE OF UNIFORMLY CONVERGENT TAYLOR SERIES.

Introduction

The main subject of the present paper is the space U_A consisting of the functions in the disc-algebra C_A whose Taylor series converge uniformly on the unit circle $\mathbb{T}$. It is clear from the definition that U_A is a proper subset of C_A. Indeed, for every subset E of $\mathbb{T}$ with the zero Lebesgue measure there is a function f in C_A whose Fourier series diverges on E (see [1], p.58). It is possible, of course, to mention other properties showing the distinction between the spaces C_A and U_A. For example, U_A is not an algebra [2].

Our main aim in this paper is the study of properties common for C_A and U_A both. Having originated in the classical papers of Banach [3], Paley [4], Rudin [5], [6] and Carleson [7], the theory of the disc-algebra was continued in the direction of our interest in the series of papers [8-14]. The things were not so good for the space U_A. But now they are looking up thanks to certain recent progress mainly connected with the papers [15-21] and diminishing the disproportion of two theories.

The authors of the present paper hope to publish in a future a survey of harmonic analysis in the spaces U_A and C_A. We collected here some new results and some new approaches to the subject which appeared during our work on the above mentioned survey.

For the more detailed exposition some notation is needed. Let m denote the usual normalized Lebesgue measure on $\mathbb{T}$ The Fourier coefficients of the integrable function f on the circle are defined by the formula

$$\hat{f}(n)=\int_{\mathbb{T}} f\cdot\bar{\zeta}^n\,dm\,,\qquad n\in\mathbb{Z}\overset{def}{=\!=}\{\ldots,-1,0,1,\ldots\}\,.$$

Then the symbol $D_n * f \overset{def}{=\!=} \sum_{k=-n}^{n}\hat{f}(k)\zeta^k$ denotes the partial sum of the Fourier series of f. By Weierstrass theorem C_A is a closed subalgebra of $C(\mathbb{T})$, the algebra of all continuous functions on the unit circle $\mathbb{T}$. So the following formula

$$C_A = \{ f \in C(\mathbb{T}) : \hat{f}(n) = 0,\ n \in \mathbb{Z}, n < 0 \}$$

will not mislead the reader. The space U_A consists of functions f in C_A satisfying

$$\lim_{n \to +\infty} \| D_n * f - f \|_\infty = 0 .$$

Here $\| f \|_\infty \stackrel{def}{=} \sup_{\zeta \in \mathbb{T}} | f(\zeta) |$ stands for the usual norm in $C(\mathbb{T})$. It is a Banach space with respect to the norm

$$\| f \|_U \stackrel{def}{=} \sup_{n \in \mathbb{Z}} \| D_n * f \|_\infty .$$

Let U_A^* denote the conjugate space for U_A. For every functional Φ the Cauchy transform of Φ is defined by

$$\mathcal{K}\Phi(z) = \sum_{n \geq 0} z^n < \zeta^n, \Phi > , \quad |z| < 1 .$$

It is clear that $\mathcal{K}\Phi$ is a holomorphic function in the open unit disc $\mathbb{D} = \{ z \in \mathbb{C} : |z| < 1 \}$. Let H^p, $p > 0$, denote the usual Hardy space in $\mathbb{D}$.

The next theorem is of prime **importance** for what follows.

THEOREM (S.A.Vinogradov [15], [16]). If $\Phi \in U_A^*$ then $\mathcal{K}\Phi \in \bigcap_{p<1} H^p$. Moreover there is a positive number $\varkappa$ such that

$$\sup_{0<r<1} m\{ \zeta \in \mathbb{T} : | \mathcal{K}\Phi(r\zeta) | > y \} \leq \frac{\varkappa}{y} \| \Phi \|_{U_A^*}$$

for every positive y and for every Φ in U_A^*.

It should be noted that the last theorem hinges on the famous Carleson's theorem [22] about the almost everywhere convergence of the Fourier series of the functions in $L^2(\mathbb{T})$. It extends Kolmogorov - Smirnov's conjugation theorem.

The contents of the section 1 are intimately connected with the Banach's theorem [3] and the Rudin - Carleson's theorem [23] on the one hand and with a recent result of Oberlin [24] on the other.

THEOREM (Oberlin [24]). Let E be a closed subset of $\mathbb{T}$ and let $mE = 0$. Then $U_A | E = C(E)$.

We give a simple proof of this theorem. It is based on Vinogradov's theorem mentioned above and on the following result whose proof is given in §2.

THEOREM 2.1. Let μ be a finite complex Borel measure on $\mathbb{T}$ and let μ_s denote its singular part in the Lebesgue decomposition of μ with respect to the measure m. Then

$$\lim_{y\to+\infty} y\cdot m\{t\in\mathbb{T} : |\mathcal{K}\mu(t)|>y\} = \frac{1}{\pi}\|\mu_s\|.$$

Here $\mathcal{K}\mu(z)$ stands for the Cauchy integral $\int_{\mathbb{T}} \frac{d\mu(t)}{t-z}$.

Our proof of the Oberlin's theorem may be extended in two directions. The first approach leads to the theorem about a simultaneous interpolation and to an axiomatic theory which differs from the Bishop's one [25]. The second generalization of the proof strengthens the Oberlin's theorem. Referring the interested reader to the section 1 we shall formulate now the simultaneous interpolation theorem for the space U_A. To do this remind a definition. A subset Λ of the set $\mathbb{N} \overset{def}{=} \{0,1,2,\dots\}$ of all non-negative integers is called a $\Lambda(2)$-set if there is a positive constant $C = C_\Lambda$ such that for every polynomial $f = \sum_{n\in\Lambda} \hat{f}(n) z^n$ with the frequencies in Λ the following inequality holds:

$$c\cdot\Big(\int_{\mathbb{T}} |f|^2 dm\Big)^{1/2} \le \int_{\mathbb{T}} |f|\, dm .$$

For a closed subset E of $\mathbb{T}$ and for Λ, $\Lambda\subset\mathbb{N}$, let Q be a linear operator defined by

$$Qf = (f|E, \hat{f}|\Lambda), \quad f\in U_A .$$

THEOREM. The equality

$$QU_A = C(E)\times \ell^2(\Lambda)$$

holds if and only if $mE=0$ and Λ is a $\Lambda(2)$-set.

In section 2 we deal with some extensions of the theorem 2.1. These results are applied in §4 to some problems of the interpolation theory in U_A. Here is one of these applications. For every square-summable function f, $f\in L^2(\mathbb{T})$, let

$$f_+(z) = \sum_{n=0}^{\infty} \hat{f}(n) z^n$$

denote the orthogonal projection of f into the Hardy class H^2. The next theorem deals with the space

$$V(K) \overset{def}{=} \{ f \in L^2(\mathbb{T}) : supp(f) \subset K,\ f_+ \in U_A \} .$$

Here K is a closed subset of $\mathbb{T}$ **and the symbol** $supp(f)$ stands for the support of the function f.

THEOREM. Let K be a closed subset of $\mathbb{T}$, let $mK > 0$ and let E be a closed subset of the set of density points for K, $mE = 0$. Let Λ be a subset of $\mathbb{N}$ with Hadamard's gaps. Let, finally, the operator Q be defined by

$$Qf \overset{def}{=} (f_+ | E,\ \hat{f} | \Lambda),\ f \in V(K) .$$

Then $QV(K) = C(E) \times \ell^2(\Lambda)$.

There is a definition of the space $V(K)$ equivalent to the previous one. If $f \in V(K)$ then the Cauchy transform $\int_K (t-z)^{-1} f(t)\, dm(t)$ is holomorphic in $\hat{\mathbb{C}} \setminus K$. So the space of Cauchy transforms $\int_K (t-z)^{-1} f(t)\, dm(t)$, $f \in V(K)$, is explicitly the space of holomorphic functions g in $\hat{\mathbb{C}} \setminus K$ such that $g | \mathbb{D} \in U_A$ and the restriction $g | \hat{\mathbb{C}} \setminus \mathbb{D}$ belongs to the Hardy class H^2 outside of the unit disc. It is not trivial that $V(K) \neq \{\mathbb{O}\}$ if $mK > 0$ and if the set K is nowhere dense on $\mathbb{T}$. The above theorem asserts that the space $V(K)$ is large if $mK > 0$.

In section 3 the properties of the space U_A are discussed in more details. We prove an analog of the F. and M. Riesz's theorem for U_A^*. Together with one theorem of A.Pełczyński this implies the existence for every E, $mE = 0$, of a linear interpolating operator $T : C(E) \longrightarrow U_A$. Being identical on the circle, the classes of interpolating sets for C_A and U_A are different in $\mathbb{D}$. We give an example of a closed set E such that $E \cap \mathbb{T} = \{1\}$, $C_A | E = C(E)$ but $U_A | E \neq C(E)$. Some sufficient conditions for $U_A | E = C(E)$ are also given in §3.

The final part (§4) of the paper is devoted to applications of theorems 2.1 and 2.4. One of these applications was already mentioned above. The second application depends highly on the following identity

$$\mathfrak{m}(U_A^\infty) = \mathfrak{m}(U_A^*)$$

also established in §4. Here $\mathfrak{m}(U_A^\infty)$ stands for the set of all multipliers of the space $U_A^\infty \overset{def}{=} \{f \in H^\infty : \|D_n * f\|_\infty = O(1)\}$ and $\mathfrak{m}(U_A^*)$ denotes the multiplier space for $\mathcal{K}U_A^\infty$. This identity allows to extract a useful information about the space $\mathfrak{m}(U_A^\infty)$. In particular, we prove that the transformation $f \longrightarrow \hat{f} \mid \Lambda$ maps the space $\mathfrak{m}(U_A^\infty)$ onto $\ell^1(\Lambda)$ for every subset Λ of $\mathbb{N}$ with Hadamard's gaps. We prove also that Blachke product whose zero set satisfies the Frostman condition (see §3 for the definition), belongs to $\mathfrak{m}(U_A^\infty)$. On the other hand, if S is an inner function in $\mathfrak{m}U_A^\infty$ then the radial limits $\lim_{r \to 1-0} S(rt) = S(t)$ exist everywhere on $\mathbb{T}$ and $|S(t)|=1$, $t \in \mathbb{T}$. This implies that S is a Blaschke product.

In conclusion we announce a theorem generalizing a recent result due to de Leeuw, Katznelson and Kahane [18]. Combining the method of [18] with the method of S.Kisljakov [19], and with our scheme of interpolation, we get our result (see the end of §4 for the formulation).

§1. An axiomatic approach to the Banach-Rudin-Carleson theorems

The Banach interpolation theorem was proved for the first time in [3]. It is the premise for the following definition.

DEFINITION. Let Λ be a subset of $\mathbb{Z}$. It is called a Banach set if for every square-summable sequence $x = (x_n)_{n\in\Lambda}$, $x \in \ell^2(\Lambda)$, there is a function f in $C(\mathbb{T})$ satisfying

$$\hat{f}(n) = x_n, \quad n \in \Lambda .$$

THEOREM (Banach [3]). A finite union of gap subsets of $\mathbb{Z}$ is a Banach set.

A Banach subset Λ can also be described as a subset of $\mathbb{Z}$ generating the Riesz basis $(\zeta^n)_{n\in\Lambda}$ in the closed span of the family $(\zeta^n)_{n\in\Lambda}$ in $L^p(\mathbb{T})$, $0<p<2$. Let $\mathcal{P}_\Lambda$ denote the set of all trigonometrical polynomials with frequencies in Λ.

THEOREM (Rudin [6]). For every p in $(0,2)$ the following conditions are equivalent:

1) Λ is a Banach set;

2) there is a positive number α_p such that

$$\alpha_p \|f\|_2 \leqslant \|f\|_p \ .$$

Here $\|f\|_p \overset{def}{=\!=} (\int |f|^p dm)^{1/p}$ stands for the usual L^p -norm. For $p=1$ this theorem is due to Banach [3]. It should be noted that after the paper [6] of Rudin the Banach sets were given the name of $\Lambda(2)$ -sets. Nevertheless we prefer to use the first term in honor of Banach. The following fundamental theorem describes interpolating sets for C_A on the circle.

THEOREM (Carleson - Rudin [23]). Let E be a closed subset of $\mathbb{T}$. Then $C_A \mid E = C(E)$ iff $mE=0$.

It was proved in [13] that if $\Lambda \subset \mathbb{N} = \{0,1,2,\dots\}$ then the concepts of Banach sets for C_A and $C(\mathbb{T})$ are identical. Moreover, Vinogradov [13] coupled Banach theorem and the theorem of Carleson - Rudin. Let $Q: C_A \longrightarrow C(E)\times \ell^2(\Lambda)$ be defined by

$$Qf = (f \mid E,\ \hat{f} \mid \Lambda).$$

THEOREM (Vinogradov [13]). The identity $Q(C_A) = C(E)\times \ell^2(\Lambda)$ holds if and only if $mE=0$ and Λ is a Banach set.

In this section we shall formulate simple axioms which imply the assertion of the above theorem for a large class of spaces $\mathcal{E}$. The leading example will be the space $\mathcal{E} = U_A$. Let $\mathcal{P}_A$ denote the set of all polynomials.

AXIOM 1. Let $\mathcal{E}$ be a Banach space of functions holomorphic in $\mathbb{D}$ such that

$$\mathcal{E}\text{-}clos\,\mathcal{P}_A = \mathcal{E}$$

$$\varlimsup_{n\to\infty} \|z^n\|_{\mathcal{E}}^{1/n} \leqslant 1 . \tag{A1}$$

It follows from this axiom that $(1-\zeta z)^{-1} \in \mathcal{E}$ if $|\zeta|<1$. Indeed,

$$\frac{1}{1-\zeta z} = \sum_{n=0}^{\infty} \zeta^n z^n$$

and the series in the right-hand side of the equality converges in $\mathcal{E}$ (see (A1)). The space $\mathcal{E}$ satisfying the axiom 1, it is possible to define the Cauchy transform

$$\mathcal{K}\Phi(\zeta) \overset{def}{=\!=} \langle (1-\zeta z)^{-1}, \Phi\rangle = \sum_{n=0}^{\infty} \zeta^n \langle z^n, \Phi\rangle, \ \Phi \in \mathcal{E}^*, \ |\zeta|<1.$$

It is clear that Cauchy transform is a one-to-one map of the space $\mathcal{E}^*$ onto $\mathcal{K}\mathcal{E}^*$.

AXIOM 2. There is a positive number $\varkappa$ such that

$$\sup_{0<r<1} m\{t\in\mathbb{T} : |\mathcal{K}\Phi(rt)|>y\} \leqslant \frac{\varkappa}{y}\|\Phi\|_{\mathcal{E}^*} \tag{A2}$$

for every $y>0$ and for every Φ in $\mathcal{E}^*$.

Axioms 1 and 2 are valid for the spaces C_A and U_A. For the space C_A this follows from the Kolmogorov's theorem [26] and for U_A from Vinogradov's theorem mentioned in §1. Here is another example. Let $(n_k)_{k\geqslant 1}$ be an increasing sequence of positive integers with Hadamard's gaps: $n_{k+1}/n_k \geqslant q$, $q>1$. Then

$$U_A(n_k) = \{f\in C_A : \lim_{k\to+\infty} \|D_{n_k} * f - f\|_\infty = 0\}.$$

Clearly $U_A \subset U_A(n_k) \subset C_A$ and (A2) is a consequence of Kolmogorov's theorem ([28], ch.XIII, 1.17).

The following technical lemma reformulates the condition (A2).

LEMMA 1.1. Let $\mathcal{E}$ be a Banach space satisfying (A1) and (A2). Then $\mathcal{K}\mathcal{E}^* \subset \bigcap_{p<1} H^p$ and moreover

$$\|\mathcal{K}\Phi\|_p \leqslant \varkappa\left(1+\frac{p}{1-p}\right)\|\Phi\|_{\mathcal{E}^*}, \quad p<1, \tag{1}$$

$$m\{t\in\mathbb{T} : |\mathcal{K}\Phi(t)|>y\} \leqslant \frac{\varkappa}{y}\|\Phi\|_{\mathcal{E}^*}. \tag{2}$$

Proof. Let a number r, $r>1$, be fixed and let $\lambda(y) = m\{t\in\mathbb{T} : |\mathcal{K}\Phi(rt)|>y\}$. Then

$$\|\mathcal{K}\Phi\|_p^p = -\int_0^{+\infty} y^p\, d\lambda(y) = p\cdot\int_0^\infty \lambda(y) y^{p-1} dy \leqslant$$

$$\leqslant p\cdot\int_0^\infty y^{p-1}\min\left\{1, \frac{\varkappa}{y}\cdot\|\Phi\|_{\mathcal{E}^*}\right\}dy = \varkappa^p\left(1+\frac{p}{1-p}\right)^p\cdot\|\Phi\|_{\mathcal{E}^*}^p.$$

It follows from this inequality that $\mathcal{K}\mathcal{E}^* \subset H^p$ if $p<1$. Therefore the radial limits

$$\mathcal{K}\Phi(\zeta) \overset{def}{=\!=} \lim_{r\to 1-0} \mathcal{K}\Phi(r\zeta)$$

exist for almost every ζ in $\mathbb{T}$. Let

$$E_r \stackrel{def}{=} \{\zeta\in\mathbb{T} : |\mathcal{K}\Phi(r\zeta)| > y\},\ 0<r\leqslant 1$$

and let $r_n \uparrow 1$. Then the set E_1 is covered by $(\bigcup_n \bigcap_{\ell\geqslant n} E_{r_\ell}) \cup \mathcal{N}$, where $\mathcal{N}$ is a subset of $\mathbb{T}$ with zero Lebesgue measure. Therefore (2) is a consequence of the obvious inequality

$$m(\bigcap_{\ell\geqslant n} E_{r_\ell}) \leqslant \frac{\varkappa}{y}\|\Phi\|_{\mathcal{E}^*}. \quad \bullet$$

Now we are in a position to formulate the main theorem of this section.

THEOREM 1.2. Let $\mathcal{E}$, $\mathcal{E}\subset C_A$, be a Banach space endowed with a norm stronger than the sup-norm and satisfying axioms 1 and 2. Let E be a closed subset of $\mathbb{T}$, $mE=0$, and let Λ be a Banach set. Then $Q(\mathcal{E}) = C(E)\times \ell^2(\Lambda)$.

It is interesting to note that known proofs of the Carleson-Rudin theorem use the F. and M. Riesz's theorem. Our proof of theorem 1.2 shows that this theorem may be avoided. The theorem 2.1 substitutes the theorem of F. and M.Riesz in our approach. This not only simplifies the proof but extends it in a more general setting. The idea of our approach appeared in connection with the recent paper [24].

PROOF OF THEOREM 1.2. It is clear that the operator $Q: \mathcal{E}\to C(E)\times\ell^2(\Lambda)$ is bounded. By the Banach theorem [29] it is onto iff the conjugate operator Q^* is an isomorphism onto its image. Some words about the notation for the duality between the spaces we consider. The duality between $C(E)\times\ell^2(\Lambda)$ and $M(E_*)\times\ell^2(\Lambda)$ is standard and is defined by

$$<(f,y),(\mu,x)> = \int_{E_*} f(\bar t)\,d\mu(t) + \sum_{n\in\Lambda} x_n y_n .$$

Here $E_* \stackrel{def}{=} \{\bar t : t\in E\}$. The Cauchy transform defines the duality between $\mathcal{E}$ and $\mathcal{E}^*$ on the dense subset of rational functions with poles outside the closed unit disc. It is easy now to compute the Cauchy transform of a functional $\Phi = Q^*(\mu,x)$, $(\mu,x)\in M(E_*)\times\ell^2(\Lambda)$:

$$\mathcal{K}\Phi(z) = <(1-z\bar t)^{-1}, Q^*(\mu,x)> = <Q(1-z\bar t)^{-1}, (\mu,x)> =$$

$$= \int_{E_*}\frac{d\mu(t)}{1-z\bar t} + \sum_{n\in\Lambda} x_n\cdot z^n .$$

The space $\mathcal{E}$ being imbedded into the disc-algebra, every finite Borel measure σ defines a continuous functional on $\mathcal{E}$:

$$f \longrightarrow \int_{\mathbb{T}} f(\bar{t})\, d\sigma(t) .$$

To simplify the notation we denote this functional by the same symbol σ . Let $\sigma = \sigma_a + \sigma_s$ be the Lebesgue decomposition of the measure σ with respect to the measure m.

LEMMA 1.3. Let $\mathcal{E}$ be the same as in theorem 1.2. Then for every p, $0<p<1$, there are positive constants a and b_p such that for every finite Borel measure σ on $\mathbb{T}$ the following inequalities hold

$$\|\sigma_s\|_{M(\mathbb{T})} \leqslant a \cdot \|\sigma\|_{\mathcal{E}^*} \tag{3}$$

$$\|\mathcal{K}\sigma_a\|_{H^p} \leqslant b_p \cdot \|\sigma\|_{\mathcal{E}^*} . \tag{4}$$

We shall give the proof of lemma later and now we shall show how to finish the proof of the theorem. Let $x = (x_n)_{n\in\Lambda}$, $x \in \ell^2(\Lambda)$, and let $\mu \in M(E_*)$. Then the formula

$$\sigma = \Big(\sum_{n\in\Lambda} x_n \zeta^n\Big)\, dm + \mu$$

defines the Lebesgue decomposition of σ . It is clear that for $\Phi = Q^*(\mu, x)$ we have $\mathcal{K}\Phi = \mathcal{K}\sigma$. It follows from the inequality (4) that

$$\Big\|\sum_{n\in\Lambda} x_n \zeta^n\Big\|_p \leqslant b_p \cdot \|\Phi\|_{\mathcal{E}^*} .$$

The set Λ being a Banach set, there is $c_\Lambda > 0$ such that

$$c_\Lambda \cdot \|x\|_{\ell^2(\Lambda)} \leqslant \Big\|\sum_{n\in\Lambda} x_n \zeta^n\Big\|_p \leqslant b_p \|\Phi\|_{\mathcal{E}^*} .$$

On the other hand it follows from (3) that

$$\|\mu\|_{M(E_*)} \leqslant a \cdot \|\Phi\|_{\mathcal{E}^*} .$$

This implies

$$\|\mu\|_{M(E_*)} + \|x\|_{\ell^2(\Lambda)} \leqslant \text{const.}\, \|\Phi\|_{\mathcal{E}^*} . \quad \bullet$$

PROOF OF LEMMA 1.3. To prove the inequality (3) it is sufficient to compare the identity

$$\lim_{y\to+\infty} y\cdot m\{t\in\mathbb{T}: |\mathcal{K}\sigma(t)|>y\} = \frac{1}{\pi}\|\sigma_s\|$$

with (2). This implies (3) with $a = \varkappa\cdot\pi$.

Lemma 1.1. shows now that

$$\|\mathcal{K}\sigma_a\|^p_{H^p} \le \|\mathcal{K}\sigma\|^p_{H^p} + \|\mathcal{K}\sigma_s\|^p_{H^p} \le \varkappa^p\Big(1+\frac{p}{1-p}\Big)\|\sigma\|^p_{\mathcal{E}^*} + \|\mathcal{K}\sigma_s\|^p_{H^p}.$$

By Kolmogorov-Smirnov theorem [27] we have

$$\|\mathcal{K}\sigma_s\|^p_{H^p} \le \frac{4}{1-p}\|\sigma_s\|^p_{M(\mathbb{T})}.$$

Therefore using (3) we get

$$\|\mathcal{K}\sigma_a\|^p_{H^p} \le b_p\|\sigma\|^p_{\mathcal{E}^*} + \frac{4}{1-p}\|\sigma_s\|^p_{M(\mathbb{T})} \le b_p\|\sigma\|^p_{\mathcal{E}^*} + c_p\cdot\|\sigma\|^p_{\mathcal{E}^*}. \quad \bullet$$

§2. An asymptotic formula for the distribution function of the boundary values of Cauchy integrals

In this section we prove theorem 2.1.,which was formulated in the introduction,and some of its generalizations. But we begin with a collection of proofs of the so-called Boole formulae (see formulae (5) and (6) below). Let

$$\mathcal{H}\mu(z) = \int_{\mathbb{T}} \frac{t+z}{t-z}\,d\mu(t)$$

be the Schwarz integral for a positive measure μ and let

$$\lambda_{\mathbb{T}}(y) = m\{t\in\mathbb{T}: |\mathcal{H}\mu(t)|>y\}, \quad y>0.$$

For a positive measure μ on $\mathbb{R}$ we consider the distribution function

$$\lambda_{\mathbb{R}}(y) = m\{x\in\mathbb{R}: |\mathcal{K}\mu(x)|>y\}.$$

Unexpectedly enough there exist explicit formulae for $\lambda_{\mathbb{T}}$ and $\lambda_{\mathbb{R}}$ *) This fact has applications in ergodic theory and an interested reader may find the literature in [30]. The formula for $\lambda_{\mathbb{T}}$ was found independently in [31]. G.Boole has found the formula for $\lambda_{\mathbb{R}}$ assuming that μ is a finite sum of point masses [32].

Boole's formulae can be written as follows:

$$\lambda_{\mathbb{T}}(y) = \frac{2}{\pi}\,\mathrm{arctg}\,\frac{y}{\|\mu\|} \tag{5}$$

*) The measure μ is assumed to be singular.

$$\lambda_{\mathbb{R}}(y)=\frac{2\|\mu\|}{y} \quad . \tag{6}$$

The proof of (6) (accordingly to Boole and Levinson [33], p.69). Let

$$\mu=\sum_{k=1}^{n} \alpha_k\cdot\delta_{(t_k)}\ ,\quad \alpha_k>0\ ,$$

where $t_1<t_2<\ldots<t_n$ and let $\{J_1,\ldots,J_n\}$ be the partitioning of the extended line $\hat{\mathbb{R}}\overset{def}{=\!=}\mathbb{R}\cup\{\infty\}$ defined by the points t_k, $k=1,\ldots,n$. Then we have for every measurable non-negative function F :

$$\int_{\mathbb{R}} F(\mathcal{K}\mu(x))\,dx=\sum_k \int_{J_k} F(\mathcal{K}\mu(x))\,dx=\int_{\mathbb{R}} F(y)\cdot\sum_k \frac{dx_k}{dy}\,dy\,.$$

Here x_k denotes the inverse function for the restriction $\mathcal{K}\mu\,|\,J_k$. It should be noted that $\mathcal{K}\mu$ increases on each J_k because $(\mathcal{K}\mu)'(x)=\int\frac{d\mu(t)}{(t-x)^2}>0$. It is clear that the numbers $x_1(y),\ldots,x_n(y)$ form a full collection of roots of the equation

$$y\cdot\prod_{k=1}^{n}(t_k-x)-\sum_{k=1}^{n}\alpha_k\prod_{j\neq k}(t_j-x)=0$$

or equivalently of

$$x^n+x^{n-1}\Big[\frac{\sum_{k=1}^{n}\alpha_k}{y}-\sum_{k=1}^{n}t_k\Big]+\ \ldots\ =0\,.$$

By Viète theorem $x_1(y)+\ldots+x_n(y)=-y^{-1}\|\mu\|+\sum_{k=1}^{n}t_k$ and therefore

$$\lambda_{\mathbb{R}}(y)=\int_{\{|\mathcal{K}\mu|>y\}}dx=\|\mu\|\cdot\int_{\{|t|>y\}}\frac{dt}{t^2}=\frac{2\cdot\|\mu\|}{y}\ .$$

To prove (6) for any singular measure we use the following limit procedure. Without loss of generality we may assume that μ has a compact support K , $\int_K dx=0$. Choosing n complementary intervals of the set K , we put discrete masses at their ends so that the μ-measure of the closed segment connecting the neighboring **points is equal to the sum masses at the end-points of this segment. Let μ_n denote the discrete measure constructed above. Then $\mathcal{K}\mu_n\to\mathcal{K}\mu$ uniformly on compacts separated from K .** ●

To prove the formula (5) (see [24], [30]) assume for the simplicity that $\|\mu\|=1$. Then $\mathcal{H}\mu(z)=1+2z\,\mathcal{K}\mu(z)$

and it is clear that every asymptotic formula for $\mathcal{H}\mu$ implies an analogous formula for $\mathcal{K}\mu$. The measure μ is non-negative and singular. So the Schwarz integral maps $\mathbb{T}$ onto $\mathbb{R}$ and the unit disc into the right half-plane. It follows that

$$\mathcal{H}\mu(z) = \frac{1 - I(z)}{1 + I(z)},$$

where I is an inner function, $I(0) = 0$. The map $t \to I(t)$, $t \in \mathbb{T}$, is measure preserving for the measure space $(\mathbb{T}, m)$. Indeed,

$$\int_{\mathbb{T}} z^n \, dm \circ I^{-1} = \int_{\mathbb{T}} I^n \, dm = \begin{cases} 1, & n = 0 \\ 0, & n \neq 0, \end{cases}$$

and therefore $m \circ I^{-1} = m$. Using this, we get

$$m\{t \in \mathbb{T} : |\mathcal{H}\mu(t)| > y\} = m\{t \in \mathbb{T} : \left|\frac{1 - I(t)}{1 + I(t)}\right| > y\} =$$

$$= m\{t \in \mathbb{T} : \left|\frac{1-t}{1+t}\right| > y\} = \frac{2}{\pi} \operatorname{arctg} y. \quad \bullet$$

Our next proof works for the case of the disc and for the upper half-plane as well. Let

$$g(z) = i\mathcal{H}\mu(z) = u(z) + iv(z),$$

where v denotes a non-negative harmonic function in $\mathbb{D}$, having zero boundary values a.e. on $\mathbb{T}$. Let $\varepsilon > 0$ be fixed. Then

$$\int_{\mathbb{T}} \left\{\frac{g}{g + i\varepsilon} + \frac{\bar{g}}{\bar{g} - i\varepsilon}\right\} dm = 2\int_{\mathbb{T}} \frac{|g|^2 + \varepsilon v}{|g + i\varepsilon|^2} \, dm.$$

By the mean-value theorem the left-hand side of the identity is equal to $2v(0)(v(0)+\varepsilon)^{-1}$. The right-hand side is equal to $2 \cdot \int_{\mathbb{T}} \frac{|g|^2}{|g|^2 + \varepsilon^2} \, dm$ because $v(\zeta) = 0$ almost everywhere on $\mathbb{T}$. Putting $a = \|\mu\|$ for the brevity, we get

$$\frac{a}{a + \varepsilon} = -\int_0^\infty \frac{y^2 \, d\lambda_{\mathbb{T}}(y)}{y^2 + \varepsilon^2}$$

or after the change of variables $y = \sqrt{t}$, $\varepsilon^2 = x$:

$$\frac{a}{a + \sqrt{x}} = \int_0^\infty \frac{d\nu(t)}{t + x},$$

where $d\nu(t) = -t\,d\lambda_{\mathbb{T}}(\sqrt{t})$ denotes a positive measure on $[0, +\infty)$. The integral transform $F(z) = \int_0^\infty \frac{d\nu(t)}{t+z}$ is the well-known Stieltjes transform. The function F is holomorphic in $\hat{\mathbb{C}} \setminus [-\infty, 0]$ and the Stieltjes inversion formula restores the measure ν :

$$\frac{d\nu}{dx}(x) = \lim_{y \to 0+} \frac{F(-x-iy) - F(-x+iy)}{2\pi i y}, \quad x \geqslant 0.$$

The equality holds in the sense of generalized functions (see [34], p.70).

A simple calculation shows that

$$d\nu(x) = \frac{a}{\pi} \cdot \frac{\sqrt{x}}{a^2 + x}\, dx$$

and therefore

$$\lambda_{\mathbb{T}}(y) = \frac{2}{\pi} \operatorname{arctg} \frac{y}{\|\mu\|}. \quad \bullet$$

THEOREM 2.1. Let μ be a finite complex Borel measure on $\mathbb{T}$ and let μ_s denote its singular part. Then

$$\lim_{y \to +\infty} y \cdot m\{t \in \mathbb{T} : |\mathcal{K}\mu(t)| > y\} = \frac{1}{\pi} \cdot \|\mu_s\|.$$

PROOF. Let $|\mu|$ denote the variation of the measure μ.

LEMMA 2.2. For every $\varepsilon > 0$ there are smooth functions p, q in $C^\infty(\mathbb{T})$ such that

$$\|\mu - p|\mu|\|_{M(\mathbb{T})} \leqslant \varepsilon, \qquad \||\mu| - q\cdot\mu\|_{M(\mathbb{T})} \leqslant \varepsilon,$$

$$\|p\|_\infty \leqslant 1, \quad \|q\|_\infty \leqslant 1.$$

PROOF OF THE LEMMA. For a positive measure ν, $\nu \in M(\mathbb{T})$, let

$$B = \{f \in L^\infty(d\nu) : \|f\|_\infty \leqslant 1\}$$

and let W denote the closure of the ball $\{f \in C^\infty(\mathbb{T}) : \|f\|_\infty \leqslant 1\}$ in the space $L^1(d\nu)$. It is easy to see that $B \subset W$ (and actually $B = W$). Indeed, if $\Phi \in L^\infty(d\nu)$ and if $|\int \Phi f\, d\nu| \leqslant 1$ for every f in W, then the identity

$$\sup\{|\int_{\mathbb{T}} \Phi f\, d\nu| : f \in C^\infty(\mathbb{T}), \|f\|_\infty \leqslant 1\} = \sup\{|\int_{\mathbb{T}} \Phi f\, d\nu| : f \in C(\mathbb{T}), \|f\|_\infty \leqslant 1\}$$

$$= \| \Phi d\nu \|_{M(\mathbb{T})} = \int_{\mathbb{T}} |\Phi| d\nu$$

shows that $|\int_{\mathbb{T}} f \Phi d\nu| \leq 1$ for $f \in B$.

Putting $\nu = |\mu|$ and $h = \frac{d\mu}{d|\mu|}$, we see that $h \in L^\infty(d|\mu|)$ and that $|h| = 1$ $|\mu|$-almost everywhere It follows that $h, \bar{h} \in B$ and therefore for a given $\varepsilon > 0$ there are smooth functions p and q such that $\|p\|_\infty \leq 1$, $\|q\|_\infty \leq 1$, $\int_{\mathbb{T}} |h - p| d|\mu| \leq \varepsilon$, $\int_{\mathbb{T}} |\bar{h} - q| d|\mu| \leq \varepsilon$. ●

The following lemma finishes the proof of the theorem.

LEMMA 2.3. Let $\mu \in M(\mathbb{T})$ and let $\lambda_\varphi(t) \stackrel{def}{=\!=} m\{\zeta \in \mathbb{T} : |\varphi(\zeta)| > t\}$, φ being a measurable function on $\mathbb{T}$. Then

$$\overline{\lim_{t \to +\infty}}\, t \cdot \lambda_{\mathcal{K}\mu}(t) \leq \overline{\lim_{t \to +\infty}}\, t \cdot \lambda_{\mathcal{K}|\mu|}(t),$$

$$\underline{\lim_{t \to +\infty}}\, t \lambda_{\mathcal{K}\mu}(t) \geq \underline{\lim_{t \to +\infty}}\, t \lambda_{\mathcal{K}|\mu|}(t).$$

PROOF OF THE LEMMA. It is sufficient to consider the first inequality. The proof of the second one is analogous to the proof of the first. Let ε be a fixed number in $(0,1)$. By lemma 2.2. we may find a function p in $C^\infty(\mathbb{T})$ such that

$$\|p\|_\infty \leq 1, \quad \|d\mu - p\, d|\mu|\|_{M(\mathbb{T})} \leq \varepsilon^2.$$

Now

$$\mathcal{K}\mu = p \cdot \mathcal{K}|\mu| + f + g,$$

where $f(z) = \mathcal{K}(p|\mu|)(z) - p(z)\mathcal{K}(|\mu|)(z) = \int_{\mathbb{T}} \frac{p(\zeta) - p(z)}{\zeta - z} d\mu(\zeta)$ and $g = \mathcal{K}(\mu - p|\mu|)$. It is clear that $f \in C(\mathbb{T})$ and therefore $\lim_{t \to \infty} t \lambda_f(t) = 0$. By the Kolmogorov's theorem we have

$$\lambda_g(t) \leq c_0 \cdot \frac{\|\mu - p|\mu|\|}{t} \leq c_0 \frac{\varepsilon^2}{t}, \; t > 0.$$

Finally, $\|p\|_\infty \leq 1$ and therefore

$$|\mathcal{K}\mu| \leq |\mathcal{K}|\mu|| + |f| + |g|.$$

This implies

$$\lambda_{\mathcal{K}\mu}(t) \leq \lambda_{\mathcal{K}|\mu|}((1-\varepsilon)t) + \lambda_f(\tfrac{\varepsilon}{2}t) + \lambda_g(\tfrac{\varepsilon}{2}t).$$

We have

$$\overline{\lim_{t\to+\infty}}\, t\lambda_{\mathcal{K}\mu}(t) \leq \overline{\lim_{t\to+\infty}}\, t\cdot\lambda_{\mathcal{K}|\mu|}((1-\varepsilon)t) + \overline{\lim_{t\to+\infty}}\, t\lambda_f(\tfrac{\varepsilon}{2}t) +$$

$$+ \overline{\lim_{t\to+\infty}}\, t\lambda_g(\tfrac{\varepsilon}{2}t) \leq \frac{1}{1-\varepsilon}\, \overline{\lim_{t\to+\infty}}\, t\lambda_{\mathcal{K}|\mu|}(t) + \frac{2c_0}{\varepsilon}\cdot\varepsilon^2 .$$

It remains only to make ε tend to zero. ●

REMARK. The assertion of lemma 2.3 may be sharpened. Let V denote a measurable subset of $\mathbb{T}$. Then

$$\overline{\lim_{t\to+\infty}}\, t\lambda_{\mathbb{1}_V\cdot\mathcal{K}\mu}(t) = \overline{\lim_{t\to+\infty}}\, t\cdot\lambda_{\mathbb{1}_V\cdot\mathcal{K}|\mu|}(t),\quad \mathbb{1}_V(\zeta) \overset{def}{=\!=} \begin{cases} 1, & \zeta\in V \\ 0, & \zeta\notin V \end{cases} . \tag{7}$$

The second theorem of this section finds its applications in delicate questions of the interpolation theory. Remind that a point ζ in the subset E of the circle is called a density point for E if

$$\lim_{\delta\to 0+} \frac{m(I_\delta(\zeta)\cap E)}{\delta} = 1 ,$$

I_δ being an arc of the circle centered at the point ζ with the arc length equal to δ.

THEOREM 2.4. Let K be a compact on $\mathbb{T}$ and let $mK>0$. Let E be a closed subset of the set of density points for K having a zero Lebesgue measure. Then for every measure μ supported on E

$$\lim_{y\to+\infty} y\cdot m\{t\in K : |\mathcal{K}\mu(t)|>y\} = \frac{1}{\pi}\|\mu\| .$$

PROOF. The equality (7) shows that we may assume the measure μ to be positive. For every positive integer j and for every $\varepsilon>0$ let $E_j = E_j(\varepsilon)$ denote the set of all points x in E such that the inequality

$$\frac{m(I\cap K)}{m(I)} \geqslant 1-\varepsilon \tag{8}$$

holds for every open arc I, $x\in I$, $mI<j^{-1}$.
The set E being a set of density points for K, it is clear that $E = \bigcup_{j\geq 1} E_j$. The set E_j is obviously closed.

LEMMA 2.5. Let $(G_n)_{n\geq 1}$ be a sequence of open sets on $\mathbb{T}$

satisfying

a) $\lim_m mG_n = 0$;

b) for each n the closure of every component of the set G_n contains a point of E_j.

Then $\displaystyle\lim_{n\to\infty} \frac{m(G_n \cap K)}{m(G_n)} \geqslant 1-\varepsilon$.

PROOF OF THE LEMMA. Let $G_n = \bigcup_{s\geqslant 1} \ell_{ns}$, where $(\ell_{ns})_{s\geqslant 1}$ is a sequence of components of the open set G_n. The condition a) of the lemma implies the inequality $\sup_{s\geqslant 1} m(\ell_{ns}) < j^{-1}$ if n is sufficiently large. By the condition b) $\operatorname{clos} \ell_{ns} \cap E_j \neq 0$ and therefore (8) implies

$$m(\ell_{ns} \cap K) \geqslant (1-\varepsilon)\, m(\ell_{ns})$$

for every s if the number n is large. The proof is finished by adding the above inequalities. ●

For $\varepsilon > 0$ we may find j such that $\|\mu - \mu | E_j\|_{M(\mathbb{T})} < \varepsilon$. Put $\mu_j \overset{def}{=\!=} \mu | E_j$ for the brevity and consider an increasing sequence $y_n \uparrow +\infty$. It is clear that the set

$$G_n = \{ t \in \mathbb{T} : |\mathcal{K}\mu_j(t)| > y_n \}$$

is open. It follows from theorem 2.1 that $\lim_n m(G_n) = 0$ and it is easy to see that the condition b) of lemma 2.5 is valid for G_n. Therefore

$$m(G_n \cap K) \geqslant (1-\varepsilon)\, m(G_n),$$

n being large, and consequently

$$\lim_{y\to+\infty} y \cdot m\{t \in K : |\mathcal{K}\mu_j(t)| > y\} \geqslant (1-\varepsilon)\frac{1}{\pi}\|\mu_j\|.$$

But it follows from

$$\lambda_{\mathbb{1}_K \cdot \mathcal{K}\mu_j}(y) \leqslant \lambda_{\mathbb{1}_K \cdot \mathcal{K}\mu}((1-\sqrt{\varepsilon})y) + \lambda_{\mathcal{K}(\mu-\mu_j)}(y\sqrt{\varepsilon})$$

that

$$(1-\sqrt{\varepsilon}) \lim_{y\to+\infty} y m\{t \in K : |\mathcal{K}\mu(t)| > y\} \geqslant (1-\varepsilon)\frac{1}{\pi}\|\mu_j\| -$$

$$- \lim_{y\to+\infty} y\, \lambda_{\mathcal{K}(\mu-\mu_j)}(y\sqrt{\varepsilon}) \geqslant (1-\varepsilon)\frac{1}{\pi}(\|\mu\| - \varepsilon) - c_0\sqrt{\varepsilon}$$

(the last inequality is a consequence of Kolmogorov - Smirnov theorem). This means (ε is arbitrary positive number) that

$$\underline{\lim}_{y\to+\infty} y\cdot m\{t\in K : |\mathcal{K}\mu(t)|>y\} \geq \frac{1}{\pi}\|\mu\|_{M(\mathbb{T})}\ .$$

On the other hand by theorem 2.1

$$\overline{\lim}_{y\to+\infty} ym\{t\in K : |\mathcal{K}\mu(t)|>y\} \leq \overline{\lim}_{y\to+\infty} ym\{t\in\mathbb{T} : |\mathcal{K}\mu(t)|>y\} = \frac{1}{\pi}\|\mu\|.\bullet$$

REMARK. For our applications the estimate

$$\lim_{y\to+\infty} ym\{t\in K : |\mathcal{K}\mu(t)|>y\} \geq \gamma\cdot\|\mu\| \tag{9}$$

will be sufficient. Therefore the assumptions on the set E may be considerably weakened. If we assume that the set E consists of points x satisfying $\frac{m(I\cap K)}{m(I)} \geq \delta$ for every interval I containing x and having a sufficiently small diameter, then (9) holds with $\gamma = \delta\cdot\pi^{-1}$.

§3. Interpolation in the space U_A

This section is connected mainly with the space U_A but it is convenient to consider two more spaces. Let $U(\mathbb{T})$ stand for the space $\{f\in C(\mathbb{T}) : \lim_{n\to\infty}\|D_n * f - f\|_\infty = 0\}$. To define the second one let $D_{k,\ell} * f$ denote the partial sum $\sum_{j=k}^{\ell}\hat{f}(j)\zeta^j$. Then

$$SU(\mathbb{T}) \overset{def}{=} \{f\in C(\mathbb{T}) : \lim_{\substack{k\to-\infty\\ \ell\to+\infty}}\|D_{k,\ell} * f - f\|_\infty = 0\}$$

and the norm in the space $SU(\mathbb{T})$ is defined by

$$\|f\|_{SU} = \sup_{(k,\ell)}\|D_{k,\ell} * f\|_\infty\ .$$

Clearly the map $f \longrightarrow \zeta\cdot f$ is an isometry of $SU(\mathbb{T})$. It is obvious that

$$\|f\|_{U_A} \leq \|f\|_{SU} \leq 2\|f\|_{U_A}\ ,\quad f\in U_A\ .$$

This section is opened by an analog of F. and M.Riesz's theorem for the space $U(\mathbb{T})^*$. Its proof uses a construction due to Oberlin [24].

Let $\mathcal{E}$ be a Banach space imbedded continuously into the space $C(\mathbb{T})$. Then for every trigonometrical polynomial p we may define a continuous functional $\Phi_p \in \mathcal{E}^*$ by the formula

$$< f, \Phi_p > = \sum_{n\in\mathbb{Z}} \hat{f}(n)\hat{p}(n).$$

Let $\mathcal{P}$ denote the set of all trigonometrical polynomials.

DEFINITION. A functional Φ, $\Phi \in \mathcal{E}^*$, is called absolutely continuous (briefly $\Phi \in \mathcal{E}^*_a$) if

$$\Phi \in \mathcal{E}^*\text{-}clos\{\Phi_p : p\in\mathcal{P}\}.$$

Using this term the classical F. and M.Riesz theorem may be formulated as follows. If $\mu \in C(\mathbb{T})^*$ and if $\int_{\mathbb{T}} \zeta^n d\mu = 0$ for $n = 1, 2, \ldots$, then $\mu \in C(\mathbb{T})^*_a$.

THEOREM 3.1. Let $\Phi \in U(\mathbb{T})^*$ and let

$$< \zeta^n, \Phi > = 0, \quad n \in \{-1, -2, \ldots\}.$$

Then $\Phi \in U^*_a(\mathbb{T})$ and moreover

$$\Phi \in clos\{\Phi_p : p \in \mathcal{P}_A\}.$$

Some preparation is needed for the proof of this theorem. Let $\overline{\mathbb{N}} = \{0, 1, \ldots, \infty\}$ denote the one-point compactification of the set $\mathbb{N}$ and let $\mathcal{S} = \mathbb{T}\times\overline{\mathbb{N}}$. For $n\in\overline{\mathbb{N}}$ let $\mathbb{T}_n$ denote the closed subset $\mathbb{T}\times\{n\}$ of the compact $\mathcal{S}$. It is clear that the mapping i

$$i(f)|\mathbb{T}_n = D_n * f, \quad n\in\overline{\mathbb{N}}, \quad f\in U(\mathbb{T}),$$

where $D_\infty * f \overset{def}{=\!=} f$, is an isometrical imbedding of $U(\mathbb{T})$ into $C(\mathcal{S})$. Let $M(\mathcal{S})$ denote the Banach space of all finite Borel measures on $\mathcal{S}$. Then $U^* = i^*(M(\mathcal{S}))$. Therefore for every Φ, $\Phi\in U^*$, there is a measure μ on $\mathcal{S}$ such that $\Phi = i^*\mu$. Denoting $\mu_n \overset{def}{=\!=} \mu|\mathbb{T}_n$, $n\in\overline{\mathbb{N}}$, we get an identity

$$<f,\Phi> = <f, i^*\mu> = \int_{\mathbb{T}_\infty} f\,d\mu_\infty + \sum_{k\geq 0}\int_{\mathbb{T}_k} D_k * f\,d\mu_k.$$

The following lemma is actually contained in [24].

LEMMA 3.2. Let $\mu \in M(S)$ and let the functional $\Phi = i^*(\mu)$ satisfy

$$<\zeta^n, \Phi> = 0, \quad n \in \{-1, -2, \dots\}.$$

Then the measure μ_∞ is absolutely continuous with respect to Lebesgue measure on $\mathbb{T}$.

PROOF. A function $D_K * f$ being a trigonometrical polynomial, we may replace every measure μ_K by its convolution with the Vallée-Poussin Kernel. So without loss of generality we may assume that $d\mu_K = h_K\, dm$, $h_K \in \mathcal{P}$. Let the symbol ν_n denote the restriction of the measure μ to the set $\bigcup_{K=0}^{n} \mathbb{T}_K$ and let $\Phi_n = i^*(\nu_n)$. Clearly Φ_n is absolutely continuous and

$$U(\mathbb{T})^*\text{-}\lim_n \Phi_n = \Phi_\infty \overset{def}{=\!=} i^*(\mu | \bigcup_{K=0}^{n} \mathbb{T}_K).$$

It follows from the condition of the lemma that

$$\int_{\mathbb{T}_\infty} \bar{\zeta}^n d\mu_\infty = -<\bar{\zeta}^n, \Phi_\infty>, \quad n \in \{1, 2, \dots\}.$$

Multiplying these equalities by z^n, $|z| < 1$, and summing over n, $n \in \{1, 2, \dots\}$ we get

$$\mathcal{K}\mu_\infty(z) = -\mathcal{K}\Phi_\infty(z) + < \mathbb{1}, \Phi_\infty > + \int_{\mathbb{T}} d\mu_\infty.$$

By Vinogradov's theorem

$$m\{t \in \mathbb{T} : |\mathcal{K}\Phi_\infty(t) - \mathcal{K}\Phi_n(t)| > y\} \leq \frac{const}{y} \| \Phi_n - \Phi_\infty \|_{U(\mathbb{T})^*}.$$

Taking into account that $\mathcal{K}\Phi_n \in \mathcal{P}_A$ and that $\lim_{n\to\infty} \| \Phi_n - \Phi_\infty \|_{U^*} = 0$, we have

$$\lim_{y \to +\infty} y \cdot m\{t \in \mathbb{T} : |\mathcal{K}\mu_\infty(t)| > y\} = 0.$$

It now follows from theorem 2.1 that μ_∞ is absolutely continuous. ●

Now we are in a position to prove theorem 3.1.

PROOF OF THEOREM 3.1. Let Φ be a functional satisfying the conditions of the theorem. There is a measure μ in $M(S)$ such that $\Phi = i^*(\mu)$ and $d\mu_K = h_K\, dm$, h_K being a trigonometrical polynomial. It follows from lemma 3.3 that the measure μ_∞ is absolutely continuous. Therefore for every $\varepsilon>0$ there is a polynomial p satisfying

$$\| d\mu_\infty - p\,dm \|_{M(\mathbb{T}_\infty)} < \varepsilon \ .$$

This implies

$$\| \Phi - \Phi_n - i^*(p\,dm) \|_{U^*(\mathbb{T})} \leqslant \varepsilon + \| \Phi_\infty - \Phi_n \|_{U(\mathbb{T})^*}$$

and it is clear now that $\Phi \in U_a^*(\mathbb{T})$.

To prove the second part of the theorem, we note that $f \longrightarrow D_n * f$ is a contractive mapping of $U(\mathbb{T})$ into itself. Duality arguments show

$$\| D_n * \Phi \|_{U^*} \leqslant \| \Phi \|_{U^*}$$

and therefore $\lim_{n\to\infty} \| D_n * \Phi - \Phi \|_{U^*} = 0$ for Φ in $clos\{\Phi_p : p \in \mathcal{P}\}$. The functional Φ being orthogonal to the monomials ζ^n, $n \in \{-1,-2,\dots\}$, it is clear that $D_n * \Phi \in \mathcal{P}_A$. ●

THEOREM 3.3. (An analog of the Lebesgue decomposition theorem). For every Φ in U^* (respectively in U_A^*) there is a unique pair (μ_s, Φ_a), where μ_s is a singular measure and $\Phi_a \in U_a^*$, such that

$$\Phi = \mu_s + \Phi_a, \quad \|\Phi\|_{U^*} = \|\mu_s\|_{M(\mathbb{T})} + \|\Phi_a\|_{U^*} \ .$$

PROOF. We shall give a proof only for the space $U(\mathbb{T})^*$ (for U_A^* it is analogous). Let $\Phi \in U(\mathbb{T})^*$ and let μ be a measure on $\mathbf{S}$ satisfying the equality $\Phi = i^*\mu$.

If $\nu \perp i(U)$, then $i^*\nu = 0$ and by lemma 3.2 $\nu_\infty \in M_a(\mathbb{T})$. It is useful to note that

$$\|\Phi\| = inf\{ \|\mu - \nu\|_{M(\mathbf{S})} : \nu \perp i(U) \} \ .$$

Let $(\mu_\infty)_s$ denote the singular part of the measure μ_∞. Then

$$\|\Phi\| = \|(\mu_\infty)_s\| + inf\{ \|\mu - (\mu_\infty)_s - \nu\|_{M(\mathbf{S})} : \nu \perp i(U) \} \ .$$

The functional $\Phi_a = i^*(\mu - (\mu_\infty)_s)$ is obviously absolutely continuous. Denoting $\mu_s = i^*\mu_\infty$ we see that

$$\Phi = \mu_s + \Phi_a \quad \text{and} \quad \|\Phi\|_{U^*} = \|\mu_s\|_{M(\mathbb{T})} + \|\Phi_a\|_{U^*} .$$

The unicity of the above decomposition is a simple consequence of theorem 2.1. ●

COROLLARY 3.4. Let E be a closed subset of $\mathbb{T}$ and let $mE = 0$. Then there is a linear interpolating operator $L : C(E) \longrightarrow U_A$ such that $\|L\| = 1$.

The proof is a combination of theorem 3.1 with Bishop's theorem ([25], ; and with a theorem due to A.Pełczyński (see [35], [36]).

In connection with this corollary a question arises whether a bounded linear interpolating operator in the Banach problem does exist. It appears that the answer is negative even in the case of the disc-algebra.

THEOREM 3.5. Let Λ be an infinite Banach set. Then the restriction operator $Q : f \longrightarrow \hat{f} | \Lambda$, $f \in C_A$, has no bounded right inverse operator.

PROOF. Suppose that this were not the case. Then we could find a bounded operator $R : \ell^2(\Lambda) \longrightarrow C(\mathbb{T})$, satisfying $QR = \mathbb{1}_{\ell^2(\Lambda)}$. It follows that $P = RQ$ is a bounded projection in $C(\mathbb{T})$:

$$(RQ)^2 = R(QR)Q = RQ .$$

Using the famous Arens-Rudin trick ([23]) we get that the projection

$$P_\Lambda \overset{def}{=\!=} \int_{\mathbb{T}} \tau_{\bar{t}} P \tau_t \, dm(t) , \qquad \tau_t f(\zeta) = f(\zeta \bar{t}) ,$$

is bounded. Simple computations show that

$$P_\Lambda \zeta^n = \zeta^n \cdot \mathbb{1}_\Lambda(n) .$$

Here $\mathbb{1}_\Lambda$ denotes the indicator of the set Λ. It follows from the Banach theorem that for every sequence a, $a \in \ell^2(\Lambda)$, there is a function g in $C(\mathbb{T})$ satisfying

$$P_\Lambda g = \sum_{n \in \Lambda} a_n \zeta^n .$$

The set Λ being infinite, there is an infinite subset Λ_0 of Λ with Hadamard's gaps. Then Λ_0 is a Sidon set (see

[37]) and by the definition of Sidon set we have a contradiction:

$$\sum_{n\in\Lambda_0} |a_n| < +\infty . \qquad \bullet$$

Up to now the problem of interpolation for subsets of the circle has been considered. It is time to discuss that problem for subsets of the unit disc. The things are more complicated in this case. It turned out that there exists a subset of $\operatorname{clos}\mathbb{D}$ which is an interpolating set for the disc-algebra but fails to be interpolating for U_A. The construction of the corresponding example is our second task in this section. The following theorem is a matter of common knowledge (see, nevertheless, the exercise 7 to ch.10 of [23]).

THEOREM. A closed subset E of $\operatorname{clos}\mathbb{D}$ is interpolating one for C_A iff

(1) $m(E\cap\mathbb{T}) = 0$ and $E\cap\mathbb{D}$ is a discrete subset of $\mathbb{D}$;

(2) $$\inf_{\lambda\in E\cap\mathbb{D}} \prod_{\xi\in E\cap(\mathbb{D}\setminus\{\lambda\})} \left|\frac{\lambda-\xi}{1-\bar{\xi}\lambda}\right| > 0 .$$

The condition (2) of the theorem is the famous Carleson condition. We refer the reader to the survey [38], [39], where one can find various reformulations of (2). Two conditions important for our construction are closely connected with the Carleson condition. These are Frostman condition

$$\sup_{t\in\mathbb{T}} \sum_{\lambda\in E} \frac{1-|\lambda|^2}{|t-\lambda|} < +\infty \qquad (\mathrm{F})$$

and the sparseness condition

$$\inf_{\lambda\neq\xi,\,\lambda,\xi\in E} \left|\frac{\lambda-\xi}{1-\bar{\xi}\lambda}\right| > 0 . \qquad (\mathrm{S})$$

It is easy to see that if a set satisfies (F) and (S) both then it satisfies the Carleson condition too.

The importance of the Frostman condition is emphasized by the following simple fact. Let $\mu = \sum_{\lambda\in E}(1-|\lambda|^2)\,\delta_\lambda$, where δ_λ is the unit mass at λ. Then the natural embedding $\mathcal{K}C_A^* \subset L^1(\mu)$ is continuous iff $E\in(\mathrm{F})$. Moreover

$$\sup_{\|f\|_{\mathcal{K}C_A^*}\leq 1} \sum_{\lambda\in E} |f(\lambda)|(1-|\lambda|^2) = \sup_{t\in\mathbb{T}} \sum_{\lambda\in E} \frac{1-|\lambda|^2}{|t-\lambda|} .$$

Some technical preparations are needed for the construction of our example. From now on we shall consider the space U_A as a closed subset of $SU(\mathbb{T})$ with the induced norm. The space $SU(\mathbb{T})$ is not an algebra but, the operator $f \to z\cdot f$ being isometry, the algebra $\mathcal{F}\ell^1(\mathbb{T}) \overset{def}{=} \{ f \in C(\mathbb{T}) : \sum_{n\in\mathbb{Z}} |\hat{f}(n)| = \|f\|_{\mathcal{F}\ell^1} < \infty \}$ is contained in the algebra of all multipliers for $SU(\mathbb{T})$ and moreover

$$\sup_{\|g\|_{SU}\le 1} \| f\cdot g\|_{SU} \le \|f\|_{\mathcal{F}\ell^1} . \tag{10}$$

Let α denote a function defined on the family of all finite subsets of $\mathbb{D}$ by the formula

$$\alpha(\Lambda) = \sup_{\|x\|_{C(\Lambda)}\le 1} \inf\{ \|f\|_{SU} : f\in U_A \,;\; f|\Lambda = x \} .$$

For a finite subset Λ let B^Λ denote the Blaschke product with the zero set Λ. If $\lambda\in\Lambda$ we put for brevity $B_\lambda^\Lambda \overset{def}{=} B^{\Lambda\setminus\{\lambda\}}$.

LEMMA 3.6. Let Λ be a finite subset of $\mathbb{D}$. Then

$$\sup_{t\in\mathbb{T}} \sum_{\lambda\in\Lambda} \frac{1-|\lambda|^2}{|t-\lambda|} \le \alpha(\Lambda)\cdot \|B^\Lambda\|_{\mathcal{F}\ell^1} .$$

The proof is based on the explicit formula for $\alpha(\Lambda)$:

$$\alpha(\Lambda) = \sup_{\|x\|_{C(\Lambda)}\le 1} \inf_{h\in U_A} \Big\| \sum_{\lambda\in\Lambda} x(\lambda) \frac{B_\lambda(z)}{B_\lambda(\lambda)} \cdot \frac{1-|\lambda|^2}{(1-\bar{\lambda} z)} + Bh \Big\|_{SU}$$

where $B \overset{def}{=} B^\Lambda$ for the sake of brevity. The equality $\|B\|_{\mathcal{F}\ell^1} = \|\bar{B}\|_{\mathcal{F}\ell^1}$ and (10) show that

$$\alpha(\Lambda)\|B\|_{\mathcal{F}\ell^1} \ge \sup_{\|x\|_{C(\Lambda)}\le 1} \inf_{h\in U_A} \Big\| \sum_{\lambda\in\Lambda} x(\lambda) \frac{1}{B_\lambda(\lambda)} \frac{1-|\lambda|^2}{\zeta-\lambda} + h \Big\|_{SU} .$$

We have by the Hahn-Banach theorem that

$$\inf_{h\in U_A} \Big\| \sum_{\lambda\in\Lambda} \frac{x(\lambda)}{B_\lambda(\lambda)} \cdot \frac{1-|\lambda|^2}{\zeta-\lambda} + h \Big\|_{SU} = \sup_{\|\Phi\|_{SU^*}\le 1,\, \Phi|U_A = 0} \Big| \sum_{\lambda\in\Lambda} \frac{x(\lambda)}{B_\lambda(\lambda)} \Big\langle \frac{1-|\lambda|^2}{\zeta-\lambda}, \Phi \Big\rangle \Big| .$$

Now let z be a fixed point of $\mathbb{D}$ and let

$$\langle f, \Phi_z \rangle = \sum_{n=1}^{\infty} \hat{f}(-n) z^{n-1} .$$

The norm of the usual orthogonal projection of the space $SU(\mathbb{T})$ onto U_A being equal to **one**, we have $\sup_{z\in\mathbb{D}} \|\Phi_z\| \le 1$. Clearly $\Phi_z | U_A \equiv 0$ and $\langle (\zeta-\lambda)^{-1}, \Phi_z \rangle = (1-\lambda z)^{-1}$. Hence

$$\|B\|_{\mathcal{F}\ell^1}\cdot\alpha(\Lambda) \geqslant \sup_{|z|\leqslant 1}\ \sup_{\|x\|_{C(\Lambda)}\leqslant 1}\left|\sum_{\lambda\in\Lambda}\frac{x(\lambda)}{B_\lambda(\lambda)}\cdot\frac{1-|\lambda|^2}{1-\bar\lambda z}\right| =$$

$$= \sup_{|z|=1}\sum_{\lambda\in\Lambda}\frac{1}{|B_\lambda(\lambda)|}\cdot\frac{1-|\lambda|^2}{|1-\bar\lambda z|} \geqslant \sup_{t\in\mathbb{T}}\sum_{\lambda\in\Lambda}\frac{1-|\lambda|^2}{|t-\lambda|}. \quad \bullet$$

Let H^∞ denote the Hardy class of all bounded functions holomorphic in $\mathbb{D}$ and let

$$U_A^\infty = \{ f\in H^\infty :\ \sup_{h\in\mathbb{Z}}\| D_h * f\|_\infty < +\infty \}.$$

Our first example exhibits the difference between the spaces H^∞ and U_A^∞ from the interpolation theoretical point of view.

EXAMPLE 1. Let $(a_k)_{k\geqslant 1}$ be an increasing sequence of positive numbers such that $\sup \frac{1-a_{k+1}}{1-a_k} < 1$, $\lim_{k\to\infty} a_k = 1$. And let $(s_k)_{k\geqslant 1}$ be a sequence of positive integers satisfying

$$\lim_{k\to+\infty}(\log s_k)(1-a_k) = +\infty .$$

Then the set $\Lambda \overset{def}{=\!=} \bigcup_{k\geqslant 1}\Lambda_k$, $\Lambda_k \overset{def}{=\!=} \{ z\in\mathbb{C} : z^{s_k} = a_k \}$, has the following properties

$$H^\infty|\Lambda = \ell^\infty(\Lambda), \quad \text{but } U_A^\infty|\Lambda \neq \ell^\infty(\Lambda).$$

PROOF. It is clear that $\Lambda\in(S)$ and it is easy to check that the measure $\sum_{\lambda\in\Lambda}(1-|\lambda|^2)\,\delta_\lambda$ is a Carleson measure [38]. This implies $H^\infty|\Lambda = \ell^\infty(\Lambda)$. We are going to prove now that $\lim_{k\to\infty}\alpha(\Lambda_k) = +\infty$ which implies obviously $U_A^\infty|\Lambda \neq \ell^\infty(\Lambda)$. To do this we fix a positive integer k and note that

$$B^{\Lambda_k}(z) = \frac{a_k - z^{s_k}}{1-a_k z^{s_k}} .$$

Clearly $\left\|\frac{z-a}{1-\bar a z}\right\|_{\mathcal{F}\ell^1} = 1+2|a|$. Therefore $\|B^{\Lambda_k}\|_{\mathcal{F}\ell^1}\leqslant 3$ and lemma 3.6. implies

$$\sup_{t\in\mathbb{T}}\sum_{\lambda\in\Lambda_k}\frac{1-|\lambda|^2}{|t-\lambda|}\leqslant 3\cdot\alpha(\Lambda_k).$$

To estimate the left-hand side of the above inequality we use a simple but efficient estimate due to V.I.Vasjunin (see [40]):

$$\sup_{t\in\mathbb{T}} \sum_{\lambda\in\Lambda_k} \frac{1-|\lambda|^2}{|t-\lambda|} \geqslant \int_{\mathbb{T}} \sum_{\lambda\in\Lambda_k} \frac{1-|\lambda|}{|t-\lambda|}\, dm(t) \geqslant \frac{1}{4\pi} \sum_{\lambda\in\Lambda_k} (1-|\lambda|)\log\frac{1}{1-|\lambda|}.$$

It follows that

$$3\,\alpha(\Lambda_k) \geqslant \frac{1}{4\pi}\cdot s_k\cdot(1-a_k^{1/s_k})\log\frac{1}{1-a_k^{1/s_k}} .$$

The condition $\lim\limits_{k\to\infty}(1-a_k)\log s_k = +\infty$ of the lemma implies therefore

$$\lim_{k\to\infty} \alpha(\Lambda_k) = +\infty . \quad \bullet$$

The next lemma is important for our second example.

LEMMA 3.7. Let n be a positive integer, let ℓ be an integer dividing n and let

$$\Lambda_n \stackrel{def}{=} \{z\in\mathbb{C} : z^n = \tfrac{1}{2}\},\quad \Lambda_{n,\ell} \stackrel{def}{=} \Lambda_n \cap \{z\in\mathbb{C} : |\arg z| \leqslant \tfrac{2\pi}{\ell}\} .$$

Then $\alpha(\Lambda_n) \leqslant 2\ell\cdot\alpha(\Lambda_{n,\ell})$.

PROOF. It is well-known that the function

$$\varphi_0(e^{it}) = \max\left(0, 1-\left|\frac{\ell t}{2\pi}\right|\right),\quad t\in[-\pi,\pi]$$

has non-negative Fourier coefficients. Therefore

$$\|\varphi_0\|_{\mathcal{F}\ell^1} = \varphi_0(1) = 1 .$$

Let $\zeta_\ell = \exp\left(\frac{2\pi i}{\ell}\right)$, and let

$$\varphi_k(t) = \varphi_0(t\bar{\zeta}_\ell^k),\quad t\in\mathbb{T},\quad k\in\{0,1,\dots,\ell-1\} .$$

Then

$$\sum_{k=0}^{\ell-1} \varphi_k = \mathbb{1}$$

on $\mathbb{T}$. We are going to build an analogous decomposition of the indicator of the set Λ_n. For this aim the functions φ_k should be replaced by their Lagrange polynomials. Recall that the Lagrange polynomial of a function f, $f\in\mathcal{F}\ell^1(\mathbb{T})$, is defined by

$$I_n(f,z) \stackrel{def}{=} \sum_{k=0}^{n-1} z^k \cdot \sum_{j\equiv k (\mathrm{mod}\, n)} \hat{f}(j) .$$

We see that the restriction of $I_n(f,z)$ on the group of

roots of unity of the power n is identical with that of the function f. Clearly

$$\|I_n(f)\|_{\mathcal{F}\ell^1} \leq \|f\|_{\mathcal{F}\ell^1} .$$

Denoting $P_k(z) \overset{def}{=\!=} I_n(\varphi_k, 2^{1/n} z)$, we have

$$\|P_k\|_{\mathcal{F}\ell^1} = \sum_{j=0}^{n} |\hat{P}_k(j)| 2^{j/n} \leq 2 .$$

Let $E_0 = \Lambda_{n,\ell}$ and let $E_k = \zeta_\ell^k \cdot E_0$, $k \in \{0, 1, \ldots, \ell-1\}$. It is easy to check that

$$supp(P_k | \Lambda_n) \subset E_k , \quad \sum_{k=0}^{\ell-1} P_k | \Lambda_n = \mathbb{1}_{\Lambda_n} .$$

For $x \in C(\Lambda_n)$ let $\{f_0, f_1, \ldots, f_{\ell-1}\}$ denote a subset of U_A such that

$$f_k | E_k = x | E_k , \quad k \in \{0, \ldots, \ell-1\} .$$

The identity

$$x = P_0 f_0 + P_1 f_1 + \ldots + P_{\ell-1} f_{\ell-1}$$

being valid on the set Λ_n, it follows that

$$inf\{\|f\|_{SU} : f \in U_A,\ f|\Lambda_n = x\} \leq \sum_{k=0}^{\ell-1} \|P_k\|_{\mathcal{F}\ell^1} \cdot inf\{\|f_k\|_{SU} : f_k \in U_A, f_k|E_k = x|E_k\}.$$

Taking supremum in the above inequality, we get

$$\alpha(\Lambda_n) \leq 2 \cdot \sum_{k=0}^{\ell-1} \alpha(E_k) = 2\ell \cdot \alpha(\Lambda_{n,\ell}) . \quad \bullet$$

THEOREM 3.8. There is an interpolating subset of $clos\,\mathbb{D}$ for the disc-algebra which is not interpolating for the space U_A.

PROOF. Let

$$J_k = \{e^{it} : \frac{2\pi}{2^{2k}} \leq t \leq \frac{2\pi \cdot 2}{2^{2k}}\} , \quad k \in \{1, 2, \ldots\}$$

be a sequence of arcs on the unit circle and let $n_k = 2^{2^{8k}}$. The set E_k is defined by

$$E_k = \{z^{n_k} = \frac{1}{2} , \quad z \cdot |z|^{-1} \in J_k\} .$$

Then the desired set E can be defined as follows:

$$E = \bigcup_{k=1}^{\infty} E_k \cup \{1\} .$$

Clearly $C_A | E = C(E)$. To prove $U_A | E \neq C(E)$ it is obviously sufficient to check that $\lim_{k \to +\infty} \alpha(E_k) = +\infty$.

Applying lemma 3.7 to the set E_k with $n = n_k$, $\ell = 2^{2k+1}$, we have

$$\alpha(\Lambda_{n_k}) \leq 2^{2k+2} \alpha(E_k) .$$

On the other hand

$$\alpha(\Lambda_{n_k}) \geq \frac{1}{4\pi} n_k \left(1 - \left(\tfrac{1}{2}\right)^{1/n_k}\right) \log \frac{1}{1 - \left(\tfrac{1}{2}\right)^{1/n_k}} \sim$$

$$\sim \frac{\log 2}{4\pi} \cdot \log \frac{n_k}{\log 2} = 2^{8k} \cdot \frac{\log 2}{4\pi} \cdot \log\left(\frac{2}{\log 2}\right)$$

and therefore $\lim_{k \to \infty} \alpha(E_k) = +\infty$. ●

It is natural now to look for conditions on the set E sufficient for the equality $U_A | E = C(E)$. Let E be a **closed** subset of $\operatorname{clos} \mathbb{D}$ and let

$$E_1 = \mathbb{T} \cap E \quad , \quad E_2 = \mathbb{D} \cap E .$$

THEOREM 3.9. The equality $U_A | E = C(E)$ holds iff $mE_1 = 0$ and $U_A^{\infty} | E_2 = \ell^{\infty}(E_2)$.

PROOF. The necessity of the first of the above conditions is obvious and the necessity of the second one is a consequence of simple limit arguments. Let us assume now the fulfilment of the conditions of the theorem and let $R_E : U_A \to C(E)$ denote the restriction operator. By the Banach theorem $U_A | E = C(E)$ iff the conjugate operator R_E^* is an isomorphism onto its own image in U_A^*. Every measure μ, $\mu \in M(E)$ being decomposed in a sum $\mu = \mu_1 + \mu_2$, where $\mu_j \overset{\text{def}}{=\!=} \mu | E_j$, $j = 1, 2$, the following formulae hold

$$\langle f, R_E^* \mu_1 \rangle = \int_E f \, d\mu_1 ; \quad \langle f, R_E^* \mu_2 \rangle = \int_{\mathbb{T}} f(\zeta) \left\{ \int_{E_2} \frac{1-|z|}{|\zeta - z|^2} d\mu(z) \right\} dm(\zeta).$$

These formulae show that the functional $R_E^* \mu_1$ corresponds to the singular measure and that $R_E^* \mu_2$ is absolutely continuous. By theorem 3.2

$$\| R_E^* \mu \|_{U_A^*} = \| \mu_1 \|_{M(\mathbb{T})} + \| R_E^* \mu_2 \|_{U_A^*} .$$

It follows from the identity $U_A^{\infty}|E_2 = \ell^{\infty}(E_2)$ that

$$\|R_E^* \mu_2\|_{U_A^*} \geq c \cdot \|\mu_2\|_{M(E_2)} ,$$

c being a positive constant. ●

The last theorem restricts our attention by the interpolation in U_A^{∞}.

DEFINITION. A subset E of $\mathbb{D}$ satisfies the weak Newman condition (briefly $E \in (w\mathcal{N})$) if

$$\sup\{(1-|\lambda|)^{-1} \cdot \sum_{|\xi| \geq |\lambda|, \xi \in E} (1-|\xi|)\} < +\infty \qquad (w\mathcal{N})$$

THEOREM 3.10. Let $E \in (\mathcal{S})$ and let $E = F \cup N$, where $F \in (\mathbb{F})$, $N \in (w\mathcal{N})$. Then $U_A^{\infty}|E = \ell^{\infty}(E)$.

The proof is based on two applications of the famous Earl's theorem [41], which were mentioned in [39]. For the sake of completeness of the exposition we reproduce the proofs here. The theorem of Earl asserts that if $\|x\|_{\infty}$ is sufficiently small then there is a Blaschke product B such that $x|E = B|E$. The zero set of B can be regarded as a small perturbation of the set E not destroying the property to be a Frostman set or a weak Newman set (see [38]). It remains to use two known results. The Blaschke product B^N with the zero set N satisfies

$$\hat{B}^N(n) = O\left(\frac{1}{n}\right), \quad n \to +\infty$$

(see [42], [39], p.18) and therefore $B^N \in U_A^{\infty}$. It was shown in [50] that B^F is a multiplier of the space of Cauchy integrals in the unit disc and that these multipliers are multipliers of the space U_A^{∞} (see also §4 below). Hence $B = B^F \cdot B^N \in U_A^{\infty}$. ●

We are finishing this section by the discussion of an interesting phenomenon which connects the real variable interpolation theory with the complex one.

The condition $-\infty < \int_{\mathbb{T}} \log|f|\, dm$ being satisfied for $f \in C_A$, $f \not\equiv 0$, the necessity of Carleson-Rudin theorem is obvious. It appears that there are spaces X, $X \subset C(\mathbb{T})$, containing functions vanishing on a set of positive measure, but nevertheless

$$X|E = C(E) \Rightarrow mE = 0 . \qquad (11)$$

The famous Menshov's theorem yields such an example.

THEOREM (Menshov [43]). Let $E = \operatorname{clos} E \subset \mathbb{T}$ and let $U(\mathbb{T}) | E = C(E)$. Then $mE = 0$.

Simple proofs of this theorem can be found in [44] and [24]. It is clear that $C_A | \mathbb{T} \not\subset U(\mathbb{T})$. Therefore the question arises whether the space X larger than disc-algebra and satisfying (11) does exist. We shall show that the space

$$C_H \overset{def}{=} \{ f \in C(\mathbb{T}) : P_+ f \in C(\mathbb{T}) \}, \quad \|f\| \overset{def}{=} \|f\|_\infty + \|P_+ \bar{f}\|_\infty ,$$

has the desired properties ($P_+ f \overset{def}{=} \sum_{k \geq 0} \hat{f}(k) \zeta^k$).

THEOREM 3.11. Let $E = \operatorname{clos} E \subset \mathbb{T}$ and let $C_H | E = C(E)$. Then $mE = 0$.

PROOF. There is a natural imbedding of the space C_H into $C(\mathbb{T} \times \{1,2\})$ defined by

$$j(f) = (f, \overline{P_+ \bar{f}}), \quad f \in C_H .$$

Then the annihilator of the subspace $j(C_H)$ is equal to the set of all pairs (μ_1, μ_2) satisfying

$$\int_{\mathbb{T}} f \, d\mu_1 + \int_{\mathbb{T}} \overline{P_+ \bar{f}} \, d\mu_2 = 0, \quad f \in C_H .$$

Putting $f = z^n$, $n \geq 1$, in this formula, we get $d\mu_1 = h\,dm$, $h \in H^1$. Let now $f = \bar{z}^n$, $n \geq 0$, then the formula implies $\hat{\mu}_1(n) + \hat{\mu}_2(n) = 0$, $n \geq 0$. It follows that $d\mu_2 = (g - h)\,dm$, $g \in \overline{H^1_0}$. Therefore

$$\|(\mu, 0)\|_{j(C_H)^*} = \inf \{ \|\mu - h\,dm\|_{M(\mathbb{T})} + \int_{\mathbb{T}} |g - h| \, dm : h \in H^1, \ g \in \overline{H^1_0} \} .$$

Assuming E to be an interpolating set for C_H with $mE > 0$ we get the existence of a positive constant c such that the following inequality holds:

$$\inf_{h \in H^1, g \in \overline{H^1_0}} \{ \int_E |f - h| \, dm + \int_{\mathbb{T} \setminus E} |h| \, dm + \int_{\mathbb{T}} |g - h| \, dm \} \geq c \cdot \int_{\mathbb{T}} |f| \, dm ,$$

f being a summable function supported on E. For every function k in $L^2(\mathbb{T})$ we put in the previous inequality $f = P_+ k | E$, $h = P_+ k$, $g = -P_- h$. Then we get

$$c \int_E |P_+ k| \, dm \leq \int_{\mathbb{T} \setminus E} |P_+ k| \, dm + \int_{\mathbb{T}} |k| \, dm .$$

If now k is the Poisson kernel at the point $1 - \varepsilon$, $\varepsilon \in (0,1)$, then $|P_+ k(\zeta)| = |\zeta - 1 + \varepsilon|^{-1}$ for ζ in $\mathbb{T}$. Denoting

$\zeta = e^{i\theta}$, we get

$$\frac{1}{|\theta|+\varepsilon} \leqslant |P_+ k(\zeta)| \leqslant \frac{3}{|\theta|+\varepsilon} .$$

Let $E^* \stackrel{def}{=} \{\theta \in (-\pi,\pi] : e^{i\theta} \in E\}$ and let us assume that 0 is a density point for E^*. Adding the integral $c \cdot \int\limits_{\mathbb{T}\setminus E} |P_+ h|\, dm$ to the both parts of (3) we get

$$c \cdot \int_{-\pi}^{\pi} \frac{d\theta}{|\theta|+\varepsilon} \leqslant 3(1+c) \int\limits_{[-\pi,\pi]\setminus E^*} \frac{d\theta}{|\theta|+\varepsilon} + 2\pi .$$

It is clear now that $\lambda(\theta) \stackrel{def}{=} \int_{-\theta}^{\theta} \mathbb{1}_{[-\pi,\pi]\setminus E^*}(t)\,dt = o(\theta),\ \theta \to 0+$, and therefore

$$\int\limits_{[-\pi,\pi)\setminus E} \frac{d\theta}{|\theta|+\varepsilon} = \int_0^{\pi} \frac{d\lambda(\theta)}{\theta+\varepsilon} = \frac{\lambda(\pi)}{\pi+\varepsilon} + \int_0^{\pi} \frac{\lambda(\theta)}{\theta+\varepsilon}\, d\theta = o\left(\log \frac{1}{\varepsilon}\right).$$

This implies the contradiction:

$$2 \cdot \log\left(1+\frac{\pi}{\varepsilon}\right) = \int_{-\pi}^{\pi} \frac{d\theta}{|\theta|+\varepsilon} = o\left(\log \frac{1}{\varepsilon}\right). \quad \bullet$$

§ 4. Two applications of the asymptotic formula

In this section we shall give two different applications of the results proved in §2. The first one is connected with the interpolation theory in the space $V(K)$. Recall that K denotes a compact of positive Lebesgue measure on the unit circle $\mathbb{T}$ and that

$$V(K) \stackrel{def}{=} \{ f \in L^2(\mathbb{T}) : supp(f) \subset K,\ f_+ \in U_A \} .$$

The norm in the Banach space $V(K)$ is defined by

$$\|f\|_{V(K)} \stackrel{def}{=} \Big(\int\limits_{\mathbb{T}} |f|^2 dm \Big)^{1/2} + \|f_+\|_{U_A} .$$

Clearly the space $V(K)$ is a closed subset of $V(\mathbb{T})$

DEFINITION. A subset Λ of $\mathbb{Z}$ is called $\Lambda(p)$-set, $p>0$, if the L^r-norms are equivalent on the space of all trigonometrical polynomials with frequencies in Λ for all r, $r \leqslant p$.

THEOREM 4.1. Let E be a closed subset of $\mathbb{T}$ contained in the set of density points for K and let $mE=0$. Let Λ be a $\Lambda(s)$-set for some $s>2$. The operator Q is defined by the formula

$$Qf \overset{def}{=\!=} (f_+|E,\ \hat{f}|\Lambda), \quad f\in V(K).$$

Then $QV(K)=C(E)\times \ell^2(\Lambda)$.

The proof of theorem 4.1 is based on a description of the conjugate space $V(K)^*$. Every functional Φ in $V(\mathbb{T})^*$ gives rise to a pair of analytic functions

$$\mathcal{K}\Phi(z) \overset{def}{=\!=} \sum_{n=0}^{\infty} <\zeta^n, \Phi> z^n, \quad |z|<1,$$

$$\mathcal{K}_-\Phi(z) \overset{def}{=\!=} \sum_{n=1}^{\infty} <\bar{\zeta}^n, \Phi> z^{-n}, \quad |z|>1.$$

Let now H^2 denote the usual Hardy class in $\mathbb{D}$ and let H^2_- be the Hardy class outside the unit disc. By Fatou's theorem these spaces may be considered as closed subspaces of $L^2(\mathbb{T})=H^2_-\oplus H^2$. Taking into account this agreement, we have

$$V(\mathbb{T}) = H^2_- \oplus U_A\,|\,\mathbb{T}.$$

This identity implies $\mathcal{K}\Phi\in \mathcal{K}U_A^*$ and $\mathcal{K}_-\Phi\in H^2_-$ for $\Phi\in V(\mathbb{T})^*$. The radial limits of $\mathcal{K}\Phi$ existing a.e., the function

$$P\Phi(\zeta) = \mathcal{K}\Phi(\zeta)+\mathcal{K}_-\Phi(\zeta)$$

is defined almost everywhere on $\mathbb{T}$. The correspondence $\Phi\to P\Phi$ is obviously one-to-one. Indeed, it follows from $\mathcal{K}\Phi(\zeta)=-\mathcal{K}_-\Phi(\zeta)$, $\zeta\in\mathbb{T}$, that $\mathcal{K}\Phi\in H^2$ by Smirnov's theorem (see [46]). Therefore $\Phi=0$. The space U_A satisfying the axiom 2, for every p, $0<p<1$, there is a positive constant c_p such that

$$\int_{\mathbb{T}} |P\Phi|^p dm \leqslant c_p\cdot \|\Phi\|^p_{V(\mathbb{T})^*}. \tag{12}$$

Let $E_*=\{\zeta\in\mathbb{T} : \bar{\zeta}\in E\}$. The next lemma is a generalization of a statement in [17], p.140-141.

LEMMA 4.2. Let $K=\operatorname{clos} K\subset\mathbb{T}$, $mK>0$ and let $\Phi\in V(\mathbb{T})^*$, $\Phi|V(K)\equiv 0$. Then $P\Phi(\zeta)=0$ for almost all ζ in K_*.

PROOF. The linear set

$$\mathcal{L}(K) \overset{def}{=\!=} \{\Phi\in V(\mathbb{T})^* : P\Phi(\zeta)=0 \text{ a.e. on } K_*\}$$

is weakly closed in $V(\mathbb{T})^*$. Indeed, the space $V(\mathbb{T})$ is obviously separable and therefore by the Banach theorem it **is** sufficient to prove that $\mathcal{L}(K)$ **is a weak**-$*$ sequentially closed subspace. To do this let $*-\lim_n \Phi_n = \Phi$, $\Phi_n \in \mathcal{L}(K)$. Then $\|\mathcal{K}\Phi_n\|_{\mathcal{K}U_A^*} = O(1)$ and $\lim_n \mathcal{K}_-\Phi_n = \mathcal{K}_-\Phi$ in the weak topology of H_-^2 . Closures of convex subsets of H_-^2 in the weak and strong topologies being identical, we get a sequence $(\Psi_n)_{n\geq 1}$ in $V(\mathbb{T})^*$ such that

$$\Psi_n = \sum_{k\geq n} \alpha_{kn}\Phi_n \quad , \quad \sum_{k\geq n} |\alpha_{kn}| \leq 1 ,$$

where $(\alpha_{kn})_{k\geq 1}$ are finite sequences of complex numbers satisfying $\lim_n \mathcal{K}_-\Psi_n = \mathcal{K}_-\Phi$ in the norm-topology of H_-^2 . In particular

$$\lim_{n\to\infty} \sum_{k\geq n} \alpha_{kn} = 1$$

and therefore $\lim_{n\to\infty} \mathcal{K}\Psi_n(z) = \mathcal{K}\Phi(z)$ for $z\in\mathbb{D}$. It is clear also that $\|\mathcal{K}\Psi_n\|_{\mathcal{K}U_A^*} = O(1)$. We have by the definition of the space $\mathcal{L}(K)$:

$$\mathcal{K}\Psi_n \mid K_* = -\mathcal{K}_-\Psi_n \mid K_*$$

which implies the convergence of the restrictions $\mathcal{K}\Psi_n \mid K_*$ in L^2 -metric. The space $\mathcal{K}U_A^*$ being continuously imbedded **into** $H^{1/2}$, the Hinchin-Ostrovskii (see [46]) theorem shows

$$\mathcal{K}\Phi + \mathcal{K}_-\Phi = 0$$

a.e. on K_*.

Now we are in a position to finish the proof of the lemma. The duality arguments show that it is sufficient to prove that $V(K)\supset \mathcal{L}(K)^\perp$. Let $f\in V(\mathbb{T})\cap\mathcal{L}(K)^\perp$. For every smooth function g with the support disjoint from K_* let functional Φ_g **be** defined by

$$<x,\Phi_g> = \int_{\mathbb{T}} x(\zeta)g(\bar{\zeta})\,dm(\zeta), \quad x\in V(\mathbb{T}).$$

It is clear that $P\Phi_g(\zeta) = g(\zeta)$, $\zeta\in\mathbb{T}$ and in particular, $P\Phi_g = 0$ a.e. on K_* . This implies $\Phi_g\in\mathcal{L}(K)$ and therefore

$$\int_{\mathbb{T}} f(\zeta)g(\bar{\zeta})\,dm = 0 .$$

The function g being arbitrary, we see that $\mathrm{supp}(f)\subset K$

(see [45]). ●

The following lemma may be found in [47].

LEMMA 4.3. Let Λ be a $\Lambda(s)$ - set for some $s, s>2$, and let $p\in(0,2)$. Then for every subset E of the circle $\mathbb{T}$, $mE>0$, there is a positive constant c_E such that

$$c_E \Big(\int_{\mathbb{T}} |f|^2 dm\Big)^{1/2} \le \Big(\int_E |f|^p dm\Big)^{1/p}$$

for every Λ -polynomial.

PROOF OF THEOREM 4.1. Let $(x,\mu)\in \ell^2(\Lambda)\times M(E_*)$. It is sufficient to check by the Banach theorem that

$$\|Q^*(\mu,x)\|_{V(K)^*} \geqslant \text{const}\ \big(\|x\|_{\ell^2(\Lambda)}+\|\mu\|_{M(E_*)}\big).$$

It follows from lemma 4.1 that

$$\|Q^*(\mu,x)\|_{V(K)^*} \geqslant \inf\{\|Q^*(\mu,x)+\theta\|_{V(\mathbb{T})^*} : \theta\in V(\mathbb{T})^*,\ P\theta|K_*=0\}.$$

Let symbol Φ denote $Q^*(\mu,x)$. Then

$$\mathcal{K}\Phi(z)=\sum_{n\in\Lambda,\,n\geqslant 0} x_n z^n+\int_{E_*}\frac{d\mu(t)}{1-\bar t z},\quad |z|<1,$$

$$\mathcal{K}_-\Phi(z)=\sum_{n\in\Lambda,\,n<0} x_n\cdot z^n,\quad |z|>1.$$

The combination of (12) for $p=1/2$, the Smirnov's theorem, and the identity $P\theta\,|\,K_*=0$ a.e., give

$$\Big(\int_{K_*}\Big|\sum_{n\in\Lambda} x_n\zeta^n\Big|^{1/2} dm\Big)^2 \le \text{const}\{\|\Phi+\theta\|_{V(\mathbb{T})^*}+\|\mu\|_{M(\mathbb{T})}\}. \tag{13}$$

The space U_A satisfies the conditions of axiom 2 and $P\theta|K_*=0$. Hence

$$\overline{\lim_{y\to+\infty}}\ y\cdot m\{t\in K_*:|P\Phi(t)|>y\}\le 4\varkappa\|\Phi+\theta\|_{V(\mathbb{T})^*}.$$

It is clear that

$$P\Phi(t)=\sum_{n\in\Lambda} x_n t^n+\int_{E_*}\frac{\zeta\, d\mu(\zeta)}{\zeta-t},\quad t\in\mathbb{T}.$$

The first sum in the right-hand side of the equality belongs obviously to $L^2(\mathbb{T})$. By theorem 2.4 we get

$$\overline{\lim_{y\to+\infty}}\ y m\{t\in K_*:|P\Phi(t)|>y\}\geqslant \frac{1}{\pi}\|\mu\|$$

and consequently

$$\|\mu\| \leqslant 4\pi\, æ \cdot \|\Phi + \theta\|_{V(\mathbb{T})^*} \ .$$

Comparing the last inequality with (13) and with the inequality of lemma 4.3 we get the desired estimate. ●

In conclusion of the section we discuss some properties of the space

$$\mathfrak{m}(U_A^\infty) = \{ f \in H^\infty : f \cdot U_A^\infty \subset U_A^\infty \}$$

of all multipliers of U_A^∞. This space is an interesting object for the investigation because the space U_A^∞ is not an algebra. Our first theorem conjoins the space $\mathfrak{m}(U_A^\infty)$ with two others, namely, with

$$\mathfrak{m}(U_A^*) \overset{def}{=\!=} \{ f \in \mathcal{K}U_A^* : f \cdot \mathcal{K}U_A^* \subset \mathcal{K}U_A^* \}$$

and with

$$\mathfrak{m}(U_A^*)_a \overset{def}{=\!=} \{ f \in \mathcal{K}U_A^* : f \cdot \mathcal{K}(U_A^*)_a \subset \mathcal{K}(U_A^*)_a \} \ .$$

Recall that $(U_A^*)_a$ stands for the closure of polynomials in the norm-topology of U_A^*.

THEOREM 4.4. The following identities hold

$$\mathfrak{m}(U_A^\infty) = \mathfrak{m}(U_A^*) = \mathfrak{m}(U_A^*)_a \ .$$

PROOF. To begin with it is useful to observe that both spaces $\mathfrak{m}(U_A^*)$ and $\mathfrak{m}(U_A^*)_a$ consist of bounded analytic functions (see a simple proof of this general fact in [51]). The first step is to prove the identity

$$\mathfrak{m}(U_A^*) = \mathfrak{m}(U_A^*)_a \ .$$

If $\varphi \in \mathfrak{m}(U_A^*)_a$ and if $f \in U_A^*$, then there is a sequence $(P_n)_{n \geqslant 0}$ of polynomials which converges to f in the weak * topology of U_A^*. Clearly $\|\varphi \cdot P_n\|_{U_A^*} = O(1)$ and by the trivial part of Banach-Steinhaus theorem we may conclude that $\varphi \cdot f \in U_A^*$.

Let now $\varphi \in \mathfrak{m}(U_A^*)$ and let $f \in (U_A^*)_a$. Again there is a sequence (P_n) of polynomials satisfying

$$\lim_n \|P_n - f\|_{U_A^*} = 0 \ .$$

It is clear that

$$\|\Phi\|_{U_A^*} \leqslant \|\mathcal{K}\Phi\|_{H^1} \overset{def}{=\!=} \sup_{0<r<1} \int_{\mathbb{T}} |\mathcal{K}\Phi(r\zeta)|\, dm(\zeta) \ .$$

and therefore $\varphi \cdot P_n \in (U_A^*)_a$. The equality

$$\varphi f = \varphi(f - P_n) + \varphi \cdot P_n$$

shows that $\varphi \cdot f \in (U_A^*)_a$.

To finish the proof of the theorem we are going now to establish the identity

$$\mathfrak{m}(U_A^*)_a = \mathfrak{m}(U_A^\infty) .$$

We shall use the natural duality between the spaces U_A^∞ and $(U_A^*)_a$ and shall write it in an antilinear way:

$$< \Phi, f > = \lim_{r \to 1-0} \int_{\mathbb{T}} \overline{\Phi(r\zeta)}\, f(r\zeta)\, dm .$$

Then it is clear that $\varphi \in \mathfrak{m}\,(U_A^*)_a$ iff the Toeplitz operator $T_{\overline{\varphi}}$ is bounded on the space $\mathfrak{m}(U_A^\infty)$. The space U_A^∞ may be considered as a closed subspace of $SU^\infty(\mathbb{T})$. Clearly the shift operator S and its conjugate operator S^* are isometric in $SU^\infty(\mathbb{T})$. Assuming the operator $T_{\overline{\varphi}}$ to be bounded, we have by a formula from [51] (for the non-Hilbertian case this formula appeared for the first time in [52]) that for a given $f \in SU(\mathbb{T})$:

$$\| S^{*n} T_{\overline{\varphi}} P_+ S^n f \|_{SU^\infty(\mathbb{T})} \leqslant C_f .$$

Moreover $\lim_{n \to \infty} S^{*n} T_{\overline{\varphi}} P_+ S^n f = \overline{\varphi} f$ in the weak $*$ topology of $L^\infty(\mathbb{T})$. Hence $\overline{\varphi} f \in SU^\infty(\mathbb{T})$ and by the standard arguments we see that operator $f \mapsto \overline{\varphi} f$ maps the space $SU^\infty(\mathbb{T})$ into itself. In particular, $\overline{\varphi} \cdot \overline{SU_A^\infty} \subset \overline{SU_A^\infty}$ and therefore $\varphi \in \mathfrak{m}(U_A^\infty)$. Assume now that $\varphi \in \mathfrak{m}(U_A^\infty)$. Then it follows from the isometric property of the shift S that $f \to \overline{\varphi} f$ is bounded on $SU^\infty(\mathbb{T})$. Clearly the projection P_+ is also bounded on $SU^\infty(\mathbb{T})$. Combining these facts we see $T_{\overline{\varphi}}$ is a bounded operator. ●

It appears that usually the space $\mathfrak{m}\, U_A^*$ is more convenient to deal with. Nevertheless we prefer a direct proof of the following theorem to one making use of theorem 4.4.

THEOREM 4.5. Let B be a Blaschke product with the zero set satisfying the Frostman condition (see §3). Then $B \in \mathfrak{m}(U_A^\infty)$.

PROOF. It was shown in [50] that $B \in \mathfrak{m}(\mathcal{K}C_A^*)$. Therefore

$$\frac{B}{1 - \overline{t} z} = \int \frac{d\mu_t(\zeta)}{1 - \overline{\zeta} z} , \quad |z| < 1, \quad |t| < 1,$$

where $\mu_t \in M(\mathbb{T})$, $\|\mu_t\| \leq const$. It is clear also that

$$\frac{\varphi(z)}{1-tz} = \sum_{n=0}^{\infty} \bar{t}^n D_n * \varphi(t) \cdot z^n .$$

Let now $f \in U_A^\infty$. Then

$$fB \cdot (1-\bar{t}z)^{-1} = \int_{\mathbb{T}} \frac{f(z)}{1-\bar{\zeta}z} d\mu_t(\zeta) = \sum_{n\geq 0} z^n \int_{\mathbb{T}} \bar{\zeta}^n D_n * f(\zeta) d\mu_t(\zeta) .$$

Comparing the coefficients of the Taylor expansions in both sides of the previous identity, we get

$$\| fB \|_{U_A^\infty} \leq \sup_{n\geq 0, t\in\mathbb{T}} \int | D_n * f | d|\mu_t| < const . \quad \bullet$$

LEMMA 4.6. Let $f \in \mathfrak{m}(U_A^*)$. Then

$$\sup_{t\in\mathbb{T}} \int_0^1 | f'(rt) | dr < +\infty .$$

PROOF. The proof is analogous to one given on [50] for the more narrow class $\mathfrak{m}(\mathcal{K}C_A^*)$. Given a function f in $\mathfrak{m}(U_A^*)$ and a point t in $\mathbb{T}$ it is possible to find a sequence of bounded Borel measures on $\mathbb{T}$ such that

$$\frac{f(z)}{t-z} = \sum_{n\geq 0} z^n \cdot \int_{\mathbb{T}} \frac{d\mu_n(\zeta)}{\zeta - z} , \qquad \sum_{n\geq 0} \| \mu_n \| \leq const .$$

To prove this assertion it is sufficient to note that the space $\mathcal{K}U_A^*$ may be considered as the space of all holomorphic functions F in $\mathbb{D}$ having a representation

$$F(z) = \sum_{n\geq 0} z^n \int \frac{d\mu_n(\zeta)}{\zeta - z} , \qquad \sum_{n\geq 0} \| \mu_n \| < +\infty .$$

Clearly

$$f'(z) = \sum_{n\geq 0} (z^n)' \int_{\mathbb{T}} \frac{t-z}{\zeta - z} d\mu_n(\zeta) + \sum_{n\geq 0} z^n \int_{\mathbb{T}} \frac{t-\zeta}{(\zeta - z)^2} d\mu_n .$$

Gathering two obvious inequalities

$$\left| \int_{\mathbb{T}} \frac{t-z}{\zeta - z} d\mu_n(\zeta) \right| \leq \| \mu_n \| , \quad \int_0^1 \int_{\mathbb{T}} \frac{|t-\zeta|}{|\zeta - rt|^2} d\mu_n(\zeta) dr \leq \frac{\pi}{2} \| \mu_n \| .$$

we have

$$\int_0^1 |f'(rt)|\,dr \leq \sum_{n\geq 0} \|\mu_n\| + \frac{\pi}{2}\sum_{n\geq 0}\|\mu_n\| < +\infty . \qquad \bullet$$

The last lemma together with theorem 4.4 imply the imbedding $\mathfrak{m}(U_A) \subset \mathfrak{m}(U_A^\infty) \subset \mathcal{H}_1^1$, where $\mathcal{H}_1^1$ denotes the Banach space of all holomorphic functions f in $\mathbb{D}$ such that

$$\iint_{\mathbb{D}} |f'(z)|\,dx\,dy < +\infty .$$

It was shown in [48] that for every subset Λ of $\mathbb{N}$ with Hadamard's gaps the transformation $f \to \hat{f}|\Lambda$ maps $\mathcal{H}^1$ onto $\ell^1(\Lambda)$. This means that there is no analog for the Banach theorem in $\mathfrak{m}(U_A)$.

The next theorem, being similar to one proved in [53], gives a d scription of some inner functions in the space $\mathfrak{m}(U_A^\infty)$.

THEOREM 4.7. Let S denote an inner function in $\mathfrak{m}(U_A^\infty)$. Then S is a Blaschke product, the radial limits $S(\zeta) = \lim_{r\to 1-0} S(r\zeta)$ exist everywhere on $\mathbb{T}$, and $|S(\zeta)|=1$, $\zeta\in\mathbb{T}$.

PROOF. The existence of radial limits follows easily by lemma 4.6. To compute $|S(t)|$ for $t\in\mathbb{T}$ we apply theorem 2.5. The function S being a multiplier, we see that $S(t-z)^{-1}\in \mathcal{K}U_A^\infty$. Therefore by theorem 3.3 we can conclude that there are a singular measure μ in $M(\mathbb{T})$ and an absolutely continuous functional Φ such that

$$\frac{S(z)}{t-z} = \int_{\mathbb{T}} \frac{d\mu(\zeta)}{\zeta - z} + \mathcal{K}\Phi(z), \qquad |z|<1 . \tag{14}$$

Let now E denote any closed arc on $\mathbb{T}$ containing the point t in its interior. Remembering that $|S(\zeta)|=1$ a.e. on $\mathbb{T}$ we get, obviously,

$$\lim_{y\to+\infty} y\cdot m\{t\in E : |\frac{S(t)}{t-z}| > y\} = \pi^{-1} .$$

Clearly

$$\lim_{y\to+\infty} y\cdot m\{t\in\mathbb{T} : |\mathcal{K}\Phi(t)| > y\} = 0 .$$

It follows now from (14) and by theorem 3.3 that the measure μ is supported by E and $\|\mu\| = 1$. The arc E being chosen arbitrarily, it follows that $\mu = \alpha\cdot\delta_{(t)}$, $|\alpha| = 1$.

Now we have

$$\lim_{r\to 1-0} S(rt) = \lim_{r\to 1-0}\frac{1-rt}{t-rt}\alpha + \lim_{r\to 1-0}(t-rt)\cdot\mathcal{K}\Phi(rt) = \alpha . \qquad \bullet$$

Our last application of the asymptotic formula deals with a generalization of a recent result due to de Leeuw, Katznelson, Kahane [18]. This result of three authors has given a positive answer to a question posed by Sidon in [54]. Shortly after appearance of [18] S.Kisljakov has strengthened the theorem in [19].

THEOREM (S.Kisljakov [19], [55]). For every square-summable non-negative sequence $(a_n)_{n\in\mathbb{N}}$ there is a function f in U_A such that

$$a_n \leqslant |\hat{f}(n)| , \quad n\in\mathbb{N} .$$

The method of the present paper gives a possibility for the conjugation of the above theorem with the Banach theorem and with the theorem due to Carleson and Rudin.

Let $\mathcal{E}$ be a Banach space satisfying the Axioms 1 and 2 (see §3).

THEOREM 4.8. Let E be a closed subset of the circle having zero Lebesgue measure and let Λ be a Banach set in $\mathbb{N}$. Then for every $\varphi\in C(E)$ and for every $a=(a_n)_{n\geqslant 0}\in \ell^2(\mathbb{N})$ there is a function f in $\mathcal{E}$ satisfying

1) $f\,|\,E=\varphi$;
2) $\hat{f}(n)=a_n$, $n\in\Lambda$;
3) $|a_n|\leqslant|\hat{f}(n)|$, $n\in\mathbb{N}$.

The proof of the theorem is based on an abstract scheme originating from [18] and on a method due to S.Kisljakov [19], [55] on the one hand and on our approach on the other.

References

1. K a t z n e l s o n Y. An introduction to harmonic analysis. New York. J.Wiley, 1968.
2. K a h a n e J.-P., K a t z n e l s o n Y. Sur les séries de Fourier uniformément convergentes. C.R.Acad.Sc.Paris, 1965, t.261, 3025-3028.
3. B a n a c h S. Über einige Eigenschaften der lacunären trigonometrischen Reihen. Studia Math., 1930, v.2, 207-220.
4. P a l e y R.E.A.C. A note on power series. J.London Math., Soc., 1932, v.7, 122-130.
5. R u d i n W. Boundary values of continuous analytic functions. Proc.Amer.Math.Soc., 1956, v.7.
6. R u d i n W. Trigonometric series with gaps. J. of Math.and Mech., 1960, v.9, N 2, 203-228.

7. C a r l e s o n L. Representations of continuous functions. Math.Zeit. 1957, v.66.

8. С т е ч к и н С.Б. Одна экстремальная задача для многочленов. Изв.АН СССР, сер.матем.,1956, т.20, № 6, 765–774.

9. Х а в и н В.П. О нормах некоторых операций в пространстве многочленов. Вестник ЛГУ, сер. матем., мех. и астр., 1959, вып.4, № 19, 47–59.

10. H e l s o n H. Conjugate series and a theorem of Paley. Pacific J.Math., 1958, v.8, 437-446.

11. R u d i n W. Fourier Analysis on groups. Int.Publ., N.-Y., 1962.

12. В и н о г р а д о в С.А. Особенности Пэли и интерполяционные теоремы Рудина-Карлесона для некоторых классов аналитических функций. Докл. АН СССР, 1968, т.178, № 3, 511 - - 514.

13. В и н о г р а д о в С.А. Интерполяционные теоремы Банаха--Рудина-Карлесона и нормы операторов вложения для некоторых классов аналитических функций. Зап.научн.семин.ЛОМИ, 1970, т.19, 6–54.

14. F o u r n i e r J.J.F. An interpolation problem for coefficients of H^∞ functions. Proc.Amer.Math.Soc., 1974, v.42, N 2.

15. В и н о г р а д о в С.А. Сходимость почти всюду рядов Фурье функций из L^2 и поведение коэффициентов равномерно сходящихся рядов Фурье. Докл. АН СССР, 1976, 230, № 3, 508--511.

16. В и н о г р а д о в С.А. Усиление теоремы Колмогорова о сопряженной функции и интерполяционные свойства равномерно сходящихся степенных рядов. Труды Математического института им. В. А. Стеклова АН СССР, т. 155, Л., "Наука, 1980.

17. Х р у щ ё в С.В. Проблема одновременной аппроксимации и стирание особенностей интегралов типа Коши. Труды Математического института им.В.А.Стеклова АН СССР, т.130, Л., "Наука", 1978, стр.124–195.

18. d e - L e e u w K., K a t z n e l s o n Y., K a h a n e J.-P. Sur les coefficients de Fourier des fonctions continues. C.R.Acad.Sci.Paris, 1977, 285, N 16, A1001-A1003.

19. К и с л я к о в С.В. Fourier coefficients of boundary values of analytic functions. Препринт ЛОМИ, Е-3-78, Л., 1978.

20. К и с л я к о в С.В. Коэффициенты Фурье граничных значений функций, аналитических в круге и в бидиске. Труды Математического института им.В.А.Стеклова АН СССР, т.155, Л., "Наука",

1980.
21. Ё р и к к е Б. Сравнение максимума модуля многочлена со взвешенной суммой модулей его коэффициентов. Вестник ЛГУ, 1978, № 1, 142-144.
22. C a r l e s o n L. On convergence and growth of partial sums of Fourier series. Acta Math.,1966,v.116,N 1-2,135-157.
23. Hoffman K. Banach spaces of analytic functions. Prentice-Hall, N.J., 1962.
24. O b e r l i n D.M. Uniformly convergent Fourier series and sets of measure zero. Preprint, 1978.
25. Gamelin T.W. Uniform algebras , Prentice-Hall, N.J., 1962.
26. К о л м о г о р о в А.Н. Sur les fonctions harmoniques conjuguées et les séries de Fourier. Fundamenta Mathematicae, 1925, v.7, 23-28.
27. С м и р н о в В.И. Sur les valeurs limites des fonctions régulières à l'intérieur d'un cercle. Журнал Ленингр.физ-мат.об-ва, 1928, 2:2.
28. Zygmund A. Trigonometric Series, Cambridge, 1959.

29. К а н т о р о в и ч Л.В., А к и л о в Г.П. Функциональный анализ в нормированных пространствах. М., "Ф.-М", 1959.
30. A a r o n s o n J. Ergodic theory for functions on the upper half-plane. Ann.Inst.Henri Poincaré, 1978, v.14, N 3, 233-253.
31. Ц е р е т е л и О.Д. О сопряженных функциях. Автореферат диссертации на соискание уч.степени доктора физ.-мат.наук, Тбилиси, 1976.
32. B o o l e G. On the comparison of transcendents, with certain applications to the theory of definite integrals. Phil.Trans. of the Royal Soc., 1857, v.147, 745-803.
33. L e v i n s o n N. Gap and density theorems.AMS Coll.Publ., 1940, v.26.
34. Б р ы ч к о в Ю.А., П р у д н и к о в А.П. Интегральные преобразования обобщенных функций. М., "Наука", 1977.
35. P e ł c z y ń s k i A. On simultaneous extension of continuous functions (A generalization of theorems of Rudin--Carleson and Bishop). Studia Math., 1964, v.24, N 3, 285--304.
36. P e ł c z y ń s k i A. Supplement to my paper "On simultaneous extension of continuous functions. Studia Math., 1964, v.25, 157-161.
37. L ó p e z J.M., R o s s K.A. Sidon sets. Lecture notes

in pure and appl.math., v.13, N.-Y., 1975.

38. В и н о г р а д о в С.А., Х а в и н В.П. Свободная интерполяция в H^∞ и в некоторых других классах функций. Зап. научн.семин.ЛОМИ, 1974, т.47, 15-54.

39. В и н о г р а д о в С.А., Х а в и н В.П. Свободная интерполяция в H^∞ и в некоторых других классах функций П. Зап. научн.семин.ЛОМИ, 1976, т.54, 12-58.

40. В а с ю н и н В.И. Циркулярные проекции множеств, встречающихся в теории интерполирования. Зап.научн.сем.ЛОМИ, 1979, т.92, 51-59.

41. E a r l T.P. On the interpolation of bounded sequences by bounded analytic functions. J.London Math. Soc., 1970, v.2, 544-548.

42. N e w m a n D.J., S h a p i r o H.S. The Taylor coefficients of inner functions. Mich.Math.J., 1962, v.9, N 3, 249-255.

43. М е н ь ш о в Д.Е. О рядах Фурье непрерывных функций. Ученый записки МГУ, 1951, вып.148, т.IУ, 108-132.

44. Х р у щ ё в С.В. Простое доказательство теоремы Меньшова об исправлении. Препринт ЛОМИ, Р-I-79, Л., 1978.

45 Bremermann H.Distributions, complex variables and Fourier transforms, Addison-Wesley, 1965.

46. П р и в а л о в И.И. Граничные свойства аналитических функций. М.-Л., ГИТТЛ, 1950.

47. М и х е е в И.М. О рядах с лакунами. Математический сборник, 1975, т.98, № 4, 538-563.

48. А н д р и а н о в а Т.Н. Коэффициенты Тэйлора с редкими номерами для функций, суммируемых по площади. Зап.научн.сем. ЛОМИ, 1976, т.65, 161-163.

49. H r u š č ë v S.V., V i n o g r a d o v S.A. Inner functions and multipliers of Cauchy type integrals. LOMI preprints, E-I-80, Leningrad, 1980.

50. В и н о г р а д о в С.А. Свойства мультипликаторов интегралов типа Коши и некоторые задачи факторизации аналитических функций. Труды Седьмой Зимней школы. "Теория функций и функциональный анализ", Драгобыч, 1974.

51. Halmos P. A Hilbert space problem book, Van Nostrand, Prentice-Hall, N.J., 1967.

52. Н и к о л ь с к и й Н.К. О пространствах и алгебрах тёплицевых матриц, действующих в ℓ^p. Сиб.матем.журн., 1966, т.6, № 1, 146-158.

53. H r u š č ë v S.V., V i n o g r a d o v S.A. Inner functi-

ons and multipliers of Cauchy type integrals. LOMI preprints, E-I-80, 3-29 (to appear in Arkiv for Mathematik).

54. S i d o n S. Einige Sätze und Fragestellungen über Fourier--Koeffizienten. Math.Zeit., I932, b.34, N 4, 477-480.

55. К и с л я к о в С.В. Коэффициенты Фурье граничных значений функций, аналитических в круге и в бидиске. Труды МИАН, I98I, т.I55, 77-94.

S.V.Hruščev, N.K.Nikol'skii, B.S.Pavlov

UNCONDITIONAL BASES OF EXPONENTIALS AND OF REPRODUCING KERNELS

CONTENTS

INTRODUCTION

The problem of expansion of a given function f defined on a finite interval I of real axis $\mathbb{R}$ in Dirichlet series with complex frequencies λ_n

$$f(x)=\sum_{n\in\mathbb{Z}} a_n e^{i\lambda_n x}$$

is the nearest analog of the well-known Fourier analysis problem[*]. In general, the family of exponentials $(e^{i\lambda_n x})_{n\in\mathbb{Z}}$ is not orthogonal in the Hilbert space $L^2(I)$ of all square-summable functions on I and apart from that, it need not be complete on that interval. Leaving aside the difficult completeness problem (i.e. the problem of completeness of exponentials in $L^2(I)$), we shall focus our attention on a more narrow question: to describe families of frequencies $(\lambda_n)_{n\in\mathbb{Z}}$ producing "well-behaved" bases $(e^{i\lambda_n x})_{n\in\mathbb{Z}}$ in the space $L^2(I)$.

The convergence problem for o r t h o g o n a l expansions with respect to a general complete orthonormal system $(\varphi_n)_{n\in\mathbb{Z}}$ in $L^2(I)$ is solved by the famous V.A.Steklov theorem: such an expansion converges in L^2 to the function being expanded. Moreover, the system $(\varphi_n)_{n\in\mathbb{Z}}$ being orthogonal, the corresponding Fourier series converges unconditionally; that is it converges to the same sum after any permutation of its terms. This, surely, remains true for any system $(\psi_n)_{n\in\mathbb{Z}}$ (a so-called R i e s z b a s i s) which can be obtained from the system $(\varphi_n)_{n\in\mathbb{Z}}$ by an invertible bounded linear transformation of $L^2(I)$.

In what follows we shall use a slightly more general notion of unconditional basis to avoid the hypothesis $\|\psi_n\|\asymp 1$ (i.e.

---

[*] To emphasize the relationship of a general problem to the classical one we shall use the set of all integers $\mathbb{Z}$ as an index set; this kind of numeration will be also highly convenient for the comparison of our results with the classical theory.

$\inf_n \|\psi_n\| > 0$, $\sup_n \|\psi_n\| < +\infty$.) and to cover by the same token the case of exponentials with frequencies whose imaginary parts are bounded from one side.

According to the definition of unconditional basis (see, for example, [7], [18]) every element x of a given space can be uniquely decomposed in an unconditionally convergent series $x = \sum_{n\in\mathbb{Z}} a_n \psi_n$. In this paper we deal, aside from one exception, with a Hilbert space where by the classical G.Köthe - O.Toeplitz theorem a complete system $(\psi_n)_{n\in\mathbb{Z}}$ forms an unconditional basis iff the following "approximate Parseval identity" holds

$$\|\sum_{n\in\mathbb{Z}} a_n\psi_n\|_2^2 \asymp \sum_{n\in\mathbb{Z}} |a_n|^2 \|\psi_n\|_2^2 .$$

So we take the following definition as one suitable to work with.

DEFINITION. A family $(\psi_n)_{n\in\mathbb{Z}}$ of non-zero vectors in a Hilbert space H is called an unconditional basis in H if

1) the family $(\psi_n)_{n\in\mathbb{Z}}$ spans the space H ;

2) there are positive constants c, C such that for every finite sequence of complex numbers $(a_n)_{n\in\mathbb{Z}}$ the following inequalities hold

$$c\sum_n |a_n|^2 \|\psi_n\|^2 \leqslant \|\sum_n a_n\psi_n\|^2 \leqslant C\sum_n |a_n|^2 \|\psi_n\|^2 .$$

Thus every Riesz basis is unconditional and conversely every unconditional basis satisfying $\|\psi_n\| \asymp 1$ is a Riesz basis.

The purpose of this paper is to describe all subsets $\Lambda = \{\lambda_n : n\in\mathbb{Z}\}$ of a half-plane $\mathbb{C}_\gamma \overset{def}{=} \{\zeta\in\mathbb{C} : \operatorname{Im}\zeta > \gamma\}$, $\gamma\in\mathbb{R}$, (or of the half-plane $\{\zeta\in\mathbb{C} : \operatorname{Im}\zeta < \gamma\} = \mathbb{C}^\gamma$) such that the family $(e^{i\lambda_n x})_{n\in\mathbb{Z}}$ forms an unconditional basis in $L^2(I)$.

The first fundamental progress in the outlined area was attained by N.Wiener [58] and by N.Wiener and R.Paley [59] in 1934. They proved that the system $(e^{i\lambda_n x})_{n\in\mathbb{Z}}$ forms a Riesz basis in $L^2(0, 2\pi)$ if $\lambda_n \in \mathbb{R}$, $n\in\mathbb{Z}$, and if $\sup_n |n - \lambda_n| < \pi^{-2}$. This result has been repeatedly revised and generalized; see the history of the question in §7 of Part I. The most exquisite formulation of the achievements mentioned above can be obtained by comparison of the theorems due to A.Ingham [41] and M.I.Kadec [10].

THEOREM. Let $\delta > 0$. Every family $(e^{i\lambda_n x})_{n\in\mathbb{Z}}$ satisfying

$$\sup_{n\in\mathbb{Z}} |\lambda_n - n| = \delta, \ \delta > 0,$$

forms a Riesz basis in $L^2(0,2\pi)$ if and only if $\delta < \frac{1}{4}$.

We obtain this theorem in Part I of the paper as a consequence of our main results.

In all papers, which have dealt with the subject discussed, it was assumed that $\sup_n |\operatorname{Im} \lambda_n| < \infty$ and the main tool of investigation was an idea stated in the remarkable book of N.Wiener and R.Paley [59]: to form a Riesz basis in $L^2(0,2\pi)$ it is sufficient for the family $(e^{i\lambda_n x})_{n\in\mathbb{Z}}$ to be close enough to the usual trigonometrical system $(e^{inx})_{n\in\mathbb{Z}}$.

One can hardly expect that such an approach to the general problem will be successful, though a result of part III (see §4) exhibits some connection between the general and the classical case.

Another point of view, also originated in [59] has been advanced by B.Ja.Levin. In his method a central role is played by an entire function of exponential type with zeros λ_n, $n \in \mathbb{Z}$ and whose width of the indicator diagram coincides with the length of the interval where our basis is considered. We shall call this entire function "a generating function for the family $(e^{i\lambda_n x})_{n\in\mathbb{Z}}$".

We are now going to state in terms of generating functions a condition sufficient for the exponentials to form a Riesz basis in $L^2(I)$.

DEFINITION. A countable subset $\Lambda = \{\lambda_n : n \in \mathbb{Z}\}$ of the complex plane $\mathbb{C}$ is named *separated* if

$$\inf_{n\neq m} |\lambda_n - \lambda_m| > 0. \qquad (S)$$

DEFINITION (B.Ja.Levin). An entire function f of exponential type is called a *sine-type function* (briefly STF) if its zero set is contained in a strip of a finite width, parallel to the real axis, and if

$$0 < \inf_{x\in\mathbb{R}} |f(x)| \leqslant \sup_{x\in\mathbb{R}} |f(x)| < +\infty .$$

The sine-type functions play an important role in the exponential bases problem and we will be returning from time to time to a discussion of their properties in the sequel.

THEOREM (B.Ja.Levin, V.D.Golovin [14], [6]). Let the generating function of a family $(e^{i\lambda_n x})_{n\in\mathbb{Z}}$ be a sine-type function with the width of the indicator diagram equal to a, $a>0$. Then $(e^{i\lambda_n x})_{n\in\mathbb{Z}}$ forms a Riesz basis in $L^2(I)$, $|I|=a$.

Some attempts have been made to unify the approaches mentioned above. The relevant result of V.È. Kacnelson [12] can be stated, broadly speaking, as follows. A transformation $\lambda_n\to\mu_n$, $n\in\mathbb{Z}$, of the zero set of a STF preserves the property to form a Riesz basis for the corresponding family of exponentials if the set $\{\mu_n : n\in\mathbb{Z}\}$ is separated and if $|\operatorname{Re}(\mu_n-\lambda_n)| < \frac{1}{4}\inf_{n\neq k}|\operatorname{Re}\lambda_n-\operatorname{Re}\lambda_k|$. The most subtle result has been proved by S.A.Avdonin [2], see §7 of the Part I below. The main tools of these papers are delicate estimates of canonical products.

The method of the present paper rests on completely different considerations. It comes from an explicit description of those families of exponentials $\mathcal{E}_\Lambda=(e^{i\lambda_n x})_{n\in\mathbb{Z}}$ which form unconditional bases in their closed linear spans $\operatorname{span}\mathcal{E}_\Lambda$ in $L^2(\mathbb{R}_+)$, $\mathbb{R}_+ \overset{def}{=\!=} (0,+\infty)$. This description is given by the famous C a r l e s o n c o n d i t i o n

$$\inf_n \prod_{k\neq n}\left|\frac{\lambda_k-\lambda_n}{\lambda_k-\overline{\lambda_n}}\right|>0; \tag{C}$$

see L.Carleson [28], H.Shapiro - A.Shields [56], V.È.Kacnelson [11], N.K.Nikol'skii - B.S.Pavlov [20]. If we deal with such a set of frequencies $\Lambda=\{\lambda_n : n\in\mathbb{Z}\}$ and if the transformation $f\to f\cdot\chi_{[0,a]}$, which, obviously, coincides with the orthogonal projection onto $L^2(0,a)$ *), is an isomorphism of $\operatorname{span}\mathcal{E}_\Lambda$ onto $L^2(0,a)$, then, clearly, the family $(e^{i\lambda_n x}\cdot\chi_{[0,a)})_{n\in\mathbb{Z}}$ will form an unconditional basis in $L^2(0,a)$.

This procedure is a chief ingredient of the proofs of all our results. It appeared for the first time in [22], and was used, in particular, in the proof of Levin - Golovin theorem. However, only five years later it became clear that these arguments lead not only to a full solution of the exponential Riesz bases problem [23], but also imply simple and transparent proofs of almost all known results in that area [25]. In the sequel, it turned out that the

---

*) We assume that the space $L^2(0,a)$ is imbedded into $L^2(\mathbb{R}_+)$ in a natural way.

sphere of applications of the described method can be considerably extended to cover unconditional bases of exponentials as well as the bases formed by reproducing kernels [19].

Our method has several advantages in comparison with those of Wiener - Paley and Levin; requiring less in what concerns the non-perturbed basis, it allows one to redistribute the difficulties more uniformly between the investigation of non-perturbed bases in $L^2(\mathbb{R}_+)$ and perturbed ones in $L^2(0,a)$, $a>0$. Moreover, under the slight additional requirement that the projection does not distort the elements of our family too much, the above geometrical reasoning can be inverted.

It should be noted that the solution of a well-known problem, originated in the papers of L.Schwartz [55] and P.Koosis [43], is reduced to the application of the described method too. This is a problem of equivalence of norms $(\int_{\mathbb{R}_+}|f|^2)^{1/2}$, $(\int_I |f|^2)^{1/2}$ on the span of exponentials $(e^{i\lambda_n x})_{n\in\mathbb{Z}}$ in $L^2(\mathbb{R}_+)$. Clearly, the norm equivalence together with Carleson's condition imply that the family $(e^{i\lambda_n x}\cdot\chi_I)_{n\in\mathbb{Z}}$ is an unconditional basis in its span.

The same procedure can be applied to the joint completeness problem of an operator and its adjoint (for dissipative operators and contractions). The application, outlined in [21] in a vague form, appears now more distinctly. The joint basis property is also discussed here. Both of them are important for the spectral theory of differential operators. They arise naturally, for example, in the investigation of the Sturm - Liouville problem containing a spactral parameter in the boundary condition:

$$-\frac{d^2u}{dx^2}=\lambda^2\rho^2(x)u;\qquad u'(0)=0,\quad u'(a)-i\lambda u(a)=0.$$

A similar problem for the Schrödinger operator has been considered by T.Regge [52] in connection with a question of resonance scattering theory.

Aside from the systematic exposition of [19], [23], [25] and the applications to the theory of differential operators, our paper contains some new results too. The exposition is developed along the following plan.

The main purpose of Part I is to apply the above mentioned approach to the exponential bases problem; to formulate all our main results including the results for the reproducing kernels; and to discuss the connections between them. Apart from that there is a series of examples here illustrating the general theory. Part I is concluded with a short survey of the history of problem.

Part II deals with a bases problem for reproducing kernels. The connections of the bases problem with Hankel operators and with the B.Sz.-Nagy - C.Foiaş functional model are discussed. The bases close to orthogonal are considered here also. In conclusion we outline an interpretation of our results in terms of the interpolation theory and investigate the bases problem in L^p, $p\neq 2$.

The next part, Part III, is devoted to some applications of our approach in the classical domain. We prove here some results concerning the perturbation theory for exponential unconditional bases. In particular, a new proofs for the theorems of S.A.Avdonin and V.È.Kacnelson are given. In section 3 of Part III we state an example, which is due to S.A.Vinogradov and V.I.Vasjunin, of a generating function bounded on $\mathbb{R}$ together with its reciprocal and such that $\lim_{n\to\infty} \operatorname{Im} \lambda_n = +\infty$ for a sequence $(\lambda_n)_{n\in\mathbb{Z}}$ of its zeros. It is also proved (following to V.I.Vasjunin) that in many cases an unconditional basis $(e^{i\lambda_n x})_{n\in\mathbb{Z}}$ in its closed span in $L^2(0,a)$ can be extended to be an unconditional exponential basis in the whole space $L^2(0,a)$. The last section of the Part, § 4, deals with the problem of equiconvergence of Fourier series with respect to the general unconditional basis $(e^{i\lambda_n x})_{n\in\mathbb{Z}}$ in $L^2(0,2\pi)$ and of those with respect to $(e^{inx})_{n\in\mathbb{Z}}$. A theorem is proved generalizing the well-known Levinson theorem [48].

Part IV is devoted to the applications of our geometrical approach to the above mentioned Regge problem. The main purpose of this part of the paper is to indicate new possibilities of the method rather than prove accomplished results. So it is linked to the preceding Parts by the method of investigation.

Completing the discussion we mention that we have tried to make the bulk of the article intelligible to anyone with basic knowledge of functional analysis and function theory.

ACKNOWLEDGEMENT. We are sincerely grateful to V.I.Vasjunin and S.A.Vinogradov for many fruitful discussions and for the permission to include in our text their remarkable examples. We are also indebted to P.Koosis for the attention to our paper and for the communication of his version of the proof of the theorem announced in [23], which we have used to improve our exposition. We are grateful to all participants of the V.P.Havin - N.K.Nikol'skii seminar who supported us during our work. At last, but not in the last instance, we are indebted to S.V.Kisliakov, V.V.Peller,

V.I.Vasjunin for their generous assistance in translating the Russian version of the paper into English.

PART I

BACKGROUND OF EXPONENTIAL BASES PROBLEM

1. Functional model

To translate our problem into the language used in B.Sz.-Nagy - C.Foias model some facts of common knowledge about the Hardy class H^2_+ in the upper half-plane $\mathbb{C}_+ \stackrel{def}{=} \{\zeta\in\mathbb{C} : \operatorname{Im}\zeta > 0\}$ are needed. The following sources [18], [33], [44], [54] contain the exhaustive information about the subject.

A function f which is analytic in $\mathbb{C}_+$ belongs to the Hardy class H^2_+ if

$$\|f\|^2 \stackrel{def}{=} \sup_{y>0} \frac{1}{2\pi}\int_{\mathbb{R}} |f(x+iy)|^2\,dx < +\infty .$$

By Fatou's theorem the space H^2_+ may be considered as a closed subspace of $L^2(\mathbb{R})$. It is convenient to define an inner product in $L^2(\mathbb{R})$ by the formula

$$(f,g) \stackrel{def}{=} \frac{1}{2\pi}\int_{\mathbb{R}} f\bar{g}\,dx .$$

A nontrivial function f in H^2_+ can be factored uniquely as the product

$$f = c\cdot B\cdot S\cdot f_e ,$$

where c is a unimodular constant, $|c|=1$; B is a Blaschke product; S is a singular inner function; and f_e is an outer function. A Blaschke product B with the zero sequence $(\lambda_n)_{n\in\mathbb{Z}}$ is an infinite product

$$B(z) = \prod_{n\in\mathbb{Z}} \varepsilon_n \frac{1-z/\lambda_n}{1-z/\bar{\lambda}_n} ,$$

where signs ε_n, $|\varepsilon_n|=1$ make each factor in the product non-negative at the point $z=i$. A well-known Blaschke condition

$$\sum_{n\in\mathbb{Z}} \frac{\operatorname{Im}\lambda_n}{|\lambda_n+i|^2} < +\infty \tag{B}$$

is the necessary and sufficient one for the Blaschke product to converge. To describe the factor S one can consider a one-point compactification $\hat{\mathbb{R}} \overset{def}{=\!=} \mathbb{R} \cup \{\infty\}$ of the real line $\mathbb{R}$. Then

$$S(z) = \exp\left\{-\frac{1}{\pi i} \int_{\hat{\mathbb{R}}} \frac{tz+1}{t-z}\, d\mu(t)\right\},$$

where μ is a non-negative finite measure on $\hat{\mathbb{R}}$ which, being restricted on $\mathbb{R}$, is singular with respect to the usual Lebesgue measure on $\mathbb{R}$. The measure μ of the full mass equal to $\pi \cdot a$, $a > 0$, supported by the point ∞ corresponds, obviously, to the exponential e^{iaz}. The product $c \cdot B \cdot S$ is called an inner function. Inner functions can be described as elements of the algebra H^∞ of all uniformly bounded and holomorphic in $\mathbb{C}_+$ functions, whose boundary values are unimodular a.e. on $\mathbb{R}$. The outer part f_e of the function f is defined by

$$f_e(z) = \exp\left\{\frac{1}{\pi i} \int_{\mathbb{R}} \frac{tz+1}{t-z} \cdot \frac{\log|f(t)|}{t^2+1}\, dt\right\}.$$

It should be noted that the same factorization property holds for all Hardy classes H^p_+, $0 < p \leqslant +\infty$ (H^p_+ consists of all functions f, analytic in $\mathbb{C}_+$ and satisfying

$$\|f\|_p^p \overset{def}{=\!=} \sup_{y>0} \int_{\mathbb{R}} |f(x+iy)|^p dx < +\infty).$$

The well-known Paley - Wiener theorem asserts that the inverse Fourier transform

$$\mathcal{F}^* f(t) = \int_{\mathbb{R}} e^{it\gamma} f(\gamma)\, d\gamma$$

is a one to one norm-preserving mapping of $L^2(\mathbb{R}_+)$ onto H^2_+. By the inversion formula we have

$$f(\gamma) = \mathcal{F} \cdot \mathcal{F}^* f(\gamma) = \frac{1}{2\pi} \int_{\mathbb{R}} \mathcal{F}^* f(t)\, e^{-i\gamma t}\, dt.$$

Let, for the time being, $\Lambda = \{\lambda_n : n \in \mathbb{Z}\}$ be a fixed subset of $\mathbb{C}_+$ and let $a > 0$. Clearly, the family $(e^{i\lambda_n x})_{n\in\mathbb{Z}}$ forms an unconditional basis in $L^2(0,a)$ iff the family $(e^{-i\bar{\lambda}_n x})_{n\in\mathbb{Z}}$ does. Let $\Lambda^* \overset{def}{=\!=} \{-\bar{\lambda}_n : n \in \mathbb{Z}\}$. The Fourier transform $\mathcal{F}^*$ maps the closed span $\mathcal{E}_{\Lambda^*}$ of the family $(e^{i\lambda x} \cdot \chi_{[0,\infty)})_{\lambda\in\Lambda^*}$ onto the subspace

$$K_B \overset{def}{=\!=} H^2_+ \ominus B H^2_+$$

in H^2_+, B being the Blaschke product for the sequence $(\lambda_n)_{n\in\mathbb{Z}}$ if it satisfies the Blaschke condition and the identically zero function otherwise. The proof of this fact rests on a simple calculation:

$$\mathcal{F}^*(e^{-i\bar\lambda\gamma}\cdot\chi_{[0,+\infty)})(z)=\int_0^\infty e^{i\gamma z}e^{-i\gamma\bar\lambda}\,d\gamma=\frac{i}{z-\bar\lambda}\,.$$

It remains only to observe that the span of the family $((z-\bar\lambda)^{-1})_{\lambda\in\Lambda}$ is equal to K_B.

The space $\mathcal{F}^*\mathcal{E}_{\Lambda^*}$ being described, we have to do the same for the space $\mathcal{F}^*L^2(0,a)$. Let $\theta^a(z)\overset{def}{=}e^{iaz}$, $a>0$. Clearly,

$$\mathcal{F}^*L^2(0,a)=\mathcal{F}^*L^2(\mathbb{R}_+)\ominus\mathcal{F}^*L^2(a,\infty)=H^2_+\ominus\theta^aH^2_+=K_{\theta^a}\,.$$

The program outlined in Introduction can be easily applied now. But it is natural to consider now a more general problem. Let θ be any inner function and let B be a Blaschke product with the sequence of zeros $(\lambda_n)_{n\in\mathbb{Z}}$. The function $k(z,\lambda)\overset{def}{=}\frac{i}{z-\bar\lambda}$ is, obviously, the reproducing kernel for H^2_+:

$$(f,k(\cdot,\lambda))=\frac{1}{2\pi i}\int_{\mathbb{R}}\frac{f(x)}{x-\lambda}\,dx=f(\lambda),\quad \operatorname{Im}\lambda>0.$$

Let P_θ be an orthogonal projection onto the subspace K_θ. Then the function $k_\theta(\cdot,\lambda)\overset{def}{=}P_\theta k(\cdot,\lambda)$ is the reproducing kernel for K_θ. Indeed, if $f\in K_\theta$ then

$$(f,k_\theta(\cdot,\lambda))=(f,P_\theta k(\cdot,\lambda))=(f,k(\cdot,\lambda))=f(\lambda).$$

Simple computations show that

$$k_\theta(z,\lambda)=i\,\frac{1-\overline{\theta(\lambda)}\,\theta(z)}{z-\bar\lambda}\,.$$

Now we are in a position to formulate **general problem of unconditional bases for reproducing kernels**:

What is to be assumed about the pair (θ,Λ) for the family $(k_\theta(\cdot,\lambda))_{\lambda\in\Lambda}$ to be an unconditional basis in K_θ?

2. Carleson condition

As it was already mentioned in Introduction, the test for the family $((z-\bar{\lambda}_n)^{-1})_{n\in\mathbb{Z}}$ to be an unconditional basis in its closed span in H^2_+ is given by the well-known Carleson condition:

$$\delta = \inf_n \prod_{k\neq n} \left|\frac{\lambda_n-\lambda_k}{\lambda_n-\bar{\lambda}_k}\right| > 0. \tag{C}$$

Clearly, (C) $\Longrightarrow$ (B) and therefore the Blaschke product

$$B=\prod_{n\in\mathbb{Z}} b_n, \quad b_n(z) \overset{def}{=\!=} \varepsilon_n \cdot \frac{1-z/\lambda_n}{1-z/\bar{\lambda}_n},$$

may be considered. Denoting $B_n \overset{def}{=\!=} B\cdot b_n^{-1}$, one may rewrite (C) in a more compact form

$$\inf_n |B_n(\lambda_n)|>0. \tag{C}$$

It is a matter of common knowledge, see for example [18], that the Carleson condition is equivalent to a purely geometrical one. Let $D(\zeta,r) \overset{def}{=\!=} \{\xi\in\mathbb{C} : |\xi-\zeta|<r\}$.

DEFINITION. A subset $\Lambda=\{\lambda_n : n\in\mathbb{Z}\}$ of $\mathbb{C}_+$ is called a *rare set* if there is a positive ε such that

$$D(\lambda_n,\varepsilon \operatorname{Im}\lambda_n)\cap D(\lambda_m,\varepsilon\operatorname{Im}\lambda_m)=\emptyset,\ m\neq n. \tag{R}$$

DEFINITION. A positive measure μ in $\mathbb{C}_+$ is called a C-*measure* if

$$\sup_{r>0,\,x\in\mathbb{R}} r^{-1}\mu(D(x,r))<+\infty. \tag{CM}$$

Then $\Lambda\in(C)$ iff $\Lambda\in(R)$ and the measure $\sum_{n\in\mathbb{Z}} \operatorname{Im}\lambda_n\cdot\delta_{\lambda_n}$ (δ_λ denotes the unit mass at λ) is a C-measure.

Here are two examples of sets satisfying (C): $\Lambda_1=\{2^n i : n\in\mathbb{Z}\}$, $\Lambda_2=\{i+n : n\in\mathbb{Z}\}$. In general, if $\Lambda\subset\{\zeta\in\mathbb{C} : 0<c<\operatorname{Im}\zeta<c^{-1}\}$ then the separation condition

$$\inf_{n\neq m} |\lambda_n-\lambda_m|>0 \tag{S}$$

and the Carleson condition are equivalent.

There is one more notion needed for the formulation of the main theorem on unconditional bases of rational fractions.

DEFINITION. A family of non-zero elements $(x_n)_{n\in\mathbb{Z}}$ of a Banach space X is called a **uniformly minimal family** if

$$\inf_n \operatorname{dist}\left(\frac{x_n}{\|x_n\|},\ \operatorname{span}(x_k : k \neq n)\right) > 0.$$

Clearly, any basis, and, in particular, any unconditional basis, forms a uniformly minimal family. The converse assertion does not hold in general but it, nevertheless, holds for the families of rational fractions in H^2_+ ; see theorem A below. Apparently, the main reason of this phenomenon is rooted in simple formulae for the dual family

$$\psi_n = \frac{2\operatorname{Im}\lambda_n}{z-\overline{\lambda}_n}\cdot\frac{B_n(z)}{B_n(\lambda_n)},\quad n\in\mathbb{Z},$$

of the family φ_n , $\varphi_n \overset{def}{=\!=} (z-\overline{\lambda}_n)^{-1}$ spanning the space

$$\operatorname{span}_{H^2_+}(\varphi_n : n\in\mathbb{Z}) = H^2_+ \ominus BH^2_+ = K_B .$$

It is an easy task to check that $\psi_n \in K_B$, $n\in\mathbb{Z}$, and that $\langle \varphi_n , \psi_k \rangle = \delta_{nk}$. The computation of the distance from $\|\varphi_n\|^{-1}\varphi_n$ to the $\operatorname{span}(\varphi_k : k\neq n)$ is now an elementary exercise: $\operatorname{dist}(\|\varphi_n\|^{-1}\varphi_n , \operatorname{span}(\varphi_k : k\neq n))$ =(by the Hahn - Banach theorem) = $\|\varphi_n\|^{-1}_{H^2_+}\cdot\|\psi_n\|^{-1}_{H^2_+} = (2\operatorname{Im}\lambda_n)^{1/2}(2\operatorname{Im}\lambda_n)^{-1/2}|B_n(\lambda_n)| = |B_n(\lambda_n)|$.

COROLLARY. For a family $\{(z-\overline{\lambda}_n)^{-1},\ n\in\mathbb{Z}\}$, $\operatorname{Im}\lambda_n > 0$, to be uniformly minimal it is necessary and sufficient that $(\lambda_n)_{n\in\mathbb{Z}} \in (C)$. ●

THEOREM A. Let $\Lambda = \{\lambda_n : n\in\mathbb{Z}\} \subset \mathbb{C}_+$. The following assertions are equivalent.

1. The family $((z-\overline{\lambda}_n)^{-1})_{n\in\mathbb{Z}}$ forms an unconditional basis in its own span in H^2_+ .
2. The family $((z-\overline{\lambda}_n)^{-1})_{n\in\mathbb{Z}}$ is uniformly minimal in H^2_+.
3. The family $(e^{i\lambda_n x}\chi_{\mathbb{R}_+})_{n\in\mathbb{Z}}$ forms an unconditional basis in its $L^2(\mathbb{R}_+)$ -span.
4. $\Lambda \in (C)$.

In such form Theorem A has been obtained by N.K.Nikol'skii and B.S.Pavlov [63], [20] (see also [61], [62]) as a consequence of a more general theory. Their proof hinges on preceding results of L.Carleson [28] and of H.Shapiro - A.Shields [56], [64] from the interpolation theory.

There are many ways to reformulate the assertions 1-4 of Theorem A and, first of all, to link these assertions to the objects fundamental for our approach. We mean the expansions in Fourier series with respect to the eigen-functions of the so-called "model semigroup" and the well-known interpolation problem $f(\lambda_n)(\operatorname{Im}\lambda_n)^{1/2} = a_n$, $(a_n)_{n\in\mathbb{Z}} \in \ell^2$ in H^2_+. We leave the discussion of these links - *for the time being* - till §5, not to be led too far from exponential bases. Note, however, that it is just the operator-theoretical approach (connected with the model semigroup) the proof of Theorem A in [20] was based upon.

Our last remark concerns the interplay between the unconditional bases property and the completeness problem for rational fractions in H^2_+. Obviously, (C) $\Longrightarrow$ (B), and therefore the uncompleteness is a necessary condition for the family $((z-\bar\lambda_n)^{-1})_{n\in\mathbb{Z}}$ to be an unconditional basis in its closed span in H^2_+.

Now we are in a position to make the first step towards the investigation of the basis property for exponentials. Namely, according to the plan stated in Introduction we are to prove that the Carleson condition (C) is necessary for exponentials to form an unconditional basis in $L^2(0,a)$. The next step will be to study the orthogonal projection P_{K_θ}, $\theta=\theta^a$. Because of the general nature of our geometrical reasoning, it is natural to deal with the general case of reproducing kernels at once; see the end of §1.

THEOREM 1. Let θ be an inner function and let $\Lambda = \{\lambda_n : n\in\mathbb{Z}\} \subset \mathbb{C}_+$.

1. If the family $(k_\theta(\cdot,\lambda_n))_{n\in\mathbb{Z}}$ is an unconditional basis in its span then $\Lambda\in(C)$.

2. If the family $(k_\theta(\cdot,\lambda_n))_{n\in\mathbb{Z}}$ is uniformly minimal and if

$$\sup_n |\theta(\lambda_n)| < 1 \tag{1}$$

then $\Lambda\in(C)$.

Leaving aside the proof of the assertion 1 till §1 of Part II, we shall give now a simple explanation of the assertion 2 of the theorem, which is sufficient for our analysis of exponential bases property. For $\theta=\theta^a$ the condition (1) implies, obviously, that $\Lambda\subset\mathbb{C}_\delta$, for some positive number δ . The role of the condition (1) in what follows becomes clear after we note that it is a necessary and sufficient condition for H^2_+-norms of the functions $(z-\bar\lambda_n)^{-1}$ and $P_\theta(z-\bar\lambda_n)^{-1} = k_\theta(\cdot,\lambda_n)$ to be comparable. If $\theta=\theta^a$ then it means

$$\Big(\int_{\mathbb{R}_+} |e^{i\lambda_n x}|^2\,dx\Big)^{1/2} \asymp \Big(\int_0^a |e^{i\lambda_n x}|^2\,dx\Big)^{1/2}, \quad n\in\mathbb{Z}. \tag{1a}$$

The statement 2 of Theorem 1 is an immediate corollary of Theorem A and the following elementary Lemma.

LEMMA *). Let L be a bounded linear operator in a Banach space X and let $(x_n)_{n\in\mathbb{Z}}$ be a sequence of non-zero vectors in X satisfying $C \overset{\text{def}}{=} \sup_n \|x_n\|\,\|Lx_n\|^{-1} < \infty$. Then the family $(x_n)_{n\in\mathbb{Z}}$ is uniformly minimal if the same holds for the family $(Lx_n)_{n\in\mathbb{Z}}$.

PROOF. If $a_k\in\mathbb{C}$, $b_k = a_k\|Lx_k\|\,\|x_k\|^{-1}$, then

$$\|Lx_n\cdot\|Lx_n\|^{-1} - \sum_{k\neq n} a_k Lx_k\| = \|x_n\|\,\|Lx_n\|^{-1}\|Lx_n\cdot\|x_n\|^{-1} - \sum_{k\neq n} b_k Lx_k\| \leqslant$$

$$\leqslant C\|L\|\,\|x_n\cdot\|x_n\|^{-1} - \sum_{k\neq n} b_k x_k\|.$$

It follows that

$$\operatorname{dist}(x_n\|x_n\|^{-1}, \operatorname{span}(x_k, k\neq n)) \geqslant (C\|L\|)^{-1}\operatorname{dist}(Lx_n\|Lx_n\|^{-1}, \operatorname{span}(Lx_k : k\neq n)) \bullet$$

To prove the statement 2 of Theorem 1 let $x_n = (z-\bar\lambda_n)^{-1}$, $L = P_\theta$. Then it follows from the equalities $\|x_n\|^2 = (2\operatorname{Im}\lambda_n)^{-1}$, $\|Lx_n\|^2 = (1-|\theta(\lambda_n)|^2)(2\operatorname{Im}\lambda_n)^{-1}$ that $\sup_n \|x_n\|\cdot\|Lx_n\|^{-1} < +\infty$. The trivial part of Theorem A $(2 \Longrightarrow 4)$ together with the Lemma imply $\Lambda = \{\lambda_n : n\in\mathbb{Z}\} \in (C)$. $\bullet$

A simple but, nevertheless, important remark is relevant now. Let $\theta = \theta^a$ for the time being. There are a few isomorphisms in $L^2(0,a)$ preserving the exponentials:

$$f(x) \longmapsto e^{i\alpha x} f(x), \quad \alpha\in\mathbb{C}$$

$$f(x) \longmapsto f(a-x)$$

$$f(x) \longmapsto \overline{f(x)}.$$

Any of these isomorphisms preserves, obviously, the property to be a uniformly minimal exponential family and the basis property as well. Using these isomorphisms we always can move a frequency

---

*) An analogous lemma may be found in [22].

set $\Lambda=\{\lambda_n : n\in\mathbb{Z}\}$ from any half-plane $\mathbb{C}_\gamma$ (or $\mathbb{C}^\gamma$), $\gamma\in\mathbb{R}$, to the half-plane $\mathbb{C}_\delta$, $\delta>0$. So the assumption (1) does not restrict the generality if we deal with the sets Λ contained in a half-plane $\mathbb{C}_\gamma$ (or $\mathbb{C}^\gamma$), $\gamma\in\mathbb{R}$.

The second step in splitting up our problem into two independent ones is made by theorem 2 below. We again not only formulate the theorem in its natural generality, but also give a special formulation (Theorem 2′) for the important case of exponentials.

THEOREM 2. Let θ be an inner function, $\Lambda=\{\lambda_n : n\in\mathbb{Z}\}\subset\mathbb{C}$, and let $\Lambda\in(1)$. Then the following statements are equivalent.

1. The family $(k_\theta(\cdot,\lambda_n))_{n\in\mathbb{Z}}$ forms an unconditional basis in K_θ.

2. a) $\Lambda\in(C)$; b) the operator $P_\theta|K_B$ maps isomorphically the space K_B onto K_θ, B being the Blaschke product for the sequence $(\lambda_n)_{n\in Z}$.

THEOREM 2′. Let $\Lambda=\{\lambda_n : n\in\mathbb{Z}\}\subset\mathbb{C}_\delta$, $\delta>0$, and let a be a positive number. The following statements are equivalent.

1. The family $(e^{i\lambda_n x}\cdot\chi_{[0,a)})_{n\in\mathbb{Z}}$ is an unconditional basis in $L^2(0,a)$.

2. a) $\Lambda\in(C)$; b) the restriction of the orthogonal projection $f\mapsto\chi_{[0,a)}\cdot f$ onto $\mathrm{span}_{L^2(\mathbb{R}_+)}(e^{i\lambda_n x} : n\in\mathbb{Z})$ is an isomorphism of the span onto $L^2(0,a)$.

It is clear from §1 that Theorem 2′ is covered by Theorem 2.

THE PROOF OF THEOREM 2. 1 $\Longrightarrow$ 2. From Theorem 1 it follows that $\Lambda\in(C)$ and therefore the family $((z-\bar\lambda_n)^{-1})_{n\in\mathbb{Z}}$ is an unconditional basis in its closed span $K_B=H^2\ominus BH^2$ by Theorem A . Using the condition $\|(z-\bar\lambda_n)^{-1}\|_{H^2_+}\asymp\|P_\theta(z-\bar\lambda_n)^{-1}\|_{H^2_+}$ implied by (1), we see that

$$\|P_\theta\sum_n a_n(z-\bar\lambda_n)^{-1}\|^2_{H^2_+}=\|\sum_n a_n k_\theta(\cdot,\lambda_n)\|^2_{H^2_+}\asymp$$

$$\asymp\sum_n|a_n|^2\cdot\|k_\theta(\cdot,\lambda_n)\|^2_{H^2_+}\asymp\sum_n|a_n|^2\|(z_n-\bar\lambda_n)^{-1}\|^2_{H^2_+}\asymp$$

$$\asymp\|\sum_n a_n(z-\bar\lambda_n)^{-1}\|^2_{H^2_+}.$$

This, clearly, implies that the map $P_\theta : K_B\longrightarrow K_\theta$ is an isomorphism.

2 $\Longrightarrow$ 1. The set Λ satisfying the Carleson condition, it follows by Theorem A that the family $((z-\bar\lambda_n)^{-1})_{n\in\mathbb{Z}}$

forms an unconditional basis in K_B. The family $k_\theta(\cdot,\lambda_n)= = P_\theta\,(z-\bar\lambda_n)^{-1}$ is now an unconditional basis in K_θ because it is assumed in the conditions of the theorem that the operator $P_\theta\,|\,K_B$ is an isomorphism. ●

Thus the unconditional basis problem for exponentials defined on a finite interval, as well as the more general problem for reproducing kernels in K_θ, is reduced to the study of the conditions of invertibility of the operator $P_\theta : K_B \to K_\theta$. We shall describe later, see §§ 3,5, all pairs of inner functions (θ_1,θ_2) such that $P_{\theta_1} : K_{\theta_2} \longrightarrow K_{\theta_1}$ is an isomorphism, and shall be especially detailed in the leading case $\theta_1 = \exp iaz$, $\theta_2 = B \overset{def}{=\!=} \prod_{n\in\mathbb{Z}} b_{\lambda_n}$. Such a description, see § 4, may be given directly in terms of the distribution of numbers $(\lambda_n)_{n\in\mathbb{Z}}$, and all known results on exponential bases in $L^2(0,a)$ can be easily derived after that.

To end this section we note that Theorems 2, 2′ can be given a form covering the case of unconditional bases in their closed linear span (i.e. not assuming the family under consideration to be complete in the whole space). Let us do this, e.g., for Theorem 2.

THEOREM 2 bis. Let θ be an inner function, let $\Lambda = =\{\lambda_n : n\in\mathbb{Z}\}\subset\mathbb{C}$, and let $\Lambda\in(1)$. Then the following statements are equivalent.

1. The family $\{k_\theta(\cdot,\lambda_n) : n\in\mathbb{Z}\}$ forms an unconditional basis in its closed linear span.

2. a) $\Lambda\in(C)$, b) the operator $P_\theta : K_B \longrightarrow K_\theta$ is left-invertible.

3. The invertibility tests for $P_\theta\,|\,K_B$: geometrical and analytical aspects

Let M and N be closed subspaces of a Hilbert space H. The invertibility of the operator $P_M\,|\,N$ means clearly that the subspaces are "close" (in a sense). Geometrically speaking this "closeness" can be expressed as the positivity of the angle $\langle N, M^\perp\rangle$ formed by subspaces N and $M^\perp \overset{def}{=\!=} H\ominus M$; a precise definition of the angle $\langle X,Y\rangle$ will be given later (§ 2, Part II; now it will be used only nominally). Note for the time being, that $\cos\langle X,Y\rangle = \|P_X\,|\,Y\|$ (see § 2, II).

The following Lemma gives simple geometrical conditions for the operator $P_\theta\,|\,K_B$ to be invertible.

LEMMA. Let M and N denote closed subspaces of a Hilbert space H. The following statements are equivalent:
1. $\mathrm{Ker}(P_M|N)=\{0\}$; 2. $M^\perp \cap N=\{0\}$; 3. $\mathrm{clos}(M+N^\perp)=H$,
4. $\mathrm{clos}\, P_N M=N$.

The following statements are equivalent: 1. $P_M|N$ is left-invertible; 2. $\|P_N|M^\perp\|<1$; 3. $0<\langle N,M^\perp\rangle$; 4. $H=M+N^\perp$.

There is no sense to burden our text with the highly standard proof of the Lemma; see however Lemma 2.1, § 2, Part II. Considering $P_M|N$ as a mapping of N into M we see that

$$(P_M|N)^*=P_N|M,$$

so that the Lemma yields the following useful conclusion.

COROLLARY. The following statements are equivalent.

1. The projection P_M maps the subspace N isomorphically onto M.
2. $\max(\|P_N|M^\perp\|,\ \|P_M|N^\perp\|)<1$.
3. $0<\langle N,M^\perp\rangle$ and $N+M^\perp=H$.
4. $\|P_N|M^\perp\|<1$, $M\cap N^\perp=\{0\}$.

We may now return to the problem of the invertibility of the operator $P_\theta|K_B$ arisen at the end of § 2. Let $\mathbb{P}_+$ be the orthogonal projection of $L^2(\mathbb{R})$ onto H^2_+, and let $\mathbb{P}_-=I-\mathbb{P}_+$.

LEMMA. $P_\theta=\theta\mathbb{P}_-\bar\theta|H^2_+$.

PROOF. It is clear that $\theta\mathbb{P}_-\bar\theta x=0$ if $x\in\theta H^2_+$. If $x\perp\theta H^2_+$, then, obviously, $\bar\theta x\perp H^2_+$ and therefore $\theta\mathbb{P}_-\bar\theta x=x$. ●

THEOREM 3. Let θ_j be an inner function, $P_{\theta_j}=P_{K_{\theta_j}}$, $K_{\theta_j}=H^2_+\ominus\theta_j H^2_+$, $j=1,2$. The following statements are equivalent.

1. The operator $P_{\theta_1}: K_{\theta_2}\longrightarrow K_{\theta_1}$ is invertible.
2. $\mathrm{dist}(\theta_1\bar\theta_2, H^\infty)<1$, $\mathrm{dist}(\theta_2\bar\theta_1, H^\infty)<1$.
3. $\mathrm{dist}(\theta_1\bar\theta_2, H^\infty)<1$, $\bar\theta_2\theta_1 H^2_-\cap H^2_+=\{0\}$.
4. $\mathrm{dist}(\theta_2\bar\theta_1, H^\infty)<1$, $\theta_2\bar\theta_1 H^2_-\cap H^2_+=\{0\}$.
5. $0<\langle\theta_2 H^2_-, \theta_1 H^2_+\rangle$, $\theta_2 H^2_-+\theta_1 H^2_+=L^2(\mathbb{R})$.
6. $0<\langle\theta_1 H^2_-, \theta_2 H^2_+\rangle$, $\theta_1 H^2_-+\theta_2 H^2_+=L^2(\mathbb{R})$.

PROOF. To use the obtained tests of invertibility of $P_M|N$, where $M=K_{\theta_1}$, $N=K_{\theta_2}$, we are to calculate the norm $\|P_{\theta_2}|K^\perp_{\theta_1}\|$:

$$\|P_{\theta_2}|K^\perp_{\theta_1}\|=\|P_{\theta_2}|\theta_1 H^2_+\|=\|\theta_2\mathbb{P}_-\bar\theta_2|\theta_1 H^2_+\|=$$

$$=\sup\left\{\left|\int_{\mathbb{R}}\bar\theta_2\theta_1 h_1\bar h_2\right| : h_1\in H^2_+,\ h_2\in H^2_-,\ \|h_i\|\leqslant 1\right\}=$$

(we use well-known properties of spaces H_+^p : $H_-^2=\{\bar f : f\in H_+^2\}$; the unit ball of H_+^1 coincides with the set $\{fg : \|f\|_{H_+^2}\le 1, \|g\|_{H_+^2}\le 1\}$; see the sources indicated at the begining of § 1)

$$=\sup\{|\int_{\mathbb{R}}\bar\theta_2\theta_1 h| : h\in H_+^1,\ \|h_1\|\le 1\}=$$

(the Hahn - Banach theorem) $=\operatorname{dist}(\bar\theta_2\theta_1, H^\infty)$.

So 1 $\Longleftrightarrow$ 2, as was to be proved. The remaining assertions can be obtained by a formal application of the corollary stated above. It is useful to note that $\theta H_-^2 = H_-^2 \oplus K_\theta$ for any inner function θ. ●

The same arguments lead to the following tests.

THEOREM 3 bis. Let the conditions of Theorem 3 be satisfied. Then the following assertions are equivalent.

1. The operator $P_{\theta_1} : K_{\theta_2}\longrightarrow K_{\theta_1}$ is left-invertible.
2. $\operatorname{dist}(\theta_1\bar\theta_2, H^\infty)<1$.
3. $0<\langle\theta_2 H_-^2, \theta_1 H_+^2\rangle$.
4. $L^2(\mathbb{R})=\theta_1 H_-^2+\theta_2 H_+^2$. ●

Any reader familiar with the Hankel operators may descry the Hankel operator $H_{\bar\theta_1\theta_2}$ at the right-hand side of the formula

$$P_{\theta_1}\,|\,K_{\theta_2}^\perp=\theta_1 \mathbb{P}_-\bar\theta_1\,|\,\theta_2 H_+^2 .$$

This connection of the bases problem with the Hankel (and Toeplitz) operators and with their spectral theory will be very useful. Remind necessary definitions.

Let $L^\infty(\mathbb{R})$ be the space of all bounded measurable functions φ on $\mathbb{R}$ with the natural norm

$$\|\varphi\|_\infty=\operatorname{ess\,sup}_{\mathbb{R}}|\varphi| .$$

DEFINITION. Let $\varphi\in L^\infty(\mathbb{R})$. The *Toeplitz operator* with the symbol φ is the operator T_φ on H_+^2 defined by

$$T_\varphi f=\mathbb{P}_+\varphi f,\quad f\in H_+^2 .$$

The *Hankel operator* H_φ with the same symbol is defined by the formula

$$H_\varphi f=\mathbb{P}_-\varphi f,\quad f\in H_+^2 .$$

The operators T_φ and H_φ are different parts of the multiplication operator

$$\varphi f = H_\varphi f + T_\varphi f, \ f \in H^2_+ . \tag{2}$$

Now we see that

$$P_{\Theta_2} | K^\perp_{\Theta_1} = \Theta_2 P_- \bar\Theta_2 | \Theta_1 H^2_+ = \Theta_2 H_{\bar\Theta_2 \Theta_1} \cdot \bar\Theta_1 | K^\perp_{\Theta_1} \tag{3}$$

and therefore

$$\| P_{\Theta_2} | K^\perp_{\Theta_1} \| = \| H_{\bar\Theta_2 \Theta_1} \|.$$

Returning to the Theorem 3, one can immediately note that it is reduced to the well-known Nehari theorem.

THEOREM (Z.Nehari [50], [54]). If $\varphi \in L^\infty(\mathbb{R})$, then

$$\| H_\varphi \| = dist(\varphi, H^\infty).$$

On the other hand the Hankel operators appearing in Theorems 3 and 3bis have unimodular symbols $\varphi = \Theta_1 \bar\Theta_2$, $\varphi = \Theta_2 \bar\Theta_1$. Then it follows from (2) that

$\| H_\varphi \| < 1$ iff T_φ is left-invertible;

$\| H_\varphi \| < 1, \| H_{\bar\varphi} \| < 1$ iff T_φ is an invertible operator.

Putting these remarks together with Theorems 3 and 3bis, we obtain the following result.

THEOREM 4. Let Θ_j be an inner function for $j=1,2$. Then the operator $P_{\Theta_1} : K_{\Theta_2} \longrightarrow K_{\Theta_1}$ is an isomorphism (respectively left-invertible) if and only if the Toeplitz operator $T_{\Theta_1 \bar\Theta_2}$ is invertible (respectively left-invertible). ●

In order to translate now the invertibility of $P_{\Theta_1} | K_{\Theta_2}$ into "the language of inner functions" Θ_1, Θ_2 (or returning to exponential and reproducing kernel bases - into the language of the Blachske product B with the zero set $\Lambda = \{\lambda_n : n \in \mathbb{Z}\}$) we can apply the invertibility criteria of the Toeplitz operator theory, and in particular A.Devinatz's - H.Widom's theorem [31], [57], [54]. For its formulation a new portion of definitions is needed.

The first deals with the Hilbert transform in $L^\infty(\mathbb{R})$. The space $L^\infty(\mathbb{R})$ being not contained in $L^2(\mathbb{R})$ it is impossible to extend the Hilbert transform (from $L^2(\mathbb{R})$) by means of the usual Cauchy integral. We shall use the conformally-invariant form to remove the singularity at infinity. Namely, we define the Hilbert transform $\tilde{v}$ of a function v, $v \in L^\infty(\mathbb{R})$ by

$$\tilde{v}(x) = \frac{1}{\pi}\,(\mathrm{v.p.})\int_{\mathbb{R}} \left\{ \frac{1}{x-t} + \frac{t}{1+t^2} \right\} v(t)\,dt .$$

The Schwarz formula

$$V(z) = \frac{1}{\pi i}\int_{\mathbb{R}} \left\{ \frac{1}{t-z} - \frac{t}{1+t^2} \right\} v(t)\,dt$$

recovers the function V by its real part v only, provided $V \in H^\infty$ and $\operatorname{Im} V(i) = 0$.

DEFINITION. A non-negative function w is called a function satisfying Helson - Szegö condition (briefly $w \in (HS)$) if there are functions u, v in $L^\infty(\mathbb{R})$ such that

$$\|v\|_\infty < \pi/2 \quad \text{and} \quad w = \exp\{u + \tilde{v}\}. \tag{HS}$$

Another form of the Helson -Szegö condition has been obtained in a remarkable paper of B.Muckenhoupt, R.Hunt and R.Wheeden [40]. Let $\mathcal{J}$ be the family of all intervals on $\mathbb{R}$.

THEOREM (R.A.Hunt, B.Muckenhoupt, R.L.Wheeden [40]). The (HS)-condition is equivalent to (A_2)-condition of Muckenhoupt:

$$\sup_{I \in \mathcal{J}} \frac{1}{|I|}\int_I w\,dx \cdot \frac{1}{|I|}\int_I w^{-1}\,dx < \infty. \tag{A_2}$$

THEOREM (A.Devinatz, H.Widom [31], [57]). A Toeplitz operator T_φ with a unimodular symbol φ ($|\varphi| = 1$ a.e.) is invertible if and only if

$$\varphi = e^{i(\tilde{u} + v + c)}, \quad \text{where } c \in \mathbb{R};\ u, v \in L^\infty(\mathbb{R}),\ \|v\|_\infty < \pi/2.$$

The next theorem combined with Theorem 4 will be a key tool for the proofs of many efficient basis tests.

THEOREM 5. Let φ be a unimodular function. The following conditions are equivalent.

1. The Toeplitz operator is invertible.
2. $\operatorname{dist}_{L^\infty}(\varphi, H^\infty) < 1$, $\operatorname{dist}_{L^\infty}(\bar{\varphi}, H^\infty) < 1$.
3. There is an outer function f, $f \in H^\infty$, satisfying

$$\| \varphi - f \|_\infty < 1.$$

4. There is a branch of the argument α of the unimodular function φ, $\varphi(x) = e^{i\alpha(x)}$, such that

$$\inf\{ \|\alpha - \tilde{v} - c\|_\infty : v \in L^\infty(\mathbb{R}),\ c \in \mathbb{R}\} < \pi/2.$$

5. There are a unimodular constant λ and an outer function h such that

$$\varphi = \lambda \cdot \frac{\bar{h}}{h}, \quad |h|^2 \in (HS) \qquad (\text{or } |h|^2 \in (A_2)\ ).$$

To obtain a list of invertibility tests for $P_\theta | K_B$ it remains only to put $\varphi = \bar{B}\theta$ in the condition of the theorem.

Referring the reader to § 2, Part II for the proof of Theorem 5, we mention that the equivalence 1 $\Longleftrightarrow$ 2 has been already proved and the equivalence 1 $\Longleftrightarrow$ 5 is a simple consequence of the A.Devinatz - H.Widom theorem.

4. Basis property of exponentials on an interval

Comparing Theorems 2,2 and 2bis with Theorems 3,3bis, 4 and 5 one can easily obtain a series of tests for the basis property mentioned in the title of the section. Nevertheless, for the convenience of the reader we formulate one of them.

Let $\widetilde{L^\infty} \overset{def}{=} \{\tilde{v} : v \in L^\infty(\mathbb{R})\}$ and let $\widetilde{L^\infty} + \mathbb{C} = \{u + c : u \in \widetilde{L^\infty},\ c \in \mathbb{C}\}$. It is useful to note that non-zero constants can not coincide with $\tilde{v}$, the harmonic continuation of $\tilde{v}$ is vanishing at the point i. For any function f defined on $\mathbb{R}$ let

$$dist_{L^\infty}(f, \widetilde{L^\infty} + \mathbb{C}) \overset{def}{=} \inf\{\|f - g\|_\infty : g \in \widetilde{L^\infty} + \mathbb{C}\},$$

assuming that $\|f - g\|_\infty = +\infty$ if $f - g \notin L^\infty(\mathbb{R})$.

Let $\Lambda \subset \mathbb{C}_\delta$, $\delta > 0$, and let B be a Blaschke product with the zero set Λ. It is easy to see that the function α_Λ defined by

$$\alpha_\Lambda(x) = 2\int_0^x \sum_{\lambda \in \Lambda} \frac{\operatorname{Im}\lambda}{|\lambda - t|^2}\,dt - ax, \quad x \in \mathbb{R},$$

is a continuous branch of argument, up to an additive constant, of the unimodular function $B\bar{\theta}^a$ on $\mathbb{R}$.

THEOREM 6. Let $\Lambda=\{\lambda_n : n\in\mathbb{Z}\}\subset\mathbb{C}_\delta$, $\delta>0$. Then the family $(e^{i\lambda_n x})_{n\in\mathbb{Z}}$ forms an unconditional basis in $L^2(0,a)$ if and only if

$$\Lambda\in(C),\quad dist_{L^\infty}(\alpha_\Lambda,\widetilde{L^\infty}+C)<\pi/2.$$

The sufficiency part of Theorem 6 is a simple consequence of Theorems 2′ ,4 and 5. We put aside the proof of the necessity till §1 of Part III where it will be proved that the function α , arising in Theorem 5 (see assertion 4 of that theorem), is automatically continuous under the conditions of Theorem 6. This will imply, obviously, $\alpha-\alpha_\Lambda\equiv const.$

The M.I.Kadec theorem can be easily obtained as a corollary of Theorem 6. The same reasonings fit in for the proof S.A.Avdonin and V.È.Kacnelson theorems as well; see §2 of Part III.

COROLLARY. Let $(\lambda_n)_{n\in\mathbb{Z}}$ be a sequence of real numbers and let $\sup_{n\in\mathbb{Z}}|n-\lambda_n|<1/4$. Then the family $(e^{i\lambda_n x})_{n\in\mathbb{Z}}$ is a Riesz basis in $L^2(0,2\pi)$.

PROOF. According to our remark on p. 229 , we may without loss of generality consider a family of frequencies $(\lambda_n+iy)_{n\in\mathbb{Z}}$, $y>0$. It is clear that the family $(e^{i(n+iy)x})_{n\in\mathbb{Z}}$ is a Riesz basis in $L^2(0,2\pi)$. This example is a good illustration for Theorem 5. Let $\varepsilon=\exp(-2\pi y)$, then

$$\frac{\theta^{2\pi}(z)-\varepsilon}{1-\varepsilon\theta^{2\pi}(z)}\ \stackrel{def}{=\!=}\ B_0(z)=\prod_{n\in\mathbb{Z}}\frac{1-\frac{z}{n+iy}}{1-\frac{z}{n-iy}}\ .$$

We may conclude therefore that

$$B_0\bar\theta^{2\pi}=\frac{\overline{1-\varepsilon\theta^{2\pi}}}{1-\varepsilon\theta^{2\pi}}\ .$$

The function $h=1-\varepsilon\theta^{2\pi}$ is outer and

$$1-\varepsilon\le\inf_{x\in\mathbb{R}}|h(x)|\le\sup_{x\in\mathbb{R}}|h(x)|\le1+\varepsilon.$$

Therefore statement 5 of Theorem 5 holds and the Toeplitz operator $T_{B\bar\theta^{2\pi}}$ is invertible by that theorem. Obviously, $\mathbb{Z}+iy\in\in(C)$. So the combination of Theorems 2′ and 4 implies among other things the Riesz basis property for the family $(e^{inx}\cdot e^{-y})_{n\in\mathbb{Z}}$ in $L^2(0,2\pi)$. The function $\alpha_{\mathbb{Z}+iy}$, up to an additive constant, is an argument of the unimodular function $B_0\bar\theta^{2\pi}$

This implies

$$\alpha_{\mathbb{Z}+iy}(x) = c + \widetilde{\log|h^2|}(x), \quad c \in \mathbb{R},$$

and $\alpha_{\mathbb{Z}+iy} \in \widetilde{L}^\infty + \mathbb{C}$. Moreover $\alpha_{\mathbb{Z}+iy} \in \mathrm{Re}\, H^\infty$ as $\log h^2 \in H^\infty$.

Now we may compare the functions $\alpha_{\mathbb{Z}+iy}$ and $\alpha_{\Lambda+iy}$. Let $\lambda_n = n + \delta_n$, $n \in \mathbb{Z}$. Then

$$\alpha_{\mathbb{Z}+iy}(x) - \alpha_{\Lambda+iy}(x) = 2\sum_{n\in\mathbb{Z}} \left\{ \int_0^x \frac{y}{(t-n)^2+y^2}\,dt - \int_0^x \frac{y}{(t-\lambda_n)^2+y^2}\,dt \right\} =$$

$$= 2\sum_{n\in\mathbb{Z}} \int_{x-\delta_n}^{x} \frac{y}{(t-n)^2+y^2}\,dt - 2\sum_{n\in\mathbb{Z}} \int_{-\delta_n}^{0} \frac{y}{(t-n)^2+y^2}\,dt .$$

It is time to remember that $\delta \overset{\mathrm{def}}{=} \sup_n |\delta_n| < 1/4$. An obvious estimate shows

$$\left|\sum_{n\in\mathbb{Z}} \int_{x-\delta_n}^{x} \frac{y}{(t-n)^2+y^2}\,dt\right| \le \sum_{n\in\mathbb{Z}} \int_{x-\delta}^{x} \frac{y}{(t-n)^2+y^2}\,dt = \int_{x-\delta}^{x} \sum_{n\in\mathbb{Z}} \frac{y}{(t-n)^2+y^2}\,dt .$$

It remains to show that the right-hand side of the equality is bounded by $\pi/4$ uniformly on $\mathbb{R}$. Very simple reasonings lead to this conclusion. The periodic function

$$t \longmapsto \sum_{n\in\mathbb{Z}} \frac{y}{(t-n)^2+y^2}$$

tends to a constant uniformly in t as $y \to +\infty$. Its integral along the interval $[0,1]$ is π . So the integral along any interval with length smaller than $1/4$ will be smaller than $\pi/4$ if y is sufficiently large.

We may, certainly, use a more formal calculation. By the Poisson summation formula

$$\sum_{n\in\mathbb{Z}} \frac{y}{y^2+(t-n)^2} = \pi \cdot \frac{1-\varepsilon^2}{1-2\varepsilon\cos 2\pi t + \varepsilon^2} ,$$

$\varepsilon = \exp(-2\pi y)$. It is clear that

$$\sup_{x\in\mathbb{R}} \int_{x-\delta}^{x} \sum_{n\in\mathbb{Z}} \frac{y}{y^2+(t-n)^2}\, dt = \pi \cdot \int_{-\delta/2}^{\delta/2} \frac{1-\varepsilon^2}{1-2\varepsilon\cos 2\pi t+\varepsilon^2}\, dt$$

and the right-hand side tends to $\pi\delta < \pi/4$ as $y \to +\infty$. ●

REMARK. See another proof of the Corollary in [19], [18] p.342.

Our next topic concerns the relationship between the bases problem and the theory of entire functions. Entire functions arise in the unconditional bases problem in a natural way. Assuming the family $(e^{i\lambda_n x})_{n\in\mathbb{Z}}$ is an unconditional basis in $L^2(0,a)$ we see that the co-dimension of $span\{e^{i\lambda_n x}\cdot\chi_{[0,a)} : n\in\mathbb{Z}\setminus\{0\}\}$ in $L^2(0,a)$ is equal to 1. By the Hahn - Banach theorem

$$\dim\{f\in L^2(0,a) : \int_0^a e^{i\lambda_n x} f(x)\,dx = 0,\ n\in\mathbb{Z}\setminus\{0\}\} = 1. \tag{4}$$

It follows that the Fourier - Laplace transform

$$\hat{f} = \int_0^a e^{izt} f(t)\,dt$$

vanishes exactly on the set $\{\lambda_n : n\in\mathbb{Z}\setminus\{0\}\}$ if the function f, $f\not\equiv 0$ belongs to the one-dimensional subspace considered in (4). Indeed, every zero μ $(\hat{f}(\mu)=0)$ not belonging to the set gives rise to a function g belonging to the subspace defined by (4) and not a scalar multiple of f. Indeed, let $\hat{f}(z)\cdot(z-\mu)^{-1} = \hat{g}(z)$, where $g = -ie^{-i\mu x}\int_0^x e^{i\mu s} f(s)\,ds$, $g\in L^2(0,a)$. So the function F_Λ,

$$F_\Lambda = \left(1-\frac{z}{\lambda_0}\right)\int_0^a e^{izt} f(t)\,dt \tag{5}$$

is an entire function of the exponential type a with the zero set $\{\lambda_n : n\in\mathbb{Z}\}$. It follows from (5) that the conjugate diagram of F_Λ is the segment $[0, ia]$ *). Let $\mathcal{E}_a$ denote the set of all entire functions of exponential type

*) The exhausting information about diagrams, and in general about the growth theory, may be found in [13], [27]. In our case a is the length of the interval on which the basis problem is considered.

with the conjugate diagram $[0, ia]$. An entire function of exponential type without zeros coincides with one exponential $\exp \lambda z$, $\lambda \in \mathbb{C}$. Therefore the functions in $\mathcal{E}_a$ are defined by their zero-set up to a multiplicative constant.

DEFINITION. Let $\Lambda \subset \mathbb{C}_+$, $a>0$. An entire function F_Λ in $\mathcal{E}_a$ is called a g e n e r a t i n g f u n c t i o n for the pair (Λ, a) if its zero set is Λ and if $F_\Lambda(0)=1$.

THEOREM 7. Let $\Lambda = \{\lambda_n : n \in \mathbb{Z}\} \subset \mathbb{C}_\delta$, $\delta>0$ and let $a>0$. The following conditions are equivalent.

1. The family $(e^{i\lambda_n x})_{n\in\mathbb{Z}}$ is a Riesz basis in $L^2(0,a)$
2. $\Lambda \in (C)$ and there is a generating function F_Λ for the pair (Λ, a) satisfying $|F_\Lambda|^2 | \mathbb{R} \in (HS)$ (or equivalently $|F_\Lambda|^2 | \mathbb{R} \in (A_2)$).

We shall give now only an idea of the proof, the details may be found in Part III. What we are to prove is the equivalence of the inclusion $|F_\Lambda|^2 | \mathbb{R} \in (HS)$ and of the invertibility of the Toeplitz operator $T_{\bar{\theta}^a B}$. By Theorem 5 (see the statements 1 and 5) the operator $T_{\bar{\theta}^a B}$ is invertible if and only if the unimodular function $\bar{\theta}^a B$ can be factored in a form $\bar{\theta}^a B = c\bar{h}h^{-1}$, $|c|=1$, $c\in\mathbb{C}$, $|h^2| \in (HS)$. This implies the equality

$$Bh = c\theta^a \bar{h} \tag{6}$$

holds a.e. on $\mathbb{R}$ for the outer function h. It follows from $|h^2| \in (HS)$ by V.I.Smirnov theorem that $h(z+i)^{-1} \in H^2_+$. The equality (6) means that the boundary values of the function Bh analytic in the upper half-plane coincide with the ones of $z \to c\,\theta^a(z)\,\overline{h(\bar{z})}$, which is, obviously, analytic in the lower half-plane. Using the inclusion $h(z+i)^{-1} \in H^2_+$ one can easily deduce that the function Bh is a restriction of an entire function F onto $\mathbb{C}_+$. Standard estimates show that $F \in \mathcal{E}_a$. The zero set of F is Λ. We see also that $|F|^2 = |h|^2$ on $\mathbb{R}$. These arguments can be easily converted.

REMARK. The Levin - Golovin theorem (see Introduction for the formulation) is an obvious corollary of Theorem 7.

Let now $\Lambda = \{\lambda_n : n \in \mathbb{Z}\} \subset \mathbb{R}$. It would be pleasant to have a test for the unconditional bases property in terms of this set only. To do this let

$$N_\Lambda(x)=\begin{cases}\operatorname{card}\Lambda\cap[0,x], & x\geq 0.\\ -\operatorname{card}\Lambda\cap[x,0), & x<0.\end{cases}$$

The function N_Λ is non-decreasing on $\mathbb{R}$. An asymptotic property of N_Λ equivalent to the unconditional bases property for the family $(e^{i\lambda_n x})_{n\in\mathbb{Z}}$ in $L^2(0,a)$ will be given in terms of the well-known class $BMO(\mathbb{R})$. The space $BMO(\mathbb{R})$ consists of locally integrable functions f on $\mathbb{R}$ satisfying

$$\|f\|_* = \sup_{I\in\mathcal{J}} \frac{1}{|I|}\int_I |f-f_I|\,dx<\infty, \quad f_I \overset{\text{def}}{=} \frac{1}{|I|}\int_I f\,dx.$$

Here $\mathcal{J}$ stands for the family of all intervals on $\mathbb{R}$. An important property of $BMO(\mathbb{R})$ is that this class as well as the class of function satisfying (A_2)-condition, has a completely different description. A function f belongs to BMO iff there are bounded measurable functions u, v such that $f = u+\tilde{v}$. This and other properties of BMO may be found in [44], [54]. If $f\in BMO$ then it follows that

$$\int_{\mathbb{R}} \frac{|f(x)|}{1+x^2}\,dx<+\infty$$

and so every function f in BMO has a harmonic continuation into $\mathbb{C}_+$:

$$\mathcal{U}_f(z)=\frac{1}{\pi}\int_R \frac{\operatorname{Im} z}{|t-z|^2} f(t)\,dt.$$

Let symbol $\mathcal{P}_\gamma$ denote the set of all f in BMO satisfying the following condition. There are a positive number y, a real number c and bounded measurable functions u, v such that

$$\mathcal{U}_f(x+iy)=c+\tilde{u}(x)+v(x);\quad x\in\mathbb{R},\quad \|v\|_\infty<\gamma.$$

THEOREM 8. Let $\lambda_n\in\mathbb{R}$, $n\in\mathbb{Z}$. Then the family $(e^{i\lambda_n x})_{n\in\mathbb{Z}}$ forms a Riesz basis in $L^2(0,a)$, $a>0$, iff

1. $\inf_{n\neq m}|\lambda_n-\lambda_m|>0$;
2. $N_\Lambda-\frac{a}{2\pi}x\in\mathcal{P}_{1/4}$.

The condition 2 of Theorem 8 defines a number a uniquely

because the linear function $x \mapsto x$ does not belong to BMO (indeed, $\int_{\mathbb{R}} \frac{|x|}{1+x^2}\, dx = +\infty$).

It is interesting to compare Theorem 8 with known theorems concerning the completeness problem. It follows from the condition

$$\int_{\mathbb{R}} \frac{|N_\Lambda(x) - ax|}{1+x^2}\, dx < +\infty$$

by the Beurling - Malliavin theorem that the family $(e^{i\lambda_n x})_{n\in\mathbb{Z}}$ is complete on any interval I , $|I| < a$; see theorem 71 in [51] . We see therefore that the conditions implying the unconditional basis property for a family of exponentials on I are considerably more restrictive than those for the completeness property.

The Kadec theorem may be also proved with the help of Theorem 8. Here is a sketch of the proof. Let $f(x) = N_{\mathbb{Z}}(x) - x,\ x \in \mathbb{R}$. Then the function $x \mapsto \mathcal{U}_f(x+iy),\ y>0$ belongs to $\widetilde{L}^\infty + \mathbb{C}$. If $\lambda_n = n + \delta_n$ and if $\sup_n |\delta_n| = \delta < 1/4$, then

$$N_{\mathbb{Z}}(x) - N_\Lambda(x) = \sum_{n\in\mathbb{Z}} \operatorname{sign}\delta_n \cdot \chi_{[n, n+\delta_n)}(x).$$

Therefore the Poisson integral of $N_{\mathbb{Z}} - N_\Lambda$ is equal to

$$\sum_{n\in\mathbb{Z}} \int_{x-\delta_n}^{x} \frac{y}{(t-n)^2 + y^2}\, dt.$$

The proof is finished as on p. 238 ●

The next result demonstrates the close relationship existing between general unconditional exponential bases and the classical orthogonal system $(e^{inx})_{n\in\mathbb{Z}}$ in $L^2(-\pi,\pi)$. Let $\Lambda = \{\lambda_n : n \in \mathbb{Z}\} \subset \mathbb{C}_+$ and let the family $(e^{i\lambda_n x})_{n\in\mathbb{Z}}$ be an unconditional basis in $L^2(-\pi,\pi)$. Let $(h_n)_{n\in\mathbb{Z}}$ be the dual family for $(e^{i\lambda_n x})_{n\in\mathbb{Z}}$ in $L^2(-\pi;\pi)$:

$$(e^{i\lambda_k x}, h_n) = \frac{1}{2\pi}\int_{-\pi}^{\pi} e^{i\lambda_k x}\cdot\overline{h_n(x)}\, dx = \begin{cases} 1, & n = k, \\ 0, & n \neq k. \end{cases}$$

Then it is possible to associate to every function f in $L^2(-\pi,\pi)$ the non-harmonic Fourier series

$$f \sim \sum_{n\in\mathbb{Z}} (f, h_n)\, e^{i\lambda_n x}$$

which, in accordance with our assumption, converges unconditionally in L^2 to the function f. However, the question of the pointwise convergence of such a non-harmonic Fourier series is interesting too. It were again R.Paley and N.Wiener who have studied the problem for the first time [59]. After that N.Levinson in his well-known book [48] has proved, assuming $\Lambda \subset \mathbb{R}$, $\sup_n |\lambda_n - n| < 1/4$, that for every function f in $L^2(-\pi,\pi)$

$$\lim_{N\to+\infty}\left\{\sum_{|n|\le N}\hat{f}(n)e^{inx}-\sum_{|n|\le N}(f,h_n)e^{i\lambda_n x}\right\}=0$$

uniformly on every compact subset of the interval $(-\pi,\pi)$. Here $\hat{f}(n)=\frac{1}{2\pi}\int_{-\pi}^{\pi}e^{-inx}f(x)\,dx$ stands for usual Fourier coefficients of f. In §4 of Part III this theorem is extended on each family $(e^{i\lambda_n x})_{n\in\mathbb{Z}}$, with $\Lambda\subset\mathbb{C}_+$, which forms an unconditional basis in $L^2(-\pi,\pi)$.

5. Hilbert space geometry of exponentials and reproducing kernels, and the spectral expansion of the model semigroup

Let us return once more to the Carleson condition (C) for the set $\Lambda=\{\lambda_n : n\in\mathbb{Z}\}$, $\lambda_n\in\mathbb{C}_+$. As we have already noted, this condition appeared originally in the papers of L.Carleson [28], W.K.Hayman [66], D.J.Newman [65] as a condition for the solvability of the interpolation problem in H^∞. H.Shapiro and A.Shields proved later that (C) is a necessary and sufficient condition for the following interpolation problem in H^2_+ to be solvable for any given sequence $(a_n)_{n\in\mathbb{Z}}$, $(a_n)_{n\in\mathbb{Z}}\in\ell^2(\mathbb{Z})$:

$$f\in H^2_+;\quad f(\lambda_n)\cdot\sqrt{2\,\mathrm{Im}\,\lambda_n}=a_n .$$

A formal solution of the problem is given by the formula

$$f(z)=\sum_{n\in\mathbb{Z}}\frac{2i\,\mathrm{Im}\,\lambda_n}{z-\bar{\lambda}_n}\cdot\frac{B_n(z)}{B_n(\lambda_n)}a_n+Bg,\quad g\in H^2_+ .$$

The series under the condition $\Lambda\in(C)$ turns out to be unconditionally convergent for every $(a_n)_{n\in\mathbb{Z}}\in\ell^2$. The solution f corresponding to $g=0$ has the minimal norm among other solutions and belongs to $K_B=H^2_+\ominus BH^2_+$.

In the paper [20] it was observed that the considered series is the Fourier series expansion with respect to the eigen-functions of the so-called model contractive semigroup. The model semigroup has been thoroughly studied in papers of B.Sz-Nagy, C.Foiaş, V.M.Adamjan, D.Z.Arov, M.G.Krein and others. The semigroup we want to deal with is defined in K_B by the formula

$$Z_t f = P_B \cdot U_t f, \quad f \in K_B, \ t>0,$$

where $U_t f(z) = exp(izt) \cdot f(z)$. The inner function B is named the characteristic function of the semigroup $(Z_t)_{t>0}$.

Spectral properties of $(Z_t)_{t>0}$ are now well-studied, see for example [18]. We mention only that the generator A of a model semigroup, $Z_t = exp(iAt)$, $t>0$ is a simple dissipative operator and its spectrum σ coincides with the spectrum of the characteristic function B.

In particular, every simple zero λ_n of B is a simple eigen-value for $A = A_B$ and the corresponding eigen-function is defined by

$$\psi_n(z) = \frac{(2 \operatorname{Im} \lambda_n)^{-1/2}}{z - \overline{\lambda}_n} \cdot \frac{B_{\lambda_n}(z)}{B_{\lambda_n}(\lambda_n)} .$$

If B is a Blaschke product then the family $(\psi_n)_{n \in \mathbb{Z}}$ of eigen-functions of A_B is complete in K_B . The dual system, being the family of eigen-functions for the conjugate operator A_B^* , is defined by

$$\varphi_n(z) = \frac{(2 \operatorname{Im} \lambda_n)^{1/2}}{z - \overline{\lambda}_n} ,$$

and $A_B^* \varphi_{\lambda_n} = \overline{\lambda}_n \varphi_{\lambda_n}$, $n \in \mathbb{Z}$.

By Theorem A the Carleson condition is a necessary and sufficient condition for $(\varphi_n)_{n \in \mathbb{Z}}$, as well as for $(\psi_n)_{n \in \mathbb{Z}}$ to form an unconditional basis in K_B.

Let now Θ denote a singular inner function and let B denote a Blaschke product. The invertibility problem for the operator $P_\theta : K_B \to K_\theta$, which is central for the unconditional basis problem, can be reformulated in terms of model operators. To do this consider the subspace $clos(K_B + K_\Theta)$ in H_+^2.

LEMMA. $clos(K_B + K_\Theta) = K_{B\Theta}$.

THE PROOF is an elementary calculation: if $f \perp K_B + K_\theta$ then by definition $f \in BH_+^2 \cap \Theta H_+^2 = B\Theta H_+^2$. ●

Let A be a model dissipative operator in $K_{B\theta}$ with a characteristic function $B\theta$ and let $(Z_t)_{t\geq 0}$:

$$Z_t f = P_K e^{izt} f = e^{iAt} f,$$

be a corresponding semigroup of contractions.

The spaces K_B and K_θ have a well-defined spectral sense.

LEMMA A. The space K_B is the subspace of discrete spectrum for A^* and the space K_θ is the subspace of singular continuous spectrum for A^*. Their orthogonal complements in $K \overset{def}{=\!=} K_{B\theta}$

$$K \ominus K_B = BK_\theta, \quad K \ominus K_\theta = \theta K_B$$

are the spaces of singular continuous spectrum and the space of discrete spectrum for the operator A respectively.

The point discrete spectrum $\sigma_d(A)$ of A coincides with $\Lambda = \{\lambda_n : n \in \mathbb{Z}\}$ and $\bar{\Lambda} = \{\bar{\lambda}_n : n \in \mathbb{Z}\} = \sigma_d(A^*)$.

If $\theta = \theta^a$ then the point ∞ belongs to the singular continuous spectrum of the both operators A and A^*.

The interested reader can find the proof of a proposition analogous to the Lemma in [18].

The spectral interpretation of the completeness problem and the unconditional bases problem requires to remind the reader one definition more.

DEFINITION (see [7], p.382). A family of vectors $(\varphi_n)_{n\in\mathbb{Z}}$, $\|\varphi_n\| \asymp 1$ in a Hilbert space is named ω-l i n e a r l y i n d e p e n d e n t if the conditions

$$\lim_{N\to\infty} \sum_{|n|\leq N} a_n \varphi_n = 0, \quad (a_n)_{n\in\mathbb{Z}} \in \ell^2$$

imply $a_n = 0$, $n \in \mathbb{Z}$.

To emphasize the spectral sense of the subspaces K_B and $\theta K_B = K \ominus K_\theta$ we shall use the following notation $E_d^* \overset{def}{=\!=} K_B$, $E_d \overset{def}{=\!=} K \ominus K_\theta$.

Let now $\varphi_n = \frac{(2 \operatorname{Im} \lambda_n)^{1/2}}{z - \bar{\lambda}_n}$ and let $\sup_{n\in\mathbb{Z}} |\theta(\lambda_n)| < 1$. Then it follows from §2 that $\|P_\theta \varphi_n\| \asymp 1$.

LEMMA. The following statements are equivalent:

1. the family $(P_\theta \varphi_n)_{n\in\mathbb{Z}}$ is complete in K_θ;
2. $K_\theta \cap K_B^\perp = \{\mathbb{O}\}$.

If $\Lambda \in (C)$ then the following statements are equivalent:

3. the family $(P_\theta \varphi_n)_{n\in\mathbb{Z}}$ is a ω-linearly independent;

4. $K_\theta^\perp \cap K_B = \{0\}$.

PROOF. 1 $\Leftrightarrow$ 2. Let $f \in K_\theta \ominus span(P_\theta \varphi_n : n \in \mathbb{Z})$. Then

$$0 = (P_\theta \varphi_n, f) = (\varphi_n, P_\theta f) = (\varphi_n, f)$$

and therefore $f \perp K_B$.

3 $\Longleftrightarrow$ 4. The family $(\varphi_n)_{n\in\mathbb{Z}}$ is a Riesz basis in K_B by Theorem A. Therefore for every f in $P_\theta K_B$ one may find a sequence $(a_n)_{n\in\mathbb{Z}}$ in $\ell^2(\mathbb{Z})$ such that

$$f = \sum_{n\in\mathbb{Z}} a_n P_\theta \varphi_n .$$

On the other hand each sum of such a form is the orthogonal projection of a function in K_B. Therefore the condition $P_\theta f = 0$, $f \in K_B$ appears to be equivalent to $\sum a_n P_\theta \varphi_n = 0$, $(a_n)_{n\in\mathbb{Z}} \in \ell^2$. But the kernel of the operator $P_\theta | K_B$ is $K_B \cap K_\theta^\perp$. ●

LEMMA. The following statements are equivalent:

1. the family $(P_\theta \varphi_n)_{n\in\mathbb{Z}}$ is complete in K_θ;

2. $K = clos\,(E_d + E_d^*)$.

If the family of eigen-functions of A (or A^*) forms an unconditional basis in its own span, then the following statements are equivalent:

3. the family $(P_\theta \varphi_n)_{n\in\mathbb{Z}}$ is ω-linearly independent;

4. $E_d \cap E_d^* = \{0\}$.

PROOF. Apply Lemma A. ●

It is easy to obtain the spectral test for the invertibility of $P_\theta : K_B \to K_\theta$.

LEMMA. The operator $P_\theta : K_B \longrightarrow K_\theta$ is invertible if and only if

a) $K = clos\,(E_d + E_d^*)$;

b) $0 < \langle E_d, E_d^* \rangle$.

The following theorem finds its application in Part IV for the case $\Theta = \Theta^a$.

THEOREM 9. Let Θ be an inner function, B be a Blaschke product. Suppose that the point spectrum $\sigma_p(A)$ of the model operator A defined in $K \overset{def}{=} K_{B\Theta}$ satisfies $\sup_{\lambda\in\sigma_p} |\Theta(\lambda)| < 1$ and let eigen-vectors $\left\{ \frac{\Theta}{z-\lambda} : \lambda \in \sigma_p(A) \right\}$ of A form an unconditional basis in their span. Then the following conditions are equivalent.

1. The operator $P_\theta : K_B \to K_\theta$ is invertible.

2. The family of reproducing kernels $\{(1-\overline{\theta(\lambda)}\theta)(z-\bar\lambda)^{-1} : \lambda \in \sigma_p(A)\}$ forms an unconditional basis in K_θ.

3. The joint family of eigen-functions for A and A^* forms an unconditional basis in K.

PROOF. The implications 1 $\Longleftrightarrow$ 2 are a simple corollary of Theorem 2. The statement 1 $\Longleftrightarrow$ 3 is implied by the spectral test of the invertibility of $P_\theta | K_B$. ●.

REMARK. Clearly

$$P_B | K_\theta = (P_\theta | K_B)^*.$$

It follows that the operator $P_\theta | K_B$ has a bounded inverse operator if and only if the subspaces of continuous singular spectrum for A, A^* span the space $K = K_{B\theta}$ and form a positive angle.

6. Bases problem in the disc and in the half-plane

In § 1 it was shown that the unconditional exponential bases problem leads to a more general one. By some reasons it is convenient to deal with the general case of reproducing kernels in the setting of Hardy classes in the unit disc $\mathbb{D} = \{\zeta \in \mathbb{C} : |\zeta| < 1\}$. The main purpose of the section is to establish the connection between the Hardy classes theory in the half-plane and that in the disc.

Let $\mathbb{T} \overset{def}{=\!=} \{\zeta \in \mathbb{C} : |\zeta| = 1\}$ denote the unit circle of the complex plane and let $L^2(\mathbb{T})$ be the Hilbert space of all square-summable functions on $\mathbb{T}$ with respect to the normalized Lebesgue measure m on $\mathbb{T}$. The Hardy class $H^2(\mathbb{D})$ is defined as the space of all holomorphic functions g in $\mathbb{D}$ satisfying

$$\sup_{0<r<1} \int_{\mathbb{T}} |g(r\zeta)|^2 \, dm(\zeta) < +\infty.$$

By Fatou's theorem the space $H^2(\mathbb{D})$ may be considered as a closed subspace of $L^2(\mathbb{T})$. Let θ be an inner function in $\mathbb{D}$ and let $K_\theta = H^2(\mathbb{D}) \ominus \theta H^2(\mathbb{D})$. The reproducing kernel for $H^2(\mathbb{D})$ being defined by $k(z,\lambda) = (1-\bar\lambda z)^{-1}$, the reproducing kernel for K_θ is equal to

$$k_\theta(z,\lambda)=\frac{1-\overline{\theta(\lambda)}\,\theta(z)}{1-\bar\lambda z}.$$

Let Λ be a subset of $\mathbb{D}$ satisfying the Blaschke condition

$$\sum_{\lambda\in\Lambda}(1-|\lambda|)<+\infty \tag{B}$$

and let B denote the Blaschke product

$$B=\prod_{\lambda\in\Lambda}\frac{\bar\lambda}{|\lambda|}\,\frac{\lambda-z}{1-\bar\lambda z}.$$

We remind that the Carleson condition for $\mathbb{D}$ has the same form as for $\mathbb{C}_+$. Namely, $\Lambda\in(C)$ if

$$\inf_{\lambda\in\Lambda}|B_\lambda(\lambda)|>0,\quad B_\lambda=B\cdot\frac{1-\bar\lambda z}{\lambda-z}.$$

It also may be split up into two parts; see [18].

THEOREM 10. Let $\Lambda\in(B)$ and let B be the Blaschke product with the zero set Λ. Let θ be an inner function in $\mathbb{D}$ satisfying $\sup_{\lambda\in\Lambda}|\theta(\lambda)|<1$. The following statements are equivalent.

1. The family $\left\{\frac{1-\overline{\theta(\lambda)}\,\theta}{1-\bar\lambda z}:\lambda\in\Lambda\right\}$ forms an unconditional basis in $K_\theta=H^2(\mathbb{D})\ominus\theta H^2(\mathbb{D})$.

2. $\Lambda\in(C)$ and the operator P_θ maps K_B isomorphically onto K_θ.

The operator $P_\theta|K_B$ is invertible iff the Toeplitz operator $T_{\bar B\theta}$ does. The tests for the last are given by an analog of Theorem 5; see §3.

In conclusion, some words about the relationship between the Hardy classes in the disc and in the half-plane. Clearly, the operator

$$Uf(x)\stackrel{def}{=}\frac{1}{\sqrt{\pi}}\,\frac{1}{x+i}\,f\left(\frac{x-i}{x+i}\right),\quad x\in\mathbb{R},$$

is an isometry of $L^2(\mathbb{T})$ onto $L^2(\mathbb{R})$. Let $\gamma(x)\stackrel{def}{=}\frac{x-i}{x+i}$, $x\in\mathbb{R}$. Then it is aesy to check that

$$U\cdot M_\varphi=M_{\varphi\circ\gamma}U,$$

where M_φ stands for the multiplication operator in L^2 and $\varphi \in L^\infty$. It follows from the equality $UH^2(\mathbb{D}) = H^2_+$ that an analogous formula holds for the Hankel and Toeplitz operators. It should be also noted that $UK_\theta = K_{\theta\circ\gamma}$ and that the operator U establishes a one-to-one correspondence between the reproducing kernels of K_θ and those of $K_{\theta\circ\gamma}$. So the unitary operator U allows one to move from the disc into the half-plane and vice versa.

The special condition $\sup_{\lambda\in\Lambda} |\theta(\lambda)| < 1$ imposed onto the pair (Λ, θ) plays the same rоle as in § 1-4: simplifying the problem it leads to the more elegant formulations. When θ is a function "with a single charged point" this condition does not constitute a real restriction, a linear fractional transformation (linear $z \mapsto z + iy$, $y>0$, when $\theta(z) = e^{iaz}$) of Λ gives a set with the required property. We give also a general criterion for the family to form an unconditional basis. But the criterion being somewhat cumbersome, we prefer not to quote it here (see § 4, Part II).

7. Some remarks concerning the history of the problem

As we already pointed out in Introduction the problem we have discussed goes back to the fundamental book of R.Paley and N.Wiener [59]. It was also mentioned that the problem of Riesz bases of exponentials, as it was posed by R.Paley and N.Wiener, has been solved by M.I.Kadec in [10]. The intermediate result with $\delta < \pi^{-1}\cdot \log 2$ was proved in [34]. The elegant proof of R.Duffin and J.Eachus may be found in the book [16], p.227. For the sake of completeness we represent here, essentially following the N. Levinson's book [48], an example of A.Ingham which shows that the constant $1/4$ in the Kadec theorem can not be increased.

EXAMPLE (A.Ingham). Let $\lambda_0 = 0$, let $\lambda_n = n - 1/4$ if $n > 0$, $n \in \mathbb{Z}$, and let $\lambda_n = -\lambda_{-n}$ if $n < 0$, $n \in \mathbb{Z}$. Then

$$L^2(0,2\pi) = \operatorname{span}(e^{i\lambda_n x} \cdot \chi_{[0,2\pi)} : n \in \mathbb{Z} \setminus \{0\}).$$

In particular, the family $(e^{i\lambda_n x})_{n\in\mathbb{Z}}$ is not minimal in $L^2(0,2\pi)$.

It is sufficient to prove that the generating function F_Λ (which d o e s exist in this case) satisfies

$$\int_{\mathbb{R}} \frac{|F_\Lambda(x)|}{1+x^2}\, dx = +\infty .$$

The last assertion as well as the existence of F_Λ is a consequence of the formula

$$z^{-1} \cdot e^{-i\pi z} \cdot F_\Lambda(z) = \prod_{n=1}^{\infty} \left(1 - \frac{z^2}{\lambda_n^2}\right) = c^{-1} \int_{-\pi}^{\pi} e^{izt} (\cos\tfrac{t}{2})^{-1/2}\, dt ,$$

$c \overset{\text{def}}{=\!=} \int_{-\pi}^{\pi} (\cos\frac{t}{2})^{-1/2}\, dt$, because the function $t \to (\cos\frac{t}{2})^{-1/2}$ does not belong to $L^2(-\pi,\pi)$, although it belongs, obviously, to $\bigcap_{p<2} L^p(-\pi,\pi)$. To prove the formula we are only to check that the zero set of $I(z) = \int_{-\pi}^{\pi} e^{izt} (\cos t/2)^{-1/2} dt$ coincides with $\{\lambda_n : n \in \mathbb{Z} \setminus \{0\}\}$. We have for $n \in \mathbb{Z}$, $n \geqslant 1$:

$$\int_{-\pi}^{\pi} e^{i\lambda_n x} (\cos\tfrac{x}{2})^{-1/2}\, dx = \int_{-\pi}^{\pi} e^{i\lambda_n x} \cdot \left(\frac{e^{ix/2} + e^{-ix/2}}{2}\right)^{-1/2} dx =$$

$$= \sqrt{2} \int_{-\pi}^{\pi} e^{inx} (1 + e^{ix})^{-1/2}\, dx = 0$$

since $(1+z)^{-1/2} \in H^1(\mathbb{D})$. Now we are going to prove that if $I(z) = 0$ and if $\operatorname{Re} z \geqslant 0$ then $z = \lambda_n$ for some n in $\mathbb{Z}$. The function $(\cos\frac{t}{2})^{-1/2}$ being even this would imply the desired conclusion. By the Taylor formula

$$(1+\zeta)^{-1/2} = \sum_{k=0}^{\infty} (-1)^k \frac{(2k-1)!!}{(2k)!!} \zeta^k , \quad |\zeta| \leqslant 1 .$$

Let now $\operatorname{Re} w \geqslant 0$ and let $\lambda \overset{\text{def}}{=\!=} w + 1/4$. Then

$$\int_{-\pi}^{\pi} e^{iwt} (\cos\tfrac{t}{2})^{-1/2} dt = \sqrt{2} \int_{-\pi}^{\pi} e^{i\lambda t} (1 + e^{it})^{-1/2}\, dt =$$

$$= 2\sqrt{2}\, \sin(\lambda\pi) \cdot \sum_{k=0}^{\infty} \frac{(2k-1)!!}{(2k)!!} \cdot \frac{1}{\lambda + k} .$$

But obviously,

$$\mathrm{Re}\sum_{k=0}^{\infty}\frac{(2k-1)!!}{(2k)!!}\cdot\frac{1}{\lambda+k}=\sum_{k=0}^{\infty}\frac{(2k-1)!!}{(2k)!!}\;\frac{k+\mathrm{Re}\,\lambda}{|\lambda+k|^2}>0$$

if $\mathrm{Re}\, w \geq 0$. ●

As R.M.Young noted in [60], the condition

$$|\lambda_n - n| < 1/4, \quad n\in\mathbb{Z}$$

is also insufficient for the family $(e^{i\lambda_n x})_{n\in\mathbb{Z}}$ to form a Riesz basis in $L^2(0,2\pi)$. This observation is based on the following theorem.

THEOREM (R.Duffin, A.Schaeffer [35]). Let $(\mu_n)_{n\in\mathbb{Z}}$ be a real sequence such that the family $(e^{i\mu_n x})_{n\in\mathbb{Z}}$ forms a Riesz basis in $L^2(0,a)$. Then there exists a positive number δ, $\delta>0$, such that any family $(e^{i\lambda_n x})_{n\in\mathbb{Z}}$, satisfying $\sup_n|\lambda_n-\mu_n|<\delta$ is also a Riesz basis in $L^2(0,a)$.

We obtain in Part III a generalization of this result.

It was B.Ja.Levin who showed the significance of the notion of generating function. Generalizing his definition of a sine type function, see the definition in Introduction, we give the following one.

DEFINITION. An entire function S of exponential type is called a generalized sine type function (briefly $S\in$ GSTF) if all its zeros are in $\mathbb{C}_\delta$ for some δ, $\delta>0$ and if above that

$$0<\inf_{x\in\mathbb{R}}|S(x)|\leq\sup_{x\in\mathbb{R}}|S(x)|<+\infty .$$

It is not a difficult task to give an example of GST function whose zeros $\Lambda=\{\lambda_n : n\in\mathbb{Z}\}$ satisfy the condition $\overline{\lim_{|n|\to\infty}}\,\mathrm{Im}\,\lambda_n=+\infty$. It appears nevertheless, and this is a subtle result due to S.A. Vinogradov, see §3 of Part III, that there is such an example satisfying in addition $\Lambda\in(C)$. In [14] B.Ja.Levin has proved that a family $(e^{i\lambda_n x})_{n\in\mathbb{Z}}$ is a basis in $L^2(0,a)$ if the set $\{\lambda_n : n\in\mathbb{Z}\}$ is separated and if it coincides with the zero set of a STF having the width of the indicator diagram equal to a. V.D.Golovin remarked later that in fact these families are Riesz bases in $L^2(0,a)$, see [5], [6]. Now the Levin - Golovin theorem is a simple concequence of Theorem 7 of the present paper, but at that time it was a fundamental step forward. V.È. Kacnelson has generalized the Levin-Golovin theorem as well as that of Kadec.

THEOREM (V.È. Kacnelson [12]). Let $(\lambda_n)_{n\in\mathbb{Z}}$ be a zero sequence of STF with the width of the indicator diagram equal to a, $a>0$. Let $(\mu_n)_{n\in\mathbb{Z}}$ be a sequence of points in $\mathbb{C}_+$ satisfying

$$\sup_n \operatorname{Im}\mu_n < +\infty, \quad |\operatorname{Re}\lambda_n - \operatorname{Re}\mu_n| \le d\rho_n,$$

where $d<1/4$ and $\rho_n = \inf_{k,k\ne n} |\operatorname{Re}\lambda_k - \operatorname{Re}\lambda_n|$. Let at last $\inf_{n\ne m} |\mu_n - \mu_m| > 0$. Then the family $(e^{i\mu_n t})_{n\in\mathbb{Z}}$ is a Riesz basis in $L^2(0,a)$.

This theorem has been strengthened by S.A.Avdonin in [2] and [3]. To formulate his results the next definition is needed.

DEFINITION. Let $\Lambda = \{\lambda_n : n\in\mathbb{Z}\}$ be a separated subset of a strip of a finite width, parallel to the real axis. A partitioning $\Lambda = \bigcup_{k\in\mathbb{Z}} \Lambda_k$ of Λ by some vertical lines into disjoint subsets Λ_k is named an A -p a r t i t i o n i n g if the distances ℓ_k between the lines bounding each group Λ_k are uniformly bounded.

THEOREM (S.A.Avdonin [2]). Let Λ be a zero set of STF with the width of the indicator diagram equal to a, $a>0$. Let $(\delta_\lambda)_{\lambda\in\Lambda}$ be a bounded family of complex numbers satisfying

$$\Big|\sum_{\lambda\in\Lambda_j} \operatorname{Re}\delta_\lambda\Big| \le d\ell_j$$

for some A-partitioning, where $d<1/4$. Suppose, that the set $\{\lambda+\delta_\lambda\}_{\lambda\in\Lambda}$ is separated. Then the family $(e^{i(\lambda+\delta_\lambda)x})_{\lambda\in\Lambda}$ forms a Riesz basis in $L^2(0,a)$.

A new proof of Kacnelson and Avdonin theorems will be given in §2 of Part III. The paper of Avdonin [2] contains also a theorem very similar to one of the corollaries of our Theorem 7. Let, for the time being, $\mathcal{M}$ denote the set of all positive functions φ defined on $[0,+\infty)$ and such that the function $\gamma(x) = x\cdot\frac{\varphi'(x)}{\varphi(x)}$ satisfies the following conditions

$$|\gamma(x)| \le a < 1/2, \quad \gamma'(x) = O\left(\frac{1}{x}\right), \quad x\to+\infty.$$

THEOREM (S.A.Avdonin [2]). Let Λ be a zero set of the entire function F with the width of the indicator diagram equal to a. Suppose that $0 < \inf_{\lambda\in\Lambda} \operatorname{Im}\lambda \le \sup_{\lambda\in\Lambda} \operatorname{Im}\lambda < +\infty$ and suppose there is a function φ in $\mathcal{M}$ satisfying

$$0 < \inf_{x \in \mathbb{R}} \frac{|F(x)|}{\varphi(x)} \leq \sup_{x \in \mathbb{R}} \frac{|F(x)|}{\varphi(x)} < +\infty .$$

Then the family $(e^{i\lambda t})_{\lambda \in \Lambda}$ is a Riesz basis in $L^2(0,a)$.

The paper [2] contains also some examples which show that the modulus $|F_\Lambda|^2 \mid \mathbb{R}$ satisfying (A_2) can, nevertheless, bahave irregularly.

The problem of unconditional exponential bases is closely connected with the completeness problem and with the spectral theory of Toeplitz operators. It is interesting to note that all machinery needed for the solution of the problem of exponential Riesz bases (as is given by Theorem 6) was ready in the early 60-ies. The papers [43], [45], were especially close to the solution. The paper [43], containing really a characterization of Blaschke products (for the upper half-plane) generating compact Hankel operators $H_{\bar{B}\theta^a}$ for every $a, a > 0$, contains also various combinations of all attributes of our description of bases. The same can be said on the paper [32] by R.Douglas and D. Sarason containing sufficient conditions of the completeness of exponentials involving invertibility of the Toeplitz operators $T_{\bar{B}\theta^a}$. Let us mention the paper [49] (indicated to one of us by P.Koosis), where one can find the trick employed in our proof of Kadec's theorem on $1/4$.

On the other hand, the idea of preservation of Riesz bases under some orthogonal projections was formulated (and used for a proof of the Levin-Golovin theorem) by one of us as early as in 1973 in the paper [22] .

And in conclusion we indicate the paper [29] where bases of reproducing kernels of spaces K_θ are studied. But these bases are very close to orthogonal (à la Wiener - Paley theorem). This causes strong restrictions imposed on the inner function θ (see also § 5 Part II below). Riesz bases (of exponentials or of reproducing kernels) are connected with the problem of free interpolation by analytic functions (at corresponding knots). Almost every work devoted to exponential bases, beginning from the book by N.Wiener and R.Paley, contains some interpolatory corollaries. One can also find such corollaries in § 7 Part II.

PART II
BASES OF REPRODUCING KERNELS

1. Carleson condition

In §1 Part I we have formulated the general problem concerning unconditional bases composed of reproducing kernels. Now we recall it:

Given a pair (θ,Λ) with θ an inner function in the disc $\mathbb{D}$ and $\Lambda\subset\mathbb{D}$, find necessary and sufficient conditions for the family

$$k_\theta(z,\lambda)=\frac{1-\overline{\theta(\lambda)}\theta(z)}{1-\bar{\lambda}z}\quad,\quad \lambda\in\Lambda$$

to be an unconditional basis of K_θ (or of the subspace of K_θ it generates).

This problem generalizes the problem concerning bases of rational fractions (and coincides with it when $\theta=B=\prod_{\lambda\in\Lambda} b_\lambda$), described in §2 Part I.

To link together the problems discussed we need a part of the well-known N.K.Bari theorem on Riesz bases (a proof may be found e.g. in [18], p.172).

THEOREM (N.K.Bari). Let $(\varphi_n)_{n\in\mathbb{Z}}$ be a family of nonzero vectors in a Hilbert space H and set $\psi_n=\varphi_n\|\varphi_n\|^{-1}$, $n\in\mathbb{Z}$. The following assertions are equivalent.

1. The family $(\varphi_n)_{n\in\mathbb{Z}}$ is an unconditional basis of H.
2. The Gram matrix $\{(\psi_n,\psi_m)\}_{n,m\in\mathbb{Z}}$ generates a continuous and invertible operator in the space $\ell^2(\mathbb{Z})$ and $H=$ $=\operatorname{span}(\varphi_n)_{n\in\mathbb{Z}}$.

We state now the main result of this section.

THEOREM 1.1. Suppose that the family $\{k_\theta(\cdot,\lambda):\lambda\in\Lambda\}$ is an unconditional basis in its closed linear span. Then $\Lambda\in(C)$.

PROOF. We shall extract all information we need from the Gram matrix $\Gamma=\{(\psi_n,\psi_m)\}_{n,\,m\in\mathbb{Z}}$ corresponding in the same way as in N.K.Bari Theorem to the family of functions

$$\varphi_n=\frac{1-\overline{\theta(\lambda_n)}\theta(z)}{1-\bar{\lambda}_n z}\quad,\quad n\in\mathbb{Z}\ ,$$

$\{\lambda_n:n\in\mathbb{Z}\}$ being an enumeration of Λ. Using the definition of the reproducing kernel, we obtain

$$(\varphi_n,\varphi_m)=\frac{1-\overline{\theta(\lambda_n)}\theta(\lambda_m)}{1-\overline{\lambda_n}\lambda_m}$$

and, in particular, $\|\varphi_n\|_2^2=(1-|\theta(\lambda_n)|^2)(1-|\lambda_n|^2)^{-1}$.
Hence

$$(\psi_n,\psi_m)=\frac{(1-|\lambda_n|^2)^{1/2}(1-|\lambda_m|^2)^{1/2}}{1-\overline{\lambda_n}\lambda_m}:\frac{(1-|\theta(\lambda_n)|^2)^{1/2}(1-|\theta(\lambda_m)|^2)^{1/2}}{1-\overline{\theta(\lambda_n)}\theta(\lambda_m)}.$$

Note that the absolute value of the divisor in the right-hand side of the last formula is less than 1:

$$\frac{(1-|z|^2)(1-|w|^2)}{|1-\overline{w}z|^2}=1-\left|\frac{w-z}{1-\overline{w}z}\right|^2.$$

Let $\{e_n\}_{n\in\mathbb{Z}}$ be the standard unit vector basis in $\ell^2(\mathbb{Z})$: $e_n(k)=0$ for $k\neq n$, $e_n(n)=1$. The fact that the Gram matrix defines a bounded operator in $\ell^2(\mathbb{Z})$ implies the inequality

$$\sum_{n\in\mathbb{Z}}|(\psi_n,\psi_m)|^2=\|\Gamma e_m\|^2\leq\|\Gamma\|^2<\infty,$$

from which it follows in view of the preceding remarks that

$$\sup_{m\in\mathbb{Z}}\sum_n\frac{(1-|\lambda_n|^2)(1-|\lambda_m|^2)}{|1-\overline{\lambda_n}\lambda_m|^2}\leq\|\Gamma\|^2<\infty.$$

But the last condition is necessary and sufficient for the measure $\sum_n(1-|\lambda_n|)\delta_{\lambda_n}$ to be a Carleson one (for the proof see [18] or [44])*).

Let us check now the rarity condition. If $(\varphi_n)_{n\in\mathbb{Z}}$ is an unconditional basis in H then the normed family $(\psi_n)_{n\in\mathbb{Z}}$ is uniformly disjoint (i.e. $\inf\{\|\psi_n-\psi_m\|:n\neq m\}>0$), and, consequently, $\sup_{n\neq m}|(\psi_n,\psi_m)|^2=\gamma<1$. In the

*) It should be noted that the Carleson condition (C), as well as the rarity condition (R) and the condition that the corresponding measure is a Carleson one may be transferred from the half-plane $\mathbb{C}_+$ to the disc $\mathbb{D}$ by means of conformal mapping. The equivalence (C) $\Longleftrightarrow$ (CM) & (R) still holds in $\mathbb{D}$, cf. §2.6 of Part I for the details.

case we examine this inequality may be rewriten as follows:

$$\sup_{n\neq m}\left(1-\left|\frac{\lambda_n-\lambda_m}{1-\overline{\lambda}_n\lambda_m}\right|^2\right)\left(1-\left|\frac{\theta(\lambda_n)-\theta(\lambda_m)}{1-\overline{\theta(\lambda_n)}\,\theta(\lambda_m)}\right|^2\right)^{-1}=\gamma<1\,.$$

This implies that $\inf_{n\neq m}\left|\frac{\lambda_n-\lambda_m}{1-\overline{\lambda}_n\lambda_m}\right|\geq 1-\gamma$ and hence $(\lambda_n)_{n\in\mathbb{Z}}$ satisfies the rarity condition (R). ●

Let P_θ be the orthogonal projection onto the space K_θ. Theorem 1.1 shows that each unconditional basis of the form $\{k_\theta(\cdot,\lambda):\lambda\in\Lambda\}$ in K_θ is necessarily the image under P_θ of some unconditional basis consisting of rational fractions (namely, the basis $\{(1-\overline{\lambda}z)^{-1}:\lambda\in\Lambda\}$ in K_B).

Let us assume now that P_θ does not distort very much the norms of the rational fractions:

$$\sup_{\lambda\in\Lambda}\|(1-\overline{\lambda}z)^{-1}\|_{H^2}\cdot\|P_\theta(1-\overline{\lambda}z)^{-1}\|_{H^2}^{-1}<\infty\,.\text{*)}$$

Since $\|(1-\overline{\lambda}z)^{-1}\|^2_{H^2}=(k(\cdot,\lambda),k(\cdot,\lambda))=(1-|\lambda|^2)^{-1}$ and $\|P_\theta(1-\overline{\lambda}z)^{-1}\|^2_{H^2}=k_\theta(\lambda,\lambda)=(1-|\theta(\lambda)|^2)(1-|\lambda|^2)^{-1}$, the last condition is equivalent to the following inequality:

$$\sup_{\lambda\in\Lambda}|\theta(\lambda)|<1\,. \tag{1}$$

This inequality means that (a) the poles of the rational fractions $(1-\overline{\lambda}z)^{-1}$, $\lambda\in\Lambda$ can accumulate only to the spectrum of θ on $\mathbb{T}$ (i.e. to the set $\{\zeta\in\mathbb{T}:\underline{\lim}_{\xi\to\zeta}|\theta(\xi)|=0\}$); and, moreover, (b) this accumulation must be in a sense nontangential with respect to the unit circle. We shall see later that the condition (a) is i m p l i e d b y t h e f a c t t h a t t h e f u n c t i o n s $\{k_\theta(\cdot,\lambda):\lambda\in\Lambda\}$ form an unconditional basis of the space they generate (see corollary 4.2 and its comments, page 268 and §6 p. 276).

THEOREM 1.2. Suppose that the pair (θ,Λ) satisfies condition (1). Then the following assertions are equivalent.

1. The family $\{k_\theta(\cdot,\lambda):\lambda\in\Lambda\}$ is an unconditional basis in K_θ (resp., in the subspace of K_θ it generates).

2. $\Lambda\in(C)$ and $P_\theta\,|\,K_B$ is an isomorphism of K_B onto K_θ (resp., of K_B onto $P_\theta(K_B)$).

PROOF follows the same lines as the proof of Theorem 2 (Part 1, §2). Here is its shortened version.

1⟹2. Theorem 1.1 implies that $\Lambda\in(C)$. In view of Theorem A (cf. Part 1, §2) the fractions $k(\cdot,\lambda)$, $\lambda\in\Lambda$ form an

*) From now on $H^2\overset{\text{def}}{=}H^2(\mathbb{D})$.

unconditional basis in K_B. Combining this with (1) we obtain that $P_\theta \mid K_B$ is an isomorphism.

Implication $2 \Longrightarrow 1$ is a consequence of Theorem A and inequality (1). ●

2. Projecting onto K_θ and Toeplitz operators

The condition " $P_\theta \mid K_B$ is an isomorphism onto its image" may be restated in geometric terms. To do this we need some notations and definitions.

Given a closed subspace M of a Hilbert space H we denote by $M^\perp$ the orthogonal complement to M and by P_M the orthogonal projection of H onto M.

By the angle between two subspaces M and N we mean a number (denoted $\langle N, M\rangle$) uniquely determined by $\langle N, M\rangle \in [0, \frac{\pi}{2}]$ and

$$\cos\langle N,M\rangle = \sup\{|(n,m)| : \|n\| = \|m\| = 1,\ n\in N,\ m\in M\}.$$

Clearly $\cos\langle N,M\rangle = \sup\{\|P_M x\| : \|x\| = 1,\ x\in N\} = \|P_M \mid N\| = \|P_N \mid M\| = \|P_N P_M\|$ and

$$\inf\{\|P_M n\|^2 : n\in N, \|n\|=1\} = 1 - \sup\{\|P_{M^\perp} n\|^2 : n\in N, \|n\|=1\} = \sin^2\langle N, M^\perp\rangle. \quad (*)$$

Let M, N be two subspaces of H with $M\cap N = \{0\}$. Define a (possibly discontinuous) projection $\mathcal{P}_{M\|N}$ on $M+N$ by

$$\mathcal{P}_{M\|N}(m+n) = m \qquad (m\in M,\ n\in N).$$

We call it the projection onto M along N $(\mathcal{P}^2 = \mathcal{P};\ \mathcal{P}\mid M = I;\ \mathcal{P}\mid N = \mathbb{O})$. It follows from the closed graph theorem that this projection is continuous if and only if $M+N$ is closed. Also we have

$$\sin\langle M,N\rangle = \inf_{x\in M} \frac{\|(I-P_M)x\|}{\|x\|} = \|\mathcal{P}_{M\|N}\|^{-1}.$$

LEMMA 2.1. Let M and N be closed subspaces of a Hilbert space H . The following assertions are equivalent.

1. $\mathrm{Ker}(P_M | N) = \{\mathbb{O}\}$.
2. $M^{\perp} \cap N = \{\mathbb{O}\}$.
3. $\mathrm{clos}(M+N^{\perp}) = H$.
4. $\mathrm{clos}\, P_N M = N$.

The following assertions are also equivalent.

1a. $P_M | N$ is an isomorphism (onto its image).
2a. $\cos \langle N, M^{\perp} \rangle < 1$.
3a. $\langle N, M^{\perp} \rangle > 0$.

Finally, $P_M | N$ is an isomorphism of N o n t o M if and only if any of the following (equivalent) conditions is satisfied.

1b $\cos \langle N, M^{\perp} \rangle < 1$; $\cos \langle N^{\perp}, M \rangle < 1$.
2b. $H = N + M^{\perp}$ and $N \cap M^{\perp} = \{\mathbb{O}\}$.
3b. $\mathrm{clos}(N+M^{\perp}) = H$, $\| \mathcal{P}_{N \| M^{\perp}} \| < +\infty$.

PROOF of the lemma is routine, but we include it for the sake of completeness.

The equivalence of the first four assertions follows immediately from the equality $(P_M : N \to M)^* = (P_N : M \to N)$ and the fact that $\mathrm{Ker}\, A = \{\mathbb{O}\} \Longleftrightarrow \mathrm{clos}\, A^* H = H$.

Implications 1a$\Longleftrightarrow$ 2a follow from the formula ($\overset{*}{1}$) and implications 2a$\Longleftrightarrow$3a are evident.

To prove the third part of the Lemma use once more the fact that $(P_M | N)^* = P_N | M$ and apply the Banach theorem (an operator is onto if and only if the conjugate operator is an isomorphic imbedding). ●

COROLLARY 2.2. Let Θ and B be inner functions. The following assertions are equivalent.

1. $P_\Theta | K_B$ is an isomorphism onto its image..
2. $\cos \langle K_B, \Theta H^2 \rangle < 1$.
3. $\| \mathcal{P}_{K_B \| \Theta H^2} \| < \infty$.

The operator P_Θ maps isomorphically K_B o n t o K_Θ iff any of the following equivalent conditions is satisfied:

1a. $\cos \langle K_B, \Theta H^2 \rangle < 1$, $\cos \langle K_\Theta, B H^2 \rangle < 1$.
2a. $H^2 = K_B + \Theta H^2$, $K_B \cap \Theta H^2 = \{\mathbb{O}\}$.
3a. $\mathrm{clos}(B H^2_- + \Theta H^2) = L^2(\mathbb{T})$, $\| \mathcal{P}_B^\Theta \| < \infty$,

where $\mathcal{P}_B^\Theta = \mathcal{P}_{BH^2_- \| \Theta H^2_-}$.

PROOF. Apply Lemma 2.1 with $N = K_B$, $M = K_\Theta$. When treating the condition 3a one needs to keep in mind that $K_B + H^2_- = B H^2_-$. ●

It is easy to compute the number $\cos \langle K_B, \Theta H^2 \rangle$ using the following well-known fact: every function g in the Hardy class H^1 can be represented in the form $g = h_1 \cdot h_2$ with

$h_1, h_2 \in H^2$ and $\|h_1\|_2^2 = \|h_2\|_2^2 = \|g\|_1$.

LEMMA 2.3. Let φ be a unimodular function on $\mathbb{T}$. Then

$$\cos\langle H_-^2, \varphi H^2\rangle = \mathrm{dist}_{L^\infty}(\varphi, H^\infty),$$

and, in particular, $\cos\langle K_B, \theta H^2\rangle = \mathrm{dist}(\bar{B}\theta, H^\infty)$.

PROOF. $\cos\langle H_-^2, \varphi H^2\rangle =$
$= \sup\{|\int \varphi h_+ \bar{h}_- \, dm| : \|h_\pm\|_2 \le 1,\ h_\pm \in H_\pm^2\} =$
$= \sup\{|\int_{\mathbb{T}} \varphi h \, dm : h \in H^1, \|h\|_1 \le 1, \hat{h}(0) = 0\} = \mathrm{dist}(\varphi, H^\infty)$.
$\cos\langle K_B, \theta H^2\rangle = \cos\langle BH_-^2, \theta H^2\rangle = \cos\langle H_-^2, \bar{B}\theta H^2\rangle$. ●

The first assertion of Lemma 2.3 essentially coincides with Z.Nehari theorem mentioned in Part I.

We have already pointed out (Part I, §4) that it is possible to obtain Theorem 5 combining well-known theorems of Helson - Szegö andDevinatz - Widom. A proof of Theorem 5 may be found in [18] or extracted from lectures [54]. However, we present here a proof of this theorem to make the exposition selfcontained. This proof is also of interest by another reason: it enables us to consider the Helson-Szegö theorem from a new view-point (as a theorem describing a special class of unimodular functions; see, however, [1] in connection with this view-point). Keeping in mind the unitary equivalence of the Toeplitz operators in the disc and in the half-plane mentioned in Part 1, §6 we shall prove the analog of Theorem 5 for $\mathbb{D}$.

To begin with, we introduce two definitions. If $v \in L^\infty(\mathbb{T})$ then $\tilde{v}$ stands for the harmonic conjugate of v $(\int_{\mathbb{T}} \tilde{v}\, dm = 0)$. From now on we assume all functions from $L^1(\mathbb{T})$ to be harmonically extended into $\mathbb{D}$, a function and its extension being denoted by the same letter. So for a real function v its harmonic conjugate $\tilde{v}$ is uniquely determined by $v + i\tilde{v} \in H^2(\mathbb{D})$ and $\tilde{v}(0) = 0$.

DEFINITION. Let h be an outer function in $H^2(\mathbb{D})$; h is said to satisfy the *Helson-Szegö condition* if there are $u, v \in L^\infty(\mathbb{T})$ with

$$|h^2| = \exp(u + \tilde{v}), \quad \|v\|_\infty < \frac{\pi}{2}. \tag{HS}$$

DEFINITION. A unimodular function φ on $\mathbb{T}$ is called a *Helson-Szegö function* if there are a constant λ , $|\lambda|=1$ and an outer function h satisfying Helson - Szegö condition, such that

$$\varphi=\lambda\frac{\bar{h}}{h}.$$

THEOREM 5D. Let φ be a unimodular function on $\mathbb{T}$. The following assertions are equivalent.

1. The Toeplitz operator T_φ is invertible.
2. $dist_{L^\infty}(\varphi, H^\infty)<1$, $dist_{L^\infty}(\bar{\varphi}, H^\infty)<1$.
3. There exists an outer function f in $H^\infty=H^\infty(\mathbb{D})$ such that $\|\varphi-f\|_\infty<1$.
4. There exists a Lebesgue measurable branch α of the argument of φ (i.e. $\varphi(\zeta)=e^{i\alpha(\zeta)}$, $\zeta\in\mathbb{T}$) satisfying

$$dist_{L^\infty}(\alpha, \widetilde{L^\infty(\mathbb{T})+\mathbb{C}})<\frac{\pi}{2}.$$

5. φ is a Helson - Szegö function.

Some details of the proof of this theorem are of independent interest, and so we begin just with them.

LEMMA 2.4. (R.Douglas [54]). Let $\varphi\in L^\infty(\mathbb{T})$, $|\varphi|=1$ a.e. Then the Toeplitz operator T_φ is an isomorphism (onto its image) if and only if $\|H_\varphi\|=dist(\varphi, H^\infty)<1$.

PROOF. If $f\in H^2$ then clearly

$$\varphi f=H_\varphi f+T_\varphi f\ ,\quad \|f\|^2=\|H_\varphi f\|^2+\|T_\varphi f\|^2,$$

and the result follows. ●

Let $0<\gamma\leq 1$. Set

$$A_\gamma=\{\zeta\in\mathbb{C}: |\arg\zeta|<\pi\gamma\}.$$

LEMMA 2.5. 1. If $F\in H^\infty$ and the essential image $F(\mathbb{T})$ of the circle $\mathbb{T}$ is contained in the angle $A_\gamma (0<\gamma\leq 1)$ then $F(\mathbb{D})\subset A_\gamma$.

2. If F is analytic in $\mathbb{D}$ and $F(\mathbb{D})\subset A_\gamma$ then F is outer and $F\in H^p$, $p<(2\gamma)^{-1}$.

PROOF. 1. Following J.B.Garnett ([44], p.632, [36], p.199-200), suppose that there exists a point z_0 in $\mathbb{D}$ with $w_0=F(z_0)\notin A_\gamma$. Construct a polynomial P so that

$$P(w_0)=1,\quad \sup_{w\in F(\mathbb{T})}|P(w)|<1/2.$$

Then $P(F(z_0)) = 1$, but boundary values of the function $P \circ F$ on $\mathbb{T}$ are almost everywhere less than $1/2$. This contradicts the maximum modulus principle.

2. Since $F(\mathbb{D}) \subset A_\gamma$, F has no zeros in $\mathbb{D}$, for otherwise 0 would be an interior point of $F(\mathbb{D})$. Consider the function $f = F^{1/2\gamma}$. Clearly $\operatorname{Re} f \geqslant 0$ in $\mathbb{D}$ and hence f is an outer function (one of numerous well-known ways to see this is as follows: if $\varepsilon > 0$, then $f + \varepsilon$ is evidently an outer function for it is bounded away from zero in $\mathbb{D}$; hence $\log|f(0)+\varepsilon| = \int_{\mathbb{T}} \log|f(\zeta)+\varepsilon|\,dm$ and it suffices to pass to limit, as $\varepsilon \to 0$, using monotone convergence theorem). Consequently the function F is also outer.

The remaining part of the second assertion is due to V.I.Smirnov and is widely known. Here is a proof. If $p < (2\gamma)^{-1}$ then there is a constant C so that $w \in A_{p\gamma} \Longrightarrow |w| \leq C \operatorname{Re} w$. Therefore $|F(\zeta)|^p \leq C \operatorname{Re} F(\zeta)^p$, $\zeta \in \mathbb{D}$, and, consequently,

$$\int_{\mathbb{T}} |F(r\zeta)|^p dm(\zeta) \leq C \int_{\mathbb{T}} \operatorname{Re} F(r\zeta)^p dm(\zeta) = C \operatorname{Re} F(0)^p. \quad \bullet$$

LEMMA 2.6. If the assertion 2 of Theorem 5D is fulfilled then the set $\{ f \in H^\infty(\mathbb{D}) : \|\varphi - f\|_\infty < 1 \}$ consists entirely of outer functions.

PROOF. Let $f, g \in H^\infty(\mathbb{D})$ and

$$\|1 - \bar{\varphi} f\|_\infty < 1, \qquad \|1 - \varphi g\|_\infty < 1.$$

These inequalities imply that all values of the functions $\bar{\varphi} f\,|\,\mathbb{T}$, $\varphi g\,|\,\mathbb{T}$ lie in A_γ for some γ, $\gamma < 1/2$ and so $fg(\mathbb{T}) \subset A_{2\gamma}$. By Lemma 2.5 fg is an outer function and hence f, g are also outer. $\bullet$

PROOF OF THEOREM 5D. $1 \Longleftrightarrow 2$ by Lemma 2.4, $2 \Longrightarrow 3$ by Lemma 2.6.

$3 \Longrightarrow 4$: Let f be an outer function with $\|\varphi - f\|_\infty = \gamma < 1$. There exists a number λ, $|\lambda| = 1$ such that

$$f\,|\,\mathbb{T} = \lambda \exp(\log|f| + i\,\widetilde{\log|f|}).$$

The values of the function $\bar{\varphi} f\,|\,\mathbb{T}$ lie in the angle $A_{(\arcsin\gamma)/\pi}$ and so there exists a unique real-valued function α with $\bar{\lambda}\varphi = \exp i\alpha$ and $\|\alpha - \widetilde{\log|f|}\|_\infty < \pi/2$

$4 \Longrightarrow 5$. If $\varphi = \exp i\alpha$ and $\alpha = c + \tilde{u} + v$ with

$c \in \mathbb{R}$, $u \in L^\infty(\mathbb{T})$, $u(0) = 0$, $\|v\|_\infty < \pi/2$, then we set $\lambda = \exp ic$ and find an outer function h from the equation

$$\widetilde{\log|h^2|} = -\tilde{u} - v + v(0).$$

We have then $\log|h^2| = -u + \tilde{v}$. Since $\|v\|_\infty < \pi/2$, Lemma 2.5 implies that $\exp(\tilde{v} - iv) \in H^2(\mathbb{D})$, hence $h \in L^2(\mathbb{T})$ and, consequently, h satisfies the Helson - Szegö condition. The formula $\varphi = \lambda \cdot \bar{h}/h$ follows from the construction.

5⟹2. Suppose $\varphi = \bar{h}/h$ with h satisfying the Helson - Szegö condition. Then $\log|h^2| = u + \tilde{v}$, $\widetilde{\log|h^2|} = \tilde{u} + v(0) - v$ and hence $\varphi = \exp(-i(\tilde{u} + v(0) - v))$, where $\|v\|_\infty < \pi/2$. Set

$$f_\varepsilon = \varepsilon e^{-iv(0)} e^{-u - i\tilde{u}}, \quad \varepsilon > 0.$$

Then $f_\varepsilon \in H^\infty$, $f_\varepsilon^{-1} \in H^\infty$. We have:

$$\|\varphi - f_\varepsilon\|_\infty = \|1 - \bar{\varphi} f_\varepsilon\|_\infty = \|1 - |f_\varepsilon| e^{-iv}\|_\infty < 1,$$

provided ε is sufficiently small, because $f_\varepsilon \in H^\infty$ and $\|v\|_\infty < \frac{\pi}{2}$. Similarly, $\|\bar{\varphi} - f_\varepsilon^{-1}\| < 1$. ●

REMARKS. 1. Lemma 2.6 and implication $3 \Longrightarrow 2$ show that the set $\{f \in H^\infty : \|\varphi - f\|_\infty < 1\}$ either does not intersect the set of outer functions or is contained in it.

2. The famous Helson-Szegö theorem stated below may be easily derived from Theorem 5D.

THEOREM (H.Helson, G.Szegö [38]). Let $w \in L^1(\mathbb{T})$, $w \geq 0$. Then the Riesz projection $\mathbb{P}_+$ $\left(\mathbb{P}_+\left(\sum_{n \in \mathbb{Z}} a_n z^n\right) \overset{def}{=} \sum_{n \geq 0} a_n z^n\right)$ is continuous in the weighted space $L^2(\mathbb{T}, w) = \{f : \int_{\mathbb{T}} |f|^2 w \, dm < \infty\}$ if and only if $w \in (HS)$.

Indeed, the assertion that $\mathbb{P}_+$ is continuous is equivalent to the assertion 2 of Theorem 5D with $\varphi = \bar{h}/h$, h being an outer function satisfying $h \in H^2$, $|h|^2 = w$. ●

Theorem 1.2 combined with Theorems 4 and 5D enables us to list many useful necessary and sufficient conditions for a family of reproducing kernels $(k_\theta(\cdot, \lambda))_{\lambda \in \Lambda}$ to be a basis of the space K_θ. To obtain criteria for such a family to be a basis in its closed linear span, Theorems 1.2 and 2 bis (Part I) and Lemma 2.4 can be used.

3. A criterion in terms of the model operators

Using the implications 1$\Longleftrightarrow$2 of Theorem 5D and a formula relating Hankel operators and the Functional model, it is possible to add to equivalent assertions 1-5 of Theorem 5D another one expressed in Functional model terms.

Let θ be an inner function and let S stand for the operator of multiplication by z in H^2 (z being the identity function: $z(\zeta) = \zeta$). Consider the model operator

$$T_\theta \overset{def}{=\!=} P_\theta S \,|\, K_\theta .$$

It is well known that this operator admits an H^∞- functional calculus:

$$f(T_\theta) = P_\theta f(S) | K_\theta , \quad f \in H^\infty .$$

We have also

$$f(T_\theta) P_\theta = \theta H_{\bar\theta f} .$$

This formula and some of its applications can be found in [18]. Substituting in it $f = B$ we obtain that $P_\theta | K_B$ is an isomorphism of K_B onto K_θ if and only if

$$\|\theta(T_B)\| < 1 \qquad \text{and} \quad \|B(T_\theta)\| < 1. \tag{2}$$

Similarly, $P_\theta | K_B$ is an isomorphism of K_B onto $P_\theta(K_B)$ if and only if

$$\|\theta(T_B)\| < 1 \tag{3}$$

(combine implications 1$\Longleftrightarrow$2 in Theorem 5D, theorem of Z.Nehari in §3 of Part 1 and Lemma 2.4).

Here is a consequence of these assertions.

THEOREM 3.1. Let $\Lambda = \{\lambda_n : n \in \mathbb{Z}\} \subset \mathbb{C}_+$. Suppose $\Lambda \in (C)$ and $\lim\limits_{n \to +\infty} \operatorname{Im} \lambda_n = +\infty$. Then for every positive number a the family of exponents $(e^{i\lambda_n x})_{n \in \mathbb{Z}}$ is an unconditional basis in the subspace of $L^2(0,a)$ it generates. The deficiency of this subspace in $L^2(0,a)$ is infinite.

This theorem is a special case of the following one.

THEOREM 3.2.. Let θ be an inner function in $\mathbb{D}$ and let $\theta = B\cdot S$ be the canonical factorization of θ. Let Λ be a subset of $\mathbb{D}$ satisfying the Carleson condition and also the condition $\lim\limits_{\lambda\in\Lambda, |\lambda|\to 1} |\theta(\lambda)| = 0$. Then the following assertions holds.

1. There exists a subset Λ' of Λ with $card(\Lambda\setminus\Lambda') < \infty$ so that the family $\{k_\theta(\cdot,\lambda): \lambda\in\Lambda'\}$ is an unconditional basis of its closed linear span.

2. If $S \neq const$ then $\Lambda' = \Lambda$ and $dim(K_\theta \ominus span\{k_\theta(\cdot,\lambda): \lambda\in\Lambda\}) = \infty$.

PROOF. Assertion 1 is almost immediate. Observe that the rational fractions $\{(1-\bar\lambda z)^{-1}: \lambda\in\Lambda\}$ form an unconditional basis and $\theta(T_{B'})^*(1-\bar\lambda z)^{-1} = \overline{\theta(\lambda)}\,(1-\bar\lambda z)^{-1}$, $\lambda\in\Lambda'$, where B' is the Blaschke product corresponding to the set Λ'. From this follows the inequality

$$\|\theta(T_{B'})\| \le const \sup_{\lambda\in\Lambda'} |\theta(\lambda)|,$$

the right-hand side of which is strictly less than 1 for an appropriate choice of Λ', $card(\Lambda\setminus\Lambda') < \infty$.

The essence of assertion 2 is given by the following argument. Set $\theta_\alpha \overset{def}{=\!=} BS^\alpha$, $\alpha > 0$. We still have $\lim\limits_{\lambda\in\Lambda, |\lambda|\to 1} |\theta_\alpha(\lambda)| = 0$. Hence an application of assertion 1 shows that for some $\Lambda'\subset\Lambda$ with $card(\Lambda\setminus\Lambda') < \infty$ the family $\{k_{\theta_\alpha}(\cdot,\lambda): \lambda\in\Lambda'\}$ forms an unconditional basis in its closed linear span. But if $\alpha' < \alpha$ then $K_{\theta_{\alpha'}} \subset K_{\theta_\alpha}$, $dim(K_{\theta_\alpha} \ominus K_{\theta_{\alpha'}}) = \infty$ and $K_{\theta_{\alpha'}} = P_{\theta_{\alpha'}} K_\theta$. The rest is contained in two elementary lemmas (the first one to be applied to $A = P_{\theta_{\alpha'}} \mid K_{\theta_\alpha}$).

LEMMA 3.3. Let X, Y be linear topological spaces and let A be a continuous linear map from X to Y. If $(x_n)_{n\ge 1}$ is a basis in $span_X\{x_n: n\ge 1\}$ and $(Ax_n)_{n\ge 1}$ is a basis in $span_Y\{Ax_n: n\ge 1\}$ then

$$codim\ span_X\{x_n: n\ge 1\} \ge dim\ Ker A.$$

PROOF. Note that A is one-to-one on the space $span_X\{x_n: n\ge 1\}$. ●

LEMMA 3.4. Let an inner function θ and two subsets Λ, Λ_1 of $\mathbb{D}$ satisfy

$$\Lambda\cap\Lambda_1 = \emptyset,\ dim(K_\theta \ominus span\{k_\theta(\cdot,\lambda): \lambda\in\Lambda\}) \ge card\,\Lambda_1$$

and suppose Λ_1 is finite. Then

$$\operatorname{span}\{k_\theta(\cdot,\lambda):\lambda\in\Lambda_1\}\cap\operatorname{span}\{k_\theta(\cdot,\lambda):\lambda\in\Lambda\}=\{\mathbb{O}\}.$$

PROOF. It is sufficient to consider the case $\operatorname{card}\Lambda_1=1$ (i.e. to check that $k_\theta(\cdot,\mu)\notin\operatorname{span}\{k_\theta(\cdot,\lambda):\lambda\in\Lambda\}$ provided $\mu\notin\Lambda$ and $\operatorname{span}\{k_\theta(\cdot,\lambda):\lambda\in\Lambda\}\neq K_\theta$) . Indeed, an induction by the number of the nonzero summands in $\sum_{\lambda_1\in\Lambda_1} c_{\lambda_1} k_\theta(\cdot,\lambda_1)$ enables us to reduce the Lemma to this particular case. But the "base of induction" we need is immediate: if $f\in K_\theta$, $f\perp k_\theta(\cdot,\lambda)$, $\lambda\in\Lambda$, $f\not\equiv 0$, and if n is the multiplicity of zero of f at a point μ , $\mu\notin\Lambda$ then the function g , $g\overset{def}{=\!=}P_+\bar b_\mu^n f=\bar b_\mu^n f$ (as earlier, $b_\mu(z)=\frac{\mu-z}{1-\bar\mu z}\frac{|\mu|}{\mu}$), belongs to K_θ, $g(\mu)\neq 0$ and $g|\Lambda\equiv 0$. This means that $k_\theta(\cdot,\mu)\notin\operatorname{span}\{k_\theta(\cdot,\lambda):\lambda\in\Lambda\}$. ●

To complete the proof of Theorem 3.2 it suffices now to verify that in the case $S\neq const$ we can take $\Lambda'=\Lambda$. But we have already established that $\dim(K_\theta\ominus\operatorname{span}\{k_\theta(\cdot,\lambda):\lambda\in\Lambda'\})=\infty$, Λ' being the set existing in virtue of assertion 1. By Lemma 3.4 the family $\{k_\theta(\cdot,\lambda):\lambda\in\Lambda\}$ is also a basis in the subspace it generates. ●

REMARK. Lemma 3.4 is a generalization of some propositions of R.Paley - N.Wiener [59] and N.Levinson [48] concerning the case $\theta(z)=\exp a\frac{z+1}{z-1}$, $a>0$ (i.e. families of exponents in $L^2(0,a)$). This lemma shows also that a family of reproducing kernels (or exponents) neither loses nor gains the property to form a basis of K_θ (or of the subspace of K_θ it generates) if a finite set of its members is replaced by a set of functions of the same sort having the same cardinality. Another consequence (also generalizing some remarks from the books just mentioned; cf.also R.Redheffer [51]): a family $\{k_\theta(\cdot,\lambda):\lambda\in\Lambda\}$ either is a minimal one or $k_\theta(\cdot,\mu)\in\operatorname{span}\{k_\theta(\cdot,\lambda):\lambda\in\Lambda\setminus\{\mu\}\}$ $\forall\mu\in\Lambda$.

Theorems 3.1, 3.2 show that tests to establish whether a family of reproducing kernels (or exponents) is a basis involving conditions(2), (3) may be exploited not only in general theory, but in some concrete questions as well. Here is one more example confirming this.

THEOREM 3.5. Let $\Lambda\subset\mathbb{C}$, $\inf_{\lambda\in\Lambda}\operatorname{Im}\lambda>-\infty$. The following assertions are equivalent.

1. The family $\{e^{i\lambda x}\chi_{(0,a)}:\lambda\in\Lambda\}$ is an unconditio-

nal basis in the subspace of $L^2(0,a)$ it generates for some a, $a>0$.

2. $\Lambda+iy\in(C)$ for $y>-\inf_{\lambda\in\Lambda}\operatorname{Im}\lambda$.

This theorem is, of course, a simple consequence of the analogous fact for the unit disc.

THEOREM 3.5D. Let $\Lambda\subset\mathbb{D}$ and let θ be an inner function with $\sup_{\lambda\in\Lambda}|\theta(\lambda)|<1$. The following assertions are equivalent.

1. There exists a positive integer n such that $\{k_{\theta^n}(\cdot,\lambda):\lambda\in\Lambda\}$ is an unconditional basis in its closed linear span.

2. $\Lambda\in(C)$.

PROOF. The implication $1\Longrightarrow 2$ follows from Theorem 1.1.

$2\Longrightarrow 1$. Let $B=\prod_{\lambda\in\Lambda}b_\lambda$. Since the fractions $(1-\bar\lambda z)^{-1}$, $\lambda\in\Lambda$ constitute an unconditional basis of the subspace they generate and since $\theta^n(T_B)^*(1-\bar\lambda z)^{-1}=\overline{\theta(\lambda)^n}(1-\bar\lambda z)^{-1}$, it follows that for n sufficiently large we have the inequality

$$\|\theta^n(T_B)\|<1.$$

Combining this with the condition (3) and Theorem 2 bis (Part I) we obtain the desired implication. ●

To clarify better the situation some links between Theorem 3.1 and an interesting paper of P.Koosis [43] (cf.also [46]) are to be pointed out. In Koosis' paper a necessary and sufficient condition is found for all operators

$$f\longmapsto\chi_{(a,+\infty)}f,\qquad a>0 \tag{4}$$

to be compact on the space $\operatorname{span}_{L^2(\mathbb{R}_+)}\{e^{i\lambda_n x}\chi_{\mathbb{R}_+}:n\in\mathbb{Z}\}$. The condition reads as follows [x]):

$$\lim_n\operatorname{Im}\lambda_n=+\infty,\quad\lim_{|x|\to\infty}\sum_n\frac{\operatorname{Im}\lambda_n}{|\lambda_n-x|^2}=0.$$

Theorem 3.1 is an easy consequence of this result, for Koosis condition is implied by its hypotheses (i.e. $(\lambda_n)_{n\in\mathbb{Z}}\in(C)$, $\lim_n\operatorname{Im}\lambda_n=+\infty$). It should be noted that under the hypotheses of Theorem 3.1 we can establish with an equal ease that all *operators of the form* (4) *are compact*. Indeed, each operator of such form is equal to

---

[x]) It is not hard to see that the same condition is equivalent to compactness of all Hankel operators $H_{\bar B\theta^a}$, $a>0$ where $\theta^a=e^{iaz}$.

$$(I-P_{\theta^a})|K_B=\theta^a P_+\overline{\theta}^a|K_B=\theta^a\cdot\theta^a(T_B)^*\ (\theta^a=\exp iaz),$$

and the operator $\theta^a(T_B)^*$ is evidently compact, for the eigenvectors of this operator form an unconditional basis and its eigenvalues tend to zero. ●

Note also that the proof of theorem 3.1 presented here is much simpler than that of Koosis' theorem. This is due to the fact that in Theorem 3.1 Λ is assumed to satisfy Carleson condition.

Similar links exist between Theorem 3.2 and the recent paper [39]. In [39] all pairs (B,θ) of inner functions with the following property are identified: θ is singular and the Hankel operator $H_{\overline{B}\theta^a}$ is compact for every positive a.

4. Unconditional bases of reproducing kernels (the general case)

Theorems 1.2 and 5D give a solution of the problem concerning unconditional basis families of reproducing kernels under the additional assumption that the pair (θ,Λ) satisfies condition (1). Now we are going to treat the general case. If condition (1) is not satisfied then (see §1) the orthogonal projection P_θ distorts rational fractions and so $P_\theta|K_B$ is no longer an isomorphic imbedding. It is natural to try to "correct" the fractions $k(\cdot,\lambda)$ by means of a non-bounded operator in such a manner that the subsequent application of P_θ should produce no distortion.

Let $G\in H^2$ and let $T_{\overline{G}}$ be the Toeplitz operator whose symbol is $\overline{G}$. If $G\notin H^\infty$ then this operator is unbounded, but in any case its domain contains H^∞. It is evident (and well-known) that

$$T_{\overline{G}}(1-\overline{\lambda}z)^{-1}=\overline{G(\lambda)}(1-\overline{\lambda}z)^{-1}.$$

Thus $T_{\overline{G}}$ compensates the distortion produced by P_θ provided

$$G(\lambda)=(1-|\theta(\lambda)|^2)^{-1/2},\quad \lambda\in\Lambda. \tag{5}$$

LEMMA 4.1. If the family $\{k_\theta(\cdot,\lambda):\lambda\in\Lambda\}$ is an unconditional basis of its closed linear span then

$$\sum_{\lambda\in\Lambda}\frac{1-|\lambda|^2}{1-|\theta(\lambda)|^2}<+\infty\,, \tag{6}$$

and there exists a solution G, $G\in H^2$ of the problem (5).

PROOF. Consider the normed reproducing kernels $x_\lambda = \frac{(1-|\lambda|^2)^{1/2}}{(1-|\theta(\lambda)|^2)^{1/2}}\,\frac{1-\overline{\theta(\lambda)}\,\theta}{1-\bar\lambda z}$, $\lambda\in\Lambda$. If $f\in K_\theta$ then

$$\sum_{\lambda\in\Lambda}\frac{1-|\lambda|^2}{1-|\theta(\lambda)|^2}\,|f(\lambda)|^2=\sum_{\lambda\in\Lambda}|(f,x_\lambda)|^2\leqslant \mathrm{const}\,\|f\|^2.$$

Setting here $f=P_\theta\mathbb{1}=1-\theta\,\overline{\theta(0)}$ and using $|f(\lambda)|\geqslant 1-|\theta(0)|>0$ we obtain (6). Since $\Lambda\in(C)$ (Theorem 1.1), by Theorem A of §2, Part I the problem (5) has a solution in H^2 if and only if the inequality (6) holds. ●

REMARK. The solution of the problem (5) in K_B is unique and is given by the following formula:

$$G(z)=\sum_{\lambda\in\Lambda}\left(\frac{1-|\lambda|^2}{1-|\theta(\lambda)|^2}\right)^{1/2}\frac{(1-|\lambda|^2)^{1/2}}{1-\bar\lambda z}\,\frac{B_\lambda(z)}{B_\lambda(\lambda)}\,. \tag{7}$$

COROLLARY 4.2. Suppose that the assumptions of Lemma 4.1 are satisfied. If, in addition, θ is a singular inner function and μ is the representing measure of θ then

$$\sum_{\lambda\in\Lambda}\left(\int_{\mathbb{T}}\frac{d\mu(\zeta)}{|\zeta-\lambda|^2}\right)^{-1}<\infty\,.$$

Indeed, $1-|\theta(\lambda)|^2=1-\exp\left(-2\int_{\mathbb{T}}\frac{1-|\lambda|^2}{|\zeta-\lambda|}\,d\mu(\zeta)\right)\leqslant 2\int_{\mathbb{T}}\frac{1-|\lambda|^2}{|\zeta-\lambda|^2}\,d\mu(\zeta)$. ●

We have already mentioned that if a family $\{k_\theta(\cdot,\lambda):\lambda\in\Lambda\}$ is an unconditional basis then $\operatorname{dist}(\lambda,\operatorname{supp}\mu)$ necessarily tends to 0 as $|\lambda|\to 1$, $\lambda\in\Lambda$ (see §6 for the proof). Corollary 4.2 shows that in the case of a purely singular inner function θ, moreover, it must tend *at least with some prescribed rapidity*, namely

$$\sum_{\lambda\in\Lambda}(\operatorname{dist}(\lambda,\operatorname{supp}\mu))^2<\infty\,.$$

THEOREM 4.3. Let $\Lambda\subset\mathbb{D}$ and let θ be an inner function. The following assertions are equivalent.

1. The family $\{k_\theta(\cdot,\lambda): \lambda\in\Lambda\}$ is a basis of K_θ (resp., of the subspace it generates).

2. $\Lambda\in(C)$ and there is a function G in H^1 so that the operator $P_\theta T_{\overline{G}}$ may be extended from the linear span of the fractions $(1-\overline{\lambda}z)^{-1}$, $\lambda\in\Lambda$ to an isomorphism of K_B onto K_θ (resp., into K_θ).

PROOF. Implication 1 $\Longrightarrow$ 2 follows from Theorem 1.1 and Lemma 4.1, for if G is the function from this lemma, then
$P_\theta T_{\overline{G}}(1-\overline{\lambda}z)^{-1}=(1-|\theta(\lambda)|^2)^{-1/2}k_\theta(\cdot,\lambda),\quad \lambda\in\Lambda;$
hence

$$\|P_\theta T_{\overline{G}}(1-\overline{\lambda}z)^{-1}\|=\|(1-\overline{\lambda}z)^{-1}\|,\quad \lambda\in\Lambda$$

and consequently $P_\theta T_{\overline{G}}$ may be extended to an isomorphism (indeed, it takes an unconditional basis to an unconditional basis and does not change the norms of its elements).

To prove that 2 $\Longrightarrow$ 1 we argue similarly to Theorem 1: the family of fractions $(1-\overline{\lambda}z)^{-1}$, $\lambda\in\Lambda$ is an unconditional basis of K_B, hence any isomorphic image of this family is also an unconditional basis; in particular so is the family $\overline{G(\lambda)}k_\theta(\cdot,\lambda)$, $\lambda\in\Lambda$. ●

REMARK. The property of the function G expressed by assertion 2 is shared by any other function F in H^1 satisfying

$$0<\inf_{\lambda\in\Lambda}\left|\frac{F(\lambda)}{G(\lambda)}\right|\leq\sup_{\lambda\in\Lambda}\left|\frac{F(\lambda)}{G(\lambda)}\right|<+\infty .$$

It is clear also that for any such F

$$0<\inf_{\lambda\in\Lambda}|F(\lambda)|(1-|\theta(\lambda)|^2)^{1/2}\leq\sup_{\lambda\in\Lambda}|F(\lambda)|(1-|\theta(\lambda)|^2)^{1/2}<+\infty .$$

Unfortunately Theorem 4.3 is too non-constructive, and the situation is unlikely to improve very much even if we try to use some concrete G (e.g. one given by (7) provided (6) is satisfied) when applying this theorem. As for the function (7), it very probably fails to be the most appropriate. For example, in the case (1) (i.e. $\sup_{\lambda\in\Lambda}(1-|\theta(\lambda)|^2)^{-1/2}<\infty$) it is natural to choose $G\equiv\mathbb{1}$ (and so we did in Theorems of sections 1-3). Some facts supporting what we have just said may be found in the next §5.

5. Orthogonal and nearly orthogonal bases of reproducing kernels

It was already mentioned that if $\Lambda\subset\mathbb{D}$ and $\operatorname{card}\Lambda>1$ then the family $\{k_\theta(\cdot,\lambda):\lambda\in\Lambda\}$ cannot be orthogonal. In some cases it is possible, however, to consider reproducing kernels with poles on the unit circle. For example let the function θ admit an analytic continuation through a point λ, $\lambda\in\mathbb{T}$. Then the kernel

$$k_\theta(z,\lambda)=\frac{1-\overline{\theta(\lambda)}\theta(z)}{1-\bar\lambda z}=\frac{\overline{\theta(\lambda)}}{\bar\lambda}\,\frac{\theta(\lambda)-\theta(z)}{\lambda-z}$$

evidently lies in $H^2(\mathbb{D})$ and, moreover, in K_θ. A criterion for the inclusion $k_\theta(\cdot,\lambda)\in H^2(\mathbb{D})$, $\lambda\in\mathbb{T}$ was obtained by P.Ahern and D.Clark [26]. Let

$$\theta(z)=z^N\prod_n\frac{\bar a_n}{|a_n|}\,\frac{a_n-z}{1-\bar a_n z}\,exp\left\{-\int_{\mathbb{T}}\frac{\zeta+z}{\zeta-z}\,d\mu(\zeta)\right\}$$

be the canonical factorization of an inner function θ and set

$$E_\theta\overset{def}{=\!=}\left\{\zeta\in\mathbb{T}:\ \sum_n\frac{1-|a_n|^2}{|\zeta-a_n|^2}+\int_{\mathbb{T}}\frac{d\mu(t)}{|t-\zeta|^2}<+\infty\right\}.$$

Roughly speaking, E_θ consists of those points at which the argument of θ is differentiable.

THEOREM (P.Ahern, D.Clark [26]). Let $\lambda\in\mathbb{T}$. Then the fraction $(1-\bar c\,\theta(z))(1-\bar\lambda z)^{-1}$ lies in $H^2(\mathbb{D})$ for some complex number c if and only if $\lambda\in E_\theta$. If $\lambda\in E_\theta$ then this c is in fact unique and is given by $c=\lim_{r\to 1-0}\theta(r\lambda)$.

It should be noted here that Frostman's theorem (cf. [30]) implies that θ has radial limits on a set wider than E_θ, namely on the set

$$\left\{\zeta\in\mathbb{T}:\ \sum_n\frac{1-|a_n|^2}{|\zeta-a_n|}+\int\frac{d\mu(t)}{|t-\zeta|}<+\infty\right\}.$$

Let now $\theta(\lambda)=\theta(\lambda')=\alpha$, $|\alpha|=1$ and assume $\lambda,\lambda'\in E_\theta$, $\lambda\neq\lambda'$. Then

$$(k_\theta(z,\lambda'),k_\theta(z,\lambda))=k_\theta(\lambda,\lambda')=\frac{1-\overline{\theta(\lambda')}\theta(\lambda)}{1-\bar\lambda'\lambda}=0. \tag{8}$$

This remarkable property was observed for the first time by D.Clark [29] and later (independently) by D.Georgijević [37]. We are going to illustrate this property by an example; to do this we pass for some time to the upper half-plane. Let $\theta = \theta^{2\pi} = e^{2\pi i z}$. Then $k_\theta(z,t) = i\,\frac{1-\overline{\theta(t)}\,\theta(z)}{z-t} \in K_\theta$ for all t in $\mathbb{R}$. Evidently $\theta(\zeta)=1$ if and only if ζ is an integer. For $\lambda = n \in \mathbb{Z}$ we have

$$k_\theta(z,n) = \frac{e^{2\pi i z}-1}{i(z-n)} = \int_0^{2\pi} e^{izt}\,e^{-int}\,dt\,,$$

and so the kernels $k_\theta(\cdot,n)$ are Fourier-Laplace transforms of the classical orthogonal system of exponents $\{e^{int}: n\in\mathbb{Z}\}$. We see (!) that the reproducing kernels $\{k_\theta(\cdot,n): n\in\mathbb{Z}\}$ form a complete orthogonal system in K_θ.

It turns out that this example may be generalized to a class of inner functions θ. The construction was performed by D.Clark [29] in connection with the investigation of spectra of one-dimensional perturbations of the model operator T_θ.

Let θ be an inner function and $\alpha\in\mathbb{T}$. Substituting θ for z in the Poisson kernel $\frac{1-|z|^2}{|\alpha-z|^2}$ we obtain a nonnegative harmonic function in the disc, which can be represented by a Poisson integral:

$$\frac{1-|\theta(z)|^2}{|\alpha-\theta(z)|^2} = \int_{\mathbb{T}} \frac{1-|z|^2}{|1-\bar{\zeta}z|^2}\,d\sigma_\alpha(\zeta), \quad |z|<1. \tag{9}$$

The measure σ_α is nonnegative and singular with respect to the Lebesgue measure since $\lim\limits_{r\to 1-0}\theta(r\zeta)\neq\alpha$ almost everywhere on $\mathbb{T}$. On the other hand it is well-known that the radial limits of the Poisson integral of a singular measure are equal to $+\infty$ almost everywhere with respect to this measure. Hence

$$\lim_{r\to 1-0}\theta(r\zeta)=\alpha, \qquad \sigma_\alpha\text{-a.e.}$$

Thus measures σ_α and σ_β are mutually singular if $\alpha\neq\beta$. The equality (9) can be given another form:

$$\frac{1-|\theta(z)|^2}{1-|z|^2} = \int_{\mathbb{T}} \left|\frac{1-\alpha\,\overline{\theta(z)}}{1-\bar{z}\zeta}\right|^2 d\sigma_\alpha(\zeta).$$

Since $\lim\limits_{r\to 1-0} k_\theta(r\zeta, z) = (1-\alpha\,\overline{\theta(z)})(1-\bar{z}\zeta)^{-1}$,

the last equality means that the restriction map $\mathcal{U}$:
$f \mapsto f|supp(\sigma_\alpha)$ from H^2 to $L^2(\sigma_\alpha)$ preserves the norms of the reproducing kernels $k_\theta(\cdot,\lambda)$, $|\lambda|<1$. In fact $\mathcal{U}$ can be extended to an isometry of K_θ onto $L^2(\sigma_\alpha)$ (see Clark [29] for the details).

Let $\Lambda \subset \mathbb{T}$. Then the family $\{k_\theta(z,\lambda): \lambda\in\Lambda\}$ is orthogonal in K_θ if and only if $\Lambda\subset E_\theta$ and $\theta|\Lambda \equiv \alpha$, $\alpha\in\mathbb{T}$. It turns out that every such orthogonal family is the family of eigenfunctions of a unitary operator U_α and that this U_α is a one-dimensional perturbation [x] of the model operator $P_\theta S|K_\theta$. The action of this unitary operator is described by the formula

$$U_\alpha f = z\Big(f-(f,K_0)\frac{K_0}{\|K_0\|^2}\Big)+ w(f,K_0)\frac{k_0}{\|k_0\|^2} ,$$

where

$$K_0 = P_\theta \mathbb{1} = 1-\overline{\theta(0)}\theta,\ k_0 = z^{-1}(\theta(z)-\theta(0)),\ w = \frac{\alpha-\theta(0)}{1-\overline{\theta(0)}\alpha} .$$

Restricting this formula to the support set of σ_α and using the fact that $\theta=\alpha$ a.e. with respect to σ_α we obtain

$$\mathcal{U}U_\alpha f = z\,\mathcal{U}f .$$

Hence U_α is equivalent to the operator of multiplication by z in $L^2(d\sigma_\alpha)$. This reasoning proves the following theorem of Clark.

THEOREM (D.Clark [29]). The space K_θ has an orthogonal basis consisting of reproducing kernels $\{k_\theta(z,\lambda):\lambda\in\Lambda\}$, $\Lambda\subset\mathbb{T}$ if and only if for some α, $\alpha\in\mathbb{T}$ the measure σ_α is purely atomic. ●

Unfortunately it is not easy to use this criterion. There exists, however, a simpler sufficient condition: if the set $\mathbb{T}\setminus E_\theta$ is at most countable then for any α, $\alpha\in\mathbb{T}$, the family $\{k_\theta(z,\lambda):\theta(\lambda)=\alpha,\ \lambda\in E_\theta\}$ is a complete orthogonal system in K_θ. This condition is also due to Clark. It is satisfied, for example, for inner functions $\theta(z) = \exp\{-\int_{\mathbb{T}}\frac{\zeta+z}{\zeta-z}d\mu(\zeta)\}$ such that the set $supp(\mu)$ is at most coun-

[x] It was the investigation of such perturbation that led D.Clark to all his results. A "vector-valued" theory of the same sort is developped in [71], [72].

table.

Using orthogonal bases consisting of reproducing kernels corresponding to points of the unit circle it is possible to construct unconditional reproducing kernel bases with members corresponding to points of $\mathbb{D}$. For example suppose that for a given θ a family $\{k_\theta(z,\lambda_n)\}_{n\in\mathbb{Z}}$ with $|\lambda_n|=1, \theta(\lambda_n)=\alpha, n\in\mathbb{Z}$ constitutes an orthogonal basis in K_θ. Choose for each n a point μ_n in $\mathbb{D}$ so close to λ_n that

$$\sum_n \| \, \|k_\theta(\cdot,\lambda_n)\|^{-1} k_\theta(\cdot,\lambda_n) - \|k_\theta(\cdot,\mu_n)\|^{-1} k_\theta(\cdot,\mu_n)\|^2 < 1 .$$

Then $(k_\theta(\cdot,\mu_n))_{n\in\mathbb{Z}}$ is clearly an unconditional basis in K_θ. This method to construct bases is, of course, merely a generalization of the Paley - Wiener method..

There exist however inner functions θ such that the space K_θ contains no reproducing kernels corresponding to points of the unit circle, but yet has an unconditional basis consisting of reproducing kernels. To construct such a θ it is sufficient to produce a Blaschke product B whose zeros form a Carleson set $\Lambda=\{\lambda_n : n\in\mathbb{Z}\}$, but

$$\sum_n \frac{1-|\lambda_n|^2}{|\zeta-\lambda_n|^2} \equiv +\infty$$

for $\zeta\in\mathbb{T}$. (Given such a B take simply $\theta=B$. Then by Theorem A the family $(k(\cdot,\lambda_n))_{n\in\mathbb{Z}}$ is an unconditional basis in K_B and by the theorem of P.Ahern - D.Clark the set E_B is empty). We shall show even that there exists a subset Λ of the unit disc such that $\Lambda\in(C)$ and

$$\sum_{\lambda\in\Lambda} \frac{1-|\lambda|}{|\zeta-\lambda|} = +\infty \tag{10}$$

for all ζ, $\zeta\in\mathbb{T}$. Let $(k_n)_{n\geq 2}$ be a sequence of positive integers with the properties

$$\sum_n k_n 2^{-n} < \infty, \quad \sum_n k_n 2^{-n} \log k_n = \infty .$$

(Take, for example, $k_n=[(n\log^2 n)^{-1}2^n]$, $[x]$ being the greatest integer less than or equal to x). For each n choose k_n equidistant points on the circle $\{\zeta\in\mathbb{C} : |\zeta|=r_n \stackrel{def}{=\!=} \stackrel{def}{=\!=} 1-2^{-n}\}$, and let Λ be the set of all choosen points. We claim that Λ has the desired properties.

The rarity condition (i.e. $D(\lambda,\varepsilon(1-|\lambda|))\cap$ $\cap D(\lambda',\varepsilon(1-|\lambda'|))=\emptyset$ for $\lambda,\lambda'\in\Lambda$, $\lambda\neq\lambda'$ and for sufficiently small ε, $\varepsilon>0$) as well as the fact that $\sum_{\lambda\in\Lambda}(1-|\lambda|)\delta_\lambda$ is a Carleson measure (i.e. $\sum_{\lambda\in Q}(1-|\lambda|)\leq$ $\leq const\cdot\ell$ for every rectangle $Q=\{\zeta: 1-\ell\leq|\zeta|<1,$ $\arg\zeta\in I\}$, $I\subset\mathbb{T}$, $|I|=\ell$, $\ell>0$) are easily checked. To verify (10) take $\zeta\in\mathbb{T}$ and estimate separately the summands with $|\lambda|=r_n$, $r_n=1-2^{-n}$. Note that for every such λ except possibly two we have $|\zeta-\lambda|\leq const|r_n\zeta-\lambda|$, and hence

$$\sum_{\lambda\in\Lambda,|\lambda|=r_n}\frac{1-|\lambda|}{|\zeta-\lambda|}\geq const\,2^{-n}\sum_{\lambda\in\Lambda,|\lambda|=r_n}\frac{1}{|r_n\zeta-\lambda|}\geq$$

$$\geq const\,2^{-n}\sum_{k=1}^{[k_n]}\frac{k_n}{k}\geq const\cdot k_n 2^{-n}\log k_n\,. \bullet$$

6. Interpolation by K_θ-functions. The H^p_- spaces

In this section we are mainly concerned with applications of our results on exponential bases and bases of reproducing kernels to interpolation theory, and with some variants of these results for the H^p- spaces.

We begin with the second subject, restricting ourselves by the case $1<p<\infty$. For these values of p the theory turns out to be a duplicate (with minor variations) of the H^2-theory already discussed, the reason being, of course, the L^p-continuity of the Riesz projection $\mathbb{P}_+$. We recall that $\mathbb{P}_+f\overset{def}{=\!=}$ $\overset{def}{=\!=}\sum_{n\geq0}\hat{f}(n)z^n$, and thus $\mathbb{P}_+$ maps $L^p=L^p(\mathbb{T})$ onto the Hardy space $H^p\overset{def}{=\!=}\{f\in L^p:\hat{f}(n)=0,\ \ n<0\}$. Setting $(H^p)_-=\{f\in L^p:\hat{f}(n)=0,\ n\geq0\}$ we see that for $p\in(1,\infty)$ L^p is the direct sum of H^p and H^p_-, and so using the duality $\langle f,g\rangle=\int_{\mathbb{T}}f\bar{g}\,dm$ we may identify (in the anti-linear manner) the conjugate space $(H^p)^*$ with $H^{p'}$, $\frac{1}{p}+\frac{1}{p'}=1$. It is clear that the main formulae of §§ 1-4 remain valid in the H^p-setting also.

Define for an inner function θ

$$K^p_\theta=(\theta H^{p'})^\perp\ ,\ \ k_\theta(\cdot,\lambda)=\frac{1-\overline{\theta(\lambda)}\,\theta}{1-\bar{\lambda}z}\,.$$

Then $P_\theta \overset{def}{=} \theta \mathbb{P}_- \bar\theta$ is the projection onto K_θ along θH^p, $H^p = K_\theta \dot{+} \theta H^p$ and k_θ the reproducing kernel of the space $K_\theta^{p'}$ (which may be naturally considered as the conjugate space $(K_\theta^p)^* = H^{p'}/(K_\theta^p)^\perp = H^{p'}/\theta H^{p'}$); indeed,

$$f(\lambda) = \langle f, k_\theta(\cdot,\lambda)\rangle \ ; \ \lambda \in \mathbb{D} \ , \ f \in K_\theta^{p'}$$

and $k_\theta(\cdot,\lambda) \in K_\theta^p \ (\lambda \in \mathbb{D})$. It is also clear that $k_\theta(\cdot,\lambda) = P_\theta(1-\bar\lambda z)^{-1}$, $\lambda \in \mathbb{D}$.

We shall be interested mainly in the case when the functions $k_\theta(\cdot,\lambda_n)$ "do not lie very far from rational fractions", i.e.

$$\sup_n |\theta(\lambda_n)| < 1. \tag{11}$$

As before we shall discuss u n c o n d i t i o n a l b a s e s of the form $\{k_\theta(\cdot,\lambda) : \lambda \in \Lambda\}$, $\Lambda \subset \mathbb{D}$ but now using the general difinition of this notion which we have mentioned in the first pages of our paper (p. 217)

LEMMA 6.1. Suppose that the family $\{k_\theta(\cdot,\lambda_n) : \lambda_n \in \Lambda\}$ is an unconditional basis of the subspace of H^p it generates, $1<p<\infty$ and assume that (11) holds. Then $\Lambda \in (C)$.

PROOF. Note that, in H^p , the Carleson condition (C) is still necessary and sufficient for the system $\{(1-\bar\lambda_n z)^{-1} : \lambda_n \in \Lambda\}$ to be uniformly minimal. It remains to apply to $A = P_\theta$, $x_n = (1-\bar\lambda_n z)^{-1}$ the lemma about the uniformly minimal families proved in § 2 Part I. ●

At this point some widely known facts concerning the geometry of families of rational functions $\{(1-\bar\lambda_n z)^{-1} : n \geq 1\}$ in the H^p-metric, $1<p<\infty$, should be recalled. (For a more detailed exposition see [8], [4], [17], [18]). One of these facts has already been used (namely, that the condition (C) is equivalent to the uniform minimality), others (to be used later on) are as follows. The Carleson condition (C) is equivalent to each of assertions listed below:

a) the family $\{(1-\bar\lambda_n z)^{-1} : \lambda_n \in \Lambda\}$ is an unconditional basis of the subspace of H^p it generates;

b) The family $\{(1-|\lambda_n|)^{1/p'}(1-\bar\lambda_n z)^{-1} : \lambda_n \in \Lambda\}$ is an unconditional basis (in its closed linear span) isomorphic to the standard unit vector basis of ℓ^p ;

c) $\mathcal{J} H^p = \ell^p$, where $\mathcal{J} f = \{f(\lambda_n)(1-|\lambda_n|)^{1/p} : \lambda_n \in \Lambda\}$

One more condition equivalent to a)-c) worth mentioning (seems to be present in the literature only in an implicit form, if at all):

d) $\mathcal{J}H^p \supset \ell^p$, i.e. any interpolation problem $\mathcal{J}f = a$ with the data a in ℓ^p has a solution in H^p.

To verify that d) is equivalent to a)-c) note that the inclusion $\mathcal{J}H^p \supset \ell^p$ and the closed graph theorem imply that the problem mentioned is not merely solvable, but is solvable with an estimate: there exists a constant C so that $\forall a \in \ell^p\ \exists f \in H^p$: $\mathcal{J}f = a$, $\|f\|_{H^p} \leq c\|a\|_{\ell^p}$. Taking as a the unit vectors of the space ℓ^p we obtain the uniform minimality of the family $\{(1-\overline{\lambda}_n z)^{-1} : \lambda_n \in \Lambda\}$ in $H^{p'}$, i.e. the Carleson condition (C). ●

Also well-known is the general duality between the problems concerning bases and interpolation, cf. [17], [18] for the details. In our setting it is expressed by the following lemma.

LEMMA 6.2. Let $\Lambda \subset \mathbb{D}$, Θ be an inner function, $1<p<\infty$. The following assertions are equivalent.

1. The family $\{k_\Theta(\cdot, \lambda_n) : \lambda_n \in \Lambda\}$ is an unconditional basis of the subspace of H^p it generates.

2. The space of restrictions $K_\Theta^{p'} \mid \Lambda$ is an ideal space (that is, from $f \in K_\Theta^{p'}$ and $|a_n| \leq |f(\lambda_n)|$, $\lambda_n \in \Lambda$ it follows that there exists a function g in $K_\Theta^{p'}$ interpolating $\{a_n\}$: $g(\lambda_n) = a_n$, $\lambda_n \in \Lambda$). ●

In fact (and this will be the essence of theorem 6.3 below) an ideal space mentioned in the lemma will turn out to be simply a weighted ℓ^p-space (just as for the problem of free interpolation in the whole space H^p). Now we mention only that the interpolation by $K_\Theta^{p'}$-functions is nothing else as the interpolation by functions analytically continuable trough the points of $\mathbb{T} \setminus \operatorname{spec} \Theta$ and satisfying some estimates in $\mathbb{C} \setminus \mathbb{T}$, cf. [73], [18]. That is why the condition $\lim\limits_{\lambda \in \Lambda, |\lambda| \to 1} \operatorname{dist}(\lambda, \operatorname{spec} \Theta) = 0$ mentioned on p. 268 is necessary for the family $\{k_\Theta(\cdot, \lambda) : \lambda \in \Lambda\}$ to form an unconditional basis (see also the next corollary and theorem 6.3).

For $p = 2$ no additional work is needed to give a precise theorem connecting interpolation and reproducing kernels bases. We present both a general assertion concerning the spaces $K_\Theta = K_\Theta^2$ and an assertion concerning the most interesting particular case $\Theta = \Theta^a$, connected with exponential bases $\{e^{i\lambda_n x} \chi_{(0,a)} : \lambda_n \in \Lambda\}$.

COROLLARY D. Let $\Lambda \subset \mathbb{D}$, Θ be an inner function. The following assertions are equivalent.

1. The family $\{k_\Theta(\cdot, \lambda_n) : \lambda_n \in \Lambda\}$ is an unconditional

basis of the subspace of H^2 it generates.

2. $\mathcal{J}_\theta K_\theta = \ell^2$, where $\mathcal{J}_\theta f = \{ f(\lambda_n)\left(\frac{1-|\lambda_n|^2}{1-|\theta(\lambda_n)|^2}\right)^{\frac{1}{2}} : \lambda_n \in \Lambda \}$.

If the condition (11) satisfied, then we have another equivalent assertion:

3. $\mathcal{J}_\theta K_\theta \supset \ell^2$.

COROLLARY $\mathbb{C}_+$. Let $a > 0$, $\Lambda \subset \mathbb{C}_\delta$, $\delta > 0$. The following assertions are equivalent.

1. The family $\{ e^{i\lambda_n x} \chi_{(0,a)} : \lambda_n \in \Lambda \}$ is an unconditional basis of the subspace of $L^2(0,a)$ it generates.

2. $\mathcal{J}_a E_a^2 = \ell^2$, where E_a^2 is the space of all entire functions of exponential type less than or equal to $a/2$ and square summable on $\mathbb{R}$, and $\mathcal{J}_a f \overset{def}{=\!=} \{ c_n f(\lambda_n) : \lambda_n \in \Lambda \}$, $c_n = (\operatorname{Im} \lambda_n)^{1/2} \exp(-1/2\, a \operatorname{Im} \lambda_n)$.

3. $\mathcal{J}_a E_a^2 \supset \ell^2$.

To check these corollaries one needs only to add to what has already been explained the (evident) fact that for any unconditional basis $\{x_n\}$ in a Hilbert space the space of Fourier coefficients $\{(x, x_n/\|x_n\|)\}$ coincides with ℓ^2. ●

Passing to the main result of this section we recall the Muckenhoupt condition (A_p) in terms of which the reproducing kernel bases in K_θ^p will be described. This condition, imposed for $1 < p < \infty$ on a positive function w on $\mathbb{T}$, looks as follows:

$$\sup_I \left(\frac{1}{mI}\int_I w\,dm\right)\left(\frac{1}{mI}\int_I w^{-\frac{1}{p-1}}\,dm\right)^{p-1} < \infty \qquad (A_p),$$

where the $\sup$ is taken over all intervals (arcs) of $\mathbb{T}$.

THEOREM 6.3. Let $1 < p < \infty$, $\Lambda \subset \mathbb{D}$, θ be an inner function in $\mathbb{D}$ and suppose that the condition (11) holds. The following assertions are equivalent.

1. The family $\{ k_\theta(\cdot, \lambda_n) : \lambda_n \in \Lambda \}$ is an unconditional basis of K_θ^p.

2. The family $\{ k_\theta(\cdot, \lambda_n)(1-|\lambda_n|)^{1/p'} : \lambda_n \in \Lambda \}$ is a basis of K_θ^p equivalent to the standard unit vector basis of ℓ^p.

3. $\Lambda \in (C)$ and the operator $P_\theta | K_B^p$ is an isomorphism of K_B^p onto K_θ^p (here $B = \prod_{\lambda_n \in \Lambda} b_{\lambda_n}$, the Blaschke product corresponding to the set Λ).

4. $\Lambda \in (C)$ and there exist real functions u and v and a real number c so that $u \in L^\infty(\mathbb{T})$ and

$$\overline{B}\Theta = exp(u+ic-i\tilde{v}),\ exp(\tfrac{p}{2}v)\in(A_p)\ . \tag{12}$$

5. If $f\in K_\theta^{p'}$, $f(\lambda_n)=0$, $(\lambda_n\in\Lambda)$ then $f\equiv 0$; and $\mathcal{J}^{p'}K_\theta^{p'} = \ell^{p'}$, where $\mathcal{J}^{p'}f = \{f(\lambda_n)(1-|\lambda_n|)^{1/p'}:\lambda_n\in\Lambda\}$.

6. $\mathcal{J}^{p'}K_\theta^{p'}\supset \ell^{p'}$ and if $f\in K_\theta^{p'}$, $f(\lambda_n)=0$ $(\lambda_n\in\Lambda)$ then $f\equiv 0$.

PROOF. It is clear that $3\Longleftrightarrow 2\Longrightarrow 1$, $2\Longleftrightarrow 5$ and that the implication $6\Longrightarrow 5$ has in fact already been proved (the reverse implication $5\Longrightarrow 6$ being evident). It remains to check that $1\Longrightarrow 2$ and $3\Longleftrightarrow 4$.

$1\Longrightarrow 2$. If $\{x_n\}$ is an unconditional basis in L^p-metric then "integrating over signs" we obtain

$$\int|\sum a_n x_n|^p \asymp \int(\sum|a_n|^2|x_n|^2)^{p/2}\ ;$$

the symbol $\asymp$ means that each of the integrals majorizes another one multiplied by a constant independent of the coefficients a_n. Setting $x_n = k_\theta(\cdot,\lambda_n)$ and taking into account the condition (11), Lemma 6.1 and the assertion b) concerning unconditional bases of rational fractions we get

$$\int_{\mathbb{T}}|a_n k_\theta(\cdot,\lambda_n)|^p \asymp \int_{\mathbb{T}}(\sum|a_n|^2|k_\theta(\cdot,\lambda_n)|^2)^{p/2}\asymp$$

$$\asymp\int_{\mathbb{T}}(\sum|a_n|^2\frac{1}{|1-\bar\lambda_n z|^2})^{p/2}\asymp\int_{\mathbb{T}}|\sum a_n\frac{1}{1-\bar\lambda_n z}|^p\asymp$$

$$\asymp\sum|a_n|^p(1-|\lambda_n|)^{-p/p'}.$$

This relation between the first and the last term just means that the assertion 2 holds.

$3\Longleftrightarrow 4$. Similary to the case $p=2$ (see § 3, Part I) the operator $P_\theta\,|\,K_B^p$ has the same metric properties as the Toeplitz operator $T_{\overline{B}\Theta}$ in the space H^p. The criterion of the form (12) for such an operator to be invertible is the subject-matter of the paper [53]. ●

Of course, the material of this section suggests some natural questions. We have skipped them in the hope that they have been noted by the reader who had the patience to reach this point. May be, the reader even knows already how to answer them.

PART III.

EXPONENTIAL BASES AND ENTIRE FUNCTIONS.

1. Generating functions, BMO and theorems 6, 7, 8.

In this section we investigate some properties of the generating functions corresponding to subsets of the upper half-plane and give the proofs of theorems 6, 7, 8. First recall some definitions from the theory of entire functions.

Let F be an entire function of exponential type. The 2π-periodic function h_F defined on $\mathbb{R}$ by the formula

$$h_F(\varphi)=\overline{\lim_{r\to\infty}}\frac{\log|F(re^{i\varphi})|}{r},\qquad \varphi\in\mathbb{R}$$

is called the indicator of F.

The indicator diagram of F is by definition the convex set G_F such that

$$h_F(\varphi)=\sup_{\zeta\in G_F}\operatorname{Re}(\zeta e^{-i\varphi}).$$

The set $G_F^*=\{\bar\zeta:\zeta\in G_F\}$ is called the conjugate diagram of F.

The background material concerning the above notions is contained in [13] (ch. I, §§15-17 and §§19-20). We have already explained the reason for our interest in the class $\mathcal{E}_a$, $a>0$ of all entire functions F of exponential type with $G_F^*=[0,ia]$ in Section 5 of Part I. More precisely we shall be interested in the subclass $\mathcal{M}_a$ of the class $\mathcal{E}_a$ consisting of functions F, $F\in\mathcal{E}_a$, satisfying the Muckenhoupt condition (A_2) on $\mathbb{R}$:

$$\sup_{I\in J}\left(\frac{1}{|I|}\int_I|F|^2dx\right)\left(\frac{1}{|I|}\int|F|^{-2}dx\right)<\infty. \qquad (A_2)$$

Here J is the set of all bounded intervals of the real axis Recall that the condition (A_2) is equivalent to the Helson-Szegö condition (HS), see Part I, § 4.

LEMMA 1.1. Let w be a positive function on the real axis satisfying the Helson-Szegö condition (HS). Then there exists a number $p=p_w$, $1<p<\infty$, such that

$$\int_{\mathbb{R}}\frac{w^p(x)}{1+x^2}\,dx<\infty.$$

PROOF. The hypothesis implies that $w = \exp(u + \tilde{v})$, where u, $v \in L^\infty(\mathbb{R})$, $\|v\|_\infty < \pi/2$. It remains to use the following well-known theorem due to A.Zygmund (see [9], ch, YII § 2, th. 2.11 (I)): if $\|v\|_\infty \leqslant 1$ and $0 < \lambda < \pi/2$, then

$$\int_{\mathbb{R}} \exp(\lambda |\tilde{v}(x)|) \frac{dx}{1+x^2} < \infty . \quad \bullet$$

We denote by $\mathcal{C}$ the set of all entire functions f of exponential type such that

$$\int_{\mathbb{R}} \frac{\log^+|f(x)|}{1+x^2}\, dx < \infty,$$

where $u^+ \overset{def}{=} max(u, 0)$. From lemma 1.1 it follows that

$$\int_{\mathbb{R}} \frac{|F(x)|^2}{1+x^2}\, dx < \infty$$

for $F \in \mathcal{M}_a$, $a > 0$. Hence by the M.Cartwright theorem (see [13], ch. Y, § 4) we may conclude that $\mathcal{M}_a \subset \mathcal{C}$.

The class $\mathcal{C}$ can be characterized as the set of all entire functions f of exponential type with $f|\mathbb{C}_+$, $f|\mathbb{C}_-$ belonging to the Nevanlinna classes in the corresponding half-planes (i.e. to the images of the usual Nevanlinna class in the unit disc $\mathbb{D}$ under the conformal mappings $\mathbb{C}_\pm \to \mathbb{D}$). Hence, if $f \in \mathcal{C}$ then

$$f|\mathbb{C}_+ = cSBf_e ,$$

where $c \in \mathbb{C}$, $|c| = 1$, f_e is an outer function in $\mathbb{C}_+$, B is the Blaschke product corresponding to the zeros of $f|\mathbb{C}_+$ and S is the quotient of two singular inner function in $\mathbb{C}_+$. Because of the analyticity of the function f on the real axis we have $S = \exp i\gamma z$, $\gamma \in \mathbb{R}$ (to see this recall the formula for the singular inner function from § 1, Part I). An analogous factorization formula holds also in the lower half--plane $\mathbb{C}_-$.

We state now a useful connection between the class $\mathcal{M}_a$ and unimodular Helson-Szegö functions on $\mathbb{R}$.

THEOREM 1.2. Let Λ ($\Lambda \subset \mathbb{C}_\delta$, $\delta > 0$) be a Blaschke set, let B denote the corresponding Blaschke product and let $\theta^a = \exp(iaz)$, $a > 0$. The following assertions are equivalent.

1. There exists a function of the class $\mathcal{M}_a$ with simple

zeros whose zero-set is Λ.

2. The restriction $B\bar{\Theta}^a|\mathbb{R}$ is a Helson-Szegö function, i.e. there exists a unimodular constant c and an outer (in $\mathbb{C}_+$) function h such that $|h^2||\mathbb{R}\in(HS)$ and

$$Bh = c\bar{h}\Theta^a, \qquad \text{a.e. on } \mathbb{R}.$$

PROOF. $1 \Rightarrow 2$. Let F be an entire function mentioned in the assertion 1 and let h be the outer function in $\mathbb{C}_+$ with $|h(x)|=|F(x)|$, $x\in\mathbb{R}$. By the definition of the class $\mathcal{M}_a$, $h_F(\pi/2)=0$ and hence the canonical factorization of F in $\mathbb{C}_+$ contains no singular inner factor, i.e.

$$F|\mathbb{C}_+ = c_+B\cdot h, \quad |c_+|=1.$$

An analogous reasoning for the half-plane $\mathbb{C}_-$ shows (take into account that $h_F(-\pi/2)=a$)

$$F|\mathbb{C}_- = c_-\Theta^a h^*, \quad |c_-|=1,$$

where $h^*(z)=\overline{h(\bar{z})}$ (the outer function in $\mathbb{C}_-$ with $|h^*(x)|=|F(x)|$, $x\in\mathbb{R}$). Comparing the last equality with the preceding one we obtain the assertion 2.

$2 \Rightarrow 1$. Let h be the function from the assertion 2. It is useful to note that $h(z+i)^{-1}\in H^2_+$ because of lemma 1.1. We define a function F on $\mathbb{C}\setminus\mathbb{R}$ by the equalities:

$$F=\begin{cases} Bh & (\text{on } \mathbb{C}_+),\\ c\,\Theta^a h^* & (\text{on } \mathbb{C}_-).\end{cases}$$

In fact, however, the function F admits an extension onto the whole plane $\mathbb{C}$ as an entire function. This is an immediate consequence of the following simple lemma.

LEMMA 1.3. Let f_+ and f_- be analytic functions in the upper and lower half-planes respectively, let Δ be an interval, and let

$$\sup_{0<y<1}\int_\Delta |f_\pm(x\pm iy)|\,dx < +\infty.$$

If $\lim\limits_{y\to 0+} f_+(x+iy) = \lim\limits_{y\to 0+} f_-(x-iy)$ a.e. on Δ then f_+ can be analytically continued through Δ, the continuation coinciding with f_-.

The proof is easy. It can be found for example in [42].

A more general theorem is proved in [48] .

To prove that $F \in \mathcal{M}_a$ we use the Cauchy formula and the fact that $|F(z)| \leq \exp\{a\,|\operatorname{Im} z|\} \cdot |h(\bar{z})|$, $\operatorname{Im} z > 0$, (this inequality follows easily from the definition of F). We have

$$|h(x)| = |F(x)| = \left| \frac{1}{2\pi i} \int\limits_{|\zeta - x| = 1} \frac{F(\zeta)}{\zeta - x}\, d\zeta \right| \leq \frac{1}{2\pi} \int\limits_{\{|\zeta - x| = 1\} \cap \mathbb{C}_+} |h(\zeta)|\,|d\zeta| + \frac{e^a}{2\pi} \int\limits_{\{|\zeta - x| = 1\} \cap \mathbb{C}_-} |h^*|\,|d\zeta|$$

for $x \in \mathbb{R}$.

Since $(z+i)^{-1} h \in H^2_+$ and the "arc length" on $\mathbb{C}_+ \cap \{|z - x| = 1\}$ is obviously a Carleson measure then from the Carleson imbedding theorem ([18]) it follows that

$$\int\limits_{\mathbb{C}_+ \cap \{|\zeta - x| = 1\}} \frac{|h(\zeta)|^2}{|\zeta + i|^2} |d\zeta| \leq \text{const} \int\limits_{\mathbb{R}} \frac{|h(x)|^2}{1 + x^2}\, dx .$$

This together with Schwarz's inequality imply that

$$|h(x)| \leq \frac{1 + e^a}{2\pi} \int\limits_{\mathbb{C}_+ \cap \{|\zeta - x| = 1\}} |h(\zeta)|\,|d\zeta| \leq$$

$$\leq \frac{1 + e^a}{2\pi} \left(\text{const} \int\limits_{\mathbb{R}} \frac{|h(x)|^2}{1 + x^2}\, dx \right)^{1/2} \left(\int\limits_{\mathbb{C}_+ \cap \{|\zeta - x| = 1\}} |\zeta + i|^2 |d\zeta| \right)^{1/2} \leq \text{const}\, |x + i| .$$

From the facts that h is real, $|B(\zeta)| < 1$ for $\operatorname{Im} \zeta > 0$ and the function $\log|\zeta + i|$ is a harmonic function in $\mathbb{C}_+$ representable as a Poisson integral it follows that

$$\log|F(\zeta)| \leq \log|h(\zeta)| = \frac{1}{\pi} \int\limits_{\mathbb{R}} \frac{\operatorname{Im} \zeta}{|\zeta - t|^2} \log|h(t)|\, dt \leq \text{const} + \log|\zeta + i|$$

for $\zeta \in \mathbb{C}_+$.

Therefore

$$|F(z)| \leq \text{const}\, |z + i| , \quad \operatorname{Im} z > 0. \tag{1}$$

Similarly we can prove that

$$|F(z)| \leq \text{const}\, |z - i| \exp\{a\,|\operatorname{Im} z|\} , \ \operatorname{Im} z < 0. \tag{2}$$

The inequalities (1) and (2) imply that F is an entire function of exponential type. Hence $F \in C$ because $|F| = |h|$ on $\mathbb{R}$. To prove that $F \in \mathcal{M}_a$ it remains to show that $G_F = [0, -ia]$. It follows from (1) and (2) that $G_F \subset [0, -ia]$. But

$$h_F\left(\frac{\pi}{2}\right)=\overline{\lim_{y\to+\infty}}\frac{\log|F(iy)|}{y}=\overline{\lim_{y\to+\infty}}\frac{\log|B(iy)|}{y}+\lim_{y\to+\infty}\frac{\log|h(iy)|}{y}=0,$$

$$h_F\left(-\frac{\pi}{2}\right)=\overline{\lim_{y\to+\infty}}\frac{\log|F(-iy)|}{y}=a+\lim_{y\to+\infty}\frac{\log|h(iy)|}{y}=a,$$

and therefore $G_F=[0,-ia]$.

The function F does not vanish on $\mathbb{C}_-$ and its zeros in $\mathbb{C}_+$ are in Λ . Let us show that F does not vanish on $\mathbb{R}$. Since the functions $|F|^2|\mathbb{R}$ and $|F^{-2}||\mathbb{R}$ satisfy the Helson-Szegö condition it follows from lemma 1.1 that

$$\int_{\mathbb{R}}\frac{1}{|F(x)|^2}\,\frac{dx}{1+x^2}<+\infty .$$

Hence F does not vanish on $\mathbb{R}$. ●

From theorem 1.2 it is easy to deduce theorem 7 stated in § 5, Part I.

THEOREM 7. Let $\Lambda=\{\lambda_n:n\in\mathbb{Z}\}\subset\mathbb{C}_\delta$, $\delta>0$, and $a>0$. The family of exponentials $\{e^{i\lambda_n x}\}_{n\in\mathbb{Z}}$ is an unconditional basis in $L^2(0,a)$ if and only if $\Lambda\in(C)$ and $F_\Lambda\in\mathcal{M}_a$.

PROOF. It is sufficient to check (see theorems 1 and 4 in part I) that $F_\Lambda\in\mathcal{M}_a$ if and only if the Toeplitz operator $T_{\bar\theta^a B}$ is invertible. By theorem 5 $T_{\bar\theta^a B}$ is invertible if and only if $\bar\theta^a B$ is a Helson-Szegö function. It remains to apply theorem 1.2. ●

One more application of theorem 1.2 permits us to p r o v e t h e n e c e s s i t y p a r t *) o f t h e o r e m 6 f o r m u l a t e d i n § 5 P a r t I. Recall the statement of this part of theorem 6. Let $\Lambda\subset\mathbb{C}_\delta$, $\delta>0$, and B be a Blaschke product with Simple zeros whose zero-set coincides with Λ and α_Λ be a continuous branch of the argument of $B\bar\theta^a$ defined by

$$\alpha_\Lambda(x)=2\int_0^x\sum_{\lambda\in\Lambda}\frac{\operatorname{Im}\lambda}{|\lambda-t|^2}\,dt-ax,\quad x\in\mathbb{R}.$$

---

*) Recall that the sufficiency of the same conditions for a family of exponentials to form an unconditional basis was already noted in §6, Part I immediately after the statement of theorem 6.

LEMMA 1.4. Let the family $\{e^{i\lambda_n x}\}_{n\in\mathbb{Z}}$ be an unconditional basis in $L^2(0,a)$, $\Lambda=\{\lambda_n, n\in\mathbb{Z}\}$. Then

$$dist(\alpha_\Lambda, \widetilde{L^\infty}+C) < \frac{\pi}{2}.$$

PROOF. If $\{e^{i\lambda_n x}\}_{n\in\mathbb{Z}}$ is an unconditional basis in $L^2(0,a)$ then the Toeplitz operator $T_{\bar\theta^a B}$ is invertible, and therefore $B\bar\theta^a$ is a Helson-Szegö function. By theorem 1.2 $Bh = c\,\theta^a\bar h$ where h is an outer part of an entire function of the class $\mathcal{M}_a$ in the upper half-plane. Since functions of the class $\mathcal{M}_a$ do not vanish on $\mathbb{R}$ the function $x\mapsto \log|h^2(x)|$ is infinitely differentiable. The Hilbert transform preserves the local smoothness and thus the function $\widetilde{\log|h^2|}$ is continuous on $\mathbb{R}$. It remains to use the fact that two continuous branches of the argument of a unimodular function differ by a constant function. ●

The generating function F_Λ is uniquely determined by its zero set (it was noticed in § 5 of part I). Moreover there exists a simple formula which expresses F_Λ in terms of Λ.

LEMMA ON ZEROS OF FUNCTIONS OF CARTWRIGHT CLASS (cf. [13], [27]). Let $F\in\mathcal{C}\cap\mathcal{E}_a$, $\Lambda=\{\lambda\in\mathbb{C}: F(\lambda)=0\}$ and let all zeros of F be simple. Then

1. $$\lim_{r\to+\infty}\frac{n_+(r)}{r}=\lim_{r\to-\infty}\frac{n_-(r)}{r}=\frac{a}{2\pi}, \tag{3}$$

where $n_+(r)\overset{def}{=}Card\{\lambda\in\Lambda: |\lambda|\le r, Re\,\lambda>0\}$, $n_-(r)\overset{def}{=}Card\{\lambda\in\Lambda: |\lambda|\le r, Re\,\lambda<0\}$;

2. There exists $$\lim_{r\to+\infty}\sum_{|\lambda|\le r,\lambda\in\Lambda}\frac{1}{\lambda}. \tag{4}$$

The proof of this lemma uses delicate methods of the theory of entire functions and we refer for the proof to the books [13], [27]. Note that the conditions (3) and (4) can be considered as simple necessary conditions on $\Lambda=\{\lambda_n: n\in\mathbb{Z}\}$ for $\{e^{i\lambda_n x}\}_{n\in\mathbb{Z}}$ to be an unconditional basis in $L^2(0,a)$. Let us suppose that Λ satisfies the conditions (3) and (4) of the lemma. Integrating by parts we obtain from (3) that $\sum_{\lambda\in\Lambda\setminus\{0\}}|\lambda|^{-2}<+\infty$. Therefore it follows from the K.Weierstrass factorization theorem (cf. [13], ch. 1, §4, lemma 3) that the infinite product $\prod_{\lambda\in\Lambda}(1-\frac{z}{\lambda})e^{z/\lambda}$ converges absolutely and uniformly on compact subsets of the complex plane. It follows from (4) that there exists a limit

$$\Pi = \text{v.p.}\prod_{\lambda\in\Lambda}\left(1-\frac{z}{\lambda}\right) \overset{def}{=\!=} \lim_{R\to+\infty}\prod_{\lambda\in\Lambda, |\lambda|\leq R}\left(1-\frac{z}{\lambda}\right).$$

It is known that $G_\Pi = [-i\frac{a}{2}, i\frac{a}{2}]$ if Λ is the zero set of a function of class $\mathcal{C}$. This implies the following formula for the generating function F_Λ

$$F_\Lambda = e^{i\frac{a}{2}z}\,\text{v.p.}\prod_{\lambda\in\Lambda}\left(1-\frac{z}{\lambda}\right).$$

Note that if $\Lambda\subset\mathbb{C}_\delta$, $\delta>0$, and Λ satisfies the Blaschke condition (B) then the infinite product in the formula for F_Λ converges if and only if there exists a limit

$$\lim_{R\to\infty}\sum_{\lambda\in\Lambda, |\lambda|\leq R}\frac{1}{\lambda}.$$

To prove this it is sufficient to use the fact that the condition (B) and the condition $\Lambda\subset\mathbb{C}_\delta$, $\delta>0$, imply the convergence of $\sum_{\lambda\in\Lambda}|\lambda|^{-2}$. This remark permits us to weaken the hypothesis of the if-part of theorem 7. The condition $F_\Lambda\in\mathcal{M}_a$ can be replaced by conditions (3), (4) and the following condition: the function

$$x\longmapsto\prod_{\lambda\in\Lambda}\left|1-\frac{x}{\lambda}\right|^2,\quad x\in\mathbb{R},$$

satisfies the Helson-Szegö condition on the real line. Together with the condition $\Lambda\in(C)$ this implies that $F_\Lambda\in\mathcal{M}_a$ where a is defined by (3).

To prove theorem 8 we need the following lemma.

LEMMA 1.5. Let $\{\lambda_n\}$ be a sequence of real numbers such that $\inf_{m\neq n}|\lambda_n-\lambda_m| \overset{def}{=\!=} 3\delta>0$ and let $y>0$. Then the function

$$x\mapsto\sum_{n\in\mathbb{Z}}\log\left(1-\frac{y^2}{(x-\lambda_n)^2+y^2}\right)$$

belongs to $BMO(\mathbb{R})$.

PROOF. Put $\Delta_n \overset{def}{=\!=} \{x\in\mathbb{R}: |x-\lambda_n|<\delta\}$, $n\in\mathbb{Z}$. By the hypothesis of the lemma $\text{dist}(\Delta_n,\Delta_m)\geq\delta$ if $n\neq m$. If $x\notin\Delta_n$ then $y^2((x-\lambda_n)^2+y^2)^{-1}\leq y^2(\delta^2+y^2)^{-1}$. There exists a number $C>0$ depending only on δy^{-1} such that $\log(1-t)\geq -Ct$ for $0\leq t\leq\frac{y^2}{\delta^2+y^2}$. Whence it follows that

$$0 \geqslant \sum_{n\in\mathbb{Z}, x\notin\Delta_n} \log\left(1-\frac{y^2}{(x-\lambda_n)^2+y^2}\right) \geqslant -C \sum_{n\in\mathbb{Z}, x\notin\Delta_n} \frac{y^2}{(x-\lambda_n)^2+y^2} .$$

It is clear that

$$\sum_{n\in\mathbb{Z}, x\notin\Delta_n} \frac{y^2}{(x-\lambda_n)^2+y^2} \leqslant \sum_{K=0}^{\infty} \sum_{\substack{n \\ 2^K\delta\leqslant|x-\lambda_n|\leqslant 2^{K+1}\delta}} \frac{y^2}{4^K\delta^2+y^2} \leqslant$$

$$\leqslant \sum_{K=0}^{\infty} \frac{1}{4^K\left(\frac{\delta}{y}\right)^2+1} \operatorname{Card}\{n\in\mathbb{Z} : 2^K\delta \leqslant |x-\lambda_n| \leqslant 2^{K+1}\delta\} \leqslant \sum_{K\geqslant 0} \frac{2^{K+1}}{4^K\left(\frac{\delta}{y}\right)^2+1} < +\infty.$$

These estimates imply that

$$-\sum_{n\in\mathbb{Z}} \log\left(1-\frac{y^2}{|x-\lambda_n|^2+y^2}\right) = u(x) + \sum_{n\in\mathbb{Z}} \log^+ \frac{\delta^2}{(x-\lambda_n)^2} ,$$

where $u \in L^\infty(\mathbb{R})$. The function $\log^+ \frac{\delta^2}{x^2}$ belongs to BMO and the distances between the supports Δ_n of its translates $\log^+ \frac{\delta^2}{(x-\lambda_n)^2}$ are at least δ. It follows that the sum $v(x) = \sum_{n\in\mathbb{Z}} \log^+ \frac{\delta^2}{(x-\lambda_n)^2}$ belongs to BMO. To prove this we use the description of BMO in terms of mean oscillations.

If $I \in \mathcal{J}$ and $|I| < \delta$ then $v|I = \log^+ \frac{\delta^2}{(x-\lambda_n)^2}\Big|I$ for some $n \in \mathbb{Z}$. If $|I| \geqslant \delta$ then

$$\frac{1}{|I|}\int_I |v(x)|\,dx \leqslant \frac{1}{|I|}\cdot\frac{|I|}{3\delta}\int_{-\delta}^{\delta} \log^+ \frac{\delta^2}{x^2}\,dx < +\infty. \quad \bullet$$

THEOREM 8. Let $\lambda_n \in \mathbb{R}$, $n \in \mathbb{Z}$. The family of exponentials $\{e^{i\lambda_n x}\}_{n\in\mathbb{Z}}$ is a Riesz basis in $L^2(0,a)$ if and only if

1. $\inf_{n\neq m} |\lambda_n - \lambda_m| > 0$;
2. $N_\Lambda - \frac{a}{2\pi} x \in \mathcal{P}_{1/4}$.

PROOF. The "only if" part. Let $\{e^{i\lambda_n x}\}_{n\in\mathbb{Z}}$ be a Riesz basis in $L^2(0,a)$. Then $\Lambda + iy \in (C)$ for any $y>0$ and so $\inf_{n\neq m} |\lambda_n - \lambda_m| > 0$. Since $\{e^{i\lambda_n x}\}_{n\in\mathbb{Z}}$ is a Riesz basis then there exists the generating function

$$F_\Lambda(z) = e^{i\frac{a}{2}z} \cdot \text{v.p.} \prod_{n\in\mathbb{Z}} \left(1-\frac{z}{\lambda_n}\right).$$

Let $F_{\Lambda+iy}$ be the generating function for the set $\Lambda+iy$. The functions F_Λ and $F_{\Lambda+iy}$ obviously satisfy

$$F_{\Lambda+iy}(z)\,F_\Lambda(-iy)=F_\Lambda(z-iy).$$

Our first purpose is to prove that the function $x\mapsto \log|F_\Lambda(x)|$ belongs to BMO. To prove this we consider the difference

$$\log|F_\Lambda(x)|^2-\log|F_{\Lambda+iy}(x)|^2=\log|F_\Lambda(-iy)|^2-\frac{a}{2}y+\sum_n \log\Big(1-\frac{y^2}{|x-\lambda_n|^2+y^2}\Big).$$

The sum on the right-hand side of the formula belongs to BMO by lemma 1.5. The function $x\mapsto\log|F_{\Lambda+iy}(x)|^2$ belongs to BMO because $|F_{\Lambda+iy}|^2|\mathbb{R}\in(HS)$ by theorem 7. Therefore $\log|F_\Lambda|^2|\mathbb{R}\in$ $\in\mathrm{BMO}$.

Let now c be a complex number such that $|c|=1$ and $F_\Lambda^2(i)c>0$. Then $f=F_\Lambda^2\cdot c$ is an outer function and

$$\log f(z)=\frac{1}{\pi i}\int_{\mathbb{R}}\frac{tz+1}{t-z}\,\frac{\log|F_\Lambda(t)|^2}{1+t^2}\,dt=iaz+\log c+2\sum_n\log\Big(1-\frac{z}{\lambda_n}\Big). \quad (5)$$

This formula enables us to compute the values of $\widetilde{\log|F_\Lambda|^2}$ on the real line. Note that

$$\lim_{y\to 0+}\log\Big(1-\frac{x+iy}{\lambda_n}\Big)=\log\Big|1-\frac{x}{\lambda_n}\Big|-\pi i\begin{cases}\chi_{[\lambda_n,+\infty)}(x) \text{ if } \lambda_n>0,\\ -\chi_{(-\infty,\lambda_n]}(x) \text{ if } \lambda_n<0.\end{cases}$$

It follows that

$$\log f(x)=\log|F_\Lambda(x)|^2+i(ax+\arg c-2\pi N_\Lambda(x)).$$

Thus

$$\widetilde{\log|F_\Lambda|^2}(x)=ax-2\pi N_\Lambda(x)+\arg c.$$

By theorem 7 $|F_\Lambda(x+iy)|^2\in(A_2)$ for any $y>0$. This implies that $N_\Lambda(x)-\frac{a}{2\pi}x\in\mathcal{P}_{1/4}$.

The "if" part. The most difficult step of the proof is to show that the generating function corresponding to Λ exists.

Suppose that the function $\alpha(x)=\frac{a}{2}x-\pi N_\Lambda(x)-c$ belongs to $\mathcal{P}_{1/4}$. Here c is a complex number such that the harmonic continuation of α to the half-plane $\mathbb{C}_+$ (we denote this continuation by the same letter α) vanishes at the point i.

It is obvious that $\alpha, \tilde{\alpha} \in BMO$. Using the fact that the Hilbert transform preserves the local smoothness it is easy to see that $\tilde{\alpha}$ is infinitely differentiable on $\mathbb{R} \setminus \Lambda$. In a neighbourhood of a point $\lambda \in \Lambda$ the following equality holds

$$-\tilde{\alpha}(x) = \log\left|1 - \frac{x}{\lambda}\right| + \beta_\lambda(x), \tag{6}$$

where β_λ is a differentiable function in a neighbourhood of λ.

Consider an outer function on $\mathbb{C}_+$

$$F = \exp(-\tilde{\alpha} + i\alpha).$$

It is easy to see that the function $f(z) = F(z)e^{-i\frac{a}{2}z}e^{ic}$ is real on $\mathbb{R}$ and differentiable (cf. (6)). By the symmetry principle f can be analytically continued into $\mathbb{C}_-$ and so f is an entire function.

Let us show that $F \in \mathcal{E}_a$. From the fact that $\alpha \in \mathcal{P}_{1/4}$ it follows that there exists a positive number d such that the restriction of $|F^2|$ to the line $\{\zeta \in \mathbb{C} : \operatorname{Im}\zeta = d\}$ satisfies the Helson-Szegö condition (HS). The equality $f(z) = \overline{f(\bar{z})}$ implies that $|F^2(\zeta)| = |F^2(\bar{\zeta})|e^{ad}$ if $\operatorname{Im}\zeta = -d$. Hence the restriction of $|F^2|$ to the line $\{\zeta : \operatorname{Im}\zeta = -d\}$ also satisfies the Helson-Szegö condition. It is also clear that $F|\mathbb{C}_+$, $F|\mathbb{C}_-$ belong to the Nevanlinna classes in $\mathbb{C}_+$ and $\mathbb{C}_-$. Therefore $F|\mathbb{C}_{-d}$ belongs to the Nevanlinna class in $\mathbb{C}_{-d}$. The inner part of F in $\mathbb{C}_{-d}$ has no singular factors because $F|\mathbb{C}_+$ is an outer function. Lemma 1.1 applied to $\mathbb{C}_{-d}$ implies that the function $(z + 2id)^{-1} \cdot F$ belongs to the Hardy class H^2 in $\mathbb{C}_{-d}$. Applying the method used in the proof of theorem 1.2 we obtain that $|F(z)| \leqslant \mathrm{const} \cdot |z + i|$ if $\operatorname{Im} z \geqslant 0$. Since

$$F(z) = \overline{F(\bar{z})}\, e^{iaz} e^{2ic}, \qquad \operatorname{Im} z < 0. \tag{7}$$

We obtain that $|F(z)| \leqslant \mathrm{const} \cdot |z - i| e^{a|\operatorname{Im} z|}$. These inequalities show that F is of exponential type. Moreover it is clear that $h_F(\pi/2) = 0$ (because $F|\mathbb{C}_+$ is outer) and that $h_F(-\pi/2) = a$ (cf. (7)). Thus $F \in \mathcal{E}_a$. Put $F^*(z) = F(z - id) \cdot F(-id)^{-1}$. It is easy to see that F^* is the generating function for $\Lambda + id$. Moreover, $|F^*|^2 | \mathbb{R}$ satisfies the Helson-Szegö condition. By theorem 7 we can conclude that $\{e^{i(\lambda_n + id)}\}_{n \in \mathbb{Z}}$ is a Riesz basis in $L^2(0, a)$. ●

REMARK. Let $\Lambda \subset \mathbb{R}$, $\inf\limits_{n \neq m} |\lambda_n - \lambda_m| > 0$ and

$\Lambda(x) - \frac{a}{2\pi} x \in \mathcal{P}_{1/4}$. Then the harmonic continuation $\mathcal{U}(x)$ of the function $\Lambda - \frac{a}{2\pi} x$ into the upper half-plane satisfies the following condition:

for any positive y there exist a real number c and $u, v \in L^\infty(\mathbb{R})$ such that

$$\mathcal{U}(x+iy) = c + \tilde{u}(x) + v(x)$$

and $\|v\|_\infty < 1/4$.

Indeed, if the above equality holds for some $y > 0$, $c \in \mathbb{R}$, $u, v \in L^\infty(\mathbb{R})$ then it follows from (2) that $|F_\Lambda(x+iy)|^2 \in (A_2)$ and so $|F_{\Lambda+iy}(x)|^2 \in (A_2)$. Since the translation $\Lambda + iy \to \Lambda + iy'$ induces an isomorphism in $L^2(0,a)$, $|F_{\Lambda+iy}(x)|^2 \in (A_2)$ for any $y > 0$. ●

If $\{e^{i\lambda x}\chi_{[0,a]} : \lambda \in \Lambda\}$ is an unconditional basis in $L^2(0,a)$ $(\Lambda \subset \mathbb{C}_\delta,\ \delta > 0)$ then, as we saw in § 2 of part II, the angle between the subspaces K_{θ^a} and K_B of H^2_+ is non--zero and they span H^2_+. Consider the subspaces $\theta_a H^2_- = H^2_- \oplus K_{\theta_a}$ and BH^2_+. Now it is possible to obtain an explicit formula for the projection $\mathcal{P}_{\theta H^2_- \| BH^2_+}$ onto θH^2_- along BH^2_+ using the generating function $F \overset{def}{=} F_\Lambda$.

THEOREM 1.6. For $\{e^{i\lambda_n x}\}_{n\in\mathbb{Z}}$ to be an unconditional basis in $L^2(0,a)$ it is necessary and sufficient that $\Lambda \in (C)$, $\theta H^2_- + BH^2_+$ is dense in $L^2(\mathbb{R})$ and the projection $\mathcal{P}_{\theta H^2_- \| BH^2_+}$ is bounded. If $F = F_\Lambda$ and $\mathcal{M}_g$ is the multiplication by g operator on $L^2(\mathbb{R})$ then

$$\mathcal{P}_{\theta H^2_- \| BH^2_+} = \mathcal{M}_F P_- \mathcal{M}_{1/F}.$$

PROOF. The first part of the theorem easily follows from corollary 2.2 of part II. It remains to prove the formula for the projection. It is easy to see that the operator $\mathcal{M}_F P_- \mathcal{M}_{1/F}$ is bounded in $L^2(\mathbb{R})$ if and only if P_- is bounded in the weighted space $L^2(|F|^2 dx)$ and this is equivalent (by the Hunt-Muckenhoupt-Wheeden theorem) to the fact that $|F|^2 \in (A_2)$. We check the formula on a dense subset of $L^2(\mathbb{R})$. Since the function $(z+i)^{-1}$ is outer $H^2_+ = \operatorname{span}(e^{iaz}(z+i)^{-1} : a \geq 0)$ by P.Lax's theorem. Denote by $\mathcal{X}$ the linear span of functions $e^{iaz}(z+i)^{-1}$, $a \geq 0$. It is clear that $|f(x)| \leq \frac{C_f}{|x+i|}$ where $C_f > 0$ if $f \in \mathcal{X}$. Since $|F|^2 \in (A_2)$

$$\int_{\mathbb{R}} \frac{|F(x)|^2}{1+x^2}\,dx + \int_{\mathbb{R}} \frac{|F(x)|^{-2}}{1+x^2}\,dx < +\infty.$$

At last by theorem 1.1.

$$F = Bh = c\bar{h}\,\theta_a$$

where h is an outer function. Let $f = Bg$ where $g \in \mathcal{X}$. We have

$$\mathcal{M}_F \mathbb{P}_- \mathcal{M}_{1/F} f = \mathcal{M}_F \mathbb{P}_- \frac{g}{h} = 0 \qquad \text{because}$$

obviously, $gh^{-1} \in H^2_+$.

If $f = \theta_a \bar{g}$ where $g \in \mathcal{X}$ then

$$\mathcal{M}_F \mathbb{P}_- \mathcal{M}_{1/F} f = \mathcal{M}_F \mathbb{P}_- \frac{1}{c\bar{h}\,\theta_a}\theta_a \bar{g} = F\frac{\bar{g}}{c\bar{h}} = f. \qquad \bullet$$

2. Theorems on perturbations of unconditional bases.

We begin this Section with the deducing the theorems of S.A.Avdonin and V.È. Kacnelson (for the statements see §7 PartI). The following lemma reduces the general case to the examination of bases of exponentials with only real frequencies.

LEMMA 2.1. Let $\Lambda = \{\lambda_n : n \in \mathbb{Z}\} \subset \mathbb{R}$ and let $(\delta_n)_{n\in\mathbb{Z}}$ be an arbitrary bounded sequence of real numbers. Further, let us assume that the set Λ is separated and let $\Lambda^* \overset{def}{=\!=} \{\lambda_n^* : n \in \mathbb{Z}\}$, $\lambda_n^* \overset{def}{=\!=} \lambda_n + i\delta_n$. Then the family of exponentials $(e^{i\lambda_n t})_{n\in\mathbb{Z}}$ forms a Riesz basis in the space $L^2(0,a)$ if and only if the family $(e^{i\lambda_n^* t})_{n\in\mathbb{Z}}$ does.

THE PROOF can be easily obtained from theorem 7. Let $y > 10 \sup_n |\delta_n|$. We shall examine the following ratio

$$|F_{\Lambda^*+iy}(x)|^2 \cdot |F_{\Lambda+iy}(x)|^{-2} = \prod_{n\in\mathbb{Z}} \frac{|\lambda_n + iy|^2}{|\lambda_n^* + iy|^2} \prod_{n\in\mathbb{Z}} \left| \frac{\lambda_n + i\delta_n + iy - x}{\lambda_n + iy - x} \right|^2 .$$

It is clear that

$$\left| \frac{\lambda_n + i\delta_n + iy - x}{\lambda_n + iy - x} \right|^2 = \frac{(\lambda_n - x)^2 + (y+\delta_n)^2}{(\lambda_n - x)^2 + y^2} = 1 + \frac{\delta_n(2y+\delta_n)}{(\lambda_n - x)^2 + y^2} .$$

Since $y > 10 \sup_n |\delta_n|$, we have

$$\frac{\delta_n(2y+\delta_n)}{(\lambda_n - x)^2 + y^2} \leqslant \frac{2\delta_n y + \delta_n^2}{y^2} < \frac{1}{2} .$$

Further, let λ_m be the point of Λ nearest to the fixed point x , $x \in \mathbb{R}$, and let $d = \inf_{k \neq n} |\lambda_k - \lambda_n|$. Then

$$\sum_{n\in\mathbb{Z}} \frac{\delta_n(2y+\delta_n)}{(\lambda_n-x)^2+y^2} \leqslant \frac{1}{2} + \sum_{n\neq m} \frac{\delta_n(2y+\delta_n)}{(\lambda_n-x)^2+y^2} \leqslant$$

$$\leqslant \frac{1}{2} + y^2 \sum_{n\neq m} \frac{1}{(\lambda_n-x)^2} \leqslant \frac{1}{2} + y^2 \sum_{k=0}^{\infty} \frac{4}{4^k d^2} \operatorname{Card}\{n : 2^{k-1}d < |\lambda_n - x| \leqslant 2^k d\} \leqslant$$

$$\leqslant \frac{1}{2} + 4y^2 \sum_{k=0}^{\infty} \frac{1}{2^k d^2} = \frac{1}{2} + \frac{8y^2}{d^2} .$$

This yields

$$\log|F_{\Lambda^*+iy}(x)|^2 - \log|F_{\Lambda+iy}(x)|^2 \in L^\infty(\mathbb{R}). \quad \bullet$$

Let $(\delta_n)_{n\in\mathbb{Z}}$ be a bounded sequence of real numbers and $\Lambda = \{\lambda_n : n \in \mathbb{Z}\} \subset \mathbb{R}$. We denote

$$\Delta_x(R) \stackrel{def}{=} \sum_{x-R\leqslant\lambda_k\leqslant x+R} \delta_k ,$$

and let $\lambda_n^* = \lambda_n + \delta_n$, $\Lambda^* = \{\lambda_n^* : n \in \mathbb{Z}\}$ is a "real perturbation" of the set Λ .

Lemma 2.1 allows to phrase the Avdonin's theorem as follows.

THEOREM 2.2. Let $\Lambda = \{\lambda_n : n \in \mathbb{Z}\}$ be a separated subset of the real line. Suppose that Λ is a zero set of a STF with the width of the indicator diagram equal to 2π . Let us assume that the set Λ^* is separated and

$$\varlimsup_{R\to+\infty} \sup_{x\in\mathbb{R}} \frac{|\Delta_x(R)|}{2R} < \frac{1}{4} .$$

Then the family $(e^{i\lambda_n^* t})_{n\in\mathbb{Z}}$ forms a Riesz basis in the space $L^2(0, 2\pi)$.

THE PROOF of the theorem in its essential features follows that of the Kadec $1/4$ -theorem expounded in Section 5, Part I.

Let F_Λ be a generating function for the set Λ . The following formula is true (see §1)

$$\log|F_\Lambda(x+iy)|^2 = -2d_{\Lambda+iy}(x) + c,$$

where $c \in \mathbb{R}$, $y > 0$. According to the definition of the STF the function $x \mapsto \log|F_\Lambda(x+iy)|^2$ $(x \in \mathbb{R})$ is bounded and therefore $d_{\Lambda+iy} \in \widetilde{L^\infty} + \mathbb{C}$. In order to use Theorem 6, let us compute the difference

$$\alpha_{\Lambda+iy}(x)-\alpha_{\Lambda^*+iy}(x)=2\sum_{n\in\mathbb{Z}}\int_0^x\left\{\frac{y}{(t-\lambda_n)^2+y^2}-\frac{y}{(t-\delta_n-\lambda_n)^2+y^2}\right\}dt=$$

$$=2\sum_{n\in\mathbb{Z}}\int_{x-\delta_n}^{x}\frac{y}{(t-\lambda_n)^2+y^2}\,dt+\mathrm{const}.$$

By the mean value theorem we have

$$\int_{x-\delta_n}^{x}\frac{y}{(t-\lambda_n)^2+y^2}\,dt=\frac{y}{(x-\lambda_n)^2+y^2}\cdot\delta_n\left(1+O\left(\tfrac{1}{y}\right)\right),\quad y\to+\infty,$$

uniformly with respect to x , $x\in\mathbb{R}$. It remains only to verify that

$$\varlimsup_{y\to+\infty}\ \sup_{x\in\mathbb{R}\setminus\Lambda}\left|\sum_{n\in\mathbb{Z}}\frac{y}{(x-\lambda_n)^2+y^2}\,\delta_n\right|<\frac{\pi}{4}.$$

If $x\in\mathbb{R}\setminus\Lambda$, then

$$\sum_{n\in\mathbb{Z}}\frac{y}{(x-\lambda_n)^2+y^2}\,\delta_n=\int_0^{+\infty}\frac{y}{t^2+y^2}\,d\Delta_x(t)=$$

$$=\int_0^{+\infty}\frac{\Delta_x(t)}{t}\cdot\frac{2yt^2}{(t^2+y^2)^2}\,dt=\int_0^{+\infty}\frac{\Delta_x(yt)}{2yt}\cdot\frac{4t^2}{(1+t^2)^2}\,dt.$$

Let R_0 be a such positive number that $\sup\limits_{x\in\mathbb{R}}\dfrac{|\Delta_x(R)|}{2R}<\dfrac{1}{4}$ if $R\geqslant R_0$. Then we have

$$\sup_{x\in\mathbb{R}}\left|\int_{R_0y^{-1}}^{\infty}\frac{\Delta_x(yt)}{2yt}\,\frac{4t^2}{(1+t^2)^2}\,dt\right|<\int_0^{\infty}\frac{t^2}{(1+t^2)^2}\,dt=\frac{\pi}{4},$$

and

$$\sup_{x\in\mathbb{R}}\left|\int_0^{R_0y^{-1}}\frac{\Delta_x(yt)}{2yt}\,\frac{4t^2}{(1+t^2)^2}\,dt\right|\leqslant\sup_{x\in\mathbb{R}}\sum_{x-R_0\leqslant\lambda_k\leqslant x+R_0}|\delta_k|\cdot\frac{1}{y}\int_0^{\infty}\frac{2t}{(1+t^2)^2}\,dt\leqslant$$

$$\leqslant\frac{1}{y}\sup_{k\in\mathbb{Z}}|\delta_k|\cdot\mathrm{Card}\{k: x-R_0\leqslant\lambda_k\leqslant x+R_0\}.$$

This expression obviously tends to zero as $y\to\infty$. ●

PROOF OF THE V.È. KACNELSON'S THEOREM. We shall deduce this theorem from Theorem 2.2. Lemma 2.1 permits to consider only real frequencies in this case also. Let Λ be a subset of the real line. Suppose that Λ is the zero set of a STF with the width of the indicator diagram equal to 2π. Let $\rho_n = \inf\{|\lambda_n - \lambda_m| : m \in \mathbb{Z}\setminus\{n\}\}$. In the Kacnelson's theorem "perturbations" δ_n were supposed to satisfy the condition

$$|\delta_n| \leqslant d\rho_n \text{ , where } 0 \leqslant d < \frac{1}{4}.$$

Lemma 2.3 (see below) shows that for zeros of a STF the sequence $(\rho_n)_{n\in\mathbb{Z}}$ must be bounded, say by a constant ρ, $\rho > 0$. But then the inequality

$$\frac{|\Delta_x(R)|}{2R} < \frac{1}{4}$$

obviously is valid, if $R \gg \rho$. ●

LEMMA 2.3. Let $\Lambda = \{\lambda_n : n \in \mathbb{Z}\}$ be a subset of the real line coinciding with the zero-set of a STF, and $\rho_n = \inf\{|\lambda_n - \lambda_m| : m \neq n\}$. Then the sequence $(\rho_n)_{n\in\mathbb{Z}}$ is bounded.

PROOF. Put $S = \prod_{n\in\mathbb{Z}} (1 - \frac{z}{\lambda_n})$ and suppose the width of the indicator diagram of S is equal to a. Then

$$\lim_{y\to\infty} \frac{S'(x+iy)}{S(x+iy)} = -i\frac{a}{2}, \tag{8}$$

uniformly with respect to x, $x \in \mathbb{R}$. A simple proof of this fact can be found in an interesting paper of B.Ja.Levin and I.V.Ostrovskii [15], containing many other useful facts conserning the structure of zero-sets of STF's (see the remark to lemma 2 on the page 89 in [15]). Computing the imaginary part of the equality

$$\frac{S'(z)}{S(z)} = (\log S(z))' = \sum_{n\in\mathbb{Z}} \frac{1}{z - \lambda_n},$$

we obtain from the formula (8):

$$\lim_{y\to\infty} \sup_{x\in\mathbb{R}} \left| \sum_{n\in\mathbb{Z}} \frac{y}{(x-\lambda_n)^2 + y^2} - \frac{a}{2} \right| = 0. \tag{9}$$

If the sequence $(\rho_n)_{n\in\mathbb{Z}}$ is unbounded, then for any R, $R > 0$, there exists a number n, $n \in \mathbb{Z}$, such that $\rho_n > 2R$. In this case the interval $(\lambda_n - R, \lambda_n + R)$ contains only one point of the set Λ. Let $x \in (\lambda_n - R, \lambda_n + R)$, $y = \sqrt{R}$. Then

$$\sum_{k\neq n} \frac{y}{(x-\lambda_k)^2 + y^2} \leqslant y \sum_{k\neq n} \frac{1}{(x-\lambda_k)^2} \leqslant$$

$$\leqslant y\sum_{m=0}^{\infty}\frac{\operatorname{Card}\{k: 2^m R\leqslant|x-\lambda_k|\leqslant 2^{m+1}R\}}{4^m R^2}\leqslant \operatorname{const}\frac{y}{R}=\operatorname{const}\frac{1}{\sqrt{R}}.$$

Since $\dfrac{y}{(x-\lambda_n)^2+y^2}\leqslant\dfrac{1}{y}=\dfrac{1}{\sqrt{R}}$, we have

$$\sup_{x\in(\lambda_n-R,\lambda_n+R)}\sum_{m\in\mathbb{Z}}\frac{y}{(x-\lambda_m)^2+y^2}\leqslant\frac{\operatorname{const}}{\sqrt{R}}$$

for $y=\sqrt{R}$. But this contradicts (9) if R is large enough.●

We consider now a "perturbation theorem" for unconditional bases of exponents in a more general setting dropping the assumption $\sup_{\lambda\in\Lambda}\operatorname{Im}\lambda<+\infty$.

Let us introduce some notation. Suppose that $\Lambda=\{\lambda_n: n\in\mathbb{Z}\}\subset\mathbb{C}_\delta$, $\delta>0$, and the exponentials $(e^{i\lambda_n t})_{n\in\mathbb{Z}}$ form an unconditional basis in the space $L^2(0,a)$. For every integer n consider the disc

$$D(\lambda_n,\delta_n)=\{\zeta\in\mathbb{C}: |\zeta-\lambda_n|\leqslant\delta_n\}.$$

We shall be interested in the restrictions to be imposed on $(\delta_n)_{n\in\mathbb{Z}}$ ensuring that any family $(e^{i\lambda_n^* t})_{n\in\mathbb{Z}}$ with $\lambda_n^*\in D(\lambda_n,\delta_n)$ forms an unconditional basis in $L^2(0,a)$. Denote by the symbol $P_z(t)$ the Poisson kernel $\frac{1}{\pi}\frac{\operatorname{Im} z}{|z-t|^2}$, $\operatorname{Im} z>0$.

THEOREM 2.4. Let $(e^{i\lambda_n x})_{n\in\mathbb{Z}}$ be an unconditional basis in the space $L^2(0,a)$ and

$$\inf_{y>0}\{\|2\pi\sum_{n\in\mathbb{Z}}\delta_n P_{\lambda_n+iy}\|_\infty+\operatorname{dist}(\alpha_{\Lambda+iy},\widetilde{L^\infty}+C)\}<\frac{\pi}{2}.$$

Suppose also that $\lambda_n^*\in D(\lambda_n,\delta_n)$, $n\in\mathbb{Z}$, and $\{\lambda_n^*: n\in\mathbb{Z}\}\in(C)$. Then the family $(e^{i\lambda_n^* x})_{n\in\mathbb{Z}}$ forms an unconditional basis in the space $L^2(0,a)$.

THE PROOF is based on Theorem 6. Let $\lambda\in\mathbb{C}_\delta$ (i.e. $D(\lambda,\delta)\subset\mathbb{C}_+$), and estimate the difference

$$\int_0^x\frac{\operatorname{Im}\lambda^*}{|\lambda^*-t|^2}dt-\int_0^x\frac{\operatorname{Im}\lambda}{|\lambda-t|^2}dt.$$

To do this we note the formula

$$\int_0^x\frac{1}{\bar\lambda-t}dt=\log\left(\frac{1}{1-\frac{x}{\bar\lambda}}\right)$$

where the right hand side is meant as principal value of the logarithm. Taking imaginary parts we obtain

$$\int_0^x \frac{\operatorname{Im}\lambda}{|\lambda-t|^2}\,dt = \log\frac{1}{|1-\frac{x}{\lambda}|} + \arg\left(\frac{1}{1-\frac{t}{\lambda}}\right).$$

Let us consider two cases. At first let $\operatorname{Re}\lambda^* = \operatorname{Re}\lambda$, $\lambda^* = \lambda + i\eta$, $|\eta| \leq \delta$. Then

$$\int_0^x \frac{\operatorname{Im}\lambda^*}{|\lambda^*-t|^2}dt - \int_0^x \frac{\operatorname{Im}\lambda}{|\lambda-t|^2}dt = -\log\left|\frac{x-\lambda^*}{x-\lambda}\right| + \arg\left(1-\frac{i}{\lambda^*}\right)^{-1} - \arg\left(1-\frac{i}{\lambda}\right)^{-1}.$$

It is not difficult to see that

$$\left|\frac{x-\lambda^*}{x-\lambda}\right|^2 = 1 + \frac{(\operatorname{Im}\lambda^*)^2-(\operatorname{Im}\lambda)^2}{|x-\lambda|^2} = 1 + \frac{\eta(2\operatorname{Im}\lambda+\eta)}{|x-\lambda|^2}.$$

Since $|\eta| = \delta < \operatorname{Im}\lambda$, we have $|\eta||2\operatorname{Im}\lambda+\eta| < 3\delta\operatorname{Im}\lambda$. Hence the convergence of the series $\sum\limits_{n\in\mathbb{Z}} \frac{\delta_n \operatorname{Im}\lambda_n}{|x-\lambda_n|^2}$ implies $\alpha_{\Lambda^*} - \alpha_\Lambda \in \widetilde{L}^\infty + \mathbb{C}$.

Now let $\operatorname{Im}\lambda^* = \operatorname{Im}\lambda$, i.e. $\lambda^* = \lambda + \eta$, $\eta \in (-\delta, \delta)$. Then

$$\int_0^x \frac{\operatorname{Im}\lambda}{|\lambda-t|^2}dt - \int_0^x \frac{\operatorname{Im}\lambda^*}{|\lambda^*-t|^2}dt = \int_0^x \frac{\operatorname{Im}\lambda}{|\lambda-t|^2}dt - \int_{-\eta}^{x-\eta} \frac{\operatorname{Im}\lambda}{|\lambda-t|^2}dt =$$

$$= \int_{x-\eta}^{x} \frac{\operatorname{Im}\lambda}{|\lambda-t|^2}dt - \int_{-\eta}^{0} \frac{\operatorname{Im}\lambda}{|\lambda-t|^2}dt.$$

Clearly,

$$\int_{x-\eta}^{x} \frac{\operatorname{Im}\lambda}{|\lambda-t|^2}dt = \frac{\eta\cdot\operatorname{Im}\lambda}{|\lambda-x|^2}\left(1+O\left(\frac{1}{\operatorname{Im}\lambda}\right)\right)$$

uniformly with respect to x, $x \in \mathbb{R}$.

Let now arbitrary perturbations λ_n^* be given. It is easy to see that any of them can be obtained in two steps: at first we shift the point λ_n along the real axis up to some point λ_n' and then along the imaginary axis up to the point λ_n^*. Taking a number y large enough, we have

$$\operatorname{dist}_{L^\infty}(\alpha_{\Lambda'+iy} - \alpha_{\Lambda+iy}, \mathbb{C}) \leq \sup_{x\in\mathbb{R}} 2\left(1+O\left(\tfrac{1}{y}\right)\right)\sum_n \delta_n P_{\lambda_n+iy}(x).$$

For the shifts along the imaginary axis the inclusion $\alpha_{\Lambda^*+iy} - \alpha_{\Lambda'+iy} \in \widetilde{L}^\infty + \mathbb{C}$ is valid. It remains to refer to Theorem 6. ●

COROLLARY 2.5. Let $\Lambda \subset \mathbb{C}_\delta$, $\delta > 0$, and suppose the family of the exponentials $(e^{i\lambda t})_{\lambda\in\Lambda}$ forms an unconditional basis in the space $L^2(0,a)$. Then there exists a number ε, $\varepsilon>0$, such that any choice of a single point λ_n^* from every disc $\Delta_n(\lambda_n,\varepsilon)$ gives rise to an unconditional basis in $L^2(0,a)$.

PROOF. Obvious. ●

Note that Corollary 2.5 is a generalization of the Duffin and Schaeffer theorem [35], cited in Section 7, Part I.

3. The set of frequencies does not lie in a strip of finite width. Complementation up to an unconditional basis.

Are there unconditional bases in the space $L^2(0,a)$ consisting of exponentials $(e^{i\lambda x})_{\lambda\in\Lambda}$, if $\sup_{\lambda\in\Lambda} \operatorname{Im}\lambda=+\infty$? The affirmative answer to this question was obtained by S.A.Vinogradov. His reasoning was improved later on by V.I.Vasjunin. One more question which naturally arises is as follows: it is possible to complement any unconditional basis of exponentials $(e^{i\lambda x})_{\lambda\in\Lambda}$ (in their linear span) up to an unconditional basis $(e^{i\lambda x})_{\lambda\in\Lambda'}$, $\Lambda'\supset\Lambda$, in the whole space $L^2(0,a)$? We do not know now (1980), whether this is true, but we shall find a sufficient condition (V.I.Vasjunin), ensuring above-mentioned possibility to "enlarge" the basis. In particular, any family $(e^{i\lambda x})_{\lambda\in\Lambda}$ under conditions $\Lambda\in(C)$, $\Lambda\subset\mathbb{C}_\delta$, $\delta>0$ and $\lim_{\lambda\in\Lambda}\operatorname{Im}\lambda=+\infty$ can be complemented up to an unconditional basis of exponentials in the whole space $L^2(0,a)$.

Before we shall formulate and prove the corresponding theorems, let us descuss some heuristic considerations. For an affirmative answer to the first formulated question it is obviously, necessary and sufficient the existence of an *interpolating Blaschke product* B (i.e. such that the set of its zeros is a Carleson set) and an outer function F such that $\|\Theta_a\bar{B}-cF\|<1$, $c\in\mathbb{C}$, $|c|=1$. But then the set $\{f\in H^\infty : \|\Theta_a\bar{B}-f\|_\infty<1\}$ consists of functions of the form cf_e, where $c\in\mathbb{C}$, $|c|=1$ and f_e is outer (see Remark 1 after Theorem 4 $\mathbb{D}$ from Section 2, Part II). Consider functions F, $F\in H^\infty$, such that the module of the difference $\Theta_a\bar{B}-F$ is a constant $\alpha>0$ on $\mathbb{R}$. It is well-known that such functions F exist if

$\alpha > dist(\Theta_a \bar{B}, H^\infty)$ (see [18] p.262 or [1]). If $|\Theta_a \bar{B} - F| = \alpha$ on $\mathbb{R}$, then $\Theta_a - BF = \alpha B^*$, where B^* is a Blaschke product. The S.A.Vinogradov's idea is to inverse this reasoning. Let us take a suitable Blaschke product B^* whose zeros form a Carleson set, and let $0<\alpha<1$. Then

$$\Theta_a - \alpha B^* = BF,$$

where F is an outer function. In this case the Toeplitz operator $T_{\Theta_a \bar{B}}$ is invertible, of course, and Theorems 2 and 3 may be used; but the main difficulty is to show that the Blaschke product B is interpolating if the product B^* is. Zeros of B can be controlled by means of Rouché's theorem. Therefore if the imaginary parts of zeros of the product B^* are unbounded, then the zeros of the product B are unbounded too. Let us turn now to the exact formulations.

Let B^* be a Blaschke product with zeros a_n, $n=1,2,\dots$, and let $b_n = b_{a_n} \overset{def}{=} \frac{z-a_n}{z-\bar{a}_n}$, $B_n^* \overset{def}{=} B^* b_n^{-1}$. For a given pair of numbers α, $\delta \in (0,1)$, consider a set $\mathcal{B}(\alpha,\delta)$ of Blaschke products B^* satisfying the following conditions:

$$\inf_{\operatorname{Im} z>0} \{|b_n(z)| + |B_n^*(z)|\} \geqslant \delta; \tag{IO}$$

$$\inf_n \operatorname{Im} a_n \geqslant \frac{24}{\alpha\delta^2}. \tag{II}$$

Let the symbol D_n denote the disc $\{\zeta\in\mathbb{C} : |b_n(\zeta)| \leqslant \frac{\delta}{2}\}$. It is clear that the set $\mathcal{B}(\alpha,\delta)$ consists of interpolating Blaschke products,whose zeros lie high enough above the real line. The less is the constant δ, the higher have zeros to lie.

THEOREM 3.1. (V.I.Vasjunin, S.A.Vinogradov). Let $B^*\in\mathcal{B}(\alpha,\delta)$, and $\log\frac{1}{\alpha} > \frac{2}{\delta^2}$. Then the function $\Theta - \alpha B^*$ has exactly one zero in each disc D_n and admits the factorization $\Theta - \alpha B^* = BF$, where the function F is outer and B is an interpolating Blaschke product. In particular the family $\{e^{i\lambda x} : B(\lambda)=0\}$ forms an unconditional basis in the space $L^2(0,1)$.

PROOF. Obviously $|B_n^*(\zeta)| \geqslant \frac{\delta}{2}$, if $\zeta\in D_n$. Hence $|B^*(\zeta)| = |b_n(\zeta)||B_n^*(\zeta)| \geqslant \frac{\delta^2}{4}$ on the boundary of the disc D_n. Let $G = \mathbb{C}_+ \setminus \bigcup_{n=1}^{\infty} D_n$. Then $|B^*(\zeta)| \geqslant \frac{\delta^2}{4}$ in G by the minimum principle.

The lower point of the disc $\{\zeta\in\mathbb{C} : |(\zeta-a)/(\zeta-\bar{a})| \leqslant \delta/2\}$ lies at the distance $\operatorname{Im} a \, \frac{1-\delta/2}{1+\delta/2} \geqslant \frac{\operatorname{Im} a}{3}$ above the real line, therefore $D_n \subset \{\zeta : \operatorname{Im}\zeta \geqslant s\}$, $s \overset{def}{=} 8/\alpha\delta^2$. Let $\zeta \in \partial D_n$. Then

$$|\Theta(\zeta)| = e^{-\operatorname{Im}\zeta} \leqslant e^{-s} < \alpha\frac{\delta^2}{8} < \alpha|B^*(\zeta)|$$

(because $e^{-s} < \frac{1}{s}$ when $s > 0$), hence by Rouché's theorem the function $\theta - \alpha B^*$ has exactly one zero in the each disc D_n, $n = 1, 2, \ldots$. This estimate shows also that the Blaschke product B has no other zeros in the half-plane $\{\zeta : \operatorname{Im}\zeta \geqslant s\}$.

Let us check now that the product B has no zeros in the strip $\{\zeta : 0 < \operatorname{Im}\zeta < \log 1/\alpha\}$. Indeed

$$|B(\zeta)F(\zeta)| = |\Theta(\zeta) - \alpha B^*(\zeta)| > e^{-\operatorname{Im}\zeta} - \alpha > 0$$

if $\operatorname{Im}\zeta < \log\frac{1}{\alpha}$.

So non-controlled zeros of B can lie only in the strip $\{\zeta : \log 1/\alpha \leqslant \operatorname{Im}\zeta < s\}$. Note that if $\lim_n \operatorname{Im} a_n = +\infty$, this strip d o e s contain infinitely many zeros (see Theorem 2.4, Part II, Section 2).

Let us suppose now that $B(\lambda) = 0$ and $\log 1/\alpha \leqslant \operatorname{Im}\lambda < s$. From the system of equations

$$e^{i\lambda} - \alpha B^*(\lambda) = 0$$
$$ie^{i\lambda} - \alpha B^{*\prime}(\lambda) = B'(\lambda)F(\lambda),$$

we have

$$|B'(\lambda)| = \frac{1}{|F(\lambda)|}\left|ie^{i\lambda} - \alpha B^{*\prime}(\lambda)\right| = \frac{\alpha}{|F(\lambda)|}\left|iB^*(\lambda) - B^{*\prime}(\lambda)\right|.$$

Let $B_\lambda = B \cdot b_\lambda^{-1}$, then $2\operatorname{Im}\lambda|B'(\lambda)| = |B_\lambda(\lambda)|$. It is useful to remember a trivial estimation

$$|f'(\lambda)| \leqslant \frac{\|f\|_\infty}{2\operatorname{Im}\lambda}.$$

Summarizing this information we obtain

$$|B_\lambda(\lambda)| = \frac{\alpha}{|F(\lambda)|}\cdot 2\operatorname{Im}\lambda \cdot |iB^*(\lambda) - B^{*\prime}(\lambda)| \geqslant$$
$$\geqslant \frac{\alpha}{1+\alpha}\left[2\operatorname{Im}\lambda|B^*(\lambda)| - 1\right] \geqslant \frac{\alpha}{1+\alpha}\left(\frac{\delta^2}{2}\log 1/\alpha - 1\right) > 0$$

because $\log 1/\alpha > 2/\delta^2$. Therefore the inequality taking part in the Carleson condition holds at every zero of B contained in the strip $\{\zeta : \log 1/\alpha \leq \operatorname{Im}\zeta < s\}$. Since the remaining zeros are in the discs D_n and B is an interpolating Blaschke product, the product B^* also is interpolating. ●

Now we shall show that refining the reasonings from the proof of the preceding theorem we can obtain that the generating function F_Λ, $\Lambda = \{\lambda : B(\lambda) = 0\}$ will be a GSTF (S.A.Vinogradov). Note that it is not difficult of course to give examples of GSTF with the zero-set contained in no strip of finite width. However, it is much more difficult to combine this property with the carlesonity. But at first we give an auxiliary definition.

Let $\mathcal{U}_\infty$ be the set of all unimodular functions φ on $\mathbb{R}$ representable in the form

$$\varphi = c\frac{\bar{h}}{h},$$

where $c \in \mathbb{T}$, h is an invertible element of the algebra H^∞ $(h \in (H^\infty)^{-1})$. It is clear that $\mathcal{U}_\infty$ is a group with respect to the pointwise multiplication of functions. It is easy to see that the mapping $(c, h) \mapsto c\bar{h}h^{-1}$ is an isomorphism of the group $\mathbb{T} \times (H^\infty)^{-1}$ onto $\mathcal{U}_\infty$.

LEMMA 3.2. Let $\Lambda \subset \mathbb{C}_\delta$, $\delta > 0$, and B be a Blaschke product with the zero set Λ. Then the generating function F_Λ is a GSTF with the width of the indicator diagram equal to a iff the function $\bar{B}\Theta_a$ belongs to $\mathcal{U}_\infty$.

THE PROOF of the lemma is provided by Theorem 1.2 and the definition of a GSTF. ●

THEOREM 3.3 (S.A.Vinogradov). There exists a set Λ, $\Lambda \subset \mathbb{C}_+$, such that $\Lambda \in (C)$, $\sup_{\lambda\in\Lambda} \operatorname{Im}\lambda = +\infty$ and F_Λ is a generalized sine-type function.

PROOF. Let $\alpha \in (0,1)$ and let B^* be an auxiliary Blaschke product, whose choice will be specified later. We find the required Blaschke product from the equation

$$\theta - \alpha B^* = Bf_e,$$

where f_e is an outer function and $\theta = e^{iz}$. Note that $f_e \in (H^\infty)^{-1}$ because $1-\alpha \leq |f_e| \leq 1+\alpha$ on $\mathbb{R}$. Since $|B^* - \alpha\theta| = |\theta - \alpha B^*|$ on $\mathbb{R}$, there exists a Blaschke product C such that

$$B^* - \alpha\theta = Cf_e.$$

The first equality yields

$$B\bar{\Theta}=\frac{1-\alpha B^*\bar{\Theta}}{f_e}$$

and the second one provides

$$1-\alpha\bar{B}^*\Theta=\bar{B}^*Cf_e.$$

Hence

$$B\bar{\Theta}=\frac{B^*}{C}\cdot\frac{\bar{f_e}}{f_e}.$$

Therefore to get the inclusion $B\bar{\Theta}\in\mathcal{U}_\infty$ we have to find a Blaschke product B^* such that $B^*\bar{C}\in\mathcal{U}_\infty$. In addition B must be interpolating. By Theorem 3.1 it is really so if $B^*\in\mathcal{B}(\alpha,\delta)$ and $\log 1/\alpha>2/\delta^2$.

Let $(a_n)_{n\geq1}$ be a sequence of zeros of the function B^*. We suppose that $\lim_n \operatorname{Im} a_n=+\infty$ and the discs $D_n=\{\zeta\in\mathbb{C}: \left|\frac{\zeta-a_n}{\zeta-\bar{a}_n}\right|\leq\delta/2\}$ do not intersect. Since $\log\frac{1}{\alpha}>\frac{2}{\delta^2}$ implies $\alpha<\delta^2/4$, we have

$$|B^*-\alpha\Theta|\geq|B^*|-\alpha\geq\frac{\delta^2}{4}-\alpha>0$$

in the domain $G\overset{def}{=}\mathbb{C}_+\setminus\bigcup_{n=1}^{\infty}D_n$. Hence the product C has no zeros in G. By Rouché's theorem the product C has exactly one zero, say c_n, in each disc D_n. The Rouché's theorem allows to control the behaviour of the points c_n as $n\to\infty$. In fact $|B_n^*(\zeta)|\geq\delta/2$ if $\zeta\in D_n$. Therefore the estimate

$$|B^*(\zeta)|=\left|\frac{\zeta-a_n}{\zeta-\bar{a}_n}\right||B^*(\zeta)|\geq\frac{\delta}{2}\left|\frac{\zeta-a_n}{\zeta-\bar{a}_n}\right|$$

is valid in D_n. On the other hand

$$\inf_{\zeta\in D_n}\operatorname{Im}\zeta=\frac{1-\delta/2}{1+\delta/2}\operatorname{Im}a_n>\frac{1}{3}\operatorname{Im}a_n.$$

Hence

$$|B^*(\zeta)|-|\alpha\Theta(\zeta)|\geq\frac{\delta}{2}\left|\frac{\zeta-a_n}{\zeta-\bar{a}_n}\right|-e^{-\frac{1}{3}\operatorname{Im}a_n},\quad\zeta\in D_n.$$

So by Rouché's theorem we have

$$\left|\frac{c_n-a_n}{c_n-\bar{a}_n}\right|\leq\frac{2}{\delta}e^{-\frac{1}{3}\operatorname{Im}a_n}.$$

Since $\left|\frac{\zeta-a}{\zeta-\bar{a}}\right|<\varepsilon$ implies $|\zeta-a|\leq\frac{2\varepsilon}{1-\varepsilon}\operatorname{Im}a$ and

$\frac{2}{\delta}e^{-1/3\,\mathrm{Im}\,a_n} \leq \frac{2}{\delta}e^{-\frac{8}{\alpha\delta^2}} \leq \frac{2}{e^8} < \frac{1}{5}$, we get

$$|c_n - a_n| \leq \frac{8}{\delta}\,\mathrm{Im}\,a_n e^{-\frac{1}{3}\mathrm{Im}\,a_n}.$$

Writing the explicit expressions for B^* and C we have

$$\frac{B^*}{C}(x) = \prod_{n=1}^{\infty}\frac{\varepsilon_n}{\varepsilon_n'}\prod_{n=1}^{\infty}\frac{1-x/a_n}{1-x/c_n}\,\overline{\prod_{n=1}^{\infty}\frac{1-x/c_n}{1-x/a_n}}\,.$$

So it is sufficient to check that the argument of the product $\prod_{n=1}^{\infty}\frac{1-x/a_n}{1-x/c_n}$ belongs to the space $\mathrm{Re}\,H^\infty \overset{def}{=} \{\mathrm{Re}\,f : f \in H^\infty\}$.
This follows from the formula

$$\log\left(\frac{1-x/a_n}{1-x/c_n}\right) = \log\frac{c_n}{a_n} + \log\left(1+\frac{c_n - a_n}{x - c_n}\right)$$

which implies that the logarithm of our product belongs to $\overline{H^\infty}$ if

$$\sum_{n=1}^{\infty}\mathrm{Im}\,a_n e^{-\frac{1}{3}\mathrm{Im}\,a_n} < +\infty.\qquad\bullet$$

REMARK. The method used for the proof of Theorem 3.3 allows in fact to obtain a stronger result. Namely, one can construct such Blaschke product B with Carleson set of zeros Λ , $\sup_{\lambda\in\Lambda}\mathrm{Im}\,\lambda = +\infty$, that the unimodular function $B\overline{\Theta}$ belongs to the subgroup of $\mathcal{U}_\infty$ consisting of functions of the form $\overline{h}/h$, $h \in \exp(H^\infty) \overset{def}{=} \{h : h = e^g,\ g \in H^\infty\}$. In this case the logarithm of the outer part of the generating function F_Λ will be uniformly bounded in the upper half-plane. To prove this it is sufficient to note that in the preceding example $\arg B^*\overline{C} \in \mathrm{Re}\,H^\infty$ and $\log f_e \in H^\infty$. Indeed

$$\log f_e = \log\frac{B^* - \alpha\Theta}{C} = \log\frac{B^*}{C} + \log(1 - \alpha\overline{B^*}\Theta)$$

(the equality holds on $\mathbb{R}$ and obviously implies $\log f_e \in L^\infty(\mathbb{R})$).$\bullet$

In conclusion let us prove the theorem on the "complementing up to a basis" mentioned at the beginning of the Section.

THEOREM 3.4 (V.I.Vasjunin). Let $(a_n)_{n\in\mathbb{Z}}$ be a Carleson sequence of points of the upper half-plane satisfying $\lim_n \mathrm{Im}\,a_n = +\infty$. Then for any positive number a the family $(a_n)_{n\in\mathbb{Z}}$ can be complemented up to such a family $(\lambda_n)_{n\in\mathbb{Z}}$ that the exponentials $(e^{i\lambda_n t})_{n\in\mathbb{Z}}$ form an unconditional basis in the space $L^2(0,a)$.

THE PROOF follows immediately from Corollary 2.5. Let us remember that by Theorem 3.1 for the Blaschke product with zeros $(a_n)_{n\in\mathbb{Z}}$ there exists a number α, $\alpha\in(0,1)$, such that

$$\theta_a-\alpha B^*=B\cdot F,$$

where B is an interpolating Blaschke product and F is an outer function. Let b_n be a zero of B, which is close to the zero a_n. Then $\lim|b_n-a_n|=0$ because $\operatorname{Im} a_n\to+\infty$ (see the application of Rouché's theorem in the proof of Theorem 3.2). Therefore by Corollary 2.5 we can return the zero b_n into the point a_n for each n, may be except for a finite set of n. But a finite set of zeros causes no difficulty because we can move them into any free place. ●

4. The equiconvergence of harmonic and non-harmonic Fourier series.

Suppose that $\Lambda=\{\lambda_n : n\in\mathbb{Z}\}\subset\mathbb{C}_\gamma$, $\gamma\in\mathbb{R}$, and the family of exponentials $(e^{i\lambda_n x})_{n\in\mathbb{Z}}$ forms an unconditional basis in the space $L^2(-\pi,\pi)$. Let $(h_n)_{n\in\mathbb{Z}}$ be the "coordinate family" (the dual sequence) for this basis:

$$(e^{i\lambda_m x},h_n)=\frac{1}{2\pi}\int_{-\pi}^{\pi}e^{i\lambda_m x}\,\overline{h_n(x)}\,dx=\begin{cases}1 & n=m,\\ 0 & n\neq m.\end{cases}$$

Then to each function f, $f\in L^2(-\pi,\pi)$ corresponds the non-harmonic Fourier series

$$\sum_{n\in\mathbb{Z}}(f,h_n)e^{i\lambda_n x}.$$

It is natural to consider together with the non-harmonic Fourier series the harmonic one:

$$f\sim\sum_{n\in\mathbb{Z}}\hat f(n)e^{inx};\qquad \hat f(n)\overset{\text{def}}{=\!=}\frac{1}{2\pi}\int_{-\pi}^{\pi}f e^{-inx}\,dx,\ n\in\mathbb{Z}.$$

The main theorem of this Section demonstrates that as to the convergence inside the interval $(-\pi,\pi)$, a non-harmonic Fourier series behaves in the same way as the corresponding harmonic one.

THEOREM 4.1. Let $\Lambda = \{\lambda_n : n \in \mathbb{Z}\} \subset \mathbb{C}_\gamma$, $\gamma \in \mathbb{R}$, and let a family of exponentials $(e^{i\lambda_n x})_{n\in\mathbb{Z}}$ form an unconditional basis in the space $L^2(-\pi,\pi)$. Then the equality

$$\lim_{R\to\infty} \sup_{|x|\le\pi} (\pi-|x|)^{1/2} \Big|\sum_{|n|\le R} \hat{f}(n) e^{inx} - \sum_{|\lambda_n|\le R} (f,h_n) e^{i\lambda_n x}\Big| = 0 \qquad (12)$$

holds for each function f , $f\in L^2(-\pi,\pi)$.

REMARKS. 1. The initial formulation of the Theorem guaranteed only the equiconvergence of the harmonic and non-harmonic Fourier series uniformly on compact subsets of the interval $(-\pi,\pi)$. A.M.Sedletskii has amiably informed one of the authors that proposition (12) was recently proved by him assuming the set of frequencies lies in a strip of finite width parallel to $\mathbb{R}$. Our method turned out to lead to this more general proposition too. The method of A.M.Sedletskii differs from ours.

We refer the interested reader to the paper [24] containing a lot of other useful facts about bases of exponentials. In particular it is shown there that it is impossible to improve the weight $(\pi-|x|)^{1/2}$ in (12).

2. Without loss of generality one may suppose that $\Lambda\subset\mathbb{C}_2$. Indeed, suppose Theorem 4.1 is proved for such sets Λ . Consider then the set of frequencies $\Lambda - iy$, $y>0$. It is clear that the dual sequence for the family of exponentials $(e^{i(\lambda_n - iy)x})_{n\in\mathbb{Z}}$ coincides with the family $(e^{-yt}h_n(t))_{n\in\mathbb{Z}}$. Then the non-harmonic Fourier series for the function f with respect to the new family is

$$f \sim e^{yt}\sum_n (f\cdot e^{-ys}, h_n) e^{i\lambda_n t}.$$

By assumption this series is equiconvergent with the Fourier series $e^{yt}\sum_n \widehat{fe^{-ys}}(n)\, e^{int}$. Let $S_N(f,t)$ denote the partial sum $\sum_{|n|\le N} \hat{f}(n)\, e^{int}$ of the Fourier series of f . Then we have

$$e^{yt} S_N(fe^{-ys}, t) - S_N(f,t) =$$

$$= e^{yt}\frac{1}{\pi}\int_{-\pi}^{\pi}\frac{\sin(N+1/2)(t-s)}{2\sin\frac{t-s}{2}} f(s)\left[e^{-ys}-e^{-yt}\right]ds = O(1). \quad \bullet$$

3. By technical reasons it is convenient to replace the partial sum in the formula (12) by the integral $\frac{1}{\pi}\int_{-\pi}^{\pi}\frac{\sin R(t-s)}{t-s} f(s)\,ds$. Simple estimations of the Dirichlet kernel show that the such replacement causes an error at most $O(1)\cdot\|f\|_2 \quad (R\longrightarrow+\infty)$.

4. Since the family of exponentials $(e^{i\lambda_n x})_{n\in\mathbb{Z}}$ forms an unconditional basis in $L^2(0,a)$, Λ is a Carleson set. Then there exists a positive number ε, so small that discs

$$D_n \overset{def}{=} \{\zeta\in\mathbb{C}_+ : |\zeta-\lambda_n|\leq\varepsilon\,\mathrm{Im}\,\lambda_n\},\quad n\in\mathbb{Z},$$

are disjoint. Let R be an arbitrary positive number, $D(0,R)=\{\zeta\in\mathbb{C}:|\zeta|<R\}$, and C_R be a closed curve forming the boundary of the domain $D(0,R)\cup\{D_n : D_n\cap D(0,R)\neq\emptyset\}$. (See the diagram below).

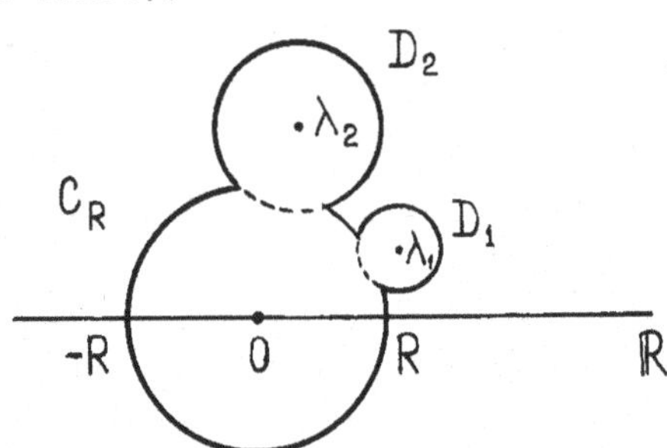

A t t h e e n d of the section w e s h a l l d e m o n s t r a t e that it is possible to replace the sum $\sum_{|\lambda_n|\leq R}(f,h_n)e^{i\lambda_n x}$ by the sum $\sum_{\lambda_n\in \mathrm{Int}\,C_R}(f,h_n)e^{i\lambda_n x}$ not violating the condition (12).

THE PROOF OF THEOREM 4.1 follows in its idea a plan, proposed by N.Levinson [48] . Though we prove a more general result, than the Levinson's one, our proof is technically simpler, because we use estimates of entire functions satisfying the condition (A_2) on $\mathbb{R}$. We have chosen the interval $(-\pi,\pi)$ instead of $(0,2\pi)$, for the sake of symmetry. Let F be the generating function for our set of frequencies. Then $G_F=[-\pi i,\pi i]$ and

$$F = \text{v.p.} \prod_{n\in\mathbb{Z}} \left(1-\frac{z}{\lambda_n}\right).$$

Clearly (see [48])

$$\frac{F}{(z-\lambda_n)F'(\lambda_n)} = \frac{1}{2\pi}\int_{-\pi}^{\pi} e^{itz}\, \overline{h_n(t)}\, dt, \quad n\in\mathbb{Z}. \qquad (13)$$

Let B be the Blaschke product with the zero-set Λ, let h be the outer part of $F|\mathbb{C}_+$ and $h^*(z) \overset{def}{=} \overline{h(\bar z)}$, $\operatorname{Im} z<0$. Then

$$F(z)=\begin{cases} B(z)e^{-\pi i z}h(z) & \text{, if } \operatorname{Im} z\geqslant 0 \\ e^{\pi i z}h^*(z) & \text{, if } \operatorname{Im} z<0. \end{cases}$$

Note that $|h|^2|\mathbb{R}$ satisfies the Helson-Szegö condition. The Blaschke product B satisfies the following condition

$$\inf_{R>0}\ \inf_{\zeta\in C_R} |B(\zeta)|>0.$$

This inequality is an immediate consequence of the Carleson condition $\inf\limits_n |B_n(\lambda_n)|>0$. Our choice of C_R is aimed just at the lower estimate of B (on C_R) .

The "algebraic" base of our proof is the following lemma due to N.Levinson which may be derived from the book [48] .

LEMMA (N.Levinson). For an arbitrary function f, $f\in L^2(-\pi,\pi)$, for any positive number R and for each t, $|t|<\pi$, the following formula holds:

$$\sum_{\lambda_n\in \operatorname{Int} C_R} (f,h_n)e^{i\lambda_n t} - \frac{1}{\pi}\int_{-R}^{R} \frac{\sin R(t-s)}{t-s} f(s)\,ds =$$

$$= \frac{1}{2\pi i}\int_{C_R} \frac{e^{i\zeta t}}{F(\zeta)}\, d\zeta \left\{ \int_{\mathbb{R}} \frac{F(x)\hat f(x)}{x-\zeta}\, dx \right\}.$$

Here $\hat f(x) = \frac{1}{2\pi}\int_{-\pi}^{\pi} e^{-ixt} f(t)\,dt$ is the Fourier transformation of f .

REMARK. Theorem 4.2 will follow from Levinson's lemma, if we prove the inequality

$$\sup_{|t|<\pi} (\pi-|t|)^{1/2} \left| \frac{1}{2\pi i}\int_{C_R} \frac{e^{i\zeta t}\, d\zeta}{F(\zeta)} \left\{ \int_{\mathbb{R}} \frac{F(x)\hat f(x)}{x-\zeta}\, dx \right\} \right| \leqslant \text{const}\cdot \|f\|_2 \qquad (14)$$

Indeed, then

$$\sup_{|t|<\pi} (\pi-|t|)^{1/2} \Big|\sum_{\lambda_n\in \mathrm{Int}\, C_R} (f,h_n)e^{i\lambda_n t} - \sum_{|n|\le R} \hat f(n) e^{int}\Big| \le \mathrm{const}\cdot \|f\|_2 .$$

It remains to note that $\mathrm{span}\{e^{i\lambda_n x}\chi_{[-\pi,\pi]} : n\in\mathbb{Z}\} = L^2(-\pi,\pi)$ and the equiconvergence holds for the exponentials $e^{i\lambda_n x}, n\in\mathbb{Z}$. ●

PROOF OF LEVINSON'S LEMMA. According to the Cauchy's formula

$$\frac{1}{2\pi i}\int_{C_R} \frac{e^{i\zeta t}}{G(\zeta)(x-\zeta)}\,d\zeta = \sum_{\lambda_n\in \mathrm{Int}\, C_R} \frac{e^{i\lambda_n t}}{G'(\lambda_n)(x-\lambda_n)} - \frac{e^{ixt}}{G(x)}\chi_{[-R,R]}(x)$$

(assuming $x\neq \pm R$). Hence

$$G(x)\frac{1}{2\pi i}\int_{C_R} \frac{e^{i\zeta t}}{G(\zeta)(x-\zeta)}\,d\zeta = \sum_{\lambda_n\in \mathrm{Int}\, C_R} \frac{e^{i\lambda_n t}G(x)}{G'(\lambda_n)(x-\lambda_n)} - e^{ixt}\chi_{[-R,R]}(x);$$

this implies, in particular, that the left part of the preceding formula belongs to $L^2(\mathbb{R})$. Computing the Fourier transform of the left and right parts and using the inversion formula we get

$$\int_{\mathbb{R}} e^{ixs}G(x)\,dx\cdot\frac{1}{2\pi i}\int_{C_R} \frac{e^{i\zeta t}}{G(\zeta)(x-\zeta)}\,d\zeta = \sum_{\lambda_n\in \mathrm{Int}\, C_R} e^{i\lambda_n t}\,\overline{h_n(s)}\,\chi_{[-\pi,\pi]}(s) - \int_{-R}^{R} e^{ix(t-s)}dx.$$

Multiplying this by $\frac{1}{2\pi} f(s)$, integrating over the interval $[-\pi,\pi]$ and interchanging the integrals we obtain

$$\int_{\mathbb{R}} \hat f(x)G(x)\Big\{\frac{1}{2\pi i}\int_{C_R}\frac{e^{i\zeta t}}{G(\zeta)(x-\zeta)}\,d\zeta\Big\}dx = \sum_{\lambda_n\in \mathrm{Int}\, C_R} (f,h_n)e^{i\lambda_n t} - \frac{1}{\pi}\int_{-\pi}^{\pi} \frac{\sin R(t-s)}{t-s} f(s)\,ds.$$

Now we need only to note that

$$\int_{\mathbb{R}} \hat f(x)G(x)\Big\{\frac{1}{2\pi i}\int_{C_R}\frac{e^{i\zeta t}}{G(\zeta)(x-\zeta)}\,d\zeta\Big\}dx = \frac{1}{2\pi i}\int_{C_R}\frac{e^{i\zeta t}}{G(\zeta)}\,d\zeta \int_{\mathbb{R}} \frac{\hat f(x)G(x)}{x-\zeta}\,dx.$$

Let us justify now the interchanging of the integrals. The function $x\mapsto |G(x)|^2$, $x\in\mathbb{R}$, satisfies the Muckenhoupt's condition (A_2) and hence $(x-\zeta)^{-1}G(x)\in L^2(\mathbb{R})$ if $\zeta\notin\mathbb{R}$. If we remove in the preceding formula the part of C_R lying in the strip $\Pi_\varepsilon \overset{\mathrm{def}}{=} \{z\in\mathbb{C} : |\mathrm{Im}\, z|\le\varepsilon\}$, the formula will follow from Fubini's theorem. Now we use an estimate which can be easily verified:

$$\Big|\frac{1}{2\pi i}\int_{C_R\cap\Pi_\varepsilon} \frac{e^{i\zeta t}}{G(\zeta)}\frac{d\zeta}{(x-\zeta)}\Big| \le C\cdot \min\Big(1, \frac{\varepsilon}{|x-R|}\Big), \quad C>0,$$

and note that according to the Sohotski's formulae the function $\int_{\mathbb{R}} \frac{\hat{f}(x)G(x)}{x-\zeta}\,dx$ is uniformly bounded on C_R and has only two points of discontinuity (jump-points) namely $\pm R$. The passage to the limit $\varepsilon \to +0$ completes the proof. ●

We shall need estimates of the Poisson's integrals $P_\zeta(W)$ of functions W satisfying the Muckenhoupt's condition (A_2) . Let us denote by h the outer function (in $\mathbb{C}_+$) satisfying $|h(x)| = W(x)$, $x \in \mathbb{R}$.

LEMMA 4.2. The following assertions are equivalent.

1) $W \in (A_2)$;

2) There exist functions u and v , $u, v \in L^\infty(\mathbb{R})$ with

$$\log W(x) = u(x) + \tilde{v}(x), \qquad \|v\|_\infty < \pi/2;$$

3) The outer function h maps the upper half-plane into an angle with vertex at the origin and with size less than $\pi/2$;

4) $\sup_{\zeta \in \mathbb{C}_+} P_\zeta(W) P_\zeta\left(\frac{1}{W}\right) < +\infty$;

5) There exists a constant C , $C > 1$, such that for $\zeta \in \mathbb{C}_+$

$$|h(\zeta)| \leqslant P_\zeta(W) \leqslant C\,|h(\zeta)|,$$

$$\frac{1}{|h(\zeta)|} \leqslant P_\zeta\left(\frac{1}{W}\right) \leqslant C\,\frac{1}{|h(\zeta)|}.$$

The proof can be found in the known paper [40] (see Theorem 2). Note that the assertion 3) of Lemma implies that the restriction of the outer function h on any line $\{\operatorname{Im} z = y\}$, $y > 0$, satisfies the Muckenhoupt's condition if the restriction of h on the real line does.

The proof of the following lemma is contained in [40] also.

LEMMA 4.3. Suppose that $W \in (A_2)$. Then there exists a constant C , $C > 0$, such that for any ζ , $\zeta \in \mathbb{C}_+$, the inequality

$$P_\zeta(W) \leqslant \frac{C}{2\operatorname{Im}\zeta} \int_{|t-\operatorname{Re}\zeta| \leqslant \operatorname{Im}\zeta} W(t)\,dt$$

is valid.

Now we prove the inequality (14). For this aim we divide the contour C_R into two parts $C_R^+ \overset{def}{=} C_R \cap \mathbb{C}_+$ and

$C_R^- \overset{def}{=} C_R \cap \mathbb{C}_-$ and prove (14) for each contour separately. Let us begin with the estimate for the boundary C_R^- (the case of C_R^+ being analogous).

If $\operatorname{Im}\zeta < 0$, then the function $(z-\zeta)^{-1}F\hat{f}$ obviously belongs to the Hardy class H^1 in the strip $\{\zeta\in\mathbb{C}: 0<\operatorname{Im}\zeta<1\}$:

$$\int_{\mathbb{R}} \frac{|F(x+iy)||\hat{f}(x+iy)|}{|x+iy-\zeta|}\,dx \leqslant \Big(\int_{\mathbb{R}} \frac{|F(x+iy)|^2}{|x+iy-\zeta|}\,dx\Big)^{1/2}\cdot\Big(\int_{\mathbb{R}}|\hat{f}(x+iy)|^2dx\Big)^{1/2} < +\infty.$$

So according to the Cauchy formula we obtain the identity

$$\int_{\mathbb{R}} \frac{F(x)\hat{f}(x)}{x-\zeta}\,dx = \int_{\mathbb{R}} \frac{F(x+i)\hat{f}(x+i)}{x+i-\zeta}\,dx,$$

from which the inequality

$$\Big|\int_{\mathbb{R}} \frac{F(x)\hat{f}(x)}{x-\zeta}\,dx\Big| \leqslant e^{2\pi}\|f\|_2\Big(\int_{\mathbb{R}} \frac{|h(x+i)|^2}{|x-(\zeta-i)|^2}\,dx\Big)^{1/2}$$

follows immediately.

Remind, that h is the outer part $F|\mathbb{C}_+$. Applying the assertion 5) of Lemma 4.2, we obtain

$$\int_{\mathbb{R}} \frac{|h(x+i)|^2}{|x-(\zeta-i)|^2}\,dx \leqslant \frac{c}{1+|\operatorname{Im}\zeta|}\,|h(2i+\bar{\zeta})|^2.$$

Hence

$$(\pi-|t|)^{1/2}\Big|\frac{1}{2\pi i}\int_{C_R^-} \frac{e^{i\zeta t}}{F(\zeta)}\,d\zeta\cdot\int_{\mathbb{R}} \frac{F(x)\hat{f}(x)}{x-\zeta}\,dx\Big| \leqslant$$

$$\leqslant \text{const}\cdot\|f\|_2\cdot(\pi-|t|)^{1/2}\int_{C_R^-} \frac{e^{|\operatorname{Im}\zeta t|}}{|h^*(\zeta)|}\cdot\frac{|h(2i+\bar{\zeta})|}{\sqrt{1+|\operatorname{Im}\zeta|}\,e^{\pi|\operatorname{Im}\zeta|}}\,|d\zeta|;$$

observing, that

$$(\pi-|t|)^{1/2}\int_{C_R^-} e^{-(\pi-|t|)|\operatorname{Im}\zeta|}\frac{|d\zeta|}{\sqrt{\operatorname{Im}\zeta}} = \int_{(\pi-|t|)C_R^-} \frac{e^{-|\operatorname{Im}\zeta|}}{\sqrt{\operatorname{Im}\zeta}}\,|d\zeta| = O(1),$$

we see, that it is sufficient to prove the inequality

$$|h(2i+\bar{\zeta})| \leqslant \text{const}\,|h^*(\zeta)|,\quad \operatorname{Im}\zeta<0. \tag{15}$$

It is obvious, that h is an entire function: $h(z)=e^{\pi iz}\cdot F(z)B^{-1}(z)$, $z\in\mathbb{C}$. Since zeroes of the product B lie in the half-plane $\mathbb{C}_2$, the function h proves to be outer in $\mathbb{C}_{-1}$. Further, if $x \in \mathbb{R}$, then

$$|h(x-i)| = \frac{e^{\pi}}{|B(x-i)|} \cdot e^{\pi} |h^*(x-i)|.$$

The function $x \longmapsto |h^*(x-i)|^2$ satisfies the condition (A_2) (see assertion 3) of the Lemma 4.2). Zeros of the product B satisfy Carleson's condition (C), outside small discs D_n, $n \in \mathbb{Z}$, lying in the half-plane $\mathbb{C}_-$, the inequality $|B(z)| \geqslant \gamma > 0$, $\operatorname{Im} z \geqslant 0$ is valid. According to the symmetry principle $B(z) = \frac{1}{\overline{B(\bar{z})}}$, $\operatorname{Im} z < 0$. Hence

$$\gamma e^{2\pi} |h^*(x-i)| \leqslant |h(x-i)| \leqslant e^{2\pi} |h^*(x-i)|$$

and, therefore, the function $x \longmapsto |h(x-i)|^2$ satisfies the condition (A_2) as well.

Consider an auxiliary function g defined in the upper half-plane:

$$g(\zeta) = h^2(\zeta - i).$$

Remembering, that $h^*(\zeta) = \overline{h(\bar{\zeta})}$, we can rewrite inequality (I5) in the following way:

$$|g(3i+\zeta)| \leqslant \text{const}\, |g(i+\zeta)|, \quad \operatorname{Im} \zeta > 0.$$

To prove this inequality let us use Lemma 4.3:

$$|g(3i+\zeta)| \leqslant P_{3i+\zeta}(|g|) \leqslant \frac{C}{2(3+\operatorname{Im}\zeta)} \int\limits_{|t-\operatorname{Re}\zeta| \leqslant 3+\operatorname{Im}\zeta} |g(t)|\, dt \leqslant \text{const} \cdot P_{i+\zeta}(|g|) \leqslant \text{const} \cdot |g(i+\zeta)|$$

(the inequality 5) of Lemma 4.2 is used in the last inequality). Thus the inequality (I5) is proved.

Some words about changes needed to estimate the contribution of the contour C_R^+ into the integral in the left-hand side of (14). Since the function $(z-\zeta)^{-1} G \hat{f}$ belongs to the Hardy class H^1 in the strip $\{\zeta : -1 \leqslant \operatorname{Im} \zeta < 0\}$ for $\zeta \in \mathbb{C}_+$, we have

$$\left| \int\limits_{\mathbb{R}} \frac{F(x) \hat{f}(x)}{x - \zeta}\, dx \right| = \left| \int\limits_{\mathbb{R}} \frac{F(x-i) \hat{f}(x-i)}{x - i - \zeta}\, dx \right| \leqslant$$

$$\leq e^{2\pi}\|f\|_2\left(\int\limits_{\mathbb{R}}\frac{|h^*(x-i)|^2}{|x-(i+\zeta)|^2}\,dx\right)^{1/2}\leq const\cdot\|f\|_2\,|h(\zeta+2i)|.$$

Since $F=Be^{-\pi i z}h$ in the half-plane $\mathbb{C}_+$ and $|B|\geq\gamma>0$ on C_R, an estimate

$$(\pi-|t|)^{1/2}\left|\frac{1}{2\pi i}\int\limits_{C_R^+}\frac{e^{i\zeta t}}{F(\zeta)}\,d\zeta\int\limits_{\mathbb{R}}\frac{F(x)\hat{f}(x)}{x-\zeta}\,dx\right|\leq$$

$$\leq\|f\|_2\cdot const\cdot\sup_{\zeta\in\mathbb{C}_+}\frac{|h(\zeta+2i)|}{|h(\zeta)|}(\pi-|t|)^{1/2}\int\limits_{C_R^+}\frac{e^{|\mathrm{Im}\,\zeta t|}}{|B(\zeta)|e^{\pi|\mathrm{Im}\,\zeta|}}\frac{|d\zeta|}{\sqrt{\mathrm{Im}\,\zeta}}\leq$$

$$\leq const\,\|f\|_{L^2(-\pi,\pi)}$$

holds.

To finish the proof of Theorem 4.1, we have to prove the assertion from Remark 4.

LEMMA 4.4. If $f\in L^2(-\pi,\pi)$, then $|(f,h_n)|\leq\leq const\sqrt{\mathrm{Im}\,\lambda_n}\,e^{-\pi\,\mathrm{Im}\,\lambda_n}\|f\|_2$.

PROOF. From the formuls (13), Parseval identity and Schwarz inequality we have

$$|(f,h_n)|\leq\|f\|_2\left\{\int\limits_{\mathbb{R}}\frac{\mathrm{Im}\,\lambda_n}{|x-\lambda_n|^2}|F(x)|^2dx\right\}^{1/2}\frac{1}{\sqrt{\mathrm{Im}\,\lambda_n}\,|F'(\lambda_n)|}.$$

Further, $F=hBe^{-\pi i z}$ in the half-plane $\mathbb{C}_+$. Hence $|F'(\lambda_n)|=|h(\lambda_n)||B'(\lambda_n)|e^{\pi\,\mathrm{Im}\,\lambda_n}$. Now we need only to note that

$$|B'(\lambda_n)|=(2\,\mathrm{Im}\,\lambda_n)^{-1}|B_n(\lambda_n)|\geq\gamma(2\,\mathrm{Im}\,\lambda_n)^{-1}$$

and applying the assertion 5) from Lemma 4.2 completes the proof. ●

For any positive number R consider the set N_R of integers n such that $\lambda_n\notin D(0,R)$ but $D(0,R)\cap D_n\neq\neq\emptyset$. Let us show that

$$\sup_R\sup_{|t|\leq\pi}\left|(\pi-|t|)^{1/2}\sum_{n\in N_R}(f,h_n)e^{i\lambda_n t}\right|\leq const\cdot\|f\|_2.$$

Due to Lemma 4.4 the inequality

$$(\pi-|t|)^{1/2}\left|\sum_{n\in N_R}(f,h_n)e^{i\lambda_n t}\right|\leq const\cdot\|f\|_2\sum_{n\in N_R}(\pi-|t|)^{1/2}\sqrt{\mathrm{Im}\,\lambda_n}\,e^{-(\pi-|t|)\mathrm{Im}\,\lambda_n}$$

is valid. The discs D_n are disjoint and their radii are proportional to the distance from the centre λ_n to the real line. Therefore the number of indices n , $n \in N_R$, with $2^k \leqslant \leqslant \operatorname{Im} \lambda_n \leqslant 2^{k+1}$, $k = 0, 1, \dots$, is uniformly bounded. We need only to prove an elementary inequality:

$$\sup_{y>0} \sum_{n \geqslant 0} (2^n y)^{1/2} e^{-2^n y} < +\infty .$$

It is clear that without loss of generality supremum in this inequality can be taken with respect to the set $\{ y : y = 2^m, m \in \mathbb{Z} \}$. Then

$$\sum_{n \geqslant 0} (2^{n+m})^{1/2} e^{-2^{n+m}} \leqslant \sum_{n=-\infty}^{\infty} 2^{n/2} e^{-2^n} < +\infty . \qquad \bullet$$

PART IY.

THE REGGE PROBLEM IN THE THEORY OF DIFFERENTIAL OPERATORS

An investigation of the completeness and bases problem for a family of eigen-functions of a differential operator containing the spectral parameter in the boundary condition is our main task in Part IY. As we shall see, the approach, which has been utilized in the preceding parts, is useful in this part as well. We intend here to demonstrate the approach in one more special situation rather than achieve results of maximal generality. Subtle results of differential operator theory form only a scenery for our exposition. So this Part can be addressed to the reader who is, possibly, for the first time, getting acquainted with the problem of eigen-function expansions.

Let $a>0$ and let ρ be a positive function on $[0,a]$. It is assumed that $\rho(x)\equiv 1$ if $x>a_\rho$ for some number a_ρ in $(0,a)$ and that

$$\max\Big(\int_0^a \rho^2 dx, \int_0^a \rho^{-2} dx\Big)<+\infty \tag{1}$$

Let

$$L^2_\rho(0,a) \stackrel{def}{=} \{ f : \|f\|^2 = \int_0^a | f(x)|^2 \rho^2(x)\, dx < +\infty \} .$$

Now the spectral problem (t h e R e g g e p r o b l e m) for a second order differential operator $L=-\rho^{-2}\dfrac{d^2}{dx^2}$ in $L^2_\rho(0,a)$ containing a spectral parameter in the boundary condition can be stated as follows. Let $\sigma(\rho)$ be the set of all complex numbers k such that the equation

$$-y''=k^2\rho^2 y\ ;\quad y'(0)=0,\ y'(a)+iky(a)=0 \tag{2}$$

has a non-zero solution $y(x,k)$. The question is - does the family $(y(x,k))_{k\in\sigma(\rho)}$ of all such solutions form a complete family or even a Riesz basis in $L^2_\rho(0,a)$? This problem came from the scattering theory on a "transparent" compact barrier for acoustic waves spreading in a medium with a constant refraction coefficient.

The plan of our exposition in Part IY is the following. In

§1 we give a brief outline of the Lax-Phillips approach to the scattering theory for wave equation. In §2 the relationship of this theory and the Regge problem is discussed. In conclusion in §3 we formulate our main result and consider an important example.

In what follows we assume the reader's familiarity with backgrounds of the theory of self-adjoint operators. A detailed exposition of the theory is given in [67], [68].

1. Lax-Phillips approach to the scattering theory.

It is well known that the wave equation for the semiinfinite string with the free end $x=0$ and the local propagation speed ρ^{-1}, $\rho(x)\equiv 1$ for $x>a_\rho$, is defined by

$$\begin{cases} \rho^2(x)u_{tt}=u_{xx} ; & u_x(0,t)=0 \\ u(x,0)=u_0(x) ; & u_t(x,0)=u_1(x),\ x\in\mathbb{R}_+ \end{cases} \tag{3}$$

The pair $\mathcal{U}(0)=(u_0,u_1)$ is called "the Cauchy data", or simply "data", of the problem (2). The evolution operator U_t of (2) transforms (by definition) the data $\mathcal{U}(0)$ into the data $\mathcal{U}(t)=(u(x,t),u_t(x,t))$ related to the moment t. A natural Hilbert space of data is a Hilbert space E of all data with finite energy:

$$E=\left\{(u_0,u_1): \int_0^\infty |u_1|^2\rho^2 dx<+\infty,\ u_0'\in L^2(\mathbb{R}_+)\right\}$$

$$\langle \mathcal{U},\mathcal{V}\rangle_E \overset{def}{=\!=} \frac{1}{2}\int_0^\infty \left\{u_{0x}\bar{v}_{0x}+\rho^2 u_1\bar{v}_1\right\}dx . \tag{4}$$

To be more precise we are to consider the space E as a space of equivalence classes identifying $\mathcal{U}$ and $\mathcal{V}$ iff $u_1=v_1$, $u_0-v_0\equiv$ const.

Let now $L^2_\rho \overset{def}{=\!=}\{f: \int_0^\infty |f|^2\rho^2 dx<+\infty\}$ be a Hilbert space with the inner product $\langle f,g\rangle_{L^2_\rho}=\int_0^\infty f\bar{g}\rho^2 dx$.

LEMMA 1.1. The operator $L=-\rho^{-2}\frac{d^2}{dx^2}$ with the domain

$$D(L)=\{f\in L^2_\rho : Lf\in L^2_\rho,\ f'(0)=0\}$$

is a self-adjoint non-negative operator in L^2_ρ .

PROOF. The set $\mathcal{D}(L)$ is dense in L^2_ρ . Indeed, if φ is any smooth function with a compact support in $(0,+\infty)$ then $\varphi\in\mathcal{D}(L)$ because according to (1) we have

$$\int_0^\infty |L\varphi|^2\rho^2 dx=\int_0^\infty \frac{|\varphi''|^2}{\rho^2}dx\leq \sup_{x\in\mathbb{R}_+}|\varphi''(x)|^2\cdot\int_{supp(\varphi)}\frac{1}{\rho^2}dx<+\infty .$$

Clearly, $f|[a_\rho,+\infty)\in L^2(a_\rho,+\infty)$, $f''|[a_\rho,+\infty)\in L^2(a_\rho,+\infty)$ if $f\in\mathcal{D}(L)$ ($\rho(x)\equiv 1$ for $x>a_\rho$) . It follows from the well-known inequality

$$\int_{a_\rho}^\infty |f'|^2 dx\leq\Big(\int_{a_\rho}^\infty |f|^2dx\Big)^{1/2}\cdot\Big(\int_{a_\rho}^\infty |f''|^2dx\Big)^{1/2}$$

(take the Fourier transform for the proof) that $\lim\limits_{x\to+\infty} f'(x)=0$ for any f in $\mathcal{D}(L)$.

It is also clear that $\int_0^x |f''(t)|dt<+\infty$ for any f in $\mathcal{D}(L)$ and for $x\in(0,+\infty)$. Indeed,

$$\int_0^x|f''(t)|dt\leq\Big(\int_0^x\frac{1}{\rho^2}|f''|^2dt\Big)^{1/2}\cdot\Big(\int_0^x\rho^2dt\Big)^{1/2}\leq\|Lf\|_{L^2_\rho}\Big(\int_0^x\rho^2dt\Big)^{1/2}<+\infty .$$

The integration by parts shows now the operator L is symmetric.

To prove $L=L^*$ it is sufficient to check that $\mathcal{D}(L^*)\subset\mathcal{D}(L)$. Let $v\in\mathcal{D}(L^*)$. Then

$$|<Lu,v>_{L^2_\rho}|=\Big|\int_0^\infty u''\cdot\bar v\,dx\Big|\leq const.\|u\|_{L^2_\rho}$$

for any u in $\mathcal{D}(L)$. This means, in particular, that

$$\Big|\int_0^\infty u''\bar v dx\Big|\leq const.\sup_{t\in supp(u)}|u(t)|\cdot\Big(\int_{supp(u)}\rho^2dt\Big)^{1/2}$$

and therefore the distribution v'' coincides with a σ-finite measure μ on $[0,+\infty)$. Then

$$\Big|\int_0^\infty u\,d\mu\Big|\leq const.\Big(\int_0^\infty|u\rho|^2dt\Big)^{1/2}$$

and therefore $d\mu = p\,dt$ is absolutely continuous with $\int_0^\infty \frac{1}{\rho x}|P|^2 dt < +\infty$. It is clear now that $\rho^{-2}\cdot v'' \in L^2_\rho$.
To prove that $v'(0) = 0$ we should only remark that

$$\int_0^\infty u''\bar{v}\,dx = -\int_0^\infty \bar{v}'\,du = \bar{v}'(0)\,u(0) - \int_0^\infty \bar{v}''u\,dx .$$

This, obviously, implies that

$$|v'(0)|\cdot|u(0)| \leq const.\|u\|_{L^2_\rho} + |<u, Lv>_{L^2_\rho}| \leq const.\|u\|_{L^2_\rho} .$$

Therefore the assumption $v'(0) \neq 0$ implies that the functional $u \mapsto u(0)$ is bounded in L^2_ρ. ●

THEOREM 1.2. The operator

$$\mathcal{L} = i\begin{pmatrix} \mathbb{O} & -I \\ L & \mathbb{O} \end{pmatrix},\ \mathcal{D}(\mathcal{L}) = \{\mathcal{U} \in E : u_0'' \in L^2(\mathbb{R}_+),\ u_0'(0)=0,\ u_1' \in L^2(\mathbb{R}_+)\}$$

is self-adjoint. The family $(U_t)_{t\in\mathbb{R}}$ of evolution operators coincides with the strongly continuous unitary group $U_t = exp(it\mathcal{L})$. For every $\mathcal{U} = (u_0, u_1)$ in $\mathcal{D}(\mathcal{L})$ the formula $\mathcal{U}(t) = U_t\mathcal{U}$ defines a function $u_0(x,t)$ satisfying (3).

PROOF. Our first task is to check that the operator $\mathcal{L}$ defined above is self-agjoint. A simple calculation shows the operator $\mathcal{L}$ is symmetric on a dense set of smooth data:

$$<\mathcal{L}\mathcal{U}, \mathcal{V}>_E = \frac{1}{2}\int_0^\infty -iu_1'\cdot\bar{v}_0'\,dx + \frac{1}{2}\int_0^\infty iu_0''\cdot\bar{v}_1\,dx =$$

$$= -\frac{1}{2}\int_0^\infty i\bar{v}_0'\,du_1 + \frac{1}{2}\int_0^\infty i\bar{v}_1\,du_0' =$$

$$= \frac{1}{2}\int_0^\infty u_1\,i\,v_0''\,dx + \frac{1}{2}\int_0^\infty iu_0'\bar{v}_1'\,dx = <\mathcal{U}, \mathcal{L}\mathcal{V}>_E .$$

Let $\mathcal{V} \in \mathcal{D}(\mathcal{L}^*)$. Then, clearly,

$$|<\mathcal{L}\mathcal{U}, \mathcal{V}>_E| \leq const.\|\mathcal{U}\|_E$$

for every $\mathcal{U}$ in $\mathcal{D}(\mathcal{L})$. Let $\mathcal{U} = (u_0, 0) \in \mathcal{D}(\mathcal{L})$. Then

$$\left|\int_0^\infty u_0''\bar{v}_1\,dx\right| \leq const.\left(\int_0^\infty |u_0'|^2 dx\right)^{1/2}$$

and therefore $v_1' \in L^2(\mathbb{R})$. Let now $\mathcal{U} = (0, u_1) \in \mathcal{D}(\mathcal{L})$.

Then it follows that

$$\left|\int_0^\infty u_1'\cdot v_0'\,dx\right| = \left|\int_0^\infty u_1''\cdot v_0\,dx\right| \le const.\,\|u_1\|_{L^2_\rho}\,.$$

Since the operator L is self-adjoint in L^2_ρ (see Lemma 1.1), we have $Lv_0\in L^2_\rho$, $v_0'(0)=0$ and therefore $v\in D(\mathcal{L})$.

By the Stone theorem the operators $U_t=exp(i\mathcal{L}t)$ are unitary and $U_{t+s}=U_tU_s$. We may now define

$$\mathcal{U}(t)\overset{def}{=\!=}U_t\,\mathcal{U}(0)\,.$$

To prove the last statement of the theorem one should only remark that $U_tD(\mathcal{L})\subset D(\mathcal{L})$ (see theorem YIII.7 of [67]) and therefore $\frac{\partial}{\partial t}\mathcal{U}(t)=i\mathcal{L}\mathcal{U}(t)$. ●

THEOREM 1.3 (Huygens principle, see [47]). Let $I=(a,b)$ be an interval on $\mathbb{R}_+$, let $x_0\in I$ and let $\mathcal{U}\in D(\mathcal{L})$ and $\mathcal{U}|I\equiv 0$. Then $\mathcal{U}(t)(x_0)=0$ for sufficiently small t .

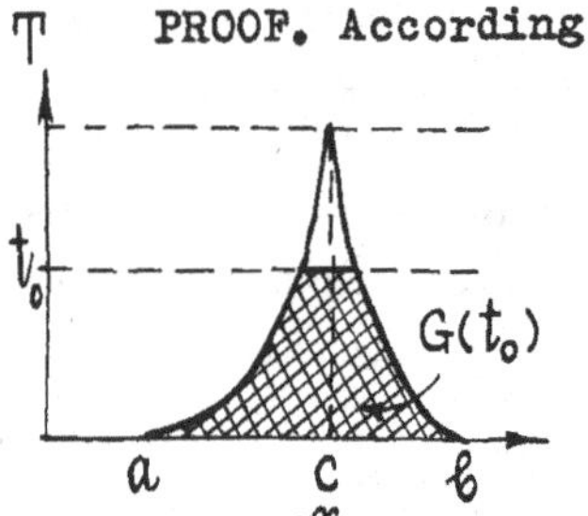

PROOF. According to (1) $\rho(x)>0$ a.e. on $\mathbb{R}_+$. So the function $\int_a^x\rho(s)\,ds$ is, obviously, strictly increasing and the function $\int_x^b\rho(s)\,ds$ decreases. A point C on the picture can be found from the equation $\int_a^c\rho(s)ds=\int_c^b\rho(s)ds$. Let $t_1(x)=\int_a^x\rho(s)\,ds$ for $a<x<c$ and let $t_2(x)=\int_x^b\rho(s)\,ds$ for $c<x<b$. At last, $T=t_1(c)=t_2(c)$. We show that the energy of the wave with the boundary values $\mathcal{U}$ vanishes in the domain $G(t_0)$. To do this let $t_1(x_1)=t_0=t_2(x_2)$. We have

$$0=\iint_{G(t_0)}\left[\rho^2(x)\,u_{tt}u_t-u_{xx}u_t\right]dx\,dt\,.$$

Clearly, for $0<t<t_0$,

$$-\int_{t_1^{-1}(t)}^{t_2^{-1}(t)}u_{xx}u_t\,dx=\int_{t_1^{-1}(t)}^{t_2^{-1}(t)}u_xu_{tx}\,dx-u_tu_x\Big|_{t_1^{-1}(t)}^{t_2^{-1}(t)}\,.$$

Therefore

$$\iint_{G(t_0)} -u_{xx}u_t\,dx\,dt = \iint_{G(t_0)} u_x u_{tx}\,dx\,dt + \int_a^{t_1^{-1}(t_0)} u_t(x,t_1(x))\,u_x(x,t_1(x))\,t_1'(x)\,dx -$$

$$- \int_{t_2^{-1}(t)}^{b} u_t(x,t_2(x))\,u_x(x,t_2(x))\,t_2'(x)\,dx .$$

But

$$\iint_{G(t_0)} \{\rho^2(x)u_{tt}\cdot u_t + u_{xt}\cdot u_t\}\,dx\cdot dt = \frac{1}{2}\iint_{G(t_0)} \frac{d}{dt}\{\rho^2 u_t^2 + u_x^2\}\,dx\,dt =$$

$$= \frac{1}{2}\int_a^{t_1^{-1}(t_0)} [\rho^2(x)u_t^2(x,t_1(x)) + u_x^2(x,t_1(x))]\,dx + \frac{1}{2}\int_{t_1^{-1}(t_0)}^{t_2^{-1}(t_0)} [\rho^2(x)u_t^2(x,t_0) + u_x^2(x,t_0)]\,dx$$

$$+ \int_{t_2^{-1}(t_0)}^{b} [\rho^2(x)\,u_t^2(x,t_2(x)) + u_x^2(x,t_2(x))]\,dx .$$

We get therefore

$$0 = \frac{1}{2}\int_{t_1^{-1}(t_0)}^{t_2^{-1}(t_0)} [\rho^2(x)\,u_t^2(x,t_0) + u_x^2(x,t_0)]\,dx +$$

$$+ \frac{1}{2}\int_{[a,t_1^{-1}(t_0)]\cup[t_2^{-1}(t_0),b]} (\rho(x)\,u_t(x,t(x)) + u_x(x,t(x)))^2\,dx . \quad \bullet$$

A remarkable property of the unitary group $(U_t)_{t\in\mathbb{R}}$ is that it has a pair of orthogonal invariant subspaces $(\mathcal{D}_+,\mathcal{D}_-)$ in E satisfying

$$U_t\,\mathcal{D}_+ \subset \mathcal{D}_+,\ t>0;\quad U_t\,\mathcal{D}_- \subset \mathcal{D}_-,\ t<0 .$$

For example, let

$$\mathcal{D}_+ = \left\{\begin{pmatrix}u_0\\u_1\end{pmatrix} : -u_1 = u_0',\ \ u_0' \in L^2(\mathbb{R}_+);\ u_0(x) \equiv const,\ x<a\right\},$$

$$\mathcal{D}_- = \left\{\begin{pmatrix}u_0\\u_1\end{pmatrix} : u_1 = u_0',\ \ u_0' \in L^2(\mathbb{R}_+);\ u_0(x) \equiv const,\ x<a\right\}.$$

Then $u_0(x+t)$ is called an *incoming wave* and $u_0(x-t)$ is called an *outgoing wave*. Clearly

$$U_t\begin{pmatrix} u_0 \\ -u_0' \end{pmatrix} = \begin{pmatrix} u_0(x-t) \\ -u_0'(x-t) \end{pmatrix}$$

if $u_0(x) \equiv const$ for $x < a$, and if $t > 0$. So we may imagine the space $\mathcal{D}_-$ as the space of incoming waves and $\mathcal{D}_+$ as the space of outgoing waves. It were P.Lax and R.Phillips who have stressed the importance of these invariant subspaces for the first time [47]. They advanced a new approach (L-Ph-approach) to the scattering theory for unitary groups which have invariant subspaces of this type [47]. Let $K \overset{def}{=\!=} E \ominus \{\mathcal{D}_+ \oplus \mathcal{D}_-\}$. The scattering matrix arising in L-Ph-approach turns out a characteristic function for the strong continuous semigroup of contractions [69]

$$Z_t \overset{def}{=\!=} P_K U_t | K, \quad t > 0 .$$

The following lemma describes the data in K.

LEMMA 1.4. Let $u \in E$. Then $u \in K$ if and only if $u_0(x) \equiv const$ for $x > a$ and $u_1(x) \equiv 0$ for $x > a$.

PROOF. Let $G_a = \{ g : \int_0^\infty |g'|^2 dx < +\infty ,\ g(x) \equiv const$ for $x < a \}$. Then clearly

$$\begin{pmatrix} g \\ 0 \end{pmatrix} = \frac{1}{2}\left\{\begin{pmatrix} g \\ -g' \end{pmatrix} + \begin{pmatrix} g \\ g' \end{pmatrix}\right\} \in \mathcal{D}_+ \oplus \mathcal{D}_- ; \quad \begin{pmatrix} 0 \\ g' \end{pmatrix} = \frac{1}{2}\left\{\begin{pmatrix} g \\ g' \end{pmatrix} - \begin{pmatrix} g \\ -g' \end{pmatrix}\right\} \in \mathcal{D}_+ \oplus \mathcal{D}_- .$$

Therefore $u \in K$ if and only if

$$< \begin{pmatrix} g \\ 0 \end{pmatrix}, u >_E = < \begin{pmatrix} 0 \\ g' \end{pmatrix}, u >_E = 0$$

for every g in G_a. The du Bois-Reymond lemma implies that this is equivalent to the statement of the lemma. ●

The semigroup $(Z_t)_{t \geqslant 0}$ is unitary equivalent to the semigroup $P_{K_S} e^{ixt} | K_S$, $t \geqslant 0$. Here S is an inner function in $\mathbb{C}_+$ and K_S stands for $H_+^2 \ominus S H_+^2$. The function S is called a c h a r a c t e r i s t i c f u n c t i o n for $(Z_t)_{t \geqslant 0}$. In the scattering theory $\bar{S}$ is known as a reflection coefficient. We shall return to its physical meaning a bit later.

It is remarkable that the unitary correspondence between

the semigroups can be given by explicit formulae. To do this we have to find a family of generalized eigen-functions for $(U_t)_{t\in\mathbb{R}}$ or equivalently for $\mathcal{L}$. In its turn this can be done with the help of so-called Jost solutions $y(x,\lambda)$:

$$-y'' = \lambda^2 \cdot \rho^2 \cdot y, \quad y(a_\rho,\lambda)=1, \quad y'(a_\rho,\lambda) = -i\lambda .$$

The existence and uniqueness of the Jost solution $y(x,\lambda)$ is implied by the standard existence theorem of the differential equations theory. Moreover, the well-known iteration method leads, obviously, to the conclusion that $\lambda \mapsto y(x,\lambda)$ is an entire function for every x in $\mathbb{R}$. Let now $a > a_\rho$. Then a Jost solution corresponding to a point a is defined by

$$y_a(x,\lambda) = e^{i\lambda(a-a_\rho)} \cdot y(x,\lambda) .$$

Clearly,

$$y_a(a,\lambda) = 1, \quad y_a'(a,\lambda) = -i\lambda$$

and $y_a(x,\lambda) = e^{-i\lambda(x-a)}$ for $x \geq a$.

It follows from the uniqueness of the Jost solution that

$$\overline{y_a(x,-\overline{\lambda})} = y_a(x,\lambda) . \tag{5}$$

A linear combination of the Jost solutions

$$\varphi_a(x,\lambda) = y_a(x,-\lambda) + \overline{S_a(\lambda)} \cdot y_a(x,\lambda)$$

satisfies the boundary condition $\varphi_a'(0,\lambda) = 0$ if

$$\overline{S_a(\lambda)} = -\frac{y_a'(0,-\lambda)}{y_a'(0,\lambda)} = -e^{-2i\lambda(a-a_\rho)} \frac{y'(0,-\lambda)}{y'(0,\lambda)} .$$

It is clear that $|S_a(\lambda)| = 1$ for $\lambda \in \mathbb{R}$ (see (5)). A simple computation shows that $\mathcal{L}\Phi_a(x,\lambda) = \lambda \Phi_a(x,\lambda)$ for

$$\Phi_a(x,\lambda) = \begin{pmatrix} 1/i\lambda \cdot \varphi_a(x,\lambda) \\ \varphi_a(x,\lambda) \end{pmatrix} .$$

Let E_0 be a dense subset of data in E which have a compact support in $\mathbb{R}_+$. We define a mapping $\mathcal{T}_-$ by the following formula. For $\mathcal{U} \in E_0$ let

$$\mathcal{T}\mathcal{U} = \frac{1}{\sqrt{2\pi}} \langle \mathcal{U}, \Phi_a(\cdot,\lambda) \rangle_E \equiv \frac{1}{2\sqrt{2\pi}} \int_0^\infty \{ u_{0x} \cdot \overline{\frac{1}{i\lambda} \varphi_{ax}} + \rho^2 u_1 \overline{\varphi}_a \} \, dx =$$

$$= \frac{1}{2\sqrt{2\pi}} \int_0^\infty (i\lambda u_0 + u_1) \rho^2 \overline{\varphi}_a \, dx .$$

THEOREM 1.5. The closure of the operator $\mathcal{T}: E_0 \to L^2(\mathbb{R})$

$$\mathcal{T}\mathcal{U} = \frac{1}{2\sqrt{2\pi}} \int_0^\infty (i\lambda u_0(x) + u_1(x)) \rho^2 \overline{\varphi_a(x,\lambda)} \, dx$$

defines an isometry of E onto $L^2(\mathbb{R})$. The following formulae hold

$$\mathcal{T}\mathcal{D}_- = H^2_- \ , \ \mathcal{T}\mathcal{D}_+ = S_a H^2_+$$

$$\mathcal{T} U_t = e^{i\lambda t} \cdot \mathcal{T} .$$

The function S_a is an inner function in $\mathbb{C}_+$ and

$$\mathcal{T}K = H^2_+ \ominus S_a H^2_+ ; \quad \mathcal{T} Z_t \mathcal{U} = P_{K_{S_a}} e^{i\lambda t} \mathcal{T}\mathcal{U}, \quad \mathcal{U} \in K .$$

PROOF. Let $\mathcal{U} \in \mathcal{D}_- \cap E_0$. Then $\mathcal{U} = (u, u')$ and $u(x) \equiv 0$ for $x \leq a$. It follows that

$$\mathcal{T}\mathcal{U} = \frac{1}{2\sqrt{2\pi}} \int_a^{+\infty} (i\lambda u + u') \{ e^{-i\lambda(x-a)} + S_a(\lambda) e^{i\lambda(x-a)} \} \, dx =$$

$$= \frac{1}{2\sqrt{2\pi}} \int_a^{+\infty} (e^{i\lambda x} u)' \{ e^{-2i\lambda x} e^{i\lambda a} + S_a(\lambda) e^{-i\lambda a} \} \, dx =$$

$$= \frac{i\lambda}{\sqrt{2\pi}} \int_a^{\infty} u e^{i\lambda(a-x)} dx = \frac{1}{\sqrt{2\pi}} \int_a^{+\infty} u' e^{i\lambda(a-x)} \, dx .$$

Hence, by the Parseval theorem and by (3) we have

$$\| \mathcal{T}\mathcal{U} \|_{L^2(\mathbb{R})} = \left(\int_a^{+\infty} |u'|^2 dx \right)^{1/2} = \| \mathcal{U} \|_E .$$

An analogous computation shows that for $\mathcal{U} \in \mathcal{D}_+ \cap E_0$

$$\mathcal{T}u=\frac{1}{2\sqrt{2\pi}}\int_a^{+\infty}(i\lambda u-u')\{e^{-i\lambda(x-a)}+S_a(\lambda)e^{i\lambda(x-a)}\}\,dx=$$

$$=S_a(\lambda)\cdot\frac{1}{\sqrt{2\pi}}\int_a^{+\infty}e^{i\lambda(x-a)}u'\,dx$$

and again we have $\|\mathcal{T}u\|_{L^2(\mathbb{R})}=\|u\|_E$. Now it follows by the Paley-Wiener theorem

$$\mathcal{T}_-\mathcal{D}_-=H^2_-\,,\qquad \mathcal{T}_-\mathcal{D}_+=S_aH^2_+\,.$$

Let now u be a smooth function in E_0 . Clearly

$$<\mathcal{L}u,\ \Phi_a(\cdot,\lambda)>_E=\lambda<u,\ \Phi_a(\cdot,\lambda)>_E$$

and consequently

$$\frac{d}{dt}<U_tu,\ \Phi_a>_E=<i\mathcal{L}u,\ \Phi_a>_E=i\lambda<U_tu,\Phi_a>_E\,.$$

Therefore the boundary condition $U_0=I$ implies

$$<U_tu,\Phi_a>_E=e^{i\lambda t}<u,\Phi_a>_E\,.$$

By theorem 1.2

$$span\,(U_t\,\mathcal{D}_-:t\in\mathbb{R})=E\,.$$

Therefore $\mathcal{T}_-$ maps the space E isometrically onto $L^2(\mathbb{R})$. It follows from $U_t\mathcal{D}_+\subset\mathcal{D}_+$ for $t>0$ that $e^{i\lambda t}\cdot S_a\cdot H^2_+\subset\subset H^2_+$ for every $t>0$. By P.Lax theorem [18] this means that S_a is an inner function $(|S_a(\lambda)|=1,\ \lambda\in\mathbb{R})$. ●

REMARK. The function S_a being a quotient of entire functions, it is clear that

$$S_a=B\cdot\Theta\,.$$

Here B denotes a Blaschke product in $\mathbb{C}_+$ whose zeros have no limit points in $\mathbb{R}$ and $\Theta(z)=exp(icz)$, $c>0$.

The transformation $\mathcal{T}_-$ is called an *incoming spectral representation* for the unitary group $(U_t)_{t\in\mathbb{R}}$. The spectral property of $\mathcal{T}_-$ means that

$\mathcal{T}_-$ transforms the group $(U_t)_{t\in\mathbb{R}}$ onto the unitary group $(e^{i\lambda t})_{t\in\mathbb{R}}$ in $L^2(\mathbb{R})$.

Let now discuss the physical meaning of the reflection coefficient $\overline{S}_a$. It is clear that

$$\mathcal{T}_-^{-1} v = \mathcal{T}_-^* v = \frac{1}{\sqrt{2\pi}} \int_{\mathbb{R}} \Phi_a(x,\lambda)\, v(\lambda)\, d\lambda \tag{6}$$

and that the evolution of the part of the "wave packet" $\mathcal{U}(x,t) = U_t \mathcal{T}_-^{-1} v$ in $\mathcal{D}_- \oplus \mathcal{D}_+$ is defined by

$$\{(\mathcal{P}_{\mathcal{D}_-} \oplus \mathcal{P}_{\mathcal{D}_+}) U_t \mathcal{T}_-^{-1} v\}(x) = \frac{1}{\sqrt{2\pi}} \int_{\mathbb{R}} e^{i\lambda t} \begin{pmatrix} \frac{1}{i\lambda} \varphi_a(x,\lambda) \\ \varphi_a(x,\lambda) \end{pmatrix} v(\lambda)\, d\lambda, \; x > a.$$

Therefore for $x > a$

$$u(x,t) = \frac{1}{\sqrt{2\pi}} \int_{\mathbb{R}} e^{i\lambda t} \begin{pmatrix} 1/i\lambda \\ 1 \end{pmatrix} e^{i\lambda(x-a)} v(\lambda)\, d\lambda +$$

$$+ \frac{1}{\sqrt{2\pi}} \int_{\mathbb{R}} e^{i\lambda t} \begin{pmatrix} 1/i\lambda \\ 1 \end{pmatrix} e^{-i\lambda(x-a)} v(\lambda)\, \overline{S_a(\lambda)}\, d\lambda =$$

$$= \Phi_{in}(x+t) + \Phi_{out}(x-t).$$

We see that the complex amplitudes $e^{-i\lambda a} \cdot v(\lambda) \cdot \begin{pmatrix} 1/i\lambda \\ 1 \end{pmatrix}$, $e^{-i\lambda a} \cdot v(-\lambda) S_a(\lambda) \begin{pmatrix} 1/-i\lambda \\ 1 \end{pmatrix}$ of the spectrum of the incoming and outgoing waves are connected with the help of reflexion coefficient.

2. The wave equation and the Regge problem.

A key to the connection between the Regge problem and the unitary group $(U_t)_{t\in\mathbb{R}}$ is given by an explicit description of the generator A of the contractive semigroup $(Z_t)_{t\geq 0}$.

THEOREM 2.1. The generator A of the semigroup $Z_t = e^{itA} = P_K e^{it\mathcal{L}} | K$ is a maximal completely dissipative operator in K. Its domain $D(A)$ is

$$\{\mathcal{U} \in K : u_0'' \in L^2(0,a);\ u_0'(0) = 0;\ u_1' \in L^2(0,a);\ u_0'(a) + u_1(a) = 0\}$$

and $A\mathcal{U} = \mathcal{L}\mathcal{U}$ for $\mathcal{U} \in K$.

PROOF. The operator A is a maximal dissipative operator, because $(Z_t)_{t\geq 0}$ is a contractive semigroup (see theorem X.48 [68]). Assuming that A has a non-trivial self-adjoint part, we see that there is a non-zero element f in K such that $Z_t f = U_t f$ for every $t > 0$. Therefore $U_t f \perp \mathcal{D}_+$ for every $t > 0$ and $f \perp U_{-t}\mathcal{D}_+$ for $t > 0$. But $E = \operatorname{span}(U_{-t}\mathcal{D}_+ : t > 0)$ and so $f = 0$.

The computation of the domain for A is a more subtle problem. Let $\mathcal{D}_0$ be the set of smooth data in K supported on compact subsets of $(0,a)$. Let $\mathcal{L}_0 = \mathcal{L} \mid \mathcal{D}_0$. Clearly, $\mathcal{L}_0$ is symmetric in K . Using Theorem 1.2 , one can easily prove that

$$D(\mathcal{L}_0^*) = \{ \mathcal{U} \in K : u_0'' \in L^2(0,a),\ \ u_0'(0) = 0,\ \ u_1' \in L^2(0,a) \}$$

and that $\mathcal{L}_0^* = \mathcal{L} \mid D(\mathcal{L}_0^*)$. Standard arguments lead to the conclusion that the deficiency indices of $\mathcal{L}_0$ are (1,1). Indeed, if, for example,

$$i\begin{pmatrix} u_0 \\ u_1 \end{pmatrix} = i\begin{pmatrix} \mathbb{O} & -I \\ L & 0 \end{pmatrix}\begin{pmatrix} u_0 \\ u_1 \end{pmatrix}$$

for $\mathcal{U}$ in $D(\mathcal{L}_0^*)$, then $u_1 = -u_0$ and $u_0 = Lu_0$. It follows that $\dim \operatorname{Ker}(i - \mathcal{L}_0^*) = 1$.

By theorem 1.3 for any $\mathcal{U}$ in $\mathcal{D}_0$ we have $U_t \mathcal{U} \in K$ if t is small enough. Therefore $\mathcal{L}_0 \subset A$ and also $\mathcal{L}_0 \subset A^*$. But then $A^* \subset \mathcal{L}_0^*$, $A = A^{**} \subset \mathcal{L}_0^*$ and therefore the domain of A is contained in $D(\mathcal{L}_0^*)$.

Let, for the time being, B denote the restriction of $\mathcal{L}$ onto the subset of data in $D(\mathcal{L}_0^*)$ satisfying the boundary condition $u_0'(a) + u_1(a) = 0$. Clearly, B is a closed operator. Moreover B is a dissipative operator in K , i.e. $\operatorname{Im}\langle B\mathcal{U}, \mathcal{U}\rangle_E \geq 0$. Indeed, for every $\mathcal{U} \in \mathcal{D}(\mathcal{L}_0^*)$

$$\langle \mathcal{L}\mathcal{U}, \mathcal{U}\rangle_E = \frac{1}{2}\int_0^a -iu_1'\overline{u}_0'\,dx + \frac{1}{2}\int_0^a -iu_0''\overline{u}_1\,dx =$$

$$= -\frac{1}{2}\int_0^a \overline{iu_0'}\,du_1 + \frac{1}{2}\int_0^a \overline{iu_1}\,du_0' = \overline{iu_0'(a)}\,u_1(a) + \overline{iu_1(a)}\,u_0'(a) -$$

$$-\frac{1}{2}\int_0^a [u_1 \cdot \overline{iu_0''} + u_0' \cdot \overline{iu_1'}]\, dx = 2i|u_1(a)|^2 + \langle \mathcal{U}, \mathcal{L}\mathcal{U} \rangle_E .$$

Therefore $\operatorname{Im} \langle \mathcal{L}\mathcal{U}, \mathcal{U} \rangle_E = |u_1(a)|^2 \geq 0$.

The operator B is a one-dimensional perturbation of $\mathcal{L}_0$. So to prove $B = A$ it is sufficient to check that $Z_t \mathcal{D}(B) \subset \mathcal{D}(B)$ (see theorem X.49 [68]), for $t > 0$. Let $\mathcal{U} \in \mathcal{D}(B)$. Then $\mathcal{P}_{\mathcal{D}_-} U_t \mathcal{U} = 0$ for $t > 0$ because $U_{-t} \mathcal{D}_- \subset \mathcal{D}_-$ and $\mathcal{U} \perp \mathcal{D}_+ \oplus \mathcal{D}_-$. This means that outside of the interval $(0, a)$ the solution $U_t \mathcal{U}(x)$ is outgoing and therefore $(U_t \mathcal{U})_0'(x) + (U_t \mathcal{U})_1(x) \equiv 0$ for $x > a$. But $(U_t \mathcal{U})_0 \in W_2^2(0, a+t)$ and $(U_t \mathcal{U})_1 \in W_1^1(0, a+t)$ and in particular, these functions are continuous. Therefore $(U_t \mathcal{U})_0'(a) + (U_t \mathcal{U})_1(a) = 0$. To finish the proof it is sufficient only to remark that the projection of $\mathcal{U} \in E$ onto K is a pair in $K \begin{pmatrix} v_0 \\ v_1 \end{pmatrix}$, $v_0(x) \equiv const$, $v_1(x) \equiv 0$ for $x > a$ which coincides with $\mathcal{U}$ on $(0, a)$. ●

REMARK. It is easy to see that for the generator $A' = -A^*$ of the conjugate semigroup the following formula holds

$$\mathcal{D}(A') = \{ \mathcal{U} \in K : u_0 \in W_2^2(0,a),\ u_1 \in W_2^1(0,a),\ u_0'(0) = 0,\ u_0'(a) - u_1(a) = 0 \}.$$

Now we are in a position to describe spectral properties of the operator A . Let $\sigma_d(B)$ denote a point spectrum of an operator B , i.e. the set of all eigen-values. Remind the reader, see lemma 1.4 , that a vector-function $\mathcal{U}$ in K is completely determined by its restriction on the interval $(0, a)$ and that $y_a(x, \lambda)$ denotes the Jost solution corresponding to a point a :

$$L y_a = \lambda^2 y_a ; \quad y_a(a, \lambda) = 1, \quad y_a'(a, \lambda) = -i\lambda .$$

THEOREM 2.2. The spectrum $\sigma(A)$ of the dissipative operator A is equal to $\sigma_d(A) \cup \{\infty\}$ and $\sigma_d(A) = \{ k \in \mathbb{C}_+ : S_a(k) = 0 \}$. The resolvent $(A - \lambda I)^{-1}$ is compact. For $k \in \sigma_d(A)$ the eigen-function $\mathcal{U}_k$ corresponding to the eigen-value k is defined by

$$\mathcal{U}_k(x) = \begin{pmatrix} \frac{1}{ik}\, y_a(x, k) \\ y_a(x, k) \end{pmatrix}, \qquad x \in [0, a] .$$

The spectrum $\sigma_d(A)$ is symmetric with respect to the imaginary axis: $\sigma_d(A) = -\overline{\sigma_d(A)}$.

PROOF. The first statement of the theorem is implied by theorem 1.5. To prove the resolvent of A is compact it is, obviously, sufficient to check that the operator $Tf = P_{K_{S_a}} \frac{1}{x+i} f$ is compact in K_{S_a} . A simple, but important formula, connecting Hankel and model operators, (see [18] , p.237) implies

$$S_a H_{\overline{S}_a \cdot \frac{1}{x+i}} = T \oplus \mathbb{O} .$$

The function S_a being holomorphic on $\mathbb{R}$, it is clear that $\overline{S}_a \cdot (x+i)^{-1} \in C_0(\mathbb{R})$ and therefore by the Hartman-Sarason theorem (see, for example, [18]) the operator $H_{\overline{S}_a \cdot (x+i)^{-1}}$ is compact.

We have by the definition of the reflection coefficient $S_a(k) = -y_a'(0,k) \cdot (y_a'(0,-k))^{-1}$ and therefore $k \in \sigma_d(A)$ iff $y_a'(0,k) = 0$ *). It follows from the definition of the Jost solution that $\mathcal{U}_k \in D(A)$. Now the proof of the equality $A\mathcal{U}_k = k\mathcal{U}_k$ is reduced to a calculation. The last statement of the theorem is an obvious consequence of (5). ●

A completely analogous result holds for the adjoint operator A^* . Clearly, $\sigma_d(A^*) = \overline{\sigma_d(A)}$. Here is a formula for the eigen-function $\mathcal{U}_k^*$, $A^*\mathcal{U}_k^* = \bar{k}\mathcal{U}_k^*$:

$$\mathcal{U}_k^* = \begin{pmatrix} \frac{1}{i\bar{k}} \, y_a(x,-\bar{k}) \\ y_a(x,-\bar{k}) \end{pmatrix} , \quad x \in [0,a] .$$

The following formulae will be useful in what follows:

$$\mathcal{U}_k + \overline{\mathcal{U}}_k^* = 2\begin{pmatrix} 0 \\ y_a(x,k) \end{pmatrix}, \quad \mathcal{U}_k - \overline{\mathcal{U}}_k^* = \frac{2}{ik}\begin{pmatrix} y_a(x,k) \\ 0 \end{pmatrix} . \qquad (7)$$

It should be remarked that

$$A^*\overline{\mathcal{U}}_k^* = -k \cdot \overline{\mathcal{U}}_k^* , \qquad k \in \sigma_d(A) .$$

THEOREM 2.3. The following are equivalent:

a) the family $\{\mathcal{U}_k, \mathcal{U}_k^* : k \in \sigma_d(A)\}$ is complete in K ;

b) the family $\{y_a(x,k) : k \in \sigma_d(A)\}$ of the eigen-func-

---

*) The operator A being dissipative, it follows $\operatorname{Im} k > 0$, otherwise we would get an eigen-value for A in $\mathbb{C}_-$.

tions for problem (2) is complete in $L^2_\rho(0,a)$.

PROOF. a)$\Longrightarrow$b) is obvious in view of (7). b)$\Longrightarrow$a). It is sufficient to check that the completeness of the family $\{y_a(x,k): k\in\sigma_d(A)\}$ in $L^2_\rho(0,a)$ implies its completeness in $W^1_2(0,a)$. We have

$$\langle f,u\rangle_1=\int_0^a f'\bar u' dx=-\int_0^a[f-f(a)]\bar u'' dx=\bar\lambda^2\int_0^a[f(x)-f(a)]\rho^2\bar u dx=$$
$$=\bar\lambda^2\langle f-f(a),u\rangle_{L^2_\rho}$$

for any u satisfying $Lu=\lambda^2 u$. ●

Henceforth we shall often assume the following technical condition is satisfied:

$$\overline{\lim_{y\to+\infty,\,S(iy)=0}}\ 2y|S'(iy)|<1. \qquad (*)$$

It should be remarked that trivial estimates using the Cauchy formula imply $2y\cdot|S'(iy)|\leq 1$.

THEOREM 2.4. Suppose the family $\{u_k,u_k^*: k\in\sigma_d(A)\}$ forms an unconditional basis in K . Then the family of the eigen--functions for the Regge problem (2) forms an unconditional basis in $L^2_\rho(0,a)$ and in $W^1_2(0,a)$ simultaneously. The converse is true if $S_a\in(*)$.

PROOF. The family of 2-dimensional subspaces spanned by the vectors u_k , $\bar u_k^*$ forms, clearly, an unconditional basis in K . Therefore the first statement of the theorem is a consequence of (7).

To prove the second one we remark the functions u_k and $\bar u_k^*$ are othogonal for $k\neq-\bar k$. Indeed,

$$k\langle u_k,\bar u_k^*\rangle=\langle Au_k,\bar u_k^*\rangle=\langle u_k,A^*\bar u_k^*\rangle=-\bar k\langle u_k,\bar u_k^*\rangle.$$

It remains to discuss the case $k=iy$, $y>0$. It follows from (*) that the angles between the vectors u_{iy} and $\bar u_{iy}^*$ are bounded away from zero. To see this we use theorem 1.5. Then the angle between $S\cdot\sqrt{y}\cdot(z-iy)^{-1}$, $(z+iy)^{-1}\sqrt{y}$ coincides with arccos $(2y|S'(iy)|)$ ●

Let $S_a=\theta^d\cdot B$, where $\theta^d=exp(idz)$, $d>0$, and B is a Blaschke product with simple zeroes in a half-plane $\mathbb{C}_\delta$ for some $\delta>0$.

THEOREM 2.5. Suppose the family $(e^{ikx})_{B(k)=0}$ forms an unconditional basis in $L^2(0,d)$. Then the family of the eigen-func -

tions for the Regge problem (2) forms an unconditional basis in $L^2_\rho(0,a)$ and in $W^1_2(0,a)$. The converse is true if $S_a \in (*)$.

PROOF. The first statement of the theorem results from theorem 2, Part I, theorem 1.5 and theorem 2.4. To prove the second statement one should simply inverse the order of theorems cited above. ●

3. Asymptotic properties of the reflexion coefficient and an example to the Regge problem.

It is assumed in this section that $\rho \in C^2[0,a_\rho]$, $\inf\limits_{0<x<a_\rho} \rho(x) > 0$ and that $\lim\limits_{x \to a_\rho - 0} \rho(x) \neq \rho(a_\rho + 0) = 1$. It follows from the formula

$$S_a(\lambda) = -e^{2i(a-a_\rho)\lambda} \cdot \frac{y'(0,\lambda)}{y'(0,-\lambda)}, \quad \operatorname{Im}\lambda \geq 0,$$

that all needed information about S_a can be extracted from the Jost solution $y(x,\lambda)$ corresponding to the point a_ρ.

We begin with an analysis of a "standard" equation

$$-y'' + \rho^{1/2}(\rho^{-1/2})'' y = \lambda^2 \rho^2 y$$

which, obviously, can be solved explicity:

$$y(x) = \rho^{-1/2}(x) \cdot e^{\pm i\lambda \int_x^{a_\rho} \rho(s)\,ds}.$$

One can easily prove that the Green function $G(x,t,\lambda)$ of the "standard" equation is defined by

$$G(x,t,\lambda) = \begin{cases} \rho^{-1/2}(x)\rho(t)^{-1/2} \cdot \dfrac{\sin\lambda \int_t^x \rho(s)\,ds}{\lambda}, & \text{if } x \leq t \\ 0, & \text{if } x > t. \end{cases}$$

Remind that by definition the Green function satisfies the equation:

$$-G'' + \rho^{1/2}(\rho^{-1/2})'' G - \lambda^2 \rho^2 G = \delta(x-t).$$

Therefore for any solution $y_0(x,\lambda)$ of the "standard" equation the solution $y(x,\lambda)$ of the integral equation

$$y(x,\lambda) = y_0(x,\lambda) + \int_x^{a_\rho} G(x,t,\lambda)\rho^{1/2}(t)(\rho^{-1/2}(t))'' y(t,\lambda)\,dt$$

satisfies (2) with the boundary conditions

$$y(a_\rho,\lambda)=y_0(a_\rho,\lambda)=1,\quad y'(a_\rho,\lambda)=y_0'(a_\rho,\lambda)=-i\lambda .$$

The following formula defines the function $y_0(x,\lambda)$:

$$y_0(x,\lambda)=\frac{\rho(a_\rho)^{\frac14}}{\rho(x)^{\frac14}}\cdot\cos\left(\lambda\int_x^{a_\rho}\rho(s)ds\right)+\frac{i\lambda-\frac{\rho'(a_\rho)}{2\rho(a_\rho)}}{[\rho(x)\cdot\rho(a_\rho)]^{\frac14}}\cdot\frac{\sin(\lambda\int_x^{a_\rho}\rho(s)ds)}{\lambda}.$$

A well-known method of iteration can be applied now to investigate the asymptotic behavior of $y'(0,\lambda)$.

Let $\Gamma(x,t,\lambda)\overset{def}{=\!=}G(x,t,\lambda)\,\rho^{1/2}(t)\,\rho^{-1/2}(t)''$. Then

$$|y_0(x,\lambda)|\leq const\exp\{|Im\lambda|\cdot\int_x^{a_\rho}\rho(s)ds\};$$

$$|\Gamma(x,t,\lambda)|\leq const\,\frac{1}{|\lambda|}\cdot\exp\{|Im\lambda|\int_x^{a_\rho}\rho(s)ds\}.$$

Let $g_0(t,\lambda)\overset{def}{=\!=}y_0(t,\lambda)$ and let

$$g_{n+1}(x,\lambda)=\int_x^{a_\rho}\Gamma(x,t,\lambda)\,g_n(t,\lambda)\,dt,\quad n\in\mathbb{Z}_+ .$$

The induction arguments imply

$$|g_n(t,\lambda)|\leq c^{n+1}\cdot\frac{(a-t)^n}{n!\cdot|\lambda|^n}\cdot\exp\{|Im\lambda|\cdot\int_t^{a_\rho}\rho(s)ds\}$$

and therefore the series

$$y(t,\lambda)=\sum_{n=0}^{\infty}g_n(t,\lambda)$$

converges to an entire function $\lambda\mapsto y(t,\lambda)$ of the exponential type:

$$y(t,\lambda)-y_0(t,\lambda)=O(1)\cdot\frac{1}{|\lambda|}\cdot\exp\{|Im\lambda|\int_t^{a_\rho}\rho(s)ds\}.$$

Let now $d\overset{def}{=\!=}\int_0^{a_\rho}\rho(s)\,ds$. A formal differentiation of the asymptotic formula gives

$$y'(0,\lambda)-y_0'(0,\lambda)=O(1)\,e^{d\cdot|Im\lambda|} .$$

The proof is given by the iteration method. A simple computation leads to the following formula:

$$y_0'(0,\lambda)=\left[\rho(a_\rho)\cdot\rho(0)\right]^{1/2}\{\lambda\sin\lambda d-i\rho(a_\rho)^{-1}\lambda\cos\lambda d\}+O(1)e^{d|Im\lambda|} . \quad (9)$$

Hence,

$$\frac{y'(0,\lambda)}{y'(0,-\lambda)}=\frac{\sin\lambda d-i\rho(a_\rho)^{-1}\cos\lambda d+O\left(\frac{1}{|\lambda|}\right)e^{d|Im\lambda|}}{\sin\lambda d+i\rho(a_\rho)^{-1}\cos\lambda d+O\left(\frac{1}{|\lambda|}\right)e^{d|Im\lambda|}} . \quad (10)$$

Clearly,

$$\sin\lambda d-i\rho(a_\rho)^{-1}\cos\lambda d=-\frac{i}{2}e^{-i\lambda d}(1+\rho^{-1}(a_\rho))\left\{e^{2i\lambda d}-\frac{\rho(a_\rho)-1}{\rho(a_\rho)+1}\right\}$$

and therefore $\sin\lambda d-i\rho(a_\rho)^{-1}\cos\lambda d$ is a sine-type function with the sequence of zeroes

$$\lambda_n^0=\frac{\pi n}{d}+\Delta+\frac{i}{2d}\log\left|\frac{1+\rho(a_\rho)}{1-\rho(a_\rho)}\right| , \quad n\in\mathbb{Z} ,$$

where $\Delta=0$ if $\rho(a_\rho)>1$ and $\Delta=\frac{\pi}{2d}$ if $\rho(a_\rho)<1$. One can easily check now that the sequence $(\lambda_n)_{n\in\mathbb{Z}}$ of the zeroes of $y'(0,-\lambda)$ satisfies

$$\lambda_n=\lambda_n^0+O\left(\frac{1}{n}\right) .$$

This implies the sequence $(\lambda_n)_{n\in\mathbb{Z}}$ is an interpolating one for H_+^∞ . It follows also from (10) that

$$\lim_{t\to+\infty}-\frac{y'(0,it)}{y'(0,-it)}=\frac{1-\rho(a_\rho)}{1+\rho(a_\rho)}$$

and therefore

$$S_a(\lambda)=e^{2i(a-a_\rho)\lambda}\cdot B(\lambda), \quad (11)$$

where B is a Blaschke product.

THEOREM 3.1. 1) If $a_\rho\leq a\leq a_\rho+d$ then the family of eigen-functions for the Regge problem (2) is complete in $L_\rho^2(0,a)$.

2) If $a=a_\rho+d$ then the family of eigen-functions forms a Riesz basis in $L_\rho^2(0,a)$.

PROOF. The function $y'(0,\lambda)$ being equal to zero at $\lambda=0$, we see that $\lambda^{-1}y'(0,\lambda)$ is an entire function. It follows from the asymptotic formula (9) that $\lambda^{-1}y'(0,\lambda)\in STF$. The width of the indicator diagram of this function is equal to $2d$.

So we see that

$$\frac{1}{\lambda} y'(0,\lambda) = e^{-i\lambda d} \cdot B \cdot h ,$$

where h is an outer function for $\mathbb{C}_+$. On the other hand, we have by (II)

$$\frac{1}{\lambda} y'(0,\lambda) = -B \frac{1}{\lambda} y'(0,-\lambda)$$

and therefore $h = -\frac{1}{\lambda} y'(0,-\lambda) e^{i\lambda d}$. This implies

$$\frac{\bar{h}}{h}(\lambda) = \frac{y'(0,\lambda)}{y'(0,-\lambda)} e^{-2i\lambda d} = -Be^{-2i\lambda d} , \quad \lambda \in \mathbb{R} .$$

It follows from $\lambda^{-1} y'(0,\lambda) \in STF$ that $|h|^2 \in (HS)$ and therefore

$$dist(\bar{\theta}^{2d} B, H^\infty_+) < 1 ,$$
$$dist(\theta^{2d} \bar{B}, H^\infty_+) < 1 .$$

We see that statement 2) of the theorem is now a simple corollary of theorem 3, Part I, theorems 2.3-2.5.

To prove statement one we should use lemma 3-bis instead of theorem 3, Part I. ●

The number $d = \int_0^{a_\rho} \rho(x)\, dx$ has a nice physical interpretation. Namely, it coincides with the time needed for the point perturbation of the end $x = 0$ of the string (2) to reach the point $x = a_\rho$ (see theorem 1.3.).

An example discussed in this section is closely related with an interesting paper [69] .

References

I. А д а м я н В.М., А р о в Д.З., К р е й н М.Г. Бесконечные ганкелевы матрицы и обобщенные задачи Каратеодори-Фейера и И.Шура. - Функц.анализ и его прил., I968, т.2, вып.4, I-I7.

2. А в д о н и н С.А. К вопросу о базисах Рисса из показательных функций в L^2. - Вестник ЛГУ, сер.матем., I974, № I3, 5-I2.

3. А в д о н и н С.А. К вопросу о базисах Рисса из показательных функций в L^2. - Зап.научн.сем.ЛОМИ, I974, т.39,I76-I77.

4. В и н о г р а д о в С.А., Х а в и н В.П. Свободная интерполяция в H^∞ и в некоторых других классах функций I. - Зап. научн.семин.ЛОМИ, I974, т.47, I5-54.

5. Г о л о в и н В.Д. О биортогональных разложениях в L^2 по линейным комбинациям показательных функций. - Зап.мех.-мат. ф-та ХГУ и Хар.мат.об-ва, I964, т.30, I8-24.

6. Г о л о в и н В.Д. Об устойчивости базиса показательных функций. - Докл. АН Арм. ССР, I963, т.36, № 2, 65-70.

7. Г о х б е р г И.Ц., К р е й н М.Г. Введение в теорию линейных несамосопряженных операторов в гильбертовом пространстве. Москва, "Наука", I965.

8. Г о ф м а н К. Банаховы пространства аналитических функций. Москва, "ИЛ", I963. (K.Hoffman, Banach spaces of analytic functions, Prentice-Hall, N.J., I962).

9. З и г м у н д А. Тригонометрические ряды. т.I. Москва, "Мир", I965. (A.Zygmund, Trigonometric Series, v.I, I959).

I0. К а д е ц М.И. Точное значение постоянной Палея-Винера. - Докл.АН СССР, I964, т.I55, № 6, I253-I254.

II. К а ц н е л ь с о н В.Э. Об условиях базисности системы корневых векторов некоторых классов операторов. - Функц.анализ и его прил., I967, т.I, вып.2, 39-5I.

I2. К а ц н е л ь с о н В.Э. О базисах из показательных функций в L^2. - Функц.анализ и его прил., I97I, т.5, вып.I, 37-47.

I3. Л е в и н Б.Я. Распределение корней целых функций. Москва, ГИТТЛ, I956.

I4. Л е в и н Б.Я. О базисах из показательных функций в L^2. - - Зап.матем.отд.ф.-м.ф-та Харьк.ун-та и Харьк.мат.об-ва, I96I, т.27, сер.4, 39-48.

I5. Л е в и н Б.Я., О с т р о в с к и й И.В. О малых возмущениях множества корней функций типа синуса. - Изв.АН СССР, сер.матем., 1979, т.43, № 1, 87-110.

16. Надь Б., Рисс Ф. Лекции по функциональному анализу. Москва, "Мир", 1979. (Riesz F., B.Sz.-Nagy, Leçons d'Analyse Fonctionnelle, Budapest, AK., 1972).

17. Никольский Н.К. Базисы из инвариантных подпространств и операторная интерполяция. - Труды Матем.ин-та им. В.А.Стеклова АН СССР, 1977, т.130, 50-123.

18. Никольский Н.К. Лекции об операторе сдвига. Москва, "Наука", 1980.

19. Никольский Н.К. Базисы из экспонент и значений воспроизводящих ядер. - Докл.АН СССР, т.252, № 6, 1316-1320.

20. Никольский Н.К., Павлов Б.С. Базисы из собственных векторов вполне неунитарных сжатий и характеристическая функция. - Изв.АН СССР, сер.матем., 1970, т.34, № 1, 90-133.

21. Павлов Б.С. О совместной полноте системы собственных функций сжатия и его сопряженного. - В сб.: Проблемы матем. физики, ЛГУ, 1971, вып.5, 101-112.

22. Павлов Б.С. Спектральный анализ дифференциального оператора с "размазанным" граничным условием. - В сб.: Проблемы матем.физики, ЛГУ, 1973, вып.6, 101-119.

23. Павлов Б.С. Базисность системы экспонент и условие Макенхоупта. - Докл.АН СССР, 1979, т.247, № 1, 37-40.

24. Седлецкий А.М. Базисы из экспонент в пространствах L^p (в печати).

25. Хрущёв С.В. Теоремы возмущения для базисов из экспонент и условие Макенхоупта. - Докл.АН СССР, 1979, т.247, № 1, 44-48.

26. Ahern P., Clark D. Radial limits and invariant subspaces. - Amer.J. of Math., 1970, v.XCII, N 2, 332-342.

27. Boas R.P. Entire Functions. New York, "AP", 1954.

28. Carleson L. An interpolation problem for bounded analytic functions. - Amer.J.Math., 1958, v.80, N 4, 921-930.

29. Clark D. One dimensional perturbations of restricted shifts. - J. anal.math., 1972, v.25, 169-191.

30. Collingwood E.F., Lohwater A.J. The theory of cluster sets. Cambridge, At the Univ.Press, 1966.

31. Devinatz A. Toeplitz operators on H^2 spaces. - - Trans.Amer.Math.Soc., 1964, v.112, 304-317.

32. Douglas R., Sarason D. A class of Toeplitz operators. - Indiana Univ.Math.J., 1971, v.20, N 10, 891-895

33. D y m H., M c K e a n H.P. Gaussian Processes, Function Theory, and the Inverse Spectral Problem. New York, "AP", I976.

34. D u f f i n R.J., E a c h u s J.J. Some notes on an expansion theorem of Paley and Wiener. - Bull.Amer.Math.Soc. I942, v.48, N I2, 850-855.

35. D u f f i n R.J., S c h a e f f e r A.C. A class of nonharmonic fourier series. - Trans.Amer.Math.Soc., I952, v.72, N 2, 34I-366.

36. G a r n e t t J.B. Two remarks on interpolation by bounded analytic functions. - Lecture Notes in Math., N 604, Berlin Springer, I977, 32-40.

37. G e o r g i j e v i č D. Bases orthogonales dans les espaces $H^p(e)$ et H^p . - C.R.Acad.Sci., Paris, I979, t.289, Ser.A, 73-74.

38. H e l s o n H., S z e g ö G. A problem in prediction theory. - Ann.Math.Pura Appl., I960, v.5I, I07-I38.

39. H r u š č ë v S.V., V o l'b e r g A. A generalization of P.Koosis interior compactness theorem. - LOMI Preprints, E-4-80, Leningrad I980.

40. H u n t R.A., M u c k e n h o u p t B., W h e e d e n R.L. Weighted norm inequalities for the conjugate function and Hilbert transform. - Trans.Amer.Math.Soc., I973, v.I76, 227-25I.

4I. I n g h a m A.E. A note on Fourier transforms. - J.London Math.Soc., I934, v.9, 29-32.

42. K a h a n e J.-P., S a l e m R. Ensembles parfaits et séries trigonométriques. Paris, Hermann, I963.

43. K o o s i s P. Interior Compact Spaces of Functions on a Half-Line. - Comm.Pure and Appl.Math., I957, v.IO, N 4, 583-6I5.

44. K o o s i s P. Introduction to H_p Spaces. London Math. Soc.Lecture Note Series 40, Cambridge Univ.Press, I980.

45. K o o s i s P. Weighted quadratic means of Hilbert transforms. - Duke Math.J., I97I, v.38, N 3, 609-634.

46. L a x P.D. Remarks on the preceeding paper. - Comm.Pure Appl.Math., I957, v.IO, N 4, 6I7-622.

47. L a x P., P h i l l i p s R. Scattering Theory, N.Y., "AP", I969.

48. L e v i n s o n N. Gap and density theorems. - Amer.Math. Soc.Coll.Publ., v.26, I940.

49. M o e l l e r J.W., F r e d e r i c k s o n P.O. A den-

sity theorem for lacunary Fourier series. - Bull. of Amer. Math.Soc., 1966, v.72, N 1, part 1, 82-86.

50. N e h a r i Z. On bounded bilinear forms. - Ann. of Math., 1957, v.65, 153-162.

51. R e d h e f f e r R.M. Completeness of sets of complex exponentials..- Advances in Math., 1977, v.24, N 1, 1-62.

52. R e g g e T. Analytic properties of the scattering matrix.- - Nuovo Cimento, 1958, v.10, N 8, 671-679.

53. R o c h b e r g R. Toeplitz Operators on Weighted H^p spaces. - Indiana Univ.Math.J., 1977, v.26, N 2, 291-298.

54. S a r a s o n D. Function theory on the unit circle. Dept. Math.Virginia Polytechnic Institute and State University. Blacksburg, Va24061, 1978.

55. S c h w a r t z L. Étude des sommes d'exponentielles réelles. Paris, Hermann, 1943.

56. S h a p i r o H.S., S h i e l d s A.L. On some interpolation problems for analytic functions. - Amer.J.Math., 1961, v.83, N 3, 513-532.

57. W i d o m H. Inversion of Toeplitz matrices III. - Notices Amer.Math.Soc., 1960, v.7, 63.

58. W i e n e r N. On the closure of certain assemblages of trigonometric functions.-Proc.Nat.Acad.Sci.,USA,1927,v.13,27.

59. W i e n e r N., P a l e y R. Fourier transforms in the complex domain.New York,AMS,1934.(русск.пер.: Винер Н.,Пэли Р.Преобразование Фурье в комплексной области,М.,"Наука",1964).

60. Y o u n g R.M. Inequalities for a perturbation theorem of Paley and Wiener. - Proc.Amer.Math.Soc., 1974, v.43, N 2, 320-322.

61. Н и к о л ь с к и й Н.К., П а в л о в Б.С. Базисы из собственных векторов, характеристическая функция и задачи интерполяции в пространстве Харди. - Докл.АН СССР, 1969, т.184, № 3, 550-553.

62. Н и к о л ь с к и й Н.К., П а в л о в Б.С. Базисы из собственных векторов вполне неунитарных сжатий. - Докл.АН СССР, 1969, т.184, № 4, 778-781.

63. Н и к о л ь с к и й Н.К., П а в л о в Б.С. Разложения по собственным векторам неунитарных операторов и характеристическая функция. - Зап.научн.семин.ЛОМИ, 1968, II, 150-203.

64. S h a p i r o H.S., S h i e l d s A.L. Interpolation in Hilbert spaces of analytic functions. - Studia math.Ser.spec., Proc.conf.funct.anal.(1960), 1963, 109-110.

65. N e w m a n D. Interpolation in H^∞. - Trans.Amer.Math.

Soc., 1959, v.92, 501-507.

66. H a y m a n W. Inteprolation by bounded functions. - Univ. de Grenoble, Annales de l'Institut Fourier, 1958, v.8, 277-290.

67. R e e d M., S i m o n B. Methods of modern mathematical physics. 1. Functional analysis.N.Y., AP,1972(русский перевод: Рид М., Саймон Б. Методы современной математической физики.I. Функциональный анализ. М., "Мир", 1977.

68. R e e d M., S i m o n B. Methods of modern mathematical physics. II: Fourier analysis, self-adjointness. N.Y.A.P. 1975(русский перевод: Рид М., Саймон Б. Методы современной математической физики.2. Гармонический анализ, самосопряженность. М., "Мир", 1978.

69. А д а м я н В.М., А р о в Д.З. Об унитарных сцеплениях полуунитарных операторов. Матем.исследования, АН Молд.ССР, I, 1966, № 2, 3-64.

70. П е к к е р М.А. Резонансы при рассеянии акустических волн со сферической неоднородностью плотности. Труды Седьмой Зимней Школы, Математическое программирование и теория операторов в линейных пространствах, Дрогобыч 1974, ЦЭМИ, М., 1976, 70-100.

71. B a l l J., L u b i n A. On a class of contraction perturbations of restricted shifts. Pacific J.Math., 1976,63, N 2.

72. F u h r m a n n P. On a class of finite dimensional contractive perturbations of restricted shifts of finite multiplicity. Isr.J.Math., 1973, 16, 162-175.

73. K r i e t e T.L., R o s e n b l u m M. Phragmén-Lindelöf theorem with applications to $M(u,v)$-functions. Pacific J. Math., 1972, 43, 175-188.

S.V.Kisliakov

WHAT IS NEEDED FOR A 0-ABSOLUTELY SUMMING OPERATOR TO BE NUCLEAR?

0. Introduction.

A famous result of Grothendieck stating that each continuous linear operator from a space $L^1(\mu)$ to a Hilbert space H is 1-absolutely summing is the origin of a considerable part of the operator ideals theory, and the present paper is also inspired by it. Besides the Grothendieck's paper [1] one can find proofs of this result e.g. in [2] or [3] , p.308. In symbols it can be written as follows: $\Pi_1(L^1(\mu), H) = \mathcal{L}(L^1(\mu), H)$ (for the notation and definitions see the end of this section), and it is known now that an analogous equality $\Pi_1(X, H) = \mathcal{L}(X, H)$ (which we call the Grothendieck equality for the space X) holds for some spaces X differing from $L^1(\mu)$, cf. [2] , [4] , [5] . It is still unknown, however, whether some particular spaces X that arise naturally in Analysis satisfy this equality. Of the greatest importance is probably the question whether we can take $X = C_A^*$, C_A being the disc algebra (cf. [6] , p.12-14 for a detailed discussion of this matter).

If $X = Y^*$ (in particular, if $X = C_A^*$) then we can try to prove the equality $\Pi_1(X,H) = \mathcal{L}(X, H)$ in the following way. Using desiriptions of operator ideals adjoint to the ideals Π_1 and $\mathcal{L}$ (cf. [3] , p.269, theorem I9.2.3 or p.276, theorem I9.4.7; cf. also [7]) we obtain that this equality is equivalent to the equality $\Gamma_\infty(H,X) = I_1(H,X)$. The latter one holds if and only if $\Gamma_1(Y,H) = I_1(Y,H)$ (here I_1 and Γ_p , $p=1,\infty$ stand for the ideals of 1-integral and L^p-factorable operators respectively)*). But since every

---

*) The fact that the conditions $\Gamma_\infty(H,X) = I_1(H,X)$ and $\Gamma_1(Y,H) = I_1(Y,H)$ are equivalent can be easily derived by passing to conjugate operators provided one knows that the equality $\Gamma_1(Y,H) = I_1(Y,H)$ implies that $\Gamma_1(Y^{**},H) = I_1(Y^{**},H)$. This implication may be proved via the local reflexivity principle ([3] , p.384) and theorem 8.7.4 at the page II5 in [3] .

1 -integral operator with values in a reflexive space is nuclear we get finally that the conditions $\Pi_1(Y^*,H) = \mathcal{L}(Y^*,H)$ and $\Gamma_1(Y,H) = N_1(Y,H)$ are equivalent.

To check the second condition it is sufficient, of course, to check the inclusion $\Gamma_1(Y,H) \subset N_1(Y,H)$ (for the reverse inclusion is evident). But Maurey's improvement of Grothendieck theorem (cf. [8] , p.II7) implies that $\Gamma_1(Y,H) \subset \Pi_0(Y,H)$ (Π_0 denotes the ideal of 0-absolutely summing operators). Therefore the space Y^* satisfies the Grothendieck equality if Y posesses the following poperty:

Every 0-absolutely summing operator from Y to a Hilbert space is 1 - nuclear (KP)

The notation (KP) is composed of the first letters of the names of the authors of [9] , because in this paper a similar scheme was applied to treat the case of translation - invariant operators. For example, it is shown in [9] that every 0-absolutely summing and translation-invariant operator from the disc algebra to the Hardy space H^2 is nuclear (cf. also [I0]). It seems to be very probable that the invariance assumption is superfluous.

CONJECTURE. The space C_A has the property (KP).

The next theorem is the strongest result which I was able to prove in support of this conjecture. Its proof inspired by the paper [9] is presented in Section 1.

THEOREM I. Every 0-absolutely summing operator from the disc algebra to ℓ^2 is p-nuclear for each $p>1$.

Note that if E and F are any Banach spaces then all classes $\Pi_r(E,F)$, $0 \leq r < 1$ coincide (cf. [8] or [3] , p.293, theorem 2I.2.9), and hence $\Pi_r(C_A,\ell^2) \subset N_p(C_A,\ell^2)$ if $p>1$ and $r<1$. However for $r=1$ the last inclusion is no longer true for the map $f \longmapsto \{\hat{f}(2^n)\}$ provides an example of an (invariant!) 1-absolutely summing operator from the disc algebra onto ℓ^2 , cf. [II] , p.20, corollary 3.I. In section 2 we construct other operators defined in a similar manner and posessing the same properties, but also an additional property of being 0-absolutely summing. Let, for example, $\mathcal{F} = \operatorname{clos}_{C(\mathbb{T})} \operatorname{span} \{z^n : n \geq 0 \text{ or } n = -2^k,\ k = 0,1,2,\dots\}$ and define an operator $\mathcal{R}$ by $\mathcal{R}f = \{\hat{f}(-2^k)\}_{k\geq 0}$.

Then (see Section 2) we have: $\mathcal{R}(\mathcal{F})=\ell^2$, $\mathcal{R}\in\Pi_0(\mathcal{F},\ell^2)$ and, moreover, $\mathcal{R}$ is translation invariant. So, by Theorem I, the space $\mathcal{F}$ is isomorphic to no factor-space of the disc algebra, though it is obtained from the latter one by adding a "small" space. We shall also discuss in Section 2 some other interesting properties of the space $\mathcal{F}$.

In section 3 we construct (among other things) a sequence of finite rank operators for which quantitative characteristics corresponding to some qualitative properties of the operator $\mathcal{R}$ described above behave in an "extreme" way. The main idea of construction of these operators is similar to that for $\mathcal{R}$ (both of them may be drafted as follows: take a "known" space, "add" to it a "small" space and then "project" onto this small space). To exploit this idea in a finite dimensional setting we need (and prove) some facts about almost Euclidean subspaces of spaces ℓ^1_n/F , F being a "not very big" subspace of ℓ^1_n (these facts are probably of independent interest). It is shown also in the same section that there is an infinite dimensional space Z with $\Pi_1(Z^*,H)=\mathcal{L}(Z^*,H)$ which does not have the property (KP). So the scheme of proving the Grothendieck equality described above does not work in general.

Finally in Section 4 it is established that $\Pi_0(C(K),Z)\subset N_1(C(K),Z)$ for every Banach space Z .

ACKNOWLEDGEMENT. I am grateful to **A.Pełczyński** for many stimulating discussions. Also I want to thank N.K.Nikol-skiĭ and V.P.Havin for reading the manuscript.

DEFINITIONS, NOTATION AND SOME STANDARD FACTS. We denote by $\mathcal{L}(X,Y)$ the set of all continuous linear operators from a quasinormed space X to a quasinormed space Y . Let X be a Banach space, $0<p<\infty$, $T\in\mathcal{L}(X,Y)$. The operator T is said to be p-absolutely summing if

$$\pi_p(T)\overset{def}{=\!=}\sup\Big\{\Big(\sum_{i=1}^{n}\|Tx_i\|^p\Big)^{1/p}:\ x_1,\dots,x_n\in X,$$

$$\sum_{i=1}^{n}|F(x_i)|^p\leqslant\|F\|^p \quad \text{for all} \quad F\in X^*;\ n=1,2,\dots\Big\}<\infty .$$

The class of all p-absolutely summing operators is denoted by Π_p . We do not reproduce here the definition of 0-absolutely summing operators (it may be found e.g. in [8]), for it is known that if X and Y are Banach spaces then (as it has already been mentioned) all classes $\Pi_r(X,Y)$, $0\leq r<1$ coincide and it is more convenient for us to deal with Π_r, $r>0$ than with Π_0.

It is well-known (cf. [3], p.232) that $T\in\Pi_p(X,Y)$ if and only if there esists a probability Radon measure μ supported by the unit ball K of the space X^* (with the topology $\sigma(X^*,X)$) so that for every x in X we have

$$\|Tx\|\leq C(\int_K|F(x)|^p d\mu(F))^{1/p} .$$

(Moreover, if $T\in\Pi_p$ then μ can be chosen in such a way that $C=\pi_p(T)$). This inequality and the measure μ are called the <u>Grothendieck-Pietsch inequality</u> and a <u>Grothendieck-Pietsch measure</u> for T. If $p\geq 1$ and A is a $\sigma(X^*,X)$-compact subset of K such that $\operatorname{clos\,conv\,eq} A=K$ (the closure in the topology $\sigma(X^*,X)$) then a Grothendieck-Pietsch measure for T may be chosen to be concentrated on A. In particular if $p\geq 1$ and X is a subspace of a space $C(S)$ then we can assume that this measure is defined on S.

We need the following version of Grothendieck-Pietsch inequality for translation invariant operators (a more general result can be found in [I2]).

THEOREM ON INVARIANT OPERATORS. Let G be a compact commutative topological group and let X be a translation-invariant subspace of $C(G)$. Let Φ be a homomorphism from G to the group of isometries of the space Y. Suppose that $p\geq 1$ and let T be a p-absolutely summing operator from X to Y satisfying $\Phi(g)(Tx)=T(L_g x)$ for $g\in G$, $x\in X$ (L_g is the rotation operator: $(L_g x)(h)=x(gh)$). Then the Haar measure of G can be taken as a Grothendieck-Pietsch measure for T.

Let (S,μ) be a measure space, $\mu(S)<\infty$, $0<p<\infty$. By $I_{p,\mu}$ we denote the formally identical operator from $L^\infty(\mu)$ to $L^p(\mu)$. If X and Y are arbitrary Banach spaces, $T\in\mathcal{L}(X,Y)$ and A, B are subspaces of X and Y so that $T(A)\subset B$ then the operator from A to B acting as the restriction of T to A is called a <u>part</u> of T. The following theorem is an immediate consequence of the Grothendieck-Pietsch inequality (cf. [3], section I7.3).

FACTORIZATION THEOREM. Let $0<p<\infty$, $T\in\mathcal{L}(X,Y)$.

Then $T \in \Pi_p(X, Y)$ if and only if $T = URV$ where U and V are linear operators and R is a part of an operator $I_{p,\mu}$.

REMARK. Of course, any Grothendieck-Pietsch measure for T may serve as a " μ " in this theorem. In particular if X is a subspace of a space $C(S)$ and $p \geq 1$ we can choose this μ to be defined on S . In this case the domain of the operator U coincides with the closure of the set X in $L^p(\mu)$. Note that the assumption $p \geq 1$ is essential here.

Let X, Y be Banach spaces, $p \geq 1$. An operator T , $T \in \mathcal{L}(X, Y)$ is said to be p-integral if there exist a measure μ and linear operators V from X to $L^\infty(\mu)$ and U from $L^p(\mu)$ to Y^{**} so that $\varkappa_Y T = U I_{p,\mu} V$ ($\varkappa_Y$ stands for the canonical imbedding operator of Y into Y^{**}). We denote the space of p-integral operators by $I_p(X, Y)$ and supply it with the norm i_p, $i_p(T) \stackrel{def}{=} \inf \|U\| \, \|\mu\|^{-1/p} \|V\|$ (the infimum is taken over all factorizations as above).

Again let $p \geq 1$ and $T \in \mathcal{L}(X, Y)$. Then T is said to be p -nuclear if T has a factorization of the form $T = U I_{p,\mu} V$, where μ is a discrete measure, $V \in \mathcal{L}(X, L^\infty(\mu))$, $U \in \mathcal{L}(L^p(\mu), Y^{**})$. The norm ν_p in the space $N_p(X, Y)$ of all p -nuclear operators is defined similarly to the norm i_p : $\nu_p(T) = \inf \|U\| \|\mu\|^{1/p} \|V\|$.

For the basic properties of the operator classes I_p and N_p see [3] .

By z we shall denote the identity mapping of the complex plane $\mathbb{C}$ onto itself as well as restrictions of this mapping to certain subsets of $\mathbb{C}$. Let m be the normalized Lebesgue measure on the unit circle $\mathbb{T}$. The disc-algebra is the subalgebra of $C(\mathbb{T})$ consisting of all continuous functions f with $\hat{f}(n) \stackrel{def}{=} \int f \bar{z}^n dm = 0$ for $n = -1, -2, \ldots$. A function f in $C(\mathbb{T})$ belongs to the disc-algebra if and only if f is a restriction to $\mathbb{T}$ of some function continuous in the closed unit disc $\bar{\mathbb{D}}$ and analytic in the open disc $\mathbb{D}$ (we reserve for this function the same notation f ; in general we often do not distinguish notationally between a harmonic function in the unit disc and its boundary values).

The norm of a function f in the space $L^p(\mu)$ (μ being a finite positive measure) will be denoted by $\|f\|_{p,\mu}$ (and simply by $\|f\|_p$ if $\mu = m$). If μ is a finite measure on the circle $\mathbb{T}$ then $H^p(\mu)$ stands for the closure of the set C_A in the space $L^p(\mu)$. We write L^p for $L^p(m)$ and H^p for $H^p(m)$. The main facts about the classes H^p may be found in [I3] . The symbol $\mathbb{P}_+$ (resp. $\mathbb{P}_-$) stands for the orthogonal projection of the space L^2 onto H^2 (resp. onto $\overline{H^2}$). Note that $\mathbb{P}_+ + \mathbb{P}_- \neq \neq id_{L^2}$. It is known that the operators $\mathbb{P}_+$ and $\mathbb{P}_-$ act in fact from L^p to L^p , $1<p<\infty$ and from L^1 to L^r , $0<r<1$ (cf. [I3]).

The space of all Radon (= Borel regular) measures on a compact space S will be denoted by $M(S)$ and we identify this space with the space $C(S)^*$.

As usual, $\mathbb{Z}$ denotes the group of integers, $\mathbb{Z}_+ \overset{def}{=} \{n\in \mathbb{Z} : n\geqslant 0\}$, $\mathbb{Z}_- \overset{def}{=} \{n\in\mathbb{Z} : n<0\}$.

The formulae in each particular section are numerated independently (if at all).

1. Operators from the disc-algebra C_A to the space ℓ^2.

As it has already been mentioned in the introduction, it is convenient for us to deal with the classes Π_r , $0<r<1$, instead of Π_0 . Fix (till the end of this section) a number r with $0<r<1$ and let $T\in\Pi_r(C_A,\ell^2)$. Our aim will be to try to prove that T is 1-nuclear (or at least p-nuclear for all $p>1$).

Since $\Pi_r\subset\Pi_1$ (cf. [3] , p.235, theorem I7.3.9), an application of Factorization theorem and of Remark after it gives a measure μ on the unit circle $\mathbb{T}$ so that T admits the following factorization:

$$C_A \xrightarrow{I_{1,\mu}|C_A} H^1(\mu) \xrightarrow{S} \ell^2 \tag{I}$$

(with $S\in\mathcal{L}(H^1(\mu),\ell^2)$). Let ν be the singular part of the measure μ with respect to m . Then $\mu = am+\nu$, $a\in L^1$. We may assume without loss of generality that $a\geqslant \mathbb{1}$ (since the replacement of a by $a+\mathbb{1}$ does not affect the factorization (I)).

Now (cf. e.g. [II] , section 2) $H^1(\mu)=H^1(am)\oplus L^1(\nu)$, $I_{1,\mu}|C_A = I_{1,am}|C_A + I_{1,\nu}|C_A$ and the operator $I_{1,\nu}|C_A$ (if considered as an operator from C_A into $L^1(\nu)$) is 1-integral. Since the space ℓ^2 is reflexive, the operator

$SI_{1,\nu} \mid C_A$ is nuclear (cf. [3], p.245, theorem 24.6.2). On the other hand it is r-absolutely summing (this follows from Maurey's extension of Grothendieck theorem, cf. [8], p. II7, théorème 94, for, $SI_{1,\nu} \mid C_A$ being nuclear, it is of the form uV with $V \in \mathcal{L}(\ell^1, \ell^2)$). So the operator $SI_{1,am} \mid C_A$ is also r-absolutely summing, and it is sufficient to prove that this operator is nuclear (or at least p-nuclear, $p > 1$).

Now we have the following situation. There is a function a, $a \in L^1(m)$, $a \geqslant \mathbb{1}$ accompanied by an operator T, $T \in \Pi_r(C_A, \ell^2)$ admitting the following factorization

$$C_A \xrightarrow{I_{1,am} \mid C_A} H^1(am) \xrightarrow{S} \ell^2. \qquad (2)$$

Let e_i^* be the coordinate functionals of the space ℓ^2. Consider for each fixed i a norm-preserving extension of the functional $S^* e_i^*$ from $H^1(am)$ to $L^1(am)$ and let h_i, $h_i \in L^\infty(am)$ be the function corresponding to this extension after we identify the spaces $L^1(am)^*$ and $L^\infty(am)$ in the standard way. We have then the following formula for S :

$$Sf = \{ \int f h_i \, a \, dm \}_{i \in \mathbb{Z}_+}, \quad f \in H^1(am).$$

Let b be an outer function satisfying $|b| = a$ a.e. (that is $b = \exp(u + i\tilde{u})$, where $u = \log a$ and $\tilde{u}$ is the harmonic conjugate of u ; the assumption $a \geqslant \mathbb{1}$ has been made to assure that such b exists, cf. [I3], p.24-25). Set $\varphi_i = \mathbb{P}_-(h_i a b^{-1})$ (note that $\sup \| h_i a b^{-1} \|_\infty < \infty$ and hence the functions φ_i are uniformly bounded in L^p for every $p < \infty$, cf. [I3]).

LEMMA I. In the situation just described the following inequality holds:

$$\Big(\int \Big(\sum_{i \in \mathbb{Z}_+} |\varphi_i|^2 \Big)^{r/2} dm \Big)^{1/r} \leqslant c \pi_r(T),$$

where c depends on r only.

PROOF. For any (say, continuous) C_A -valued function f

on $\mathbb{T}$ we have

$$\left(\int \|Tf(\zeta)\|_{\ell^2}^{r} dm(\zeta)\right)^{1/r} \leqslant \pi_r(T) \sup\left\{\left(\int |\langle f(\zeta),\mu\rangle|^r dm(\zeta)\right)^{1/r} : \mu\in M(\mathbb{T}),\quad \|\mu\|\leqslant 1\right\}.$$

(To prove this note that if $f=\sum_{k=1}^{n} x_k \chi_{\ell_k}$ with $x_k\in C_A$ and $\mu e_k = n^{-1}$ for all k then this inequality applied to f is just the inequality defining r-absolutely summing operators). Letting $b_\lambda(z)=b(\lambda z)$ $0<\lambda<1$ and substituting $f(\zeta)=(b_\lambda(z)(1-\rho\bar\zeta z))^{-1}$ with λ,ρ fixed, $0<\lambda,\ \rho<1$ we obtain:

$$\left(\int\left(\sum_{i\in\mathbb{Z}_+}\left|\int h_i(z)\frac{a(z)}{b_\lambda(z)}\frac{1}{1-\rho\bar\zeta z}\,dm(z)\right|^2\right)^{r/2} dm(\zeta)\right)^{1/r}\leqslant$$

$$\leqslant \pi_r(T)\sup\left\{\left(\int\left|\int\frac{1}{1-\rho\bar\zeta z}\frac{d\mu(z)}{b_\lambda(z)}\right|^r dm(\zeta)\right)^{1/r} : \mu\in M(\mathbb{T}), \|\mu\|\leqslant 1\right\}.$$

In view of the Kolmogorov-Smirnov theorem (cf. [13] , p.57) the right-hand side of this inequality is less than or equal to the number $c_r\pi_r(T)\sup\{\|b_\lambda^{-1}\mu\| : \mu\in M(\mathbb{T}),\ \|\mu\|\leqslant 1\}\leqslant c_r\pi_r(T)$ (since $|b|\geqslant 1$ in the unit disc $\mathbb{D}$; here c_r does not depend on λ and ρ).

By Fatou lemma and the fact that $\lim_{\lambda\to 1} b_\lambda(z)=b(z)$ a.e. we can pass to limit as $\lambda\to 1$ and get

$$\left(\int\left(\sum_{i\in\mathbb{Z}_+}\left|\int h_i\frac{a(z)}{b(z)}\frac{1}{1-\rho\bar\zeta z}\,dm(z)\right|^2\right)^{r/2} dm(\zeta)\right)^{1/r}\leqslant c_r\pi_r(T).$$

Since for every ζ the function $z\longmapsto(1-\rho z\bar\zeta)^{-1}$ is in H^2, the integral under the modulus sign coincides with the integral $\int\varphi_i(z)(1-\rho\bar\zeta z)^{-1}dm(z)$. Now using the Cauchy formula and the fact that the function $\overline{\varphi_i(z)}$ is analytic in $\mathbb{D}$ we can rewrite the last inequality in the following way:

$$\Big(\int\Big(\sum_{i\in\mathbb{Z}_+}|\varphi_i(\rho\zeta)|^2\Big)^{r/2}dm(\zeta)\Big)^{1/r}\leqslant c_r\pi_r(T)\,.$$

To complete the proof it suffices to pass to limit as $\rho\to 1$ (applying the Fatou lemma once more). ●

COROLLARY. Under the assumtions of Lemma I we have

$$\Big(\sum_{i\in\mathbb{Z}_+}\|\varphi_i\|_r^2\Big)^{1/2}\leqslant c_r\pi_r(T)\,.$$

PROOF. Apply Lemma I and the reverse Minkowski inequality in the space $L^{r/2}$. ●

REMARK. Lemma I implies immediately the result of the paper [9] mentioned in the introduction. Indeed, let T , $T\in\Pi_r(C_A, H^2)$ be a translation-invariant operator. If we identify the spaces H^2 and ℓ^2 by means of the Fourier transform then by Theorem on invariant operators (see the introduction) the operator T admits a factorization of the form (2) with $a=\mathbb{1}$ and $h_n=c_n\bar{z}^n$ (c_n are some constants). Lemma I gives now that $\sum|c_n|^2<\infty$, and this implies easily that the operator T is nuclear.

PROOF OF THEOREM I (for the statement see the introduction). Let $V\in\Pi_r(C_A,\ell^2)$ and suppose that V has a factorization of the form (2). Then for every p , $p>1$, V has also the following factorization:

$$C_A\xrightarrow{I_{p,am}|C_A}H^p(am)\xrightarrow{J}H^1(am)\xrightarrow{S}\ell^2$$

(J is the formal identity operator). Fix (till the end of the proof) a number p, $p>1$. By [II] , section 2 the operator $I_{p,am}|C_A$ is p-integral, hence it is sufficient to prove that the operator SJ is compact (cf. [3] , p.346, Proposition 24.6.3).

Suppose to the contrary that this operator is not compact. Then there exists a bounded sequence $\{x_n\}$ in $H^p(am)$ ($\|x_n\|_{p,am}\leqslant M$, say) as well as a positive number C so that $SJx_n\to 0$ weakly but $\|SJx_n\|\geqslant C$.

Passing to a subsequence we can assume that the sequence $\{SJx_n\}$ is equivalent to the unit vector basis of the space ℓ^2. Let $X = \mathrm{clos}_{\ell^2}\,\mathrm{span}\{SJx_n : n \in \mathbb{Z}_+\}$ and denote by P the orthogonal projection from ℓ^2 onto X and by Q the isomorphism between X and ℓ^2 which takes the vector SJx_n to the n-th unit vector e_n, $n = 1, 2, \ldots$. Finally, set $T = QPV$.

This operator T clearly has a factorization of the form (2) (and from now on " S " will stand for the operator which arises in the factorization of T, but not of V). Moreover, the operator T has the following property: there is a sequence $\{x_n\}$ in $H^1(am)$ such that in fact $x_n \in H^p(am)$, $\|x_n\|_{p,am} \leqslant M$ for all n and the equality $Sx_n = e_n$ holds. We shall show that this is impossible.

Given the factorization of T of the form (2) we construct the functions h_i, φ_i and b just in the same way as it was explained after the formula (2). Take a $K > 0$ and let b_K be the outer function with $|b_K| = \min(a, K)$ a.e. It is an easy consequence of well-known properties of the operator of harmonic conjugation that $b_K \longrightarrow b$ in measure (the Lebesgue one is meant) as $K \longrightarrow \infty$. Since the functions h_i are uniformly bounded a.e., we obtain

$$\left| \int (h_n x_n a b^{-1} b_K - h_n x_n a)\, dm \right| \leqslant$$

$$\leqslant \left(\int |h_n x_n|^p a\, dm \right)^{1/p} \left(\int |b^{-1} b_K - 1|^{p'} a\, dm \right)^{1/p'} \leqslant$$

$$\leqslant \mathrm{const} \left(\int |b^{-1} b_K - 1|^{p'} a\, dm \right)^{1/p'} \xrightarrow[K \to \infty]{} 0,$$

because $|b^{-1} b_K| \leqslant 1$ and $b^{-1} b_K \longrightarrow 1$ in measure. But the definition of the functions h_n and the equality $Sx_n = e_n$, $n = 0, 1, \ldots$ imply that $\int h_n x_n a = 1$. Now the preceding estimation shows that if K is a sufficiently large number (which from now on will be fixed) then for all n we have

$$\left| \int h_n x_n a b^{-1} b_K\, dm \right| > \frac{1}{2}.$$

Since $x_n b_k \in H^p$ and $h_n a b^{-1} \in H^\infty$, we obtain that

$$\int h_n x_n a b^{-1} b_k \, dm = \int \mathbb{P}_-(h_n a b^{-1}) x_n b_k \, dm = \int \varphi_n x_n b_k \, dm .$$

Hence

$$2^{-1} < \int |\varphi_n| |x_n| |b_k| \, dm \leqslant K \int |\varphi_n| |x_n| \, dm \leqslant$$

$$\leqslant K \|\varphi_n\|_{p'} \|x_n\|_p \leqslant K \|\varphi_n\|_{p'} \|x_n\|_{p,am} \leqslant KM \|\varphi_n\|_{p'} ,$$

and so $\|\varphi_n\|_{p'} \geqslant (2KM)^{-1}$. Fix now a number s with $p' < s < \infty$. As it has been already mentioned we have the inequality $\sup_n \|\varphi_n\|_s < \infty$. Together with the inequality $\|\varphi_n\|_{p'} \geqslant (2KM)^{-1}$ this implies (in view of the fact that the function $q \mapsto \log\|g\|_q$ is log-convex) that there exists a positive number δ such that $\|\varphi_n\|_r \geqslant \delta$ for all n. But this contradicts the Corollary to Lemma I. ●

2. <u>An example: the spaces $\mathcal{F}\Lambda$</u>.

The proof of Theorem I depends heavily on the fact that the Riesz projection $\mathbb{P}_-$ acts from L^1 to L^r, $r < 1$ (see the proof of Lemma I). It should be noted that some analogues of Theorem I are probably valid for some spaces other than the disc algebra but **subjected** to the following condition: a projection is to be "assigned" to such a space and this projection must behave as the Riesz projection does. Of course the proof of Theorem I does not work in this general setting, for besides of the $(L^1\text{-}L^r)$-continuity of $\mathbb{P}_-$ some very specific techniques of the theory of analytic functions in the unit disc have been involved (for example this proof does not work for the spaces $C^{(\ell)}(\mathbb{T}^n)$, $n > 2$, $\ell \geqslant 1$). But if we restrict ourselves to <u>translation invariant</u> operators only, no additional arguments except those based on the $(L^1 - L^r)$ -continuity mentioned above are needed (cf. [9], [10]; see also Remark after Lemma I).

In this section we show that a "small distorsion" of the space C_A (which affects, however, this $(L^1 - L^r)$ -continuity) may cause that no analogue of Theorem I is true for such a "distorted" space (even if we restrict ourselves to translation-invariant operators).

Let Λ be an infinite Hadamard lacunary subset of $\mathbb{Z}_-$ (i.e. $\inf \left|\frac{a_{n+1}}{a_n}\right| > 1$, where $\{a_n\}$ is the enumeration of Λ according to the magnitudes of moduli of its terms). Denote by $\mathcal{F}_\Lambda$ the closed linear span of the set $\{z^n : n \in \Lambda \cup \mathbb{Z}_+\}$ in the space $C(\mathbb{T})$. Define an operator R from $\mathcal{F}_\Lambda$ to $\ell^2(\Lambda)$ by the formula $Rf = \{\hat{f}(\lambda)\}_{\lambda\in\Lambda}$.

THEOREM 2. The operator R is 0-absolutely summing and $R(\mathcal{F}_\Lambda) = \ell^2(\Lambda)$ (hence R is noncompact and therefore it cannot be p-nuclear for any p).

REMARK. The equality $R(\mathcal{F}_\Lambda) = \ell^2(\Lambda)$ is probably known. However, I was not able to find an appropriate reference.

PROOF. Fix a number r, $0<r<1$. We shall show that $R \in \Pi_r(\mathcal{F}_\Lambda, \ell^2(\Lambda))$ (then automatically $R \in \Pi_0$, as it was already mentioned). Let X be the closure of the set $\mathcal{F}_\Lambda$ in the space L^r. In the paper [I4] it is shown that X is the direct sum of the space H^r and the space Z, $Z = \operatorname{clos}_{L^r} \operatorname{span}\{z^n : n\in\Lambda\}$. Let P be the projection from X onto Z whose kernel is H^r. It is well known (cf. e.g. [I5]) that there exists an isomorphism Φ between Z and $\ell^2(\Lambda)$ which takes the functions z^n, $n\in\Lambda$ to the unit vectors of $\ell^2(\Lambda)$. Clearly $R = \Phi P(I_{r,m}|\mathcal{F}_\Lambda)$ and it remains to apply Factorization theorem from the Introduction.

To prove the equality $R(\mathcal{F}_\Lambda) = \ell^2(\Lambda)$ it is sufficient to check that the operator R^* is an isomorphism between the spaces $\ell^2(\Lambda)$ and $T^*(\ell^2(\Lambda))$. Let $\mathcal{F}_\Lambda^\perp = \{\mu \in M(\mathbb{T}) : \int f\,d\mu = 0$ for all f in $\mathcal{F}_\Lambda\}$. By F. and M. Riesz theorem (cf. [I3]) $\mu \in \mathcal{F}_\Lambda^\perp$ if and only if $\mu = gm$ with $g \in H^1$ and $\hat{g}(n) = 0$ for $n \in (-\Lambda)$ and $n=0$ (the set all such g's will be denoted by Y). Let us identify in the canonical way the spaces $\mathcal{F}_\Lambda^*$ and $M(\mathbb{T})/\mathcal{F}_\Lambda^\perp$. Then the fact we are to check may be restated as follows: if $x = \{x_n\}_{n\in\Lambda} \in \ell^2(\Lambda)$ and $g \in Y$ then

$$\Big(\sum_{n\in\Lambda} |x_n|^2\Big)^{1/2} \leqslant C \Big\| \sum_{n\in\Lambda} x_n z^n + g \Big\|_1 ;$$

where C is an absolute constant. But this is just the well-known Paley's inequality, cf. [I5], vol.2, Ch. I2, §7. ●

COROLLARY I. The space $\mathcal{F}_\Lambda$ is not isomorphic to any quotient space of the space C_A. ●

COROLLARY 2. The space C_A is an uncomplemented subspace of $\mathcal{F}_\Lambda$.

PROOF. Let $X = clos_{C(\mathbb{T})}\, span\{z^n : n \in \Lambda\}$. Suppose to the contrary that there is a projection P from $\mathcal{F}_\Lambda$ onto C_A . Then the operator Q defined by the formula $Q = = \int S_{\zeta^{-1}} P S_\zeta \, dm(\zeta)$ (S_ζ is the rotation operator, $(S_\zeta f)(z) = f(\zeta z),\ \zeta \in \mathbb{T}$)) is a translation-invariant projection from $\mathcal{F}_\Lambda$ onto C_A . Since the kernel of Q is equal to X , the spaces X and $\mathcal{F}_\Lambda / C_A$ are isomorphic. But this is not true for X is isomorphic to ℓ^1 (cf. [I5]) and Y to ℓ^2 (by Theorem 2). ●

In the introduction we have mentioned that the operator p from C_A to ℓ^2 defined by the formula $p(f) = \{\hat{f}(2^n)\}_{n \in \mathbb{Z}_+}$, is 1 -absolutely summing and onto. It was noted by A.Pełczyński and P.Wojtaszczyk that the Grothendieck theorem is an easy consequence of this result (cf. [II] , p. 20 for the details). The operator R **enables us to derive a** stronger result with an equal ease.

COROLLARY 3 (Maurey, [8]). If $0 < r < 1$ then $\Pi_r(\ell^1, \ell^2) = \mathcal{L}(\ell^1, \ell^2)$.

PROOF. Since $R(\mathcal{F}_\Lambda) = \ell^2(\Lambda)$, any operator from ℓ^1 to $\ell^2(\Lambda)$ can be factored through R . ●

We pass now to a discussion of some linear topological properties of the spaces $\mathcal{F}_\Lambda$ which are <u>similar</u> to properties of the disc-algebra. First of all we mention that it is not hard to verify that the most of the facts concerning the spaces C_A and C_A^* proved in Sections 4, 7, 8 of [II] and stated in purely linear-topological terms have natural analogues for the spaces $\mathcal{F}_\Lambda$. (To check that these analogues are true it is sufficient in some cases to repeat an appropriate proof from [II] with minor changes; in some other cases one needs to combine a known result concerning the disc-algebra or its conjugate space with a simple argument based on the fact that $\mathcal{F}_\Lambda$ differs from C_A by a "small" space). We state here explicitly only one proposition of this sort.

PROPOSITION I. Every L^1 -factorable operator from the space $\mathcal{F}_\Lambda$ to a separable conjugate space is compact.

To prove this apply directly Remark at the page 27 in [II]. ●

By this proposition R is an example of an operator of the class Π_0 (and hence of the class Π_1) which does not factor through an $L^1(\mu)$-space. The question if such operators exist had been open not very long ago (now, however, many other examples are known). For some additional information on this subject see also Remark at the end of Section 3.

The rest of this section will be devoted to a proof of the fact that (just as in the case of the disc algebra) for every Banach space Y and every p, $p>1$ the following equality holds: $\Pi_p(\mathcal{F}_\Lambda, Y) = I_p(\mathcal{F}_\Lambda, Y)$. In contrast to the facts mentioned in the preceding paragraphs the proof of a quantitative variant of this equality (as stated in Theorem 3 below) is rather involved compared with the proof of the analogous result for the disc algebra in [II].

Let X be a Banach space and $p>1$. Set

$$k_p(X) = \sup\{ i_p(T) : T\in \Pi_p(X,Y),\ \pi_p(T)\le 1,\ Y \text{ arbitrary}\}.$$

These numeric characteristics of X were introduced in [II], p.56 and it is known that $k_p(C_A) \asymp p$ for $p\ge 2$ and $k_p(C_A)\asymp (p-1)^{-1}$ for $1<p\le 2$, cf. [I6] or [II], p.I6, Theorem 2.4. The spaces $\mathcal{F}_\Lambda$ have the same property.

THEOREM 3. There are two constants a and b (depending on the set Λ only) so that $ap\le k_p(\mathcal{F}_\Lambda)\le bp$ for $p\ge 2$ and $a(p-1)^{-1}\le k_p(\mathcal{F}_\Lambda)\le b(p-1)^{-1}$ for $1<p\le 2$. In particular $k_p(\mathcal{F}_\Lambda)<\infty$ if $1<p<\infty$.

REMARK. Let X and Y be Banach spaces and let $\dim X<\infty$. Set $\gamma_Y(X) \overset{def}{=} \inf\{\|S\|\|T\| : S\in\mathcal{L}(X,Y),\ T\in\mathcal{L}(Y,X),\ TS = id_X\}$. In [II], Section 9 some lower estimates of the constants $\gamma_{C_A}(\ell^p_{(n)})$, $p<\infty$ were obtained (they show in particular that these constants grow at least as fast as n^α as n tends to infinity, α being a number dependent on p). The unique property of the space C_A used in [II] to prove these estimates was the fact that the constants $k_p(C_A)$ behave as described above. Hence the same estimates are valid for the spaces $\mathcal{F}_\Lambda$.

To prove Theorem 3 we need a lemma which is probably of independent interest also. Its proof will be postponed till the end of this section.

LEMMA 2. Let μ be a nonnegative finite measure on the unit circle $\mathbb{T}$. Then there exists a finite measure λ on $\mathbb{T}$ with the following properties: (1) $\lambda\ge\mu$, $\|\lambda\|\le C\|\mu\|$, C being an absolute constant; (2) there exists an operator Q so that Q projects $L^p(\lambda)$ onto $H^p(\lambda)$

simultaneously for all p's and $\|Q\|_{p,\lambda} \leq d(p-1)^{-1}$ for $1<p\leq 2$, $\|Q\|_{p,\lambda} \leq dp$ for $p\geq 2$ (d is an absolute constant). Moreover, the orthogonal projection from $L^2(\lambda)$ onto $H^2(\lambda)$ may be taken to coincide with $Q \mid L^2(\lambda)$.

REMARK. A weakened version of lemma 2 was used in [II] (though it was not stated explicitly there) to prove that $k_p(C_A) \asymp p$ for $p\geq 2$ and $k_p(C_A) \asymp (p-1)^{-1}$ for $p\leq 2$. To be more precise, a projection of $L^p(\lambda)$ onto $H^p(\lambda)$ for some $\lambda \geq \mu$ was also constructed in [II] , but in contrast to lemma 2 this projection "depended on p " (i.e. it was defined by different formulae for different p's ; this formulae did not agree even on $L^\infty(\lambda)$). However the proof of this fact involved only estimates of the Riesz projection $\mathbb{P}_+$ in L^p , and to prove Lemma 2 analogous estimates in <u>weighted</u> spaces $L^p(am)$ are needed. If we argue similarly to the proof of Theorem 3 below but use only this weakened version of Lemma 2 we shall be able to prove a proposition differing from Theorem 3 only by a weaker estimate of $k_p(\mathcal{F}_\Lambda)$ for $1<p\leq 2$ (namely, $k_p(\mathcal{F}_\Lambda) \leq b(p-1)^{-3/2}$ instead of $k_p(\mathcal{F}_\Lambda) \leq$ $\leq b(p-1)^{-1}$). But to establish the "right" estimate Lemma 2 in its full strength seems to be inavoidable.

PROOF OF THEOREM 3. First of all note that the norm of the natural (i.e. translation-invariant) projection $\mathbb{P}$ of L^p onto the space $\operatorname{clos}_{L^p} \operatorname{span}\{z^n : n\in\Lambda\}$ is less than or equal to $c(p-1)^{-1/2}$ if $1<p\leq 2$ and $cp^{1/2}$ if $p>2$ (c depends on Λ only), cf. [I5] , vol.1, Ch.5, §8 . Evidently, the **same numbers are majorants for the norms** $i_p(R)$.

To begin with let us prove the lower estimates of the numbers $k_p(\mathcal{F}_\Lambda)$. Let X_p be the closure of the set $\mathcal{F}_\Lambda$ in L^p and let j be the formally identical operator from $\mathcal{F}_\Lambda$ to X_p (i.e. $j = I_{p,m} \mid \mathcal{F}_\Lambda$). Then $\pi_p(j) \leq 1$. On the other hand it follows from the results of the paper [I2] that $i_p(j)$ is equal to the norm of the translation invariant projection from L^p onto X_p , that is to the number $\|\mathbb{P}_+ + \mathbb{P}\|_{L^p\to L^p}$. Hence $k_p(\mathcal{F}_\Lambda) \geq \|\mathbb{P}_+\|_{L^p\to L^p} - \|\mathbb{P}\|_{L^p\to L^p}$, and to get the desired lower bound for $k_p(\mathcal{F}_\Lambda)$ it is sufficient to apply the above remark about $\|\mathbb{P}\|_{L^p\to L^p}$ and the known estimates from below for $\|\mathbb{P}_+\|_{L^p\to L^p}$, cf. [I3] , [I5] .

We pass to the upper estimates. Let Y be a Banach space, $1<p<\infty$ and $V\in\Pi_p(\mathcal{F}_\Lambda, Y)$, $\pi_p(V) = 1$. Set

$S = V | C_A$ and let μ, $\mu \in M(\mathbb{T})$ be a Grothendieck-Pietsch measure for S, $\|\mu\| = 1$. Then the operator S admits a factorization $S = \Gamma \circ (I_{p,\mu} | C_A)$ with $\Gamma \in \mathcal{L}(H^p(\mu), Y)$, $\|\Gamma\| \leqslant 1$. Applying Lemma 2 to the measure μ we get a measure λ and an operator Q with the properties listed in this lemma, and this enables us to factor the operator S in the following manner:

$$C_A \xrightarrow{J} L^\infty(\lambda) \xrightarrow{I_{p,\lambda}} L^p(\lambda) \xrightarrow{Q} H^p(\lambda) \xrightarrow{\Phi} H^p(\mu) \xrightarrow{\Gamma} Y ; \tag{I}$$

here J and Φ are formal identity operators. Since the space $L^\infty(\lambda)$ is metrically injective we can extend the operator J to an operator J_1 from $\mathcal{F}_\Lambda$ to $L^\infty(\lambda)$ with $\|J_1\| = \|J\|$. Let $W = \Gamma \Phi Q I_{p,\lambda} J_1$. Then $W \in \mathcal{L}(\mathcal{F}_\Lambda, Y)$ and W is clearly an extension of S satisfying $i_p(W) \leqslant$ $\leqslant \|Q\|_p \, i_p(I_{p,\lambda}) \leqslant \|Q\|_p \|\lambda\|^{1/p} \leqslant \|Q\|_p \, c^{1/p} \|\mu\| \leqslant$ $\leqslant c^{1/p} \|Q\|_p$ (c is the constant from Lemma 2). Since $(V-W) | C_A = \mathbb{O}$, we have $V - W = \mathcal{D} R$ with $\|\mathcal{D}\| \leqslant h(\|V - W\|)$ (h depending on Λ only).

Consider first the case $p \geqslant 2$. Then $i_p(V-W) \leqslant$ $\|\mathcal{D}\| i_p(R) \leqslant \|\mathcal{D}\| i_2(R) = \|\mathcal{D}\| \leqslant h(\|V\| + \|W\|) \leqslant h(\pi_p(V) + i_p(W)) \leqslant$ $\leqslant h(1 + c^{1/p} \|Q\|_p)$, hence $i_p(V) \leqslant i_p(W) +$ $+ i_p(V-W) \leqslant b_p$, and we are done.

If $p \leqslant 2$ we have to consider one more factorization of S :

$$C_A \xrightarrow{J} L^\infty(\lambda) \xrightarrow{I_{2,\lambda}} L^2(\lambda) \xrightarrow{Q} H^2(\lambda) \xrightarrow{\Psi} H^p(\mu) \xrightarrow{\Gamma} Y, \tag{2}$$

with Ψ being the formal identity operator. The crucial point is that J and Γ stand here for the <u>same</u> operators as in (I) and, moreover, $\Gamma \Phi Q I_{p,\lambda} = \Gamma \Psi Q I_{2,\lambda}$. Now the definition of the operator W together with the factorization (2) give that $\|W\| \leqslant \|\Gamma\| \|\Psi\| \|Q\|_2 \|I_{\lambda,2}\| \|J_1\| \leqslant$ $\leqslant d_2$, an absolute constant. Of course the inequality $i_p(W) \leqslant c^{1/p} \|Q\| \leqslant d c^{1/p} (p-1)^{-1}$ obtained from the factorization (I) is also true. Now we have: $i_p(V) \leqslant$ $\leqslant i_p(W) + \|\mathcal{D}\| \, i_p(R) \leqslant d c^{1/p} (p-1)^{-1} + (\|V\| + \|W\|) \, i_p(R) \leqslant$

$$\leq dc^{1/p}(p-1)^{-1}+(1+d_2)\,c(p-1)^{-1/2}\leq b(p-1)^{-1}. \qquad \bullet$$

PROOF OF LEMMA 2. Let $\mu = am+\nu$, where ν is singular with respect to m and $a\in L^1(m)$. Without loss of generality we can assume that $a(\zeta)\geqslant \varepsilon$ for some positive ε. Since $H^p(\mu)=H^p(am)\oplus L^p(\nu)$ (the ℓ^p -direct sum is meant, cf. e.g. [II], p. I4) it is sufficient to treat the case $\mu=am$ and to find for such a μ a measure λ having the properties listed in lemma 2 and an additional property of being absolutely continuous with respect to the Lebesgue measure m.

In the case $\mu=am$ the operator Q will be defined in terms of the Riesz projection $\mathbb{P}_+$ considered as an operator in a weighted space $L^p(\omega m)$. For conditions on ω ensuring that $\mathbb{P}_+$ acts in such a space see [I7], [I8], [I9]. For us the following special case of results obtained in [I7], [I8], [I9] will suffice.

Let ω be a nonnegative summable function on the circle. Then ω is said to satisfy the Muckenhoupt condition (A_1) if there exists a number c so that

$$\omega^*(x)\leqslant c\,\omega(x), \quad x\in\mathbb{T}$$

(ω^* is the Hardy-Littlewood maximal function:

$$\omega^*(x)\overset{def}{=\!=}\sup\Big\{\frac{1}{|I|}\int_I \omega\,dm : \ I \ \text{an arc}, \ x\in I\Big\}.$$

THEOREM ON WEIGHTED SPACES. If a function ω satisfies the condition (A_1) then the operator $\mathbb{P}_+$ acts from $L^p(\omega m)$ to $L^p(\omega m)$ if $1<p<\infty$ and for $1<p\leqslant 2$ we have the inequality $\|\mathbb{P}_+\|_{p,\omega m}\leqslant C_1(p-1)^{-1}$ with a constant C_1 depending only on c from the condition (A_1).

REMARK I. Let $r>1$, $g\in L^r$, $g>0$. Then it is well-known that there is a function G satisfying the conditions $G^*\leqslant C_r G$, $\|G\|_r\leqslant C_r\|g\|_r$ and $G\geqslant g$ (a construction: set $g_0=g$, $g_{n+1}=g_n^*$ for $n\geqslant 0$ and $G=\sum_{n=0}^{\infty}\delta^n g_n$, where δ is a small positive number).

REMARK 2. If a function ω satisfies the condition (A_1) and $0<\alpha<1$ then the function ω^α also satisfies the condition (A_1) <u>with the same constant</u> c :

$$(\omega^\alpha)^*(x) = \sup\left\{\frac{1}{|I|}\int_I \omega^\alpha\, dm : x\in I\right\} \leqslant$$

$$\leqslant \sup\left\{\left(\frac{1}{|I|}\int_I \omega\, dm\right)^\alpha : x\in I\right\} = (\omega^*(x))^\alpha \leqslant c^\alpha \omega(x)^\alpha \leqslant c\,\omega(x)^\alpha .$$

Now we continue the proof of Lemma 2. Given a measure μ, $\mu = am$ as above, find a function b so that $b \geqslant a$, $b^{1/2}$ satisfies the condition (A_1) with a constant independent on a and $\int b\, dm \leqslant d \int a\, dm$, d also being an absolute constant (this is possible in view of Remark I: set $g = a^{1/2}$, then $g \in L^2$; starting by this g construct a G as in Remark I and set $b = G^2$). Let v be an outer function with $|v| = b^{1/2}$ (such a v exists, for $a(\zeta) \geqslant \varepsilon$). Define an operator Q by the formula $Qf = v^{-1}\,\mathbb{P}_+(vf)$.

This operator is evidently an orthogonal projection of the space $L^2(bm)$ onto $H^2(bm)$ since the mapping $f \mapsto vf$ is a unitary operator of $L^2(bm)$ onto L^2 and it maps $H^2(bm)$ onto H^2. Hence $\|Q\|_{p,bm} = \|Q\|_{q,bm}$ if $p^{-1}+q^{-1} = 1$ and it is sufficient to estimate the number $\|Q\|_{p,bm}$ for $1 < p \leqslant 2$ only. But for such a p we have

$$\|Qf\|_{p,bm} = \left(\int |v^{-1}\mathbb{P}_+(vf)|^p b\, dm\right)^{1/p} = \left(\int |\mathbb{P}_+(vf)|^p b^{1-\frac{p}{2}}\, dm\right)^{1/p} .$$

Since $0 \leqslant 1-p/2 \leqslant 1/2$ if $1 < p \leqslant 2$, for every such p the function $b^{1-p/2}$ satisfies the condition (A_1) with the constant c independent of p (see Remark 2). Applying now Theorem on weighted spaces we obtain:

$$\|Qf\|_{p,bm} \leqslant \|\mathbb{P}_+\|_{p,b^{1-p/2}m} \|vf\|_{p,b^{1-p/2}m} \leqslant$$

$$\leqslant c_1(p-1)^{-1}\left(\int |vf|^p b^{1-p/2}\, dm\right)^{1/p} \leqslant c_1(p-1)^{-1}\left(\int |f|^p b\, dm\right)^{1/p} . \quad \bullet$$

3. Almost Euclidean subspaces and some finite dimensional examples.

In this section we restrict ourselves to spaces over the reals, but the results are valid for complex spaces as well and only minor changes of the proofs are needed. We begin with the statement of the main result of this section.

THEOREM 4. There are three constants A_1, A_2 and A_3 with the following properties. For each positive integer n there exist an n-dimensional Banach space X_n and an operator T_n from X_n to ℓ^2 so that: (a) if S is an arbitrary operator from X_n^* to ℓ^2 then $\pi_1(S) \leq A_1 \|S\|$; (b) $\sup_n \pi_0(T_n) < \infty$ and $\|T_n^* x\| \geq A_2 \|x\|$ for every x in $(T_n(X_n))^*$; (c) $\nu_1(T_n) \geq A_3 \sqrt{n}$.

REMARK. The second inequality in (b) means that the operators T_n are "uniformly with respect to n onto their images". So they are in a sense analogous to the operator $\mathcal{R}$ of the preceding section. Note also that the inequality (c) is the best (or rather the worst) possible, for if V is an operator of rank n then $\nu_1(V) \leq \sqrt{n}\, \pi_2(V) \leq \sqrt{n}\, \pi_0(V)$ (to prove this set $Y = \operatorname{Im} V$, write $\nu_1(V) \leq \pi_2(V)\, \pi_2(id_Y)$ and use the fact that $\pi_2(id_Z) = \sqrt{m}$ provided $\dim Z = m$, cf. [3], p.286, Theorem 20.2.4 and p.385, Theorem 28.2.4).

We need some facts concerning "almost Euclidean" subspaces of finite dimensional Banach spaces. Let $(E, |\cdot|)$ be an n-dimensional Euclidean space and let $\|\cdot\|$ be a norm on E satisfying the inequality $\|x\| \leq |x|$, $x \in E$. Set $B_1 = \{x \in E : \|x\| \leq 1\}$, $B_2 = \{x \in E : |x| \leq 1\}$ and suppose that $\left(\frac{vol\, B_1}{vol\, B_2}\right)^{1/n} \leq D$. The following result is proved in [20], though it is not stated there explicitly (cf. also [21]).

THEOREM ON ALMOST EUCLIDEAN SUBSPACES. Let $1 \leq k < n$. Then there are two orthogonal (in the Euclidean space $(E, |\cdot|)$) **subspaces** G_1, G_2 **of the space** E **so that** $\dim G_1 = k$, $\dim G_2 = n-k$ **and for** x **in** $G_1 \cup G_2$ **the following inequality holds:** $|x| \leq C(D, \frac{k}{n}) \|x\|$. Here $C(\cdot, \cdot)$ is a function of two real variables which does not depend on the triple $(E, |\cdot|, \|\cdot\|)$ and is bounded on each of the sets $\{(s,t) : 1 \leq s \leq B, \alpha \leq t \leq \beta\}$ $(B > 1;\ 0 < \alpha < \beta < 1)$.

Let $0 < p < \infty$. Set $L_n^p = L^p(\mu)$ where μ is a measure on the set $\{1, 2, \dots, n\}$ defined by the formula $\mu(\{j\}) = n^{-1}$, $1 \leq j \leq n$.

LEMMA 3. Take in the situation described above $E = L_n^2$ and choose the norm $\|\cdot\|$ to be the norm of the space L_n^1. Then

$$\sup_n \left(\frac{vol\, B_1}{vol\, B_2}\right)^{1/n} \overset{def}{=\!=} D < \infty$$

(B_1 and B_2 are as above the balls corresponding to the norms $\|\cdot\|$ and $|\cdot|$).

PROOF. This fact follows easily from the known formulae for the volumes of the balls B_1 and B_2 . See e.g. [20] . ●

This lemma shows that for the triples $(L^2_n, \|\cdot\|_{2,\mu}, \|\cdot\|_{1,\mu})$ the constant in the inequality from Theorem on almost Euclidean subspaces does not grow as n tends to infinity provided the ratio $\frac{k}{n}$ is bounded away from 0 and from 1 . The next lemma is crucial for the proof of Theorem 4. It shows that the same is true for appropiate quotient spaces L^2_n / F with the induced norms. This fact seems to be of independent interest.

Let $0<\beta<1$ and let F be an m-dimensional subspace of the space L^2_n with $\frac{m}{n} \leq \beta$. Set $E = L^2_n / F$ and let $|\cdot|$ and $\|\cdot\|$ be the norms on E induced by the norms of the spaces L^2_n and L^1_n . Set as above $B_1 = \{x \in E : \|x\| \leq 1\}$, $B_2 = \{x \in E : |x| \leq 1\}$. It is clear that the norm $|\cdot|$ is Euclidean and that $\|\cdot\| \leq |\cdot|$.

LEMMA 4. Under these hypotheses we have

$$\left(\frac{vol\, B_1}{vol\, B_2}\right)^{\frac{1}{n-m}} \leq D_1 ,$$

where D_1 depends on the constant β only.

PROOF. Let G be the orthogonal complement of the space F in L^2_n and let P be the orthogonal projection onto the subspace G . If we identify E and G in a natural way then the ball B_2 will "coincide" with the set $U = \{x \in G : \|x\|_{2,\mu} \leq 1\}$ and the ball B_1 will "coincide" with the set $V = P(\{x \in L^2_n : \|x\|_{1,\mu} \leq 1\})$. For every subset A of the set $\{1, \dots, n\}$ define $W_A = \{x \in L^2_n : \|x\|_{1,\mu} \leq 1,\ x(j) = 0$ for $j \notin A\}$. We shall show that

$$V = \bigcup \{P(W_A) : card\, A = n-m\} . \tag{I}$$

Let us prove first that once the formula (I) is established, the lemma easily follows. Let **vol** be the $(n-m)$-dimensional volume defined by the Euclidean structure of the space L^2_n . Then we have

$$\mathrm{vol}\,V \leq \sum_{\mathrm{card}A=n-m} \mathrm{vol}\,P(W_A) \leq \sum_{\mathrm{card}A=n-m} \mathrm{vol}\,W_A = C_n^{n-m}\,\mathrm{vol}\,W_B ,$$

where $B=\{1,2,\dots,n-m\}$. Consequently

$$\left(\frac{\mathrm{vol}\,V}{\mathrm{vol}\,U}\right)^{\frac{1}{n-m}} \leq 2^{\frac{n}{n-m}}\left(\frac{\mathrm{vol}\,W_B}{\mathrm{vol}\,U}\right)^{\frac{1}{n-m}} \leq 2^{\frac{1}{1-\beta}}\left(\frac{\mathrm{vol}\,W_B}{\mathrm{vol}\,U}\right)^{\frac{1}{n-m}} .$$

Set $U_B=\{x:\|x\|_{2,\mu}\leq 1,\ x(j)=0 \text{ for } j\notin B\}$. Then $\mathrm{vol}\,U=\mathrm{vol}\,U_B$. If x is an arbitrary vector satisfying $x(j)=0$ for $x\notin B$ then $x\in W_B$ if and only if

$$\frac{1}{n-m}\sum_{j=1}^{n-m}|x(j)| \leq \frac{n}{n-m}$$

and $x\in U_B$ if and only if

$$\left(\frac{1}{n-m}\sum_{j=1}^{n-m}|x(j)|^2\right)^{1/2} \leq \left(\frac{n}{n-m}\right)^{1/2} .$$

Now Lemma 3 implies that

$$\left(\frac{\mathrm{vol}\,W_B}{\mathrm{vol}\,U}\right)^{\frac{1}{n-m}} = \left(\frac{\mathrm{vol}\,W_B}{\mathrm{vol}\,U_B}\right)^{\frac{1}{n-m}} \leq \frac{n}{n-m}\left(\frac{n-m}{n}\right)^{1/2} D \leq \left(\frac{1}{1-\beta}\right)^{1/2} D ,$$

and hence

$$\left(\frac{\mathrm{vol}\,B_2}{\mathrm{vol}\,B_1}\right)^{\frac{1}{n-m}} \leq 2^{\frac{1}{1-\beta}}\,(1-\beta)^{-1/2} D .$$

We pass now to the proof of the formula (1). Let $\{e_j\}_{1\leq j\leq n}$ be the unit vector basis of the space L_n^1 : $e_j(k)=0$ for $k\neq j$, $e_j(j)=n$. For every x satisfying $\|x\|_{1,\mu}\leq 1$ there exist numbers ε_j , $\varepsilon_j=\pm 1$ so that x lies in the convex hull of the vectors $\varepsilon_1 e_1, \varepsilon_2 e_2, \dots, \varepsilon_n e_n, \mathbb{O}$. Let $y_j=P(\varepsilon_j e_j)$, $j=1,\dots,n$; $y_{n+1}=\mathbb{O}$. Since the range of P is $(n-m)$ -dimensional, Carathéodory theorem (cf. [22]) implies that we can choose $n-m+1$ vectors $z_1,\dots,z_{n-m+1}$ from $y_1,\dots,y_{n+1}$ so that $Px=$

$= \alpha_1 z_1 + \ldots + \alpha_{n-m+1} z_{n-m+1}$ where $\alpha_j \geq 0$, $\Sigma \alpha_j = 1$.

Now if $\alpha_j = 0$ for some j then Px is clearly in $\bigcup_{\mathrm{card}\, A = n-m} P(W_A)$. Let us show that if $\alpha_j > 0$ for all j then we can replace the above formula for x by another one in such a way that one of the vectors z_j in this new formula is $\mathbb{0}$. There are numbers $\beta_1, \ldots, \beta_{n-m+1}$ at least one of which is non-zero so that $\sum_{j=1}^{n-m+1} \beta_j z_j = \mathbb{0}$. We may assume that $\sum_{j=1}^{n-m+1} \beta_j \leq \mathbb{0}$. Define a number γ by $\gamma = \sup \{ \delta : \delta > 0$, $\alpha_j + \delta \beta_j > 0$ for all $j \}$ and set $\rho_j = \alpha_j + \gamma \beta_j$. Then $\rho_j \geq 0$, $\sum_j \rho_j \leq 1$, $Px = \sum_{j=1}^{n-m+1} \rho_j z_j + (1 - \sum_{j=1}^{n-m+1} \rho_j) \mathbb{0}$ and at least one of the numbers ρ_j is equal to zero. ●

REMARK. Let F be an m-dimensional subspace of the space L_n^2 , $m/n > \alpha > 0$. Since the intersection of F with the unit ball of the space L_n^1 is contained in the orthogonal projection of this ball onto F , the same proof shows that

$$\left(\frac{\mathrm{vol}(\{x \in F : \|x\|_{1,\mu} \leq 1\})}{\mathrm{vol}(\{x \in F : \|x\|_{2,\mu} \leq 1\})} \right)^{\frac{1}{m}} \leq D_2 ,$$

with D_2 depending on α only.

PROOF OF THEOREM 4. Theorem on almost Euclidean subspaces and lemma 3 imply that the space L_{2n}^2 contains two subspaces G_n and H_n with the following properties: (i) $\dim G_n = [\frac{n}{2}]$; (ii) $H_n = G_n^{\perp}$; (iii) the inequality

$$\|x\|_{2,\mu} \leq C \|x\|_{1,\mu} \tag{2}$$

holds for every x in $G_n \cup H_n$ with a constant C independent of n and x . Let φ be the quotient map of L_{2n}^2 onto the space L_{2n}^2 / G_n and let $|||\cdot|||$ be the norm in this quotient space induced by the norm $\|\cdot\|_{1,\mu}$. Using again Theorem on almost Euclidean subspaces and Lemma 4 we can find a subspace F_n of the space H_n so that $\dim F_n = n - [\frac{n}{2}]$ and $\|x\|_{2,\mu} \leq C_1 |||\varphi(x)|||$ for all x in F_n (C_1 being independent of n and x).

Let $E_n = \mathrm{span}(G_n \cup F_n)$ and let P_n be the orthogonal projection from E onto F_n . Then for each x in E_n we have

$$\|P_n x\|_{1,\mu} \leq \|P_n x\|_{2,\mu} \leq C_1 |||\varphi(P_n x)||| = C_1 |||\varphi(x)||| \leq C_1 \|x\|_{1,\mu} .$$

Thus if we replace the L^2_n norm in E_n by the L^1_n norm the projections P_n remain bounded uniformly in n. From this it follows easily that the norms $\|\cdot\|_{1,\mu}$ and $\|\cdot\|_{2,\mu}$ are equivalent on E_n uniformly with respect to n, since they evidently are on G_n and F_n separately. Using the log-convexity of L^p-norms we get the following:

if $x \in E_n$ then $\|x\|_{2,\mu} \leq C_2 \|x\|_{1/2,\mu}$, C_2 does not depend on n and x (3)

Now we are in a position to define the spaces X_n we want to construct. Let X_n be equal to E_n as a set and endow it with the L^∞_{2n}-norm. Clearly $\dim X_n = n$. The inequality $\pi_1(S) \leq A_1 \|S\|$ ($S \in \mathcal{L}(X^*_n, \ell^2)$, A_1 an absolute constant) follows from the finite dimensional version of the main result of papers [4], [5], since for every n the space $X^\perp_n$,

$$X^\perp_n \overset{def}{=\!=} \{ g \in L^1_{2n} : \langle x, g \rangle = 0 \quad \forall x \in X_n \}$$

is contained (as a set) in H_n and therefore for every $x \in X^\perp_n$ the inequality (2) holds (and C in this inequality does not depend on n).

Let J_n be the formal identity operator from X_n to E_n (the latter being supplied with the norm $\|\cdot\|_{2,\mu}$) and set $T_n = P_n J_n$. It follows from (3) that the projections P_n are uniformly bounded by the constant C_2 in $L^{1/2}(\mu)$-norms, so $\pi_{1/2}(T_n) \leq C_2$ and hence $\sup_n \pi_0(T_n) < \infty$. To prove the inequality $\|T^*_n x\| \geq A_2 \|x\|$, $x \in F_n = (T_n(X_n))^*$ note that it is equivalent to the following one:

$$\|x + y\|_{1,\mu} \geq A_2^{-1} \|x\|_{2,\mu}, \quad x \in F_n, \; y \in X^\perp_n .$$

But this inequality does hold for some A_2 independent of n since $F_n \perp X^\perp_n$ (in the space L^2_{2n}), $\operatorname{span}(F_n \cup X^\perp_n) = H_n$ and for x in H_n the estimate (2) holds.

It remains to show that $\nu_1(T_n) \geq A_3 \sqrt{n}$. Let I_n be the formal identity operator from F_n to X_n (F_n is supplied as above with the $L^2(\mu)$-norm). Then $\|I_n\| \leq \sqrt{2n}$ and

$$\frac{n}{2} \leq \dim F_n = \operatorname{trace} T_n I_n \leq \nu_1(T_n) \|I_n\| ,$$

and we are done. ●

COROLLARY. Let $X = (X_1 \oplus X_2 \oplus \dots)_{c_0}$. Then X does not have the property (KP) but $\Pi_1(X^*, \ell^2) = \mathcal{L}(X^*, \ell^2)$.

PROOF. The first assertion is evident. The second one is true in virtue of Remarque 8 in [5], since the inequality (2) holds for all x in $X_n^{\perp}$ uniformly in n. ●

This corollary shows that the scheme of proving the Grothendiek equality for a space Y^* described in the introduction does not work in general. Note that it is possible to construct an example demonstrating this using only Lemma 3 and avoiding Lemma 4.

REMARK. Let γ_1 be the standard norm in the ideal of L^1 - factorable operators. Then $\nu_1(T_n) \leq c\gamma_1(T_n)$, c being independent of n. (For if $T_n = AB$, $B \in \mathcal{L}(X_n, L^1(\lambda))$, $A \in \mathcal{L}(L^1(\lambda), F_n)$ then $\nu_1(T_n) \leq \pi_2(A)\pi_2(B)$; but $\pi_2(A) \leq \pi_1(A) \leq c'\|A\|$ in view of Grothendieck theorem and $\pi_2(B) \leq c''\|B\|$ in view of the results proved in [4], [5]). Hence $\gamma_1(T_n) \geq const\sqrt{n}$, $rank\, T_n \asymp n$ and $\sup_n \pi_0(T_n) < \infty$. In [23] a sequence of finite rank operators with analogous properties was constructed, but the operators T_n have an advantage of being "uniformly onto their images".

4. Operators with an arbitrary range.

Up to this section we discussed the problem of nuclearity of 0-absolutely summing operators with values in a Hilbert space, but the analogous problem for operators with an arbitrary range might also be of interest. The next theorem answers a question posed by A.Pełczyński.

THEOREM 5. For every compact topological space S and every Banach space Z the following inclusion holds: $\Pi_0(C(S), Z) \subset N_1(C(S), Z)$.

PROOF. We need a lemma which is probably of independent interest.

LEMMA 5. Let Z be a Banach space, μ a finite measure and X a subspace of $L^1(\mu)$. Denote by Y the closure of X in the space $L^r(\mu)$ $(0 < r < 1)$ and let j be the formal identity operator from $L^1(\mu)$ to $L^r(\mu)$. If W is any continuous linear operator from Y to Z then **the operator** $W \circ (j|X)$ is weakly compact.

Theorem 5 follows easily from this lemma. Indeed, let $V \in \Pi_r(C(S), Z)$. Then factorization theorem (see the Introduction) shows that V **may be included into the following commutative diagram:**

$$\begin{array}{ccccc}
L^{\infty}(\mu) & \xrightarrow{I_{1,\mu}} & L^{1}(\mu) & \xrightarrow{j} & L^{2}(\mu) \\
\cup & & \cup & & \cup \\
E & \xrightarrow{I_{1,\mu}|E} & X & \xrightarrow{j|X} & Y \\
\uparrow i & & & & \downarrow W \\
C(S) & & \xrightarrow{V} & & Z
\end{array}$$

Here μ is a finite measure, i and W are some continuous linear operators, E is a subspace of $L^{\infty}(\mu)$, $X = clos_{L^1(\mu)} E$, $Y = clos_{L^2(\mu)} E$, j is the formal identity operator from $L^1(\mu)$ to $L^2(\mu)$. By Lemma 5 the operator $W \circ (j|X)$ is weakly compact. Since the operator $(I_{1,\mu}|E) \circ i$ is 1-absolutely summing and its domain is the space $C(S)$, this operator is 1-integral (cf. [3], p.233, Proposition 17.3.5). It remains to use the fact that if A is a weakly compact operator, $B \in I_1$ and the composition AB is well-defined then AB is 1-nuclear (cf. [3], p.345, Theorem 24.6.2). ●

PROOF OF LEMMA 5. Denote the composition $W \circ (j|X)$ by T. Let $\{x_n\}$ be an arbitrary bounded sequence in X. We shall show that the sequence $\{Tx_n\}$ has a weak accumulation point. First of all we claim that there is a subsequence $\{y_k\}$ of the sequence $\{Tx_n\}$ accompanied by a sequence $\{e_k\}$ of mutually disjoint μ-measurable sets so that the set $\{y_k(1-\chi_{e_k}) : k=1,2,\dots\}$ is relatively weakly compact in $L^1(\mu)$. To verify this claim one needs to repeat a part of the proof of Theorem 5 in [24]. Here is a sketch of the argument. Set $\eta(\varepsilon) = \sup\{\int_e |x_n| d\mu : \mu e \leq \varepsilon, n=1,2,\dots\}$ and let $\delta = \lim_{\varepsilon \to 0} \eta(\varepsilon)$. If $\delta = 0$ there is nothing to prove. If $\delta > 0$ choose a sequence of positive integers $\{n_k\}$ and a sequence of sets $\{e_k\}$ so that $n_1 < n_2 < n_3 < \dots$, $me_k \to 0$ and $\int_{e_k} |x_{n_k}| d\mu \to \delta$. Setting $y_k = x_{n_k}$ we get all we need except may be the disjointness of the sequence $\{e_k\}$, and so on.

Now let $u_k = y_k \chi_{e_k}$, $v_k = y_k(1-\chi_{e_k})$. We shall assume (as we may) that $v_k \to v$ weakly in $L^1(\mu)$ and $\mu e_k \leq 2^{-k}$. By separation theorem there exists a sequence $\{z_n\}$ in $L^1(\mu)$ so that $z_n \to v$ in norm and $z_n \in conv\{v_k : k \geq n\}$. For each n write z_n as $\sum \alpha_k v_k$ with $\{\alpha_k\}$ a finite sequence, $\alpha_k \geq 0, \sum \alpha_k = 1$ and set $w_n = \sum \alpha_k u_k$. Then $\mu(supp\, w_n) \to$

$\longrightarrow 0$ and $\sup_n \| w_n \|_{1,\mu} < \infty$. Consequently $\lim_n \| j w_n \|_{r,\mu} = 0$, and since $z_n + w_n \in X$ it follows that $jv \in Y$. Let us show that $T y_k \longrightarrow W(jv)$ weakly.

Note that the above argument is applicable to any subsequence $\{v_{k_s}\}$ of the sequence $\{v_k\}$. It follows that for every such subsequence

$$W(jv) \in \operatorname{clos}_Y \operatorname{conv}\{T y_{k_s} : s \geq m\}, \quad m = 1, 2, \ldots . \tag{1}$$

Suppose that $F(T y_k) \not\longrightarrow F(W(jv))$ for some F in Y^* . Then $F(T y_{k_s}) \longrightarrow a \neq F(W(jv))$ for some sequence $\{k_s\}$ of indices. Hence for m sufficiently large we have

$$F(W(jv)) \notin F(\operatorname{clos}_Y \operatorname{conv}\{T y_{k_s} : s \geq m\}) .$$

But this contradicts the formula (1). ●

It is natural to ask now whether Theorem 1 stated in the introductión and proved in Section 1 remains valid for operators from the disc algebra to some non-Hilbert spaces E . The answer is very likely to be positive (and, moreover, some parts of the proof may be repeated almost without changes for certain non-Hilbert spaces E). However now when I am writing this I cannot formulate any satisfactory conjecture on this matter. The only thing I want to mention is that some facts concerning arbitrary spaces E can be derived from theorem 1. Here is an example.

THEOREM 6. Let E be an infinite dimensional Banach space and let $T \in \Pi_0(C_A, E)$. Then $T(C_A) \neq E$.

PROOF. Suppose to the contrary that $T(C_A) = E$. Since $\Pi_0 \subset \Pi_2$, the operator T factors through a Hilbert space and hence E is also a Hilbert space. It remains to apply theorem 1. ●

References.

1. A.G r o t h e n d i e c k. Résumé de la théorie métrique des produits tensoriels topologiques. Bol.Soc.Mat.São Paulo, 1956, 8, 1-79.
2. J.L i n d e n s t r a u s s, A.P e ł c z y ń s k i. Absolutely summing operators in $\mathcal{L}_p$ -spaces and their applications. Studia Math., 1968, 29, 275-326.
3. A.P i e t s c h. Operator ideals. Berlin, 1978.
4. С.В.К и с л я к о в. О пространствах с "малым" аннулятором.

Зап. научн. семин. ЛОМИ, 1976, 65, 192-195.

5. G.P i s i e r. Une nouvelle classe d'espaces de Banach vérifiant le théorème de Grothendieck. Ann.Inst.Fourier, 1978, 28, N 1, 69-90.
6. Н.К.Н и к о л ь с к и й, В.П.Х а в и н, С.В.Х р у щ ё в (составители и редакторы). 99 нерешённых задач линейного и комплексного анализа. Зап. научн. семин. ЛОМИ, 1978, 81.

7. Y.G o r d o n, D.R.L e w i s, J.R.R e t h e r f o r d. Banach ideals of operators with applications. J.Funct.Anal., 1973, 14, 85-129.
8. B.M a u r e y. Théorèmes de factorisation pour les opérateurs linéaires à valeurs dans les espaces L^p . Astérisque, 1974, 11.
9. S.K w a p i e ń, A.P e ł c z y ń s k i. Remarks on absolutely summing translation invariant operators from the disc algebra and its dual. Michigan J.Math., 1978, 25, 173-181.
10. S.K w a p i e ń, A.P e ł c z y ń s k i. Absolutely summing operators and translation invariant spaces of functions on compact abelian groups. Inst.Mat.PAN, preprint N 162, Warszawa, 1978.
11. A.P e ł c z y ń s k i. Banach spaces of analytic functions and absolutely summing operators. AMS Regional Conference Series in Mathematics, 30, Providence, 1977.
12. A.P e ł c z y ń s k i. p-integral operators commuting with group representations and examples of quasi p-integral operators which are not p-integral. Studia Math., 1969, 33, 63-70.
13. P.L.D u r e n. Theory of H^p spaces. Academic Press, New York and London, 1970.
14. С.В.К и с л я к о в. О рефлексивных подпространствах пространства C_A^* . Функц. анализ и его прилож., 1979, 13, № 1, 21-30.
15. A.Z y g m u n d. Trigonimetric series. Vol. 1,2. Cambridge, 1959.
16. B.S.M i t j a g i n, A.P e ł c z y ń s k i. On the nonexistence of linear isomorphisms between Banach spaces of analytic functions of one and several complex variables. Studia Math., 1975, 56, 85-96.
17. B.M u c k e n h o u p t. Weighted norm inequalities for Hardy

maximal function. Trans.Amer.Math.Soc., 1972, 165, 207-226.

18.R.A.H u n t, B.M u c k e n h o u p t, R.L.W h e e d e n. Weighted norm inequalities for the conjugate function and Hilbert transform. Trans.Amer.Math.Soc., 1973, 176, 227-251.

19.R.R.C o i f m a n, C.F e f f e r m a n. Weighted norm inequalities for maximal functions and singular integrals. Studia Math., 1974, 51, N 3, 241-250.

20.S.S z a r e k. On Kashin's almost euclidean orthogonal decomposition of ℓ_1^n . Bull.Acad.Polon.Sci., sér.sci.math., astr. et phys., 1978, 26 N 8, 691-694.

21.S.S z a r e k, N.T o m c z a k - J a e g e r m a n n. On nearly euclidean decompositions for some classes of Banach spaces. Compos.Math., 1980, 40, N 3, 367-385.

22.L.D a n z e r, B.G r u n b a u m, V.K l e e. Helly's theorem and its relatives. "Convexity", Proc.sympos.pure math., vol.7, Amer.Math.Soc., Providence, 1963.

23.T.F i g i e l, S.K w a p i e ń, A.P e ł c z y ń s k i. Sharp estimates for the constants of local unconditional structure of Minkowski spaces. Bull.Acad.Polon.Sci., sér. sci math., astr. et phys., 1977, 25, N 11, 1221-1226.

24.M.I.K a d e c, A.P e ł c z y ń s k i. Bases, lacunary sequences and complemented subspaces in the spaces L^p . Studia Math., 1861/62, 21, 161-176.

N.G.Makarov, V.I.Vasjunin

A MODEL FOR NONCONTRACTIONS AND STABILITY OF THE CONTINUOUS SPECTRUM

0. Introduction.

By the continuous spectrum*) of an operator L we mean the difference $\sigma_c(L) \overset{def}{=} \sigma(L) \setminus \sigma_0$, where σ_0 is the set of all isolated points of $\sigma(L)$ for which the corresponding Riesz projections are of finite rank. Below, we consider only the operators with the continuous spectrum on the circle $\mathbb{T}$, $\mathbb{T}=\{\zeta : \zeta \in \mathbb{C}, |\zeta|=1\}$. Under the perturbation of L by some compact operator K , the continuous spectrum will either remain unchanged or fill the whole disc; the latter is possible, of course, only if $\sigma_c(L)=\mathbb{T}$ (see Lemma 5.2 [8] ; a typical example: if $Lf \overset{def}{=} zf$, $f \in L^2(\mathbb{T})$, $Kf \overset{def}{=} -(2\pi i)^{-1}\int_{\mathbb{T}} f dz$, then $(L+K)(z-\zeta)^{-1}=\zeta(z-\zeta)^{-1}$ for each ζ, $|\zeta|<1$, and $\sigma_c(L+K)=\{\zeta : \zeta \in \mathbb{C}, |\zeta| \leq 1\}$). In the first case we say that $\sigma_c(L)$ is stable under the perturbation K . When is $\sigma_c(L)$ stable under any perturbation of rank one? under any perturbation of an arbitrary but fixed crossnormed Schatten's ideal γ? (We denote below by γ_p the standard Schatten's ideals (cf. [8]); e.g. γ_1 is the trace class ideal, γ_∞ is the ideal of all compact operators). The answers to these questions are determined by the "thinness" of the spectrum $\sigma(L)$, by the possibility to "draw" the resolvent's pole from the infinity inside the disc. The character of an appropriate "thinness" depends both on the

*) We are not sure to have chosen an apt abjective for the introduced spectrum. "Essential" may fit better, the more so it would directly generalize the Weyl's definition [7] to arbitrary operators. However, some other set is meant by "essential spectrum" in many papers; we do not argue with the generally accepted terminology.

nature of the operator L itself and on the class of perturbations.

It should be noted that the stability of the continuous spectrum under finite rank perturbations is known in Weinstein-Aronszain theory [19], [20] (which gives explicit formulae of the perturbation theory for the point spectrum) as "the regular case" (see [21], Ch.IY, § 6). It is important to distinguish the regular case from the singular one and the main result of Section 7 can be considered just as a criterion of nonsingularity in terms of the corresponding characteristic function and of the spectral measure. The principal tool to derive this criterion is a functional model for noncontractions.

The functional model by B.Sz.-Nagy and C.Foiaş plays an important role in Operator Theory. The success of the model theory makes it highly desirable to adjust the model not to contractions only - as is the case with its classical version - but to wider operator classes defined by some kind of closeness to unitary operators.

Various approaches to the construction of such models are known. Ch.Davis and C.Foiaş [1] generalized the classical construction [2], using the J-unitary dilation of an arbitrary operator L. On account of geometrical complications in the study of spaces with indefinite metrics, severe restrictions on the characteristic function were to be imposed. Without such restrictions, even in the case of finite defects, difficulties in the usage of the J-unitary dilation become considerable. There proved to be more constructive the S.N.Naboko's approach, used in the study of nondissipative operators [3].

The method of S.N.Naboko consists in the construction, for a given operator L, of an auxilliary dissipative operator T, and in the study of L in the Sz.-Nagy - Foiaş model of T. It should be mentioned that the idea to use the model of an auxilliary operator in the study of an operator, which does not admit such a model, can be found in other works as well. Thus, for example, in refs. [5], [6], [4] the unitary operators, which arise as perturbations of model constructions, are studied.

The purpose of the present paper is to apply the S.N.Naboko's method for the generalization of the Sz.-Nagy - Foiaş model to the case of "nearly unitary" operators. The direct conformal transplantation of the Naboko's model from the halfplane into the unit disc meets numerous technical difficulties. That is why we prefer not to use the results of ref. [3], but we frequently explore its ideas in our parallel construction. We also display an application of the obtained model to an operator problem concerning the behaviour of the continuous spectrum under some class of perturbations.

Let L be an arbitrary operator in a Hilbert space H. We represent the operator L as a special perturbation of some auxiliary completely nonunitary (c.n.) contraction T, "distorting" the action of T only on its defect subspace

$$L = T + \Omega_* A_0 \Omega^*. \tag{0.1}$$

Here $\Omega : E \longrightarrow H$ and $\Omega_* : E_* \longrightarrow H$ are the isometrical injections of Hilbert spaces E and E_* onto the defect subspaces $\mathfrak{D}_T$ and $\mathfrak{D}_{T*}$ of the operator T (for the definition of subspaces $\mathfrak{D}_T$ and $\mathfrak{D}_{T*}$, see section 1), and the bounded operator A_0 acts from E into E_*.

For a given operator L there are many ways of choosing its representation in the from (0.1). In contrast to S.N.Naboko, we don't fix any definite way but reserve the right to choose those T and A_0 which are the most convenient in this and that concrete situation. Next we replace the c.n. contraction T by its Sz.-Nagy - Foiaş model. From now on, H is the model space, H is contained in $\mathcal{H}$, the space of the minimal unitary dilation Z of the operator T, and

$$T = P_H Z | H$$

(P_H is the orthogonal projection of $\mathcal{H}$ onto H). In the model representation of T, we study the operator L, thus obtaining its functional model. This model allows to reduce the questions concerning the spectral structure of L to the study of some operator-valued analytic functions. The latter turn out to be closely connected with θ_L - the characteristic function of L : they participate in a factorization of θ_L. In terms of these operator-valued functions we describe the spectrum of L and write out formulae for its resolvent.

It is clear that the less the operator L and the auxiliary contraction T differ one from another, the better the constructed model "works". In the process of our study we impose some restrictions on the connection between T and L. One of them requires the existence of a bounded operator A satisfying the condition

$$A_o = \Delta_{*o} A \Delta_o , \qquad (0.2)$$

where $\Delta_0 = (I - \theta^*(0)\theta(0))^{1/2}$, $\Delta_{*0} = (I - \theta(0)\theta^*(0))^{1/2}$, $\theta = \theta_T$ - the characteristic function of T .

It turns out that any operator L can be represented in the form (0.1) with (0.2) satisfied. Moreover, for the completely nonunitary operators L (i.e. for the operators unitary on no reducing subspace), we present two canonical ways to obtain such a representation. The restriction (0.2) not only allows to avoid some technical difficulties but also gives a possibility to study more profoundly the spectral structure of L . In particular under the condition (0.2), we define N_a - the absolutely continuous subspace, and N_s - the singular subspace of the operator L ; $N_a, N_s \subset H$; we indicate the unitary operator $Z|\mathcal{N}$ ($\mathcal{N}$ is a subspace of $\mathcal{H}$ invariant under Z), whose quasi-affine transform is the restriction $L|N_a$.

As usual, the application of the model is most effective in the case where analytic operator-functions admit scalar multiples. Certainly, it takes place now not for any operator L , but only for those sufficiently "near" to unitary operators. For example, this is the case, if the operator $I - L^* L$ is of the trace class, and the spectrum $\sigma(L)$ does not fill the unit disc $\mathbb{D}$.

In the "scalar-multiple" case we study the condition for the spectral measure of the unitary operator $Z|\mathcal{N}$ to vanish on some set Σ of positive Lebesgue measure (in the following, we denote this measure by $|\cdot|$). It occurs, that this condition can be expressed in terms of boundary values of the characteristic function θ_L , and of the spectral measure of the unitary part of L . We apply the obtained criterion to the study of the stability of the continuous spectrum. The last part of the paper is devoted to this problem.

Few facts about the stability of the continuous spectrum are known. The bilateral shift provides the simplest example of an operator with the continuous spectrum unstable under rank one perturbations ([9] , problem 144). Moreover, N.K.Nikol'skii showed [10] that, for the unitary operator $\mathcal{U}$, $\sigma_c(\mathcal{U})$ is stable under perturbations of rank one (or of trace class - no difference) if and only if $\mathcal{U}$ does not contain the shift. The problem concerning γ_∞-stability was completely solved in ref.

[11] : in our situation (L is an arbitrary operator with $\sigma_c(L) \subset \mathbb{T}$) $\sigma_c(L)$ is γ_∞-stable if and only if $\sigma_c(L) \neq \mathbb{T}$.

In the present paper we study the stability of the continuous spectrum of the operators, which admit the representation (0.1) with the aforementioned analytic operator-valued functions having scalar multiples. For such operators, the rank one stability is equivalent to the γ_1-stability, and the corresponding criterion consists in the vanishing of the spectral measure of the unitary operator $z|\mathcal{N}$ on some set Σ, $|\Sigma|>0$. This criterion can be formulated in terms of the invariant subspaces and in terms of operator algebras as well.

Finally, we mention one very special case of the above assertion. Let S be the bilateral shift presented as the multiplication by z on the space $L^2(\mathbb{T})$. Let $K=(\cdot, f)g$ be an arbitrary rank one operator, f , $g \in L^2(\mathbb{T})$. Then there exists an operator of rank one K_1 , $K_1=(\cdot, f_1)g_1$, such that $\sigma(S+K+K_1) \supset \mathbb{D}$. The latter is equivalent to the identity $\det(I+(S-\lambda)^{-1}(K+K_1)) \equiv 0$, $\lambda \in \mathbb{D}$, or

$$1+\int\frac{g\bar{f}}{e^{it}-\lambda}+\int\frac{g_1\bar{f}_1}{e^{it}-\lambda}+\left(\int\frac{g\bar{f}}{e^{it}-\lambda}\right)\left(\int\frac{g_1\bar{f}_1}{e^{it}-\lambda}\right)-\left(\int\frac{g\bar{f}_1}{e^{it}-\lambda}\right)\left(\int\frac{g_1\bar{f}}{e^{it}-\lambda}\right)\equiv 0$$

(the integration is performed with respect to the normalized Lebesgue measure on the circle $\mathbb{T}$).

We would like to know how to prove the solvability (relative to f_1 and g_1) of the last equation for any f and g . Even in this simple case the application of the model preserves all its main difficulties and the outward complicacy, but it is only the model that makes it possible for the authors to solve the problem under consideration.

We are grateful to N.K.Nikol'skii and S.N.Naboko for helpful discussions and reading the manuscript.

1. The model of Szökefalvi-Nagy and Foias

We begin by briefly recalling some of the well-known facts about the model for contractions (i.e., for linear operators T satisfying $\|T\| \leqslant 1$) of a Hilbert space. Our approach is slightly different from the traditional one. We practically remove all references to the functional aspect of the model, using only the geometry of the unitary dilation space. Our point of view is close, in this sense, to the R.G.Douglas exposition in the survey [12]. For the reader's convenience a "correspondence table "of the main initial objects and formulae of the model is provided: on the left-hand side of the page we write them down in the symmetric form due to B.S.Pavlov [13], on the right-hand side - in the original form due to B.Sz.-Nagy and C.Foiaş [2]. It should be noted at once that operators in the model notation are often determined only on some dense subset, but we consider them being boundedly extended to all of the Hilbert space, if possible.

We now proceed to set forth some standard rules for notation. $\mathcal{L}(X,Y)$ is the space of all bounded linear operators acting from X into Y, $\mathcal{L}(X) \overset{\text{def}}{=} \mathcal{L}(X,X)$. $H^p(X)$ is the Hardy space of X-valued functions on the unit disc $\mathbb{D}$ of the complex plane $\mathbb{C}$ (see [14] ; [2] , V, §1) $H^p(X)$ is embedded in a natural way into the X-valued Lebesgue space $L^p(X)$ defined on the circle $\mathbb{T}$. $h_\lambda : L^2(X) \to X$ denotes the operator of evaluation at $\lambda: f \mapsto \hat{f}(\lambda)$, where $\hat{f}$ is the harmonic continuation of f, $\lambda \in \mathbb{D}$. We use the same letters to denote functions and operators of multiplication by these functions in the corresponding functional spaces, and also to denote constants and functions identically equal to these constants. The letter z denotes the identity mapping on the circle $\mathbb{T}$ (and thus denotes the operator of multiplication by z). Everywhere I is the identical operator.

Let E and E_* be separable Hilbert spaces; and let θ, $\theta \in H^\infty(\mathcal{L}(E,E_*))$ be a contraction-valued function satisfying

$$\|\theta_0 e\| < \|e\|$$

for each nonzero vector e in E, $\theta_0 \overset{\text{def}}{=} \theta(0)$. We set

$$\Delta = (I - \theta^*\theta)^{1/2}, \quad \Delta \in L^\infty(\mathcal{L}(E));$$

$$\Delta_* = (I - \theta\theta^*)^{1/2}, \quad \Delta_* \in L^\infty(\mathcal{L}(E_*));$$

$$\Delta_0 = (I - \theta_0^*\theta_0)^{1/2}, \quad \Delta_0 \in \mathcal{L}(E);$$

$$\Delta_{*0} = (I - \theta_0\theta_0^*)^{1/2}, \quad \Delta_{*0} \in \mathcal{L}(E_*).$$

Since θ is purely contractive, the kernels of the operators Δ_0 and Δ_{*0} are trivial. Hence the operators Δ_0^{-1} and Δ_{*0}^{-1} are defined correctly on the dense subsets $\Delta_0 E$ and $\Delta_{*0} E_*$ respectively.

The symbol $\begin{pmatrix} L^2(E_*) \\ \operatorname{clos} \Delta L^2(E) \end{pmatrix}$

is used to denote the orthogonal sum of the indicated Hilbert spaces. An element of this sum is represented by the column of two functions. To introduce the Hilbert space denoted below by

$$L^2\begin{pmatrix} I & \theta^* \\ \theta & I \end{pmatrix},$$

take the space of pairs $\left\{\binom{f}{f_*} : f \in L^2(E),\ f_* \in L^2(E_*)\right\}$, endow it with the seminorm p :

$$\left[p\binom{f}{f_*}\right]^2 = (f + \theta^* f_*, f)_{L^2(E)} + (\theta f + f_*, f_*)_{L^2(E_*)},$$

take its quotient space with respect to the kernel of this seminorm, and, at last, complete the obtained prehilbert space.

Now we are able to describe the model space. To do this, we set the following notation.

Pavlov's form		Sz.-Nagy -Foias' form
$\mathcal{H} = L^2\begin{pmatrix} I & \theta^* \\ \theta & I \end{pmatrix}$	‖	$\mathcal{H} = \begin{pmatrix} L^2(E_*) \\ \operatorname{clos} \Delta L^2(E) \end{pmatrix}$
	$\pi : L^2(E) \to \mathcal{H}$	
$\pi = \begin{pmatrix} I \\ 0 \end{pmatrix}$, $\pi^* = (I, \theta^*)$	‖	$\pi = \begin{pmatrix} \theta \\ \Delta \end{pmatrix}$, $\pi^* = (\theta^*, \Delta)$
	$\pi_* : L^2(E_*) \to \mathcal{H}$	
$\pi_* = \begin{pmatrix} 0 \\ I \end{pmatrix}$, $\pi_*^* = (\theta, I)$	‖	$\pi_* = \begin{pmatrix} I \\ 0 \end{pmatrix}$, $\pi_*^* = (I, 0)$

The operators π and π_* imbed isometrically $L^2(E)$ and $L^2(E_*)$ respectively into $\mathcal{H}$:

$$\pi^* \pi = I, \quad \pi_*^* \pi_* = I.$$

Observe also that

$$\pi_*^* \pi = \theta, \quad \pi^* \pi_* = \theta^*.$$

Since $\|(\pi - \pi_* \theta) f\|_{\mathcal{H}} = \|\Delta f\|_{L^2(E_*)}$ for each function

$f \in L^2(E)$, the equality

$$\tau\Delta = \pi - \pi_* \theta$$

determines an isometry

$$\tau : \operatorname{clos} \Delta L^2(E) \longrightarrow \mathcal{H}$$

$$\tau = \begin{pmatrix} I \\ -\theta \end{pmatrix} \Delta^{-1}, \ \tau^* = (\Delta, 0) \ \Bigg\| \ \tau = \begin{pmatrix} 0 \\ I \end{pmatrix}, \ \tau^* = (0, I).$$

Similarly, the equality $\tau_* \Delta_* = \pi_* - \pi\theta^*$ determines

$$\tau_* : \operatorname{clos} \Delta_* L^2(E_*) \longrightarrow \mathcal{H}$$

$$\tau_* = \begin{pmatrix} -\theta^* \\ I \end{pmatrix} \Delta_*^{-1}, \ \tau_*^* = (0, \Delta_*) \ \Bigg\| \ \tau_* = \begin{pmatrix} \Delta_* \\ -\theta^* \end{pmatrix}, \ \tau_*^* = (\Delta_*, -\theta).$$

It is clear that

$$I = \pi\pi^* + \tau_*\tau_*^* = \pi_*\pi_*^* + \tau\tau^*;$$

$$\tau^*\pi = \Delta, \ \tau^*\pi_* = 0, \quad \tau_*^*\pi = 0, \ \tau_*^*\pi_* = \Delta_* .$$

Next we set

$$H \overset{\text{def}}{=} \{x : x \in \mathcal{H}, \ \pi^* x \perp H^2(E), \pi_*^* x \in H^2(E_*)\}.$$

If we denote by P_+ and P_- the orthogonal projections of L^2 onto H^2 and onto its orthogonal complement then the projection of $\mathcal{H}$ onto the subspace H can be written down as

$$P_H = I - \pi P_+ \pi^* - \pi_* P_- \pi_*^* .$$

On $\mathcal{H}$ acts the unitary operator z . We call the c.n. contraction $T : H \longrightarrow H$, defined by

$$T \overset{\text{def}}{=} P_H z \,|\, H ,$$

the model contraction and we call H the model space of operator T . Every c.n. contraction is unitarily equivalent to some model contraction. The proof of this fact and an explicit construction of the model can be found in [2], [12], [15].

The operator $\mathcal{Z}$ on $\mathcal{H}$ is the minimal unitary dilation for the operator T. The geometry of the space of dilation is completely determined by the isometries π and π_*; the relation $\pi_*^*\pi=\theta$ provides the translation into the analytic language.

Now we return to the study of the **model** operator. It can be expressed in the form

$$T=\{z-\pi h_0^* h_0 \pi^* z\}|H.$$

Its adjoint is

$$T^*=P_H\bar{z}|H=\{\bar{z}-\bar{z}\pi_* h_0^* h_0 \pi_*^*\}|H.$$

We call $D_T \overset{def}{=} (I-T^*T)^{1/2}$ and $D_{T^*} \overset{def}{=} (I-TT^*)^{1/2}$ the defect operators, and we call the closures of their images $\mathfrak{D}_T$ and $\mathfrak{D}_{T^*}$ the defect subspaces. It is easy to check that

$$D_T^2=\omega\omega^*|H, \qquad D_{T^*}^2=\omega_*\omega_*^*|H,$$

where

$$\omega=\bar{z}(\pi-\pi_*\theta_0)h_0^* : E\longrightarrow\mathcal{H}, \tag{1.1}$$

$$\omega_*=(\pi_*-\pi\theta_0^*)h_0^* : E_*\longrightarrow\mathcal{H}. \tag{1.2}$$

$$\omega=\bar{z}\begin{pmatrix} I \\ -\theta_0\end{pmatrix}, \quad \omega_*=\begin{pmatrix} -\theta_0^* \\ I\end{pmatrix} \Bigg\| \quad \omega=\bar{z}\begin{pmatrix} \theta-\theta_0 \\ \Delta\end{pmatrix}, \quad \omega_*=\begin{pmatrix} I-\theta\theta_0^* \\ -\Delta\theta_0^*\end{pmatrix}.$$

Since $\omega^*\omega=\Delta_0^2$ and $\omega_*^*\omega_*=\Delta_{*0}^2$, the relations

$$\Omega\Delta_0=\omega, \quad \Omega_*\Delta_{*0}=\omega_*$$

define the isometrical embeddings

$$\Omega:E\longrightarrow H, \qquad \Omega_*:E_*\longrightarrow H$$

onto the defect subspaces $\mathfrak{D}_T$ and $\mathfrak{D}_{T^*}$.

The images of the operators $\omega,\omega_*,\Omega,\Omega_*$ lie in H, so

their adjoints $\omega^*, \omega_*^*, \Omega^*, \Omega_*^*$ vanish on $\mathcal{H} \ominus H$.
Since

$$h_0 \pi_*^* z \,|\, H = 0 \qquad \text{and} \qquad h_0 \pi^* | H = 0,$$

we have

$$\omega^* | H = h_0 \pi^* z | H, \quad \omega_*^* | H = h_0 \pi_*^* | H,$$

and

$$\Omega^* | H = \Delta_0^{-1} h_0 \pi^* z | H, \quad \Omega_*^* | H = \Delta_0^{-1} h_0 \pi_*^* | H. \tag{1.3}$$

The operators T and T^* can be expressed in the form

$$T = \{ z(I - \Omega\Omega^*) - \Omega_* \theta_0 \Omega^* \} | H \tag{1.4}$$
$$T^* = \{ \bar{z}(I - \Omega_* \Omega_*^*) - \Omega \theta_0^* \Omega_*^* \} | H.$$

Indeed, for example,

$$T = z - \pi h_0^* h_0 \pi^* z | H = z - \pi h_0^* \Delta_0 \Omega^* | H.$$

From

$$z\omega + \omega_* \theta_0 = (\pi - \pi_* \theta + \pi_* \theta_0 - \pi \theta_0^* \theta_0) h_0^* = \pi \Delta_0^2 h_0^*,$$

we infer that

$$z\Omega + \Omega_* \theta_0 = \pi \Delta_0 h_0^*,$$

$T = z - (z\Omega + \Omega_* \theta_0)\Omega^*$, as claimed. The established expressions for T and T^* separate the strictly contractive parts of the operators, which map one defect subspace into another, and its isometrical parts which are the unitary transformations between the orthogonal complements of $\mathcal{D}_T$ and $\mathcal{D}_{T^*}$.

In the sequel the following facts about the operator $(I - \lambda T^*)^{-1}$ will be of great utility.

We always assume that $\lambda \in \mathbb{D}$.

PROPOSITION **1.1.**

$$(I - \lambda T^*)^{-1} = (z - \lambda)^{-1} (z - \lambda \pi_* h_0^* h_\lambda \pi_*^*) | H, \tag{1.5}$$

$$\Omega_*^* (I - \lambda T^*)^{-1} = \Delta_{*0}^{-1} h_\lambda \pi_*^* | H, \tag{1.6}$$

$$(I - \lambda T^*)^{-1} \omega = (z - \lambda)^{-1} [\pi - \pi_* \theta(\lambda)] h_0^*. \tag{1.7}$$

PROOF. $(I-\lambda T^*)x=(1-\lambda\bar{z})x+\lambda\bar{z}\pi_*h_0^*h_0\pi_*^*x.$
Therefore
$$z\pi_*^*(I-\lambda T^*)x=(z-\lambda)\pi_*^*x+\lambda h_0^*h_0\pi_*^*x.$$
Since $\pi_*^*x\in H^2(E)$ for each $x\in H$, we have
$$\lambda h_\lambda\pi_*^*(I-\lambda T^*)x=\lambda h_0\pi_*^*x.$$
Thus $(I-\lambda T^*)x=(1-\lambda\bar{z})x+\lambda\bar{z}\pi_*h_0^*h_\lambda\pi_*^*(I-\lambda T^*)x,$
whence
$$x=(z-\lambda)^{-1}(z-\lambda\pi_*h_0^*h_\lambda\pi_*^*)(I-\lambda T^*)x,$$
and (1.5) is proved. The formulae (1.3) and (1.1) together with (1.5) readily imply (1.6) and (1.7). ●

REMARK 1.2. The formula (1.6) yields the inclusion
$$h_\lambda\pi_*^*H\subset\Delta_{*0}E_*.$$
In the same fashion one can obtain the formulae
$$\Omega(I-\bar{\lambda}T)^{-1}=\Delta^{-1}h_\lambda\pi^*z\,|H=\frac{1}{\lambda}\Delta_0^{-1}h_\lambda\pi^*|H,$$
and the inclusion
$$h_\lambda\pi^*H\subset\Delta_0E. \quad ●$$

Consider next an analytic operator-valued function which maps the unit disc into $\mathcal{L}(E,E_*)$:
$$\vartheta(\lambda)=\lambda\Omega_*^*(I-\lambda T^*)^{-1}\Omega.$$

We have
$$\theta=\theta_0+\Delta_{*0}\vartheta\Delta_0 \tag{1.8}$$
which follows by computations:
$$\Delta_{*0}\vartheta(\lambda)\Delta_0=\lambda h_\lambda\pi_*^*\Omega\Delta_0=\lambda h_\lambda\pi_*^*\omega=$$
$$=\lambda h_\lambda\bar{z}\pi_*^*(\pi-\pi_*\theta_0)h_0^*=$$
$$=\lambda h_\lambda\bar{z}(\theta-\theta_0)h_0^*=\theta(\lambda)-\theta_0.$$

The formula (1.8) provides the following immediate corollary.

COROLLARY 1.3. For every contraction-valued analytic function θ, the operators $\theta(\lambda)$ map Δ_0E into $\Delta_{*0}E$. ●

The next equalities can be easily deduced from (1.8)

$$\pi^*\Omega_* = \Delta_0 \vartheta^* h_0^*, \tag{1.9}$$

$$\pi_*^*\Omega_* = \Delta_{*0}(I - \vartheta\theta_0^*) h_0^*, \tag{1.10}$$

$$\pi^* z\Omega = \Delta_0(I - \vartheta^*\theta_0) h_0^*, \tag{1.11}$$

$$\pi_*^* z\Omega = \Delta_{*0}\vartheta h_0^*. \tag{1.12}$$

By (1.8) we conclude that the function $\Delta_{*0}\vartheta\Delta_0$ belongs to $H^\infty(\mathcal{L}(E, E_*))$. Furthermore, (1.9), (1.12) provides

$$\vartheta\Delta_0 \in \mathcal{L}(E, H^2(E_*)), \quad \Delta_{*0}\vartheta \in \mathcal{L}(E, H^2(E_*)),$$

whereas the function ϑ itself does not necessarily belong even to the class H^1.

We are closing this Section by some more simple formulae which will prove useful below:

$$T\Omega = -\Omega_*\theta_0 .$$

$$T^*\Omega_* = -\Omega\theta_0^* .$$

$$D_T = \Omega\Delta_0\Omega^* .$$

$$D_{T^*} = \Omega_*\Delta_{*0}\Omega_*^* .$$

2. Model representation of an arbitrary bounded operator. Spectrum and resolvent

The purpose of this Section is to begin the study of an arbitrary operator L on the model space of an auxiliary c.n. contraction T, which is connected with L by the relation (0.1)

$$L = T + \Omega_* A_0 \Omega^*, \quad A_0 \in \mathcal{L}(E, E_*) \tag{0.1}$$

Any operator can be represented in such a form, and in many fashions. The only restriction on the choice of an auxiliary contraction imposed by (0.1) is that the defect subspaces of T should contain those of L, and, in addition, the equality $L|\mathfrak{D}_T^\perp =$

$=T|\mathfrak{D}_T^{\perp}$ should be valid.

Representation (0.1) for a given operator L being fixed, we use the Sz.-Nagy – Foiaş model of the contraction T to study the action of L on the model space. We consider H, T, Ω, Ω_* having the same meaning as that in Section 1. Hence

$$L=\{z(I-\Omega\Omega^*)+\Omega_* B\Omega^*\}|H,$$
$$L^*=\{\bar{z}(I-\Omega_*\Omega_*^*)+\Omega B^*\Omega_*^*\}|H, \tag{2.1}$$

where $B \overset{def}{=\!=} A_0-\Theta_0$. Thus, we obtain the model representation of the operator L.

As is mentioned in Introduction, the less is the difference between the operators L and T, the more the model is adjusted for the analysis of the operator L. However, no additional restrictions on the relation between L and T are still imposed in this Section. We study the spectrum $\sigma(L)$ of the operator L in the general setting. Let ϑ_1 be an analytic $\mathcal{L}(E,E_*)$ -valued function, defined by

$$\vartheta_1 \overset{def}{=\!=} \vartheta-(I-\vartheta\Theta_0^*)B=\vartheta(\Delta_0^2+\Theta_0^*A_0)-A_0+\Theta_0$$

LEMMA 2.1.

$$\vartheta_1(\lambda)=\Delta_{*0}^{-1}h_\lambda\pi_*^*(z\Omega-\Omega_* B)$$
$$\vartheta_1(\lambda)\Omega^*|H=\Omega_*^*(I-\lambda T^*)^{-1}(\lambda I-L) \tag{2.2}$$

PROOF. Using (1.12) and (1.10) we verify the first equality:

$$\Delta_{*0}^{-1}h_\lambda\pi_*^*(z\Omega-\Omega_* B)=\Delta_{*0}^{-1}h_\lambda[\Delta_{*0}\vartheta-\Delta_{*0}(I-\vartheta\Theta_0^*)B]h_0^*=$$
$$=\vartheta(\lambda)-(I-\vartheta(\lambda)\Theta_0^*)B=\vartheta_1(\lambda).$$

The second equality follows from the first one, (1.6) and (2.1):

$$\Omega_*^*(I-\lambda T^*)^{-1}(\lambda I-L)=\Delta_{*0}^{-1}h_\lambda\pi_*^*[(\lambda-z)+(z\Omega-\Omega_* B)\Omega^*]|H=$$
$$=\vartheta_1(\lambda)\Omega^*|H. \quad \bullet$$

LEMMA 2.2. The number λ is in the point spectrum $\sigma_p(L)$ of the operator L if and only if $\mathcal{K}er\,\vartheta_1(\lambda)\neq\{0\}$. If this is the case, then the associated eigenspace is equal to

$$E_\lambda \overset{def}{=\!=} (z-\lambda)^{-1}(z\Omega-\Omega_* B)\,\mathrm{Ker}\,\vartheta_1(\lambda).$$

PROOF. Let x be an eigenvector of L :

$$0=(L-\lambda I)x=(z-\lambda)x-(z\Omega-\Omega_* B)\Omega^* x.$$

Since $x\neq 0$, $\Omega^* x\neq 0$. Let us verify that $\Omega^* x\in \mathcal{K}er\,\vartheta_1(\lambda)$:

$$\Delta_{*0}\vartheta_1(\lambda)\Omega^* x=h_\lambda \pi_*^*(z\Omega-\Omega_* B)\Omega^* x=h_\lambda\pi_*^*(z-\lambda)x=0.$$

Consequently, $\mathcal{K}er\,\vartheta_1(\lambda)\neq\{0\}$, and

$$x=(z-\lambda)^{-1}(z\Omega-\Omega_* B)\Omega^* x\in E_\lambda.$$

To complete the proof it remains to show that $E_\lambda\subset H$, and $E_\lambda\subseteq \mathcal{K}er(L-\lambda I)$. Let $e\in\mathcal{K}er\,\vartheta_1(\lambda)$, $x_\lambda \overset{def}{=\!=} (z-\lambda)^{-1}(z\Omega-\Omega_* B)e$. Then

$$\pi_*^* x_\lambda=(z-\lambda)^{-1}\Delta_{*0}\vartheta_1 e\in H^2(E_*),$$

$$\pi^* x_\lambda=(z-\lambda)^{-1}\pi^*(z\Omega-\Omega_* B)e=$$
$$=\frac{\bar z}{1-\lambda\bar z}[\Delta_0(I-\vartheta^*\theta_0)-\Delta_0\vartheta^* B]e\perp H^2(E).$$

Hence $x_\lambda\in H$.

$$(L-\lambda I)x_\lambda=(z-\lambda)x_\lambda-(z\Omega-\Omega_* B)\Omega^* x_\lambda=$$
$$=(z\Omega-\Omega_* B)(e-\Omega^* x_\lambda)=0,$$

because

$$\Omega^* x_\lambda=\Delta^{-1}h_0\pi^* z x_\lambda=$$

$$=\Delta^{-1}h_0(1-\lambda\bar z)^{-1}\Delta_0[(I-\vartheta^*\theta_0)-\Delta_0\vartheta^* B]e=$$

$$=[(I-\vartheta^*(0)\theta_0)-\Delta_0\vartheta^*(0)B]e=e$$

since $\vartheta(0)=0$. ●

LEMMA 2.3. $\lambda\notin\sigma(L)$ if and only if the operator $\vartheta_1(\lambda)$ is invertible. In this case

$$(L-\lambda I)^{-1}=(z-\lambda)^{-1}\{I-(z\Omega-\Omega_* B)\vartheta_1^{-1}(\lambda)\Delta_{*0}^{-1}h_\lambda\pi_*^*\}\,|\,H.$$

PROOF. We first prove the "only if" assertion. Let the operator $\vartheta_1(\lambda)$ be invertible. Denote by M the operator standing in the right-hand side of the equality in the statement. We claim that M is both right and left inverse for the operator $(L-\lambda I)$. For any $x \in H$, by Lemma 2.1 and (1.9), (1.11) we have

$$\pi_*^* M x = (z-\lambda)^{-1}[\pi_*^* x - \Delta_{*0}\vartheta_1 h_0^* \vartheta_1(\lambda)^{-1}\Delta_{*0}^{-1} h_\lambda \pi_*^* x] \in H^2(E_*) \tag{2.3}$$

$$\pi^* M x = \bar{z}(1-\lambda\bar{z})^{-1}[\pi^* x - \Delta_0(I-\vartheta^*\theta_0-\vartheta^* B)h_0^*\vartheta_1(\lambda)^{-1}\cdot$$
$$\cdot\Delta_{*0}^{-1} h_\lambda \pi_*^* x] \perp H^2(E)$$

hence $MH \subset H$. Since

$$h_\lambda \pi_*^*(L-\lambda I) = -h_\lambda \pi_*^*(z\Omega-\Omega_* B)\Omega^*|H =$$
$$= -\Delta_{*0}\vartheta_1(\lambda)\Omega^*|H,$$

we have

$$M(L-\lambda I) = I - \{(z-\lambda)^{-1}(z\Omega-\Omega_* B)\Omega^* +$$
$$+(z-\lambda)^{-1}(z\Omega-\Omega_* B)\vartheta_1(\lambda)^{-1}\Delta_{*0}^{-1}\Delta_{*0}\vartheta_1(\lambda)\Omega^*\}|H = I.$$

On the other hand, (2.2) shows that

$$\Omega^* M = \Delta_0^{-1} h_0 z \pi^* M = -\vartheta_1(\lambda)^{-1}\Delta_{*0}^{-1} h_\lambda \pi_*^*|H;$$

whence

$$(L-\lambda I)M = I - (z\Omega-\Omega_* B)\vartheta_1(\lambda)^{-1}\Delta_{*0}^{-1} h_\lambda \pi_*^* +$$
$$+(z\Omega-\Omega_* B)\vartheta_1(\lambda)^{-1}\Delta_{*0}^{-1} h_\lambda \pi_*^*|H = I.$$

For the converse, suppose that $\lambda \notin \sigma(L)$ and set

$$\varphi_1(\lambda) \overset{def}{=\!=} \Omega^*(\lambda I-L)^{-1}(I-\lambda T^*)\Omega_*.$$

By Lemma 2.1

$$\vartheta_1(\lambda)\varphi_1(\lambda) = \vartheta_1(\lambda)\Omega^*(\lambda I-L)^{-1}(I-\lambda T^*)\Omega_* =$$
$$=\Omega_*^*(I-\lambda T^*)^{-1}(\lambda I-L)(\lambda I-L)^{-1}(I-\lambda T^*)\Omega_* = I.$$

By Lemma 2.2 $\mathcal{K}er\ \vartheta_1(\lambda) = \{0\}$, and we infer from the equality

$$\vartheta_1(\lambda)[\varphi_1(\lambda)\vartheta_1(\lambda)-I] = [\vartheta_1(\lambda)\varphi_1(\lambda)-I]\vartheta_1(\lambda)=0$$

that $\varphi_1(\lambda)\,\vartheta_1(\lambda) = I$, completing the proof. ●

We introduce next another three operator-valued holomorphic functions to describe the spectrum of the operator L outside the unit disc and the spectrum and the resolvent of the adjoint operator L^* :

$$\vartheta_2 \overset{def}{=} \vartheta - B(I-\theta_0^*\vartheta) = (A_0\theta_0^* + \Delta_{*0}^2)\vartheta - A_0 + \theta_0$$

$$\vartheta_3 \overset{def}{=} (I-\theta_0^*\vartheta) - B^*\vartheta = I - A_0^*\vartheta$$

$$\vartheta_4 \overset{def}{=} (I-\vartheta\theta_0^*) - \vartheta B^* = I - \vartheta A_0^*.$$

By a computation similar to that in Lemma 2.1, the following equalities can be established

$$\vartheta_2^*(\lambda) = \Delta_0^{-1} h_\lambda \pi^* (\Omega_* - z\Omega B^*)$$

$$\Omega_*\vartheta_2(\lambda) = (\lambda I - L)(I-\lambda T^*)^{-1}\Omega \qquad (2.4)$$

$$\vartheta_3^*(\lambda) = \Delta_0^{-1} h_\lambda \pi^* (z\Omega - \Omega_* B)$$

$$\Omega\vartheta_3(\lambda) = (I-\lambda L^*)(I-\lambda T^*)^{-1}\Omega$$

$$\vartheta_4(\lambda) = \Delta_{*0}^{-1} h_\lambda \pi_*^* (\Omega_* - z\Omega B^*)$$

$$\vartheta_4(\lambda)\Omega_*^*|H = \Omega_*^*(I-\lambda T^*)^{-1}(I-\lambda L^*).$$

If $\lambda \notin \sigma(L)$. then

$$\vartheta_2(\lambda)^{-1} = \Omega^*(I-\lambda T^*)(\lambda I - L)^{-1}\Omega_*.$$

If $\lambda^{-1} \notin \sigma(L^*)$ then

$$\Omega\vartheta_3(\lambda)^{-1} = (I-\lambda T^*)(I-\lambda L^*)^{-1}\Omega, \qquad (2.5)$$

$$\vartheta_4(\lambda)^{-1}\Omega_*^*|H = \Omega_*^*(I-\lambda L^*)^{-1}(I-\lambda T^*). \qquad (2.6)$$

We state without proof some more formulae involving the functions ϑ_i

$$\pi_*^* L = \{z\pi_*^* - \Delta_{*0}\vartheta_1 h_0^* \Omega^*\}|H$$

$$\pi^* L = \{z\pi^* - \Delta_0 \vartheta_3^* h_0^* \Omega^*\}|H$$

$$\pi_*^* L^* = \{\bar{z}\pi_*^* - \bar{z}\Delta_{*0}\vartheta_4 h_0^* \Omega_*^*\}|H \tag{2.7}$$

$$\pi^* L^* = \{\bar{z}\pi^* - \bar{z}\Delta_0 \vartheta_2^* h_0^* \Omega_*^*\}|H .$$

We summarize the established results in

THEOREM 2.4.

I. 1) $\lambda \in \sigma_p(L) \Longleftrightarrow \mathrm{Ker}\, \vartheta_1(\lambda) \neq \{0\}$,

$$\mathrm{Ker}\,(L-\lambda I) = (z-\lambda)^{-1}(z\Omega - \Omega_* B)\,\mathrm{Ker}\,\vartheta_1(\lambda).$$

2) $\bar{\lambda} \in \sigma_p(L^*) \Longleftrightarrow \mathrm{Ker}\, \vartheta_2^*(\lambda) \neq \{0\}$,

$$\mathrm{Ker}\,(L^* - \bar{\lambda} I) = (1-\bar{\lambda}z)^{-1}(\Omega_* - z\Omega B^*)\,\mathrm{Ker}\,\vartheta_2^*(\lambda).$$

3) $\bar{\lambda}^{-1} \in \sigma_p(L) \Longleftrightarrow \mathrm{Ker}\, \vartheta_3^*(\lambda) \neq \{0\}$,

$$\mathrm{Ker}\,(I - \bar{\lambda} L) = (1-\bar{\lambda}z)^{-1}(z\Omega - \Omega_* B)\,\mathrm{Ker}\,\vartheta_3^*(\lambda).$$

4) $\lambda^{-1} \in \sigma_p(L^*) \Longleftrightarrow \mathrm{Ker}\, \vartheta_4(\lambda) \neq \{0\}$,

$$\mathrm{Ker}\,(I - \lambda L^*) = (z-\lambda)^{-1}(\Omega_* - z\Omega B^*)\,\mathrm{Ker}\,\vartheta_4(\lambda)$$

II. 1) $\lambda \notin \sigma(L) \Longleftrightarrow \vartheta_1(\lambda)$ is invertible $\Longleftrightarrow$ $\vartheta_2(\lambda)$ is invertible.

In this case,

$$(L-\lambda I)^{-1} = (z-\lambda)^{-1}[I - (z\Omega - \Omega_* B)\vartheta_1(\lambda)^{-1}\Delta_{*0}^{-1} h_\lambda \pi_*^*]|H,$$
$$(L^* - \bar{\lambda} I)^{-1} = (1-\bar{\lambda}z)^{-1}[z - (\Omega_* - z\Omega B^*)\vartheta_2^*(\lambda)^{-1}\Delta_0^{-1} h_\lambda \pi^* z]|H.$$

2) $\bar{\lambda}^{-1} \notin \sigma(L) \Longleftrightarrow \vartheta_3(\lambda)$ is invertible $\Longleftrightarrow$ $\vartheta_4(\lambda)$ is invertible.

In this case,

$$(I - \bar{\lambda} L)^{-1} = (1-\bar{\lambda}z)^{-1}[I - (z\Omega - \Omega_* B)\vartheta_3^*(\lambda)^{-1}\Delta_0^{-1} h_\lambda \pi^*]|H,$$

$$(I-\lambda L^*)^{-1}=(z-\lambda)^{-1}[z-\lambda(\Omega_*-z\Omega B^*)\vartheta_4(\lambda)^{-1}\Delta_{*0}^{-1}h_\lambda\pi_*^*]|H \quad \bullet$$

3. Special choice of an auxiliary contraction

The work with an arbitrary model representation of an operator L can meet some technical difficulties. We have indirectly come across one: in general, the functions ϑ_i need not be bounded. As is already mentioned, some kind of difficulties can be avoided by choosing an adequate auxiliary contraction in the representation (0.1). For example, the following restriction imposed on the representation (0.1) enables us to obtain bounded operator-functions:

There exists an operator $A\in\mathcal{L}(E,E_*)$,

such that $\Delta_{*0}A\Delta_0=A_0$. (0.2)

Any bounded operator admits a representation satisfying this restriction. Moreover, for completely nonunitary operators we describe in this section two standard choices of an auxiliary contraction. How to deal with unitary operators will be shown in Section 5.

Now we introduce some more notations. If $M\in\mathcal{L}(X,Y)$, then $|M|\overset{def}{=}(M^*M)^{1/2}$, $M=V_M|M|$ - polar decomposition of M , $D_M=|I-M^*M|^{1/2}$, $J_M=\operatorname{sign}(I-M^*M)$.

We consider $\operatorname{sign}0=1$, i.e. $J_M|\operatorname{Ker}(I-M^*M)=I$.

$\chi_M^{\pm}\overset{def}{=}\frac{1}{2}(I\pm J_M)$, and χ_M^0 denotes the orthogonal projection onto $\operatorname{Ker}D_M$.

Till now the same letter "L" served both for the operator and for its unitarily equivalent model. For a while this agreement will be broken.

Let M be a completely nonunitary operator in a separable Hilbert space. We wish to construct its representations in the model spaces of two preferable (in some sense) contractions T_0 and T_L .

REPRESENTATION IN THE MODEL T_0 . Set $M_0=M\chi_M^0$.

M_0 is a c.n. partial isometry. Let T_0 be its Sz.-Nagy - Foias model and let $\mathcal{U}$ be a unitary operator which provides an equivalence between T_0 and M_0 : $T_0 = \mathcal{U}^* M_0 \mathcal{U}$.
Thus, the operator

$$L \overset{def}{=} T_0 + \Omega_* A_0 \Omega^*,$$

$$A_0 = \Omega_*^* \mathcal{U}^* M (I - \chi_M^0) \mathcal{U} \Omega,$$

yields a model for the operator M in the model space of T_0.

For this representation, we have

$$\theta_0 = 0, \quad \vartheta = \theta, \quad \Delta_0 = I, \quad \Delta_{*0} = I.$$

$$\vartheta_1 = \vartheta_2 = \theta - A_0, \quad \vartheta_3 = I - A_0^* \theta, \quad \vartheta_4 = I - \theta A_0^*.$$

It is clear that (0.2) is valid $(A = A_0)$, and the functions ϑ_i are bounded. Such a model is convenient for the study of operators with finite defect indices (i.e. with $\dim \mathfrak{D}_M < \infty$, and $\dim \mathfrak{D}_{M^*} < \infty$).

REPRESENTATION IN THE MODEL T_L.

Consider the operator $M\varphi(|M|)$ where

$$\varphi(t) = \begin{cases} 1, & t \in [0,1] \\ t^{-2}, & t \in (1, \infty) \end{cases}$$

PROPOSITION 3.1. $M\varphi(|M|)$ is a completely nonunitary contraction.

PROOF. **The operator** $M\varphi(|M|)$ **is a contraction because**

$$I - |M\varphi(|M|)|^2 = I - |M|^2 \varphi^2(|M|) = D_M^2 \varphi(|M|) \geqslant 0. \tag{3.1}$$

To prove the complete nonunitarity of $M\varphi(|M|)$, suppose that $M\varphi(|M|)$ has a reducing subspace on which it is unitary, that is, the defect operators of $M\varphi(|M|)$ are zero on this subspace. Since the operators $\varphi(|M|)$ and $\varphi(|M^*|)$ are invertible, we infer from (3.1) that D_M and D_{M^*} are also zero on this subspace, and the operators $|M|$ and $|M^*|$ are equal to the identity on it. Since $\varphi(1) = 1$, the operators M and $M\varphi(|M|)$ coincide on it, as well as their adjoints do. Thus, the chosen subspace reduces M , and the restriction of M onto this subspace is unitary. Since M is completely nonunitary, this can happen only if the subspace is trivial. ●

Let T_L denote the Sz.-Nagy - Foiaş model of the operator $M\varphi(|M|)$. Let $\mathcal{U}$ be again an operator providing a unitary equivalence between T_L and $M\varphi(|M|)$, $T_L=\mathcal{U}^*M\varphi(|M|)\mathcal{U}$. Then

$$\theta_0=-\Omega_*^*T_L\Omega=-\Omega_*^*\mathcal{U}^*M\varphi(|M|)\mathcal{U}\Omega.$$

Define

$$L=T_L+\Omega_*A_0\Omega^*,\quad A_0=\Omega_*^*\mathcal{U}^*M[I-\varphi(|M|)]\mathcal{U}\Omega.$$

PROPOSITION 3.2. $\mathcal{U}L\mathcal{U}^*=M$.

PROOF. Since $\Omega\Omega^*$ is the projection onto the defect subspace of T_L, we have

$$\mathcal{U}\Omega\Omega^*\mathcal{U}^*=I-\chi_M^0.$$

Similarly,

$$\mathcal{U}\Omega_*\Omega_*^*\mathcal{U}^*=I-\chi_{M^*}^0.$$

Hence

$$\mathcal{U}L\mathcal{U}^*=\mathcal{U}T_L\mathcal{U}^*+\mathcal{U}^*\Omega_*A_0\Omega^*\mathcal{U}=$$

$$=M\varphi(|M|)+(I-\chi_{M^*}^0)M[I-\varphi(|M|)](I-\chi_M^0)=$$

$$=M\varphi(|M|)+M(I-\chi_M^0)[(I-\varphi(|M|)](I-\chi_M^0)=$$

$$=M\varphi(|M|)+M[I-\varphi(|M|)]=M,$$

because

$$[I-\varphi(|M|)]\chi_M^+=0,\qquad \chi_M^0=\chi_M^+\chi_M^0 \quad\bullet$$

If, as usual, $B\overset{def}{=\!=}A_0-\theta_0$, then

$$B=\Omega_*^*\mathcal{U}^*M\mathcal{U}\Omega.$$

This implies (since now we write $\chi^\pm,\chi_*^\pm$ instead of $\chi_B^\pm,\chi_{B^*}^\pm$)

$$\theta_0=-B\varphi(|B|)=-B(\chi^++|B|^{-2}\chi^-) \tag{3.2}$$

$$A_0=B[I-\varphi(|B|)]=BD_B^2|B|^{-2}\chi^-. \tag{3.3}$$

By the notation $|B|^{-2}\chi^-$ we mean the function $t\mapsto$ $\mapsto t^{-2}\chi_{(1,\infty)}(t)$ of the operator $|B|$, which is correctly defined, though the operator $|B|^{-2}$ may not exist. Instead of $\varphi(|B|)$ we shall write $\chi^+ + |B|^{-2}\chi^-$. By (3.2) we obtain

$$\Delta_0 = D_B(\chi^+ + |B|^{-1}\chi^-),$$

$$\Delta_{*0} = D_{B^*}(\chi^+_* + |B^*|^{-1}\chi^-_*).$$

Since

$$\Delta_{*0} B\chi^- \Delta_0 = B\Delta_0^2\chi^- = \\ = BD_B^2 |B|^{-2}\chi^- = A_0$$

we have $A = B\chi^-$, and the condition (0.2) is satisfied.

REMARK 3.3. A given contraction T does not uniquely determine the operator L satisfying $T = T_L$. Indeed, a model contraction T being fixed, we can choose an arbitrary orthogonal projection χ^- of E , with only two restrictions imposed:

1) $\chi^-|\theta_0| = |\theta_0|\chi^-$,

2) the operator $(I-\chi^-) + |\theta_0|\chi^-$ is invertible. Then we define in a standard way $L = T + \Omega_* A_0 \Omega^*$, where

$$A_0 = B + \theta_0, \quad B = -\theta_0[(I-\chi^-) + |\theta_0|\chi^-]^{-2}$$

and it is easily seen that $T = T_L$ ●

Now we return to the general case, supposing that (0.2) is valid. Instead of functions ϑ_i we introduce bounded functions θ_i defined by

$$\theta_1 = \theta(I + \theta_0^* A) - A$$
$$\theta_2 = (I + A\theta_0^*)\theta - A$$
$$\theta_3 = (I + A^*\theta_0) - A^*\theta$$
$$\theta_4 = (I + \theta_0 A^*) - \theta A^*$$

New functions are connected with the former ones by the relations

$$\theta_1\Delta_0 = \Delta_{*0}\vartheta_1, \quad \Delta_{*0}\theta_2 = \vartheta_2\Delta_0,$$
$$\Delta_0\theta_3 = \vartheta_3\Delta_0, \quad \theta_4\Delta_{*0} = \Delta_{*0}\vartheta_4. \tag{3.4}$$

We can describe the spectra $\sigma(L)$ and $\sigma(L^*)$ in ferms of θ_i in the same fashion as it has been done in Theorem 2.4. in terms of ϑ_i.

THEOREM 3.4.

I. 1) $\lambda \in \sigma_p(L) \Leftrightarrow \mathcal{K}er\, \theta_1(\lambda) \neq \{0\}$,

$$\mathcal{K}er\,(L-\lambda I) = (z-\lambda)^{-1}(\pi h_0^* - \omega_* A)\,\mathcal{K}er\,\theta_1(\lambda).$$

2) $\bar\lambda \in \sigma_p(L^*) \Leftrightarrow \mathcal{K}er\, \theta_2^*(\lambda) \neq \{0\}$,

$$\mathcal{K}er\,(L^*-\lambda I) = (1-\bar\lambda z)^{-1}(\pi_* h_0^* - z\omega A^*)\,\mathcal{K}er\,\theta_2^*(\lambda).$$

3) $\bar\lambda^{-1} \in \sigma_p(L) \Leftrightarrow \mathcal{K}er\, \theta_3^*(\lambda) \neq \{0\}$,

$$\mathcal{K}er\,(I-\bar\lambda L) = (1-\bar\lambda z)^{-1}(\pi h_0^* - \omega_* A)\,\mathcal{K}er\,\theta_3^*(\lambda).$$

4) $\lambda^{-1} \in \sigma_p(L^*) \Leftrightarrow \mathcal{K}er\, \theta_4(\lambda) \neq \{0\}$,

$$\mathcal{K}er\,(I-\lambda L^*) = (z-\lambda)^{-1}(\pi_* h_0^* - z\omega A^*)\,\mathcal{K}er\,\theta_4(\lambda).$$

II. 1) $\lambda \notin \sigma(L) \Leftrightarrow \theta_1(\lambda)$ is invertible $\Leftrightarrow \theta_2(\lambda)$ is invertible.

In this case

$$(L-\lambda I)^{-1} = (z-\lambda)^{-1}[I-(\pi h_0^* - \omega_* A)\,\theta_1(\lambda)^{-1} h_\lambda \pi_*^*]\,|H,$$

$$(L^*-\bar\lambda I)^{-1} = (1-\bar\lambda z)^{-1}[z-(\pi_* h_0^* - z\omega A^*)\,\theta_2^*(\lambda)^{-1} h_\lambda \pi^* z]\,|H,$$

$$\theta_1(\lambda)^{-1} = \theta_0^* + \omega^*(\lambda I - L)^{-1}\omega_*(I + A\theta_0^*),$$

$$\theta_2(\lambda)^{-1} = \theta_0^* + (I+\theta_0^* A)\,\omega^*(\lambda I - L)^{-1}\omega_*.$$

2) $\bar\lambda^{-1} \notin \sigma(L) \Leftrightarrow \theta_3(\lambda)$ is invertible $\Leftrightarrow \theta_4(\lambda)$ is invertible.

In this case

$$(I-\bar\lambda L)^{-1} = (1-\bar\lambda z)^{-1}[I-(\pi h_0^* - \omega_* A)\,\theta_3^*(\lambda)^{-1} h_\lambda \pi^*]\,|H,$$

$$(I-\lambda L^*)^{-1} = (z-\lambda)^{-1}[z-\lambda(\pi_* h_0^* - z\omega A^*)\,\theta_4(\lambda)^{-1} h_\lambda \pi_*^*]\,|H,$$

$$\Theta_3(\lambda)^{-1} = I + \lambda A^* \omega_*^* (I - \lambda L^*)^{-1} \omega,$$
$$\Theta_4(\lambda)^{-1} = I + \lambda \omega_*^* (I - \lambda L^*)^{-1} \omega A^*.$$

OUTLINE OF THE PROOF. The proof of the theorem repeats the reasoning of Section 2. We have to use the equality (3.4) and the following identities

$$z\Omega - \Omega_* B = (\pi h_0^* - \omega_* A)\Delta_0$$
$$\Omega_* - z\Omega B^* = (\pi_* h_0^* - z\omega A^*)\Delta_{*0}.$$

The only thing which differs from what we have had in Section 2 is the existence of the operators $\Theta_i(\lambda)^{-1}$ **if** $\lambda \notin \sigma(L)$ **(or** $\bar{\lambda}^{-1} \notin \sigma(L)$**).** Let us verify the formula for $\Theta_1(\lambda)^{-1}$.

$$\Theta_1(\lambda)[\Theta_0^* + \omega^*(\lambda I - L)^{-1}\omega_*(I + A\Theta_0^*)] =$$

$$= I + [\Theta(\lambda)\Theta_0^* - I + \Theta_1(\lambda)\omega^*(\lambda I - L)^{-1}\omega_*](I + A\Theta_0) = I,$$

because of

$$\Theta_1(\lambda)\omega^*(\lambda I - L)^{-1}\omega_* = \Delta_{*0}\vartheta_1(\lambda)\Omega^*(\lambda I - L)^{-1}\omega_* =$$

$$= \Delta_{*0}\Omega_*^*(I - \lambda T^*)^{-1}(\lambda I - L)(\lambda I - L)^{-1}\omega_* =$$

$$= h_\lambda \pi_*^* \omega_* = I - \Theta(\lambda)\Theta_0^*.$$

Here we have used the formulae (3.4), (2.2), (1.6) and (1.2).

$$[\Theta_0^* + \omega^*(\lambda I - L)^{-1}\omega_*(I + A\Theta_0^*)]\Theta_1(\lambda) =$$
$$= I + [\Theta_0^*\Theta(\lambda) - I + \omega^*(\lambda I - L)^{-1}\omega_*\Theta_2(\lambda)](I + \Theta_0^* A) = I$$

because of

$$\omega^*(\lambda I - L)^{-1}\omega_*\Theta_2(\lambda) = \omega^*(\lambda I - L)^{-1}\Omega_*\vartheta_2(\lambda)\Delta_0 =$$

$$= \omega^*(\lambda I - L)^{-1}(\lambda I - L)(I - \lambda T^*)^{-1}\Omega\Delta_0 =$$

$$= h_0 \pi^* z(z - \lambda)^{-1}[\pi - \pi_*\Theta(\lambda)]h_0^* = I - \Theta_0^*\Theta(\lambda).$$

Here we have used the formulae (3.4), (2.4), (1.7) and (1.3). ●

In the end of this section we state the expressions for the functions θ_i for the model with $T=T_L$. We have

$$I+\theta_0^* A = I + A^* \theta_0 = \chi^+,$$

$$I+A\theta_0^* = I+\theta_0 A^* = \chi_*^+,$$

hence

$$\theta_1 = \theta\chi^+ - B\chi^-,$$

$$\theta_2 = \chi_*^+\theta - \chi_*^- B,$$

$$\theta_3 = \chi^+ - \chi^- B^*\theta,$$

$$\theta_4 = \chi_*^+ - \theta B^*\chi_*^-.$$

4. Factorization of characteristic function

We define **the characteristic function of the operator** L by

$$\theta_L(\lambda) \overset{\text{def}}{=\!=} \Omega_*^* [-LJ_L + \lambda D_{L^*}(I-\lambda L^*)^{-1} D_L]\Omega$$

$\theta_L(\lambda)$ is a $\mathcal{L}(E, E_*)$ - valued function, holomorphic whenever $\bar{\lambda}^{-1} \notin \sigma(L)$.

Our definition differs somewhat from the generally accepted one, for the spaces E and E_* can be "wider" than the defect subspaces of L . Thus, the function θ_L depends on the model representation of the operator L.

PROPOSITION 4.1. θ_L is J -contractive in the following sense:

$$J_B - \theta_L^* J_{B^*} \theta_L \geqslant 0 , \qquad J_{B^*} - \theta_L J_B \theta_{L^*}^* \geqslant 0.$$

PROOF. Since

$$I - \Omega_* \Omega_*^* = \chi_{T^*}^\circ , \qquad \chi_{T^*}^\circ D_{L^*} = 0,$$

$$\chi_{T^*}^\circ L J_L \Omega = T\chi_T^\circ \Omega = 0,$$

we have

$$\Omega_* \theta_L(\lambda) = [-LJ_L + \lambda D_{L^*}(I-\lambda L^*)^{-1} D_L]\Omega,$$

from which follows

$$J_B-\theta_L^*(\lambda)J_{B^*}\theta_L(\lambda)=J_B-\theta_L^*(\lambda)\Omega_*^*J_{L^*}\Omega_*\theta_L(\lambda)=$$
$$=\Omega^*\{J_L-[-L^*J_{L^*}+\bar{\lambda}D_L(I-\bar{\lambda}L)^{-1}D_{L^*}]J_{L^*}[-LJ_L+$$
$$+\lambda D_{L^*}(I-\lambda L^*)^{-1}D_L]\}\Omega=$$
$$=\Omega^*D_L(I-\bar{\lambda}L)^{-1}\{(I-\bar{\lambda}L)(I-\lambda L^*)+\bar{\lambda}L(I-\lambda L^*)+$$
$$+\lambda(I-\bar{\lambda}L)L^*-|\lambda|^2(I-LL^*)\}(I-\lambda L^*)^{-1}D_L\Omega=$$
$$=(1-|\lambda|^2)\Omega^*D_L(I-\bar{\lambda}L)^{-1}(I-\lambda L^*)^{-1}D_L\Omega\geqslant 0.$$

In the similar way we verify that

$$J_{B^*}-\theta_L(\lambda)J_B\theta_L(\lambda)^*=(1-|\lambda|^2)\Omega_*^*D_{L^*}(I-\lambda L^*)(I-\bar{\lambda}L)^{-1}D_{L^*}\Omega_*\geqslant 0$$ ●

Now we exhibit the factorization of the characteristic function. We consider at first an arbitrary representation $L=T+\Omega_*A_0\Omega^*$. In this general case the factorization will contain some noninvertible constant operators in both sides of the equality. Next, treating the case $T=T_L$, we get rid of this constants.

THEOREM 4.2.

$$\theta_L(\lambda)D_BJ_B=D_{B^*}\vartheta_4^{-1}(\lambda)\vartheta_1(\lambda),$$
$$J_{B^*}D_{B^*}\theta_L(\lambda)=\vartheta_2(\lambda)\vartheta_3(\lambda)^{-1}D_B.$$

PROOF. We shall use the formulae

$$\theta_L(\lambda)D_BJ_B=D_{B^*}\Omega_*^*(I-\lambda L^*)^{-1}(\lambda I-L)\Omega,$$
$$J_{B^*}D_{B^*}\theta_L(\lambda)=\Omega_*^*(\lambda I-L)(I-\lambda L^*)^{-1}\Omega D_B.$$

For example, the checking of the first formula proceeds as follows

$$\theta_L(\lambda)D_BJ_B=\Omega_*^*[-D_{L^*}L+\lambda D_{L^*}(I-\lambda L^*)^{-1}(I-L^*L)]\Omega=$$
$$=D_{B^*}\Omega_*^*(I-\lambda L^*)^{-1}[-(I-\lambda L^*)L+\lambda(I-L^*L)]\Omega=$$
$$=D_{B^*}\Omega_*^*(I-\lambda L^*)^{-1}(\lambda I-L)\Omega.$$

So from (2.5) and Lemma 2.1 we have

$$\theta_L(\lambda)D_B J_B = D_{B^*}\Omega_*^*(I-\lambda L^*)^{-1}(I-\lambda T^*)(I-\lambda T^*)^{-1}(\lambda I - L)\Omega =$$

$$= D_{B^*}\vartheta_4(\lambda)^{-1}\Omega_*^*(I-\lambda T^*)^{-1}(\lambda I - L)\Omega = D_{B^*}\vartheta_4(\lambda)^{-1}\vartheta_1(\lambda).$$

Similarly from (2.4) and (2.3) follows

$$J_{B^*}D_{B^*}\theta_L(\lambda) = \Omega_*^*(\lambda I - L)(I-\lambda T^*)^{-1}(I-\lambda T^*)(I-\lambda L^*)^{-1}\Omega D_B =$$

$$= \Omega_*^*(\lambda I - L)(I-\lambda T^*)^{-1}\Omega\,\vartheta_3(\lambda)^{-1}D_B = \vartheta_2(\lambda)\vartheta_3(\lambda)^{-1}D_B. \bullet$$

We can simplify the factorization presented in Theorem 4.2 in the case where $T = T_L$. In order to do this we introduce additional operator-valued functions $\widetilde{\theta}_i$ which differ from the functions θ_i by invertible operators.

$$\widetilde{\theta}_1 \stackrel{def}{=} \theta_1(\chi^+ - |B|^{-1}\chi^-) = \theta\chi^+ + V\chi^-,$$

$$\widetilde{\theta}_2 \stackrel{def}{=} (\chi_*^+ - |B^*|^{-1}\chi_*^-)\theta_2 = \chi_*^+\theta + \chi_*^- V_*^*,$$

$$\widetilde{\theta}_3 \stackrel{def}{=} (\chi^+ + |B|^{-1}\chi^-)\theta_3 = \chi^+ - \chi^- V^*\theta,$$

$$\widetilde{\theta}_4 \stackrel{def}{=} \theta_4(\chi_*^+ + |B^*|^{-1}\chi_*^-) = \chi_*^+ - \theta V_*\chi_*^-.$$

Here $V = V_B, V_* = V_{B^*}$ are isometrical operators from the polar decompositions $B = V_B|B|$, $B^* = V_{B^*}|B^*|$.

Since

$$\widetilde{\theta}_1 = \Delta_{*0}\vartheta_1 D_B^{-1}J_B, \quad \widetilde{\theta}_2 = J_{B^*}D_{B^*}^{-1}\vartheta_2\Delta_0,$$

$$\widetilde{\theta}_3 = D_B^{-1}\vartheta_3\Delta_0, \quad \widetilde{\theta}_4 = \Delta_{*0}\vartheta_4 D_{B^*}^{-1},$$

Theorem 4.2 provides

COROLLARY 4.3. $\theta_L(\lambda) = \widetilde{\theta}_4(\lambda)^{-1}\widetilde{\theta}_1(\lambda) = \widetilde{\theta}_2(\lambda)\widetilde{\theta}_3(\lambda)^{-1}$ $\bullet$

As follows from Theorem 2.4, the factors in the latter factorizations are responsible for the spectrum of L inside and outside the circle $\mathbb{T}$ respectively.

In the sequel we shall need a relation between the operators $J_B - \theta_2^* J_{B^*}\theta_L$ and $I - \theta_T^*\theta_T$. We consider the case $T = T_L$.

PROPOSITION 4.4.

$$\Delta^2 = \Theta_3^*(J_B - \Theta_L^* J_{B^*}\Theta_L)\Theta_3,$$

$$\Delta_*^2 = \Theta_4(J_{B^*} - \Theta_L J_B \Theta_L^*)\Theta_4^*.$$

PROOF.

$$\Theta_3^*(J_B - \Theta_L^* J_{B^*}\Theta_L)\Theta_3 = \Theta_3^* J_B\Theta_3 - \Theta_2^* J_B \Theta_2 =$$

$$= (\chi^+ - \theta V \chi^-)(\chi^+ + \chi^- V^*\theta) - (\theta^*\chi_*^+ + V_*\chi_*^-)(\chi_*^+\theta - \chi_*^- V_*^*) =$$

$$= \chi^+ - \theta^*\chi_*^-\theta - \theta^*\chi_*^+\theta + \chi^- = I - \theta^*\theta = \Delta^2$$

The second equality can be proved similarly. ●

5. The absolutely continuous subspace.

From now on we deal only with model representations of an arbitrary operator L for which the condition (0.2) is valid. In this setting we define the absolutely continuous (N_a) and singular (N_s) subspaces of L. For unitary operators and c.n. contractions our concept agrees with the standard ones (see, e.g., [13] for a discussion on absolutely continuous subspace of c.n. contractions). In the present and in the next sections several different descriptions of the subspaces N_a and N_s are given. Some of them essentially use the model language. The others appeal to the boundary behaviour of the resolvent not involving the model representation at all. Our exposition in this section often follows the approach of S.N.Naboko [3].

Let us denote $\mathcal{N} = \mathrm{Ker}[(I + A\theta_0^*)\pi_*^* - A\pi^*]$.

PROPOSITION 5.1. $\mathcal{N} = \{x : (L-\mu)P_H(z-\mu)^{-1}x = P_H x$ for any μ, $\mu \in \mathbb{C}\setminus\mathbb{T}\}$.

PROOF. We have successively

$$P_H - (L-\mu)P_H(z-\mu)^{-1} = (P_H z - LP_H)(z-\mu)^{-1};$$

$$P_H z - LP_H = P_H z P_{H^\perp} - (L-T)P_H;$$

the first summand:

$$P_H z P_{H^\perp} = (I - \pi P_+ \pi^* - \pi_* P_- \pi_*^*) z \pi_* P_- \pi_*^* =$$

$$= (\pi_* P_+ - \pi P_+ \theta^*) z P_- \pi_*^* = \omega_* h_0 \pi_*^* z;$$

the second summand:

$$(L-T)P_H = \omega_* A \omega^* P_H = \omega_* A \omega^* = \omega_* A h_0 (\pi^* - \theta_0^* \pi_*^*) z.$$

Consequently,

$$P_H - (L-\mu) P_H (z-\mu)^{-1} = \omega_* h_0 z (z-\mu)^{-1} [(I + A\theta_0^*)\pi_*^* - A\pi^*], \quad (5.I)$$

and

$$\mathcal{N} \subset \mathcal{K}er\, [P_H - (L-\mu) P_H (z-\mu)^{-1}].$$

To prove the converse, suppose that

$$P_H x = (L-\mu) P_H (z-\mu)^{-1} x \qquad \text{for any } \mu,\ \mu \in \mathbb{C},\ |\mu| \neq 1,$$

and check that the function

$$f = [(I + A\theta_0^*)\pi_*^* - A\pi^*] x$$

is identically zero. In fact, since $\mathcal{K}er\, \omega_* = 0$, the equality (5.I) implies that the Cauchy integral of the function f vanishes on $\mathbb{C} \setminus \mathbb{T}$; hence $f \equiv 0$. ●

DEFINITION. The subspace $N_a \overset{\text{def}}{=\!=} \operatorname{clos} P_H \mathcal{N}$ is called the absolutely continuous subspace of the operator L.

REMARK 5.2. Define N_{*a} as the absolutely continuous subspace of the operator L^* in its model with the auxiliary contraction T^*. Then in the representation of L on the model space of T we have:

$$N_{*a} = \operatorname{clos} P_H \mathcal{N}_*$$

$$\mathcal{N}_* = \mathcal{K}er\, [(I + A^* \theta_0) \pi^* - A^* \pi_*^*].$$

To make sure of it, let $\hat{\mathcal{H}}, \hat{z}, \hat{H}$ etc. denote the objects having the same meaning in the model of L^* as $\mathcal{H}, z, H$ etc. in the model of L (in particular, $\hat{L} = L^*$). Note that there exists a unitary operator from $\hat{\mathcal{H}}$ onto $\mathcal{H}$ which transforms $\hat{z}$ into $\bar{z}$, $\hat{E}$ into E_*, $\hat{E}_*$ into E,

$\hat{\pi}$ into $\pi_* \bar{z} c$, $\hat{\pi}_*$ into $\pi \bar{z} c$, $\hat{\theta}$ into θ_{T^*} , $\hat{H}$ into H , $\hat{L}$ into L^* , $\hat{A}$ into A^* etc. ($c: L^2(X) \to L^2(X)$ is defined by $(cf)(\zeta) = f(\bar{\zeta})$, θ_{T^*} is the characteristic function of T^* , $\theta_{T^*}(\lambda) = \theta^*(\bar{\lambda})$).
It is easy to see that the equality

$$(I + \hat{A}\hat{\theta}_0^*)\hat{\pi}_*^*\hat{x} = \hat{A}\hat{\pi}^*\hat{x}$$

corresponds to the equality

$$(I + A^*\theta_0)\pi^* x = A\pi_*^* x.$$

The aforementioned operator can be expressed in the B.S.Pavlov's form by

$$\bar{z} c \begin{pmatrix} 0 & I \\ I & 0 \end{pmatrix} : L^2 \begin{pmatrix} I & \theta_{T^*}^* \\ \theta_{T^*} & I \end{pmatrix} \to L^2 \begin{pmatrix} I & \theta^* \\ \theta & I \end{pmatrix}. \qquad \bullet$$

PROPOSITION 5.3.

$$P_H \mathcal{N} = \{ y : \pi^* y = \theta_3^* f_- ,\ f_- \in H_-^2(E);\ \pi_*^* y = \theta_1 f_+ ,\ f_+ \in H^2(E) \},$$

$$P_H \mathcal{N}_* = \{ y : \pi^* y = \theta_2^* f_- ,\ f_- \in H_-^2(E_*);\ \pi_*^* y = \theta_4 f_+ ,\ f_+ \in H^2(E_*) \}.$$

PROOF. Let us check, for example, the first formula. Let $x \in \mathcal{N}$, i.e.

$$A\pi^* x = (I + A\theta_0^*)\pi_*^* x; \tag{5.2}$$

and let $y = P_H x$. Then

$$\pi^* y = \theta_3^* P_- (\pi^* - \theta_0^* \pi_*^*) x,$$

$$\pi_*^* y = \theta_1 P_+ (\theta_0^* \pi_*^* - \pi^*) x.$$

Indeed,

$$\pi^* y = \pi^* (I - \pi P_+ \pi^* - \pi_* P_- \pi_*^*) x =$$

$$= P_- \pi^* x - \theta^* P_- \pi_*^* x =$$

$$= [I - (\theta^* - \theta_0^*) A] P_- \pi^* x - [\theta^* - (\theta^* - \theta_0^*)(I + A\theta_0^*)] P_- \pi_*^* x =$$

$$= \Theta_3^* P_- (\pi^* - \Theta_0^* \pi_*^*) x .$$

We get the expression for $\pi_*^* y$ analogously.

Conversely, if $f_- \in H_-^2(E)$, $f_+ \in H^2(E)$, and y is such that $\pi^* y = \Theta_3^* f_-$, $\pi_*^* y = \Theta_1 f_+$, then we take the element

$$x = y + \pi_* A f_- - \pi (I + \Theta_0^* A) f_+ \tag{5.3}$$

and verify that $x \in \mathcal{N}$ and $P_H x = y$.

The latter is valid because $\pi_* H_-^2(E) \perp H$ and $\pi H^2(E) \perp H$. To check that $x \in \mathcal{N}$, let us compute $\pi^* x$ and $\pi_*^* x$:

$$\pi^* x = \pi^* y + \Theta^* A f_- - (I + \Theta_0^* A) f_+ =$$

$$= (\Theta_3^* + \Theta^* A) f_- - (I + \Theta_0^* A) f_+ = (I + \Theta_0^* A)(f_- - f_+)$$

$$\pi_*^* x = \pi_*^* y + A f_- - \Theta (I + \Theta_0^* A) f_+ =$$

$$= A f_- + [\Theta_1 - \Theta (I + \Theta_0^* A)] f_+ = A (f_- - f_+)$$

hence (5.2) holds. ●

S.N.Naboko in ref. [3] provides the description of the absolutely continuous subspace in terms of the "smoothness" of the resolvent. In the proposition to follow we present a similar result for the model with $T = T_L$.

PROPOSITION 5.4. Let L be a completely nonunitary operator. For the model with $T = T_L$,

$$P_H \mathcal{N} = \{ y : y \in H , \quad \exists H\text{-valued functions } x \text{ and } x' \text{ such that } y = (L - \lambda) x(\lambda) = (I - \bar\lambda L) x'(\bar\lambda),$$

$$D_L x \in H^2(H), \ D_L x' \in H^2(H) \} .$$

PROOF. Since the operators D_B and Δ_0 are similar, the conditions $D_L x \in H^2(H)$, $D_L x' \in H^2(H)$ are equivalent to the conditions $\omega^* x \in H^2(E)$, $\omega^* x' \in H^2(E)$. The formulae (2.7) imply the following identities:

$$h_\lambda \pi_*^* (L - \lambda) = -\Theta_1(\lambda) \omega^* | H ,$$

$$h_\lambda \pi^* (I - \bar\lambda L) = \bar\lambda \Theta_3^*(\lambda) \omega^* | H . \tag{5.4}$$

Therefore, if x and x' are such as in the statement of the theorem, we set $f_+(\lambda) = -\omega^* x(\lambda)$ and $f_-(\lambda) = \bar{\lambda}\omega^* x'(\lambda)$. Then the equalities (5.4) provide $\pi_*^* y = \theta_1 f_+$, $\pi^* y = \theta_3^* f_-$, i.e. $y \in P_H \mathcal{N}$, according to the previous proposition.

Conversely, if $\pi_*^* y = \theta_1 f_+$, $f_+ \in H^2(E)$ and $\pi^* y = \theta_3^* f_-$, $f_- \in H^2_-(E)$, then it is natural (see the formulae for the resolvent from Theorem 3.4) to set

$$x(\lambda) = (z-\lambda)^{-1}[y - (\pi h_0^* - \omega_* A) f_+(\lambda)],$$

$$x'(\lambda) = (1-\lambda z)^{-1}[y - (\pi h_0^* - \omega_* A) f_-(\bar{\lambda})].$$

Direct computations provide that $x(\lambda) \in H$, $x'(\lambda) \in H$ for any $\lambda \in \mathbb{D}$ and $(L-\lambda)x(\lambda) = (I-\bar{\lambda}L)x'(\bar{\lambda}) = y$. In addition, $\omega_* x(\lambda) = -f_+(\lambda)$ and $\omega_* x'(\lambda) = \lambda^{-1} f_-(\bar{\lambda})$, i.e. by the note at the beginning of the proof, $D_L x \in H^2(E)$ and $D_L x' \in H^2(E)$. ●

The next proposition is also adopted from [3] (cf. also [2] ch.II, §3).

PROPOSITION 5.5. If the spectrum $\sigma(L)$ does not cover the unit disc, the operator

$$Q \overset{def}{=\!=} P_H | \mathcal{N} : \mathcal{N} \to N_a$$

is a quasi-affinity intertwining $L | N_a$ and $z | \mathcal{N}$ (cf. [2] for terminology).

PROOF. The equality $Qz|\mathcal{N} = LQ$ follows from Proposition 5.I. Thus the only thing still to be checked is that $\operatorname{Ker} Q = \{0\}$. It is not difficult to verify the following formulae:

$$A\pi^* Q = \theta_4^* P_- \pi_*^* | \mathcal{N},$$
$$(I + A\theta_0^*)\pi_*^* Q = -\theta_2 P_+ \pi^* | \mathcal{N}.$$

If $x \in \operatorname{Ker} Q$, then $\theta_4^* P_- \pi_*^* x = 0$ and $\theta_2 P_+ \pi^* x = 0$. The operators $\theta_4(\lambda)$ are invertible in a neighbourhood of the origin, hence $P_- \pi_*^* x = 0$. Since $\sigma(L)$ does not cover the unit disc, $\theta_2(\lambda)$ are also invertible for some open set of λ's, so $P_+ \pi_* x = 0$. Therefore $x \in H$, and $x = Qx = 0$. ●

DEFINITION. The subspace N_s, $N_s \overset{def}{=} H \ominus N_a$, is called the singular subspace of the operator L.

Similarly, $N_{*s} = H \ominus N_a$.

Obviously, the subspaces N_a and N_s are invariant under rational functions (with poles off $\sigma(L)$) of the operator L. The study of these subspaces will be continued in the next section.

6. Scalar multiples.

As usual, the efficiency of the model increases considerably if the corresponding analytic operator valued functions admit scalar multiples. In this section we study operators allowing a model representation in which the bounded functions θ_i $(i=1,2,3,4)$ have scalar multiples. The latter means the existence of the bounded analytic operator--valued functions Υ_i and nonzero bounded scalar functions δ_i, such that

$$\theta_i \Upsilon_i = \delta_i I, \quad \Upsilon_i \theta_i = \delta_i I.$$

This representation will be said to satisfy the condition (SM). We present here the results, having in mind the model's application in the next section.

It is known (see [2], ch. YIII, § I) that the characteristic function of a contraction T admits scalar multiple if $I - T^* T \in \mathfrak{S}_1$ and the spectrum $\sigma(T)$ does not cover the whole unit disc. The thing is that in this case the difference between $\theta(\lambda)$ and some constant unitary operator is of trace class for any λ (so $\det \theta(\lambda)$ exists; see [8]), and $\theta(\lambda)$ is invertible for some λ (thus $\det \theta(\lambda)$ is not identically zero). We use this observation to prove

THEOREM 6.1. Let L be a bounded operator such that $I - T^* T \in \mathfrak{S}_1$, and $\mathbb{D} \not\subset \sigma(L)$. Then L has a model representation satisfying (SM).

PROOF. Since L is the orthogonal sum of the unitary and completely nonunitary operators, it is sufficient to prove the theorem separately for unitary and for c.n.operators.

The case of c.n. operators. We claim that (SM) is satisfied for the representation in the model space of the operator $T = T_L$. Let us check, for example, the existence of a scalar multiple for θ_1. Note that it suffices to find a scalar multiple for the function $\chi_*^+\theta\chi^+ : H^2(\chi^+E)\to H^2(\chi_*^+E_*)$. In fact, if there exists a function $\Upsilon^+\in H^\infty(\mathcal{L}(E_*,E))$ such that

$$\Upsilon^+=\chi^+\Upsilon^+=\Upsilon^+\chi_*^+$$

and

$$\Upsilon^+\theta\chi^+=\delta\chi^+,\quad \chi_*^+\theta\Upsilon^+=\delta\chi_*^+$$

then the function

$$\Upsilon_1\overset{\text{def}}{=\!=}\Upsilon^+ + |B|^{-2}\chi^-B^*\theta\Upsilon^+ - \delta|B|^{-2}\chi^-B^*$$

satisfies

$$\Upsilon_1\theta_1=\delta I,\quad \theta_1\Upsilon_1=\delta I.$$

But the operator $\chi_*^+\theta(\lambda)\chi^+ : \chi^+E\longrightarrow\chi_*^+E_*$ is invertible, provided $\lambda\notin\sigma(L)$ $((\chi_*^+\theta(\lambda)\chi^+)^{-1}=\chi^+\theta_1(\lambda)^{-1}\chi_*^+)$, and the difference $\chi^+-V_B^*\chi_*^+\theta(\lambda)\chi^+$ belongs to the trace class, where V_B is the isometrical factor in the polar decomposition $B=V_B|B|$. Thus the function $\chi_*^+\theta\chi^+$ admits a scalar multiple, and the assertion follows.

T h e c a s e o f u n i t a r y o p e r a t o r s. Let U be a unitary operator in a separable Hilbert space H. Then there exists an orthogonal decomposition of H:

$$H=\bigoplus_{k=1}^{k_0}H_k,\qquad k_0\leqslant\infty$$

such that the subspaces H_k reduce U, and the operators $U|H_k$ have simple spectra, i.e.

$$H_k=\operatorname{span}(U^n x_k : n\in\mathbb{Z})$$

for some x_k, $x_k\in H_k$.

Let R be a trace class selfadjoint operator on a Hilbert space E, $\dim E=k_0$, and $0<R<I$. Let Φ be an isometrical embedding of E into H, such that $\Phi E= =\operatorname{span}\{x_k\}$, and Φ^*x_k are the eigenvectors of R. Then the operator

$$T=U(I-\Phi R\Phi^*)$$

is a c.n. contraction. Indeed,

$$I-T^*T=\Phi(2R-R^2)\Phi^*>0,$$

and let H_0 be a reducing subspace on which T is unitary. Then $H_0 \cap H_K$ also reduces. Since the vector x_K belongs to the defect subspace of $T|H_K$ and $T|H_0 \cap H_K$ is unitary, $x_K \perp H_0 \cap H_K$. Hence $T|H_0 = \mathcal{U}|H_0$, and $\mathcal{U}^n x_K \perp \perp H_0 \cap H_K$ for any n, $n \in \mathbb{Z}$. Thus $H_0 \cap H_K = \{0\}$, and $H_0 = \{0\}$, proving the complete nonunitarity of T.

So we can assume that H is the model space of T. As

$$D_T = \Phi(2R-R^2)^{1/2}\Phi^*, \qquad D_{T*} = \mathcal{U}\Phi(2R-R^2)^{1/2}\Phi^*\mathcal{U}^*,$$

we have

$$\Theta(\lambda) = \Omega_*^* \mathcal{U}\Phi[R-I-\lambda(2R-R^2)^{1/2}\Phi^*\mathcal{U}^*(I-\lambda T^*)^{-1}\Phi(2R-R^2)^{1/2}]\Phi^*\Omega.$$

Since the characteristic function is determined up to constant unitary factors, we can choose

$$E = E_*, \quad \Omega = \Phi, \quad \Omega_* = \mathcal{U}\Phi.$$

Note that in this case $\Theta(\lambda) + I \in \gamma_1$ for all λ's. We have also

$$\Theta_0 = \Theta_0^* = R - I,$$

$$\Delta_0 = \Delta_{*0} = (2R - R^2)^{1/2},$$

$$A_0 = \Phi^*\mathcal{U}^*(\mathcal{U} - T)\Phi = R.$$

Hence (0.2) is valid and

$$A = (2I - R)^{-1}.$$

Finally, the analytic operator-functions

$$\Theta_1 = -\Theta_4 = (\Theta - I)A,$$

$$\Theta_2 = -\Theta_3 = A(\Theta - I)$$

admit scalar multiples. Indeed, the function $\frac{1}{2}(I - \Theta)$ is contractive. It is invertible at the origin, and $I - \frac{1}{2}(I - \Theta(\lambda)) = \frac{1}{2}(I + \Theta(\lambda)) \in \gamma_1$ for any λ; so the assertion follows. ●

Now we turn to the study of operators admitting model representation in which (SM) is valid. This assumption remains unchanged up to the end of the section. First we give a new description of the absolutely continuous and singular subspaces.

PROPOSITION 6.2.

$$\mathcal{N}_*^{\perp} = \operatorname{Ker}[\delta_1 \Upsilon_3^* \pi^* - \bar{\delta}_3 \Upsilon_1 \pi_*^*],$$

$$\mathcal{N}^{\perp} = \operatorname{Ker}[\delta_4 \Upsilon_2^* \pi^* - \bar{\delta}_2 \Upsilon_4 \pi_*^*]$$

PROOF. We shall verify only the first formula. Let $\delta_1 \Upsilon_3^* \pi^* x = \bar{\delta}_3 \Upsilon_1 \pi_*^* x$. Then

$$\delta_1 \bar{\delta}_3 \Delta \tau^* x = \delta_1 \bar{\delta}_3 (\pi^* - \theta^* \pi_*^*) x =$$
$$= \delta_1 \theta_3^* \Upsilon_3^* \pi^* x - \bar{\delta}_3 \theta \theta_1 \Upsilon_1 \pi_*^* x =$$
$$= \bar{\delta}_3 (\theta_3^* - \theta^* \theta_1) \Upsilon_1 \pi_*^* x = \bar{\delta}_3 \Delta^2 (I + \theta_0^* A) \Upsilon_1 \pi_*^* x,$$

hence

$$\delta_1 \tau^* x = \Delta (I + \theta_0^* A) \Upsilon_1 \pi_*^* x.$$

If $y \in \mathcal{N}_*$, that is $(I + A^* \theta_0) \pi^* y = A^* \pi_*^* y$, then

$$(x, \bar{\delta}_1 y) = (\tau^* x, \bar{\delta}_1 \tau^* y) + (\pi_*^* x, \bar{\delta}_1 \pi_*^* y) =$$
$$= ((I + \theta_0^* A) \Upsilon_1 \pi_*^* x, \Delta \tau^* y) + (\pi_*^* x, \bar{\delta}_1 \pi_*^* y) =$$
$$= (\Upsilon_1 \pi_*^* x, (I + A^* \theta_0)(\pi^* - \theta^* \pi_*^*) y) + (\Upsilon_1 \pi_*^* x, \theta_1^* \pi_*^* y) = 0.$$

Hence x is orthogonal to $\bar{\delta}_1 \mathcal{N}_*$.

Note that $\operatorname{clos} \bar{\delta}_1 \mathcal{N}_* = \mathcal{N}_*$. Indeed, let $y \in \mathcal{N}_*$ and $y \perp \bar{\delta}_1 \mathcal{N}_*$. Define $\varphi \overset{\text{def}}{=} \|\pi^* y\|_E^2 + \|\tau_*^* y\|_{E_*}^2$. Since $z \mathcal{N}_* = \mathcal{N}_*$, we have $y \perp z^n \bar{\delta}_1 \mathcal{N}_*$ for any integer n, i.e. $0 = (z^n \bar{\delta}_1 y, y)_{\mathcal{H}} = \int z^n \bar{\delta}_1 \varphi \quad \forall n$. Thus $\bar{\delta}_1 \varphi = 0$, and since $\delta_1 \in H^\infty$, $\varphi = 0$, and $y = 0$. We have proved that $\mathcal{N}_*^{\perp} \supset \operatorname{Ker}[\delta_1 \Upsilon_3^* \pi^* - \bar{\delta}_3 \Upsilon_1 \pi_*^*]$.

To prove the converse inclusion it suffices to check that

$$[\delta_1 \Upsilon_3^* \pi^* - \bar{\delta}_3 \Upsilon_1 \pi_*^*]^* L^2(E) \subset \mathcal{N}_*.$$

Let $x = (\bar{\delta}_1 \pi \Upsilon_3 - \delta_3 \pi_* \Upsilon_1^*) g$, $g \in L^2(E)$. Then

$$[(I + A^* \theta_0) \pi^* - A^* \pi_*^*] x = \bar{\delta}_1 [(I + A^* \theta_0) - A^* \theta] \Upsilon_3 g -$$
$$- \delta_3 [(I + A^* \theta_0) \theta^* - A^*] \Upsilon_1^* g = (\bar{\delta}_1 \theta_3 \Upsilon_3 - \bar{\delta}_3 \theta_1^* \Upsilon_1^*) g = 0. \quad \bullet$$

COROLLARY 6.3.

$$\mathcal{N} = \operatorname{clos}(\bar{\delta}_4 \pi \Upsilon_2 - \delta_2 \pi_* \Upsilon_4^*) L^2(E_*)$$

$$\mathcal{N}_* = \operatorname{clos}(\bar{\delta}_1 \pi \Upsilon_3 - \delta_3 \pi_* \Upsilon_1^*) L^2(E). \quad \bullet$$

COROLLARY 6.4.

$$N_S = \operatorname{Ker}(\delta_1 \Upsilon_3^* \pi^* - \bar{\delta}_3 \Upsilon_1 \pi_*^*) \cap H,$$

$$N_{*S} = \operatorname{Ker}(\delta_4 \Upsilon_2^* \pi^* - \bar{\delta}_2 \Upsilon_4 \pi_*^*) \cap H. \quad \bullet$$

In the sequel we shall use the boundary values of the resolvent (in the weak operator topology). In terms of such values one more description of the singular subspace is provided. Thus we ascertain that the definitions of the absolutely continuous and singular subspaces of the operator L coincide in all models of L satisfying (SM).

For arbitrary elements x and y of the space H consider two functions:

$$\mathcal{F}_{x,y}^+(\lambda) \equiv \mathcal{F}^+(\lambda) \overset{\text{def}}{=} ((L-\lambda I)^{-1}x, y), \quad \lambda \notin \sigma(L),$$

$$\mathcal{F}_{x,y}^-(\lambda) \equiv \mathcal{F}^-(\lambda) \overset{\text{def}}{=} ((L-\bar{\lambda}^{-1} I)^{-1}x, y), \quad \bar{\lambda}^{-1} \notin \sigma(L).$$

PROPOSITION 6.5.

$$\mathcal{F}^+(\lambda) = ((z-\lambda)^{-1}x, y)_{\mathcal{H}} - \lambda^{-1}((I+\Theta_0^* A)\Theta_1(\lambda)^{-1} h_\lambda \pi_*^* x, h_\lambda \pi^* y)_E,$$

$$\mathcal{F}^-(\lambda) = -\bar{\lambda}((1-\bar{\lambda} z)^{-1}x, y)_{\mathcal{H}} - \bar{\lambda}(A\Theta_3^*(\lambda)^{-1} h_\lambda \pi^* x, h_\lambda \pi_*^* y)_{E_*}.$$

PROOF. To prove these equalities we make use of the formulae for the resolvent from Theorem 3.4:

$$\mathcal{F}^+(\lambda) = ((z-\lambda)^{-1}x, y)_{\mathcal{H}} - ((z-\lambda)^{-1}(\pi h_0^* - \omega_* A)\Theta_1(\lambda)^{-1} h_\lambda \pi_*^* x, y)_{\mathcal{H}}$$

$$= ((z-\lambda)^{-1}x, y)_{\mathcal{H}} - (\Theta_1(\lambda)^{-1} h_\lambda \pi_*^* x, [(I+A^*\Theta_0)h_0\pi^* - A^* h_0 \pi_*^*](\bar{z}-\bar{\lambda})^{-1} y)_E.$$

Since

$$h_0(\bar{z}-\bar{\lambda})^{-1}\pi_*^* y = 0, \quad h_0(\bar{z}-\bar{\lambda})^{-1}\pi^* y = \bar{\lambda}^{-1} h_\lambda \pi^* y,$$

we obtain the required formula. The computation of $\mathcal{F}^-(\lambda)$ can be performed in the same manner. ●

COROLLARY 6.6. Under the condition (SM) the functions $\mathcal{F}_+$ and $\overline{\mathcal{F}_-}$ are meromorphic in $\mathbb{D}$ and of bounded characteristics.

Indeed, the first summands are of Nevanlinna class being the Cauchy integrals, and the assertion is true for the second summands because $\Theta_i^{-1} = \delta_i^{-1}\Upsilon_i$. ●

This corollary implies the existence of nontangential boundary values of the functions $\mathcal{F}^{\pm}$ a.e. on $\mathbb{T}$. It is the difference $\mathcal{F}^+(\zeta) - \mathcal{F}^-(\zeta)$, $\zeta \in \mathbb{T}$, that is important. Applying the Privalov theorem about boundary values of Cauchy integrals, we obtain a.e. on $\mathbb{T}$:

$$\zeta(\mathcal{F}^+ - \mathcal{F}^-) = (\pi_*^* x, \pi_*^* y)_{E_*} + (\tau^* x, \tau^* y)_E -$$
$$-\delta_1^{-1}((I + \theta_0^* A)\Upsilon_1 \pi_*^* x, \pi^* y)_E + \bar{\delta}_3^{-1}(A\Upsilon_3^* \pi^* x, \pi_*^* y)_{E_*}.$$

Taking into account connections between τ, τ_* and π, π_* we can rewrite this expression in the following forms:

$$\zeta(\mathcal{F}^+ - \mathcal{F}^-) = ([\tau^* - \delta_1^{-1}\Delta(I + \theta_0^* A)\Upsilon_1 \pi_*^*]x, [\tau^* + \delta_3^{-1}\Delta\Upsilon_3 A^* \pi_*^*]y)_E \quad (6.1)$$

or

$$\zeta(\mathcal{F}^+ - \mathcal{F}^-) = ([\tau_*^* + \bar{\delta}_4^{-1}\Delta_*\Upsilon_4^* A\pi^*]x, [\tau_*^* - \bar{\delta}_2^{-1}\Delta_*(I + \theta_0 A^*)\Upsilon_2^* \pi^*]y)_{E_*} \quad (6.2)$$

PROPOSITION 6.7.

$$N_s = \{x : x \in H, \mathcal{F}_{x,y}^+(\zeta) = \mathcal{F}_{x,y}^-(\zeta) \text{ a.e. } \zeta \in \mathbb{T}, \text{ for any } y, y \in H\},$$

$$N_{*s} = \{y : y \in H, \mathcal{F}_{x,y}^+(\zeta) = \mathcal{F}_{x,y}^-(\zeta) \text{ a.e. } \zeta \in \mathbb{T}, \text{ for any } x, x \in H\}.$$

PROOF. Let $x \in N_s$, then (see the proof of Proposition 6.2) $\delta_1 \tau^* x = \Delta(I + \theta_0^* A)\Upsilon_1 \pi_*^* x$. Hence the formula (6.1) yields $\mathcal{F}^+ = \mathcal{F}^-$ a.e. on $\mathbb{T}$.

Conversely, suppose that $\mathcal{F}^+ = \mathcal{F}^-$ a.e. on $\mathbb{T}$ for any $y \in H$, and set $y = \omega e$, $e \in E$. We have

$$\tau^* y = \bar{z}\Delta e, \quad \pi_*^* y = \bar{z}(\theta - \theta_0)e$$

and

$$\delta_3 \tau^* y + \Delta\Upsilon_3 A^* \pi_*^* y = \bar{z}\Delta\Upsilon_3 e.$$

Therefore we obtain from the formula (6.1):

$$0=\delta_1\bar{\delta}_3(\mathcal{F}^+-\mathcal{F}^-)=\left(\left[\delta_1\Delta\tau^*-\Delta^2(I+\theta_0^*A)\Upsilon_1\pi_*^*\right]x,\Upsilon_3 e\right)_E=$$

$$=\left(\delta_1\pi^*x-\left[\theta^*\theta_1+\Delta^2(I+\theta_0^*A)\right]\Upsilon_1\pi_*^*x,\Upsilon_3 e\right)_E=$$

$$=\left(\delta_1\pi^*x-\theta_3^*\Upsilon_1\pi_*^*x,\Upsilon_3 e\right)_E=\left(\left[\delta_1\Upsilon_3^*\pi^*-\bar{\delta}_3\Upsilon_1\pi_*^*\right]x,e\right)_E$$

So $x\in\operatorname{Ker}\left[\delta_1\Upsilon_3^*\pi^*-\bar{\delta}_3\Upsilon_1\pi_*^*\right]\cap H=N_s$.

Analogously can be checked the expression for N_{*s}. ●

In conclusion of this Section we study the spectral structure of the unitary operator $Z|\mathcal{N}$. Let $\mathcal{E}_Z$ denote its spectral measure. We set the question when the Lebesgue measure is absolutely continuous with respect to the measure $\mathcal{E}_Z$; i.e. when does the operator $Z|\mathcal{N}$ contain the bilateral shift. Remind that we consider the condition (SM) being satisfied.

PROPOSITION 6.8. Let Σ be a measurable subset, $\Sigma\subset\mathbb{T}$. Then the following assertions are equivalent

1) $\mathcal{E}_Z(\Sigma)=0$.
2) The operators $\theta(\zeta)$ are unitary for a.e. ζ, $\zeta\in\Sigma$.
3) The operators $\theta(\zeta)$ are isometrical for a.e. ζ, $\zeta\in\Sigma$.

PROOF. Note first that $\mathcal{E}_Z(\Sigma)=0$ iff $\chi_\Sigma x=0$ for any $x\in\mathcal{N}$

1) $\Longrightarrow$2). If $x\in\mathcal{N}$, we have $\chi_\Sigma\tau^*x=0$ and $\chi_\Sigma\tau_*^*x=0$. Due to Corollary 6.3 we can take $x=(\bar{\delta}_4\pi\Upsilon_2-\delta_2\pi_*\Upsilon_4^*)g_*$, $g_*\in L^2(E_*)$. For such x's $\tau^*x=\bar{\delta}_4\Delta\Upsilon_2 g_*$, $\tau_*^*x=-\delta_2\Delta_*\Upsilon_4^*g_*$. Hence $\chi_\Sigma\Delta=0$ and $\chi_\Sigma\Delta_*=0$, that is the operators $\theta(\zeta)$ are unitary for a.e. $\zeta\in\Sigma$.

2) $\Longrightarrow$3) is obvious.

3)$\Longrightarrow$1). By Corollary 6.3 it suffices to show that the vectors $x=\chi_\Sigma(\bar{\delta}_4\pi\Upsilon_2-\delta_2\pi_*\Upsilon_4^*)g_*$ are zero for every g_*, $g_*\in L^2(\Sigma_*)$. But $\pi_*^*x=\Upsilon_4^*A\chi_\Sigma\Delta^2\Upsilon_2 g_*$ and $\tau^*x=\bar{\delta}_4\chi_\Sigma\Delta\Upsilon_2 g_*$; hence if $\Delta(\zeta)=0$ for a.e. ζ, $\zeta\in\Sigma$, then $\|x\|^2=\|\pi_*^*x\|^2+\|\tau^*x\|^2=0$, that is $x=0$, ending the proof. ●

REMARK 6.9. Let the representation of a c.n. operator L in the model of $T=T_L$ satisfy the condition (SM). Then by Corollary 4.3 the characteristic function θ_L has boundary values (in the strong operator topology) $\theta_L(\zeta)$, for a.e. ζ on $\mathbb{T}$. From Proposition 4.4 and the previous proposition, it follows that in this case $\mathcal{E}_z(\Sigma)=0$ iff the operators $\theta_L(\zeta)$ are $\mathcal{J}$ -unitary (equivalently, $\mathcal{J}$ -isometrical) for a.e. ζ , $\zeta\in\Sigma$.

REMARK 6.10. For a unitary operator $\mathcal{U}$, our concept of the absolutely continuous subspace coincides with the traditional one, which is expressed in terms of the spectral measure $\mathcal{E}_{\mathcal{U}}$. Indeed, the subspace N_a reduces $\mathcal{U}$, thus $\mathcal{U}|N_a$ is unitary and, by Proposition 5.5, is unitarily equivalent to $z|\mathcal{N}$ (see § 3, ch.II, [2]). The latter is absolutely continuous as a part of the minimal unitary dilation of c.n. contraction. Hence, $\mathcal{U}|N_a$ is absolutely continuous, and it remains to establish that the spectral measure of the restriction of $\mathcal{U}$ onto the subspace $N_s=N_a^{\perp}$ is singular. Proposition 6.7 shows that this is true. ●

Two previous remarks provide the following corollary of Proposition 6.8 (see Theorem 6.1):

COROLLARY 6.11. If $I-L^*L\in\mathfrak{S}_1$ and $\mathbb{D}\not\subset\sigma(L)$, then the operator $z|\mathcal{N}$ does not contain the bilateral shift if and only if there exists a set Σ , $|\Sigma|>0$, such that the operators $\theta_L(\zeta)$ are $\mathcal{J}$ -unitary ($\mathcal{J}$ -isometrical) for a.e. ζ , $\zeta\in\Sigma$, and such that $\mathcal{E}(\Sigma)=0$, where $\mathcal{E}$ is the spectral measure of the unitrary part of the operator L . ●

7. The stability of the continuous spectrum.

The last section of the paper is devoted to the study of the continuous spectrum of the operators admitting the model in which the functions θ_i have scalar multiples. Our study is performed entirely on such a model. In Introduction we stated the problem of the stability of the continuous spectrum and discussed the definitions involved. Now we would like to note that the stability of $\sigma_c(L)$ is closely related with the properties of the lattices of invariant subspaces of the operator L and of its resolvent $R_\lambda \overset{\text{def}}{=} (L-\lambda I)^{-1}$, $\lambda\notin\sigma(L)$. (We denote them $\operatorname{Lat} L$ and $\operatorname{Lat} R_\lambda$ respectively). The following general lemma is true for an arbitrary operator in a Banach space.

LEMMA 7.1.*) Let $\sigma_c(L)\subset\mathbb{T}$. The following assertions are equivalent.

1) $\sigma_c(L)$ is stable under rank-one perturbations,
2) $\operatorname{Lat} L\supset \operatorname{Lat} R_\lambda$ for any λ , $\lambda\notin\sigma(L)$,
3) $\operatorname{Lat} L\supset \operatorname{Lat} R_\lambda$ for some λ , $\lambda\in\mathbb{D}\setminus\sigma(L)$.

PROOF. $3\Rightarrow 1$. Suppose that L admits a perturbation

*) Suggested by N.K.Nikol'skii.

$K = (\cdot, x)y$ of rank one such that $\mathbb{D} \subset \sigma(L+K)$. We must show that $\operatorname{Lat} L \subset \operatorname{Lat} R_{\lambda_0}$ for any $\lambda_0 \in \mathbb{D} \setminus \sigma(L)$. Fix λ_0. The operator $L + K - \lambda I$ is not invertible for any $\lambda \in \mathbb{D}$. The identity

$$L + K - \lambda I = (L - \lambda I)(I + R_\lambda K), \qquad \lambda \notin \sigma(L)$$

provides that $\det(I + R_\lambda K) \equiv 0$ in some neighbourhood of λ_0. Since

$$\det(I + R_\lambda K) = 1 + (R_\lambda y, x) =$$

$$= 1 + \sum_{n=0}^{\infty} (\lambda - \lambda_0)^n (R_{\lambda_0}^{n+1} y, x),$$

we have

$$\left.\begin{aligned} &(R_{\lambda_0} y, x) = -1, \\ &(R_{\lambda_0}^n y, x) = 0, \quad n \geqslant 2. \end{aligned}\right\} \tag{7.1}$$

These equalities imply that the invariant subspace $\operatorname{span}\{R_{\lambda_0}^n y : n \geqslant 2\}$ of the operator R_{λ_0} does not belong to $\operatorname{Lat} L$. 1 ⇒ 2. Now assume that there exists a point λ_0, $\lambda_0 \notin \sigma(L)$, and a subspace G such that $G \in \operatorname{Lat} R_{\lambda_0}$, $G \notin \operatorname{Lat} L$. Without loss of generality, we can consider the subspace G to have the form $G = \operatorname{span}\{R_{\lambda_0}^n y : n \geqslant 2\}$, Ly not belonging to G. By the Hahn-Banach theorem there is a vector x for which (7.1) holds. Converting the reasoning of the first half of the proof, we obtain that $\sigma(L + (\cdot, x)y) \supset \mathbb{D}$, i.e. $\sigma_c(L)$ is unstable under a perturbation of rank one. ●

Unfortunately, we are not able to present a similar geometrical interpretation of the stability of $\sigma_c(L)$ even for the case of perturbation of rank two. We do not know, in particular, whether the rank-one stability implies the rank-two stability or γ_1 - stability. On the other hand, the second assertion in the above Lemma ($\operatorname{Lat} L \supset \operatorname{Lat} R_\lambda$) follows from the formally stronger one: $L \in \mathcal{R}(R_\lambda)$ ($\mathcal{R}(\cdot)$ denotes the weak-closed algebra generated by the operators I and $(\cdot)$). It is also unknown whether these conditions are equivalent. The affirmative answer to this question would follow from the well-known conjecture:

$$\mathcal{R}(A) = \{B : AB = BA,\ \operatorname{Lat} A \subset \operatorname{Lat} B\}$$

where A, B are arbitrary bounded operators (see [16]).

We can give answers to the questions stated above for the operators studied in the previous Section. We assume since now that L admits a model representation in which the functions θ_i have scalar multiples. We remind that the class of such operators contains all operators with $I - L^* L \in \gamma_1$, and $\mathbb{D} \not\subset \sigma(L)$.

THEOREM 7.2. If the operators $\theta(\zeta)$ are isometrical on some set $\sum$ $(\zeta \in \sum)$ of positive measure, $\sum \subset \mathbb{T}$, then $\sigma_c(L)$ is stable under the trace class perturbations.

The proof is preceded by the following

LEMMA 7.3. Let K' and K'' be operators of the Hilbert-Schmidt ideal γ_2 . Then for a.e. ζ , $\zeta \in \mathbb{T}$, the function $\lambda \mapsto \| K''(L-\lambda)^{-1} K' \|_{\gamma_2}$ is bounded in a Stolz angle with vertex at ζ . The same is true for the function $\lambda \mapsto \| K''(L-\bar{\lambda}^{-1})^{-1} K' \|_{\gamma_2}$.

PROOF. Let $K'' = \sum_{k} \alpha''_k (\cdot, y''_k) x''_k$ be the Schmidt decomposition of K'' , where $\{x''_k\}$, $\{y''_k\}$ are orthonormal families in H , and $\sum |\alpha''_k|^2 < \infty$. Similarly, let $K' = \sum_{\ell} \alpha'_\ell (\cdot, y'_\ell) x'_\ell$. Then

$$\| K''(L-\lambda I)^{-1} K' \|^2_{\gamma_2} = \sum_{m} \| K''(L-\lambda I)^{-1} K' y'_m \|^2 =$$

$$= \sum_{k,m} |\alpha''_k|^2 |\alpha'_m|^2 |((L-\lambda I)^{-1} x'_m, y''_k)|^2.$$

We extend K' and K'' to all of $\mathcal{H}$ by setting them zero on $\mathcal{H} \ominus H$. The extended operators are denoted by $\mathcal{K}'$ and $\mathcal{K}''$ respectively.

$$\| \mathcal{K}''(z-\lambda)^{-1} \mathcal{K}' \|^2_{\gamma_2} = \sum_{k,m} |\alpha''_k|^2 |\alpha'_m|^2 |((z-\lambda)^{-1} x'_m, y''_k)|^2.$$

Proposition 6.5 provides

$$\| K''(L-\lambda)^{-1} K' \|^2_{\gamma_2} \le 2 \| \mathcal{K}''(z-\lambda)^{-1} \mathcal{K}' \|^2_{\gamma_2} +$$

$$+ 2|\lambda|^{-2} \sum_{k,m} |\alpha''_k|^2 |\alpha'_m|^2 |((I+\theta_0^* A)\theta_1(\lambda)^{-1} h_\lambda \pi_*^* x'_m, h_\lambda \pi^* y''_k)|^2.$$

The function $\| \mathcal{K}''(z-\lambda)^{-1} \mathcal{K}' \|^2_{\gamma_2}$ is bounded in a Stolz angle with vertex at ζ for a.e. ζ , $\zeta \in \mathbb{T}$, as $(z-\lambda)^{-1}$

is the resolvent of the unitary operator z, and $\mathcal{K}', \mathcal{K}'' \in \gamma_2$ (cf. [IO]). We wish to prove the same for

$$\sum_{k,m} |\alpha''_k|^2 |\alpha'_m|^2 |((I+\theta_0^* A)\Upsilon_1(\lambda) h_\lambda \pi_*^* x'_m, \; h_\lambda \pi^* y''_k)|^2.$$

This sum is not greater than

$$(\sum_m |\alpha'_m|^2 \|(I+\theta_0^* A)\Upsilon_1(\lambda) h_\lambda \pi_*^* x'_m\|_E^2)(\sum_k |\alpha''_k|^2 \|h_\lambda \pi^* y''_k\|_E^2) \leqslant$$

$$\leqslant const\, (\sum_m |\alpha'_m|^2 \|h_\lambda \pi_*^* x'_m\|_{E_*}^2)(\sum_k |\alpha''_k|^2 \|h_\lambda \pi^* y''_k\|_E^2).$$

To finish the proof, it suffices to invoke the following

SUBLEMMA. Let $\beta_k \geqslant 0$, $\sum \beta_k < \infty$, $f_k \in H^2(E)$, $\|f_k\|_{H^2(E)} \leqslant 1$. Then the function $\lambda \mapsto \sum \beta_k \|f_k(\lambda)\|_E^2$ is bounded in a Stolz angle with vertex at ζ for a.e. ζ, $\zeta \in \mathbb{T}$.

PROOF. If $f \in H^2(E)$, then $\lambda \mapsto \|f(\lambda)\|_E^2$ is subharmonic. Hence

$$\|f_k(re^{i\varphi})\|_E^2 \leqslant \frac{1}{2\pi}\int_0^{2\pi} \|f_k(e^{it})\|_E^2 P_r(\varphi-t)\,dt,$$

P_r denotes the Poisson kernel,

$$\sum \beta_k \|f_k(re^{i\varphi})\|_E^2 \leqslant \frac{1}{2\pi}\int_0^{2\pi} \sum \beta_k \|f_k(e^{it})\|_E^2 P_r(\varphi-t)\,dt =$$

$$= \frac{1}{2\pi}\int_0^{2\pi} h(e^{it}) P_r(\varphi-t)\,dt,$$

where $h(e^{it}) \overset{def}{=\!=} \sum \beta_k \|f_k(e^{it})\|_E^2$. By hypothesis, h is summable. Therefore the angular boundedness follows from the properties of the Poisson integral. ●●

PROOF OF THEOREM 7.2 essentially repeats the reasoning of [IO] . Assume that there exists an operator K, $K \in \gamma_1$, such that $\sigma(L+K) \supset \mathbb{D}$. Since $L-\lambda I+K=(L-\lambda I)(I+R_\lambda K)$, the operators $I+R_\lambda K$ are not invertible for any λ, $\lambda \in \mathbb{D} \setminus \sigma(L)$. Let

$$d(\mu) \overset{def}{=\!=} det_2(I+R_\mu K), \quad \mu \in \mathbb{C} \setminus \sigma(L),$$

where det_2 denotes the regularized determinant (see [8] , ch.IY for the definition and properties). The function d is analytic in $\mathbb{C} \setminus \sigma(L)$, and $d(\lambda) \equiv 0$ in $\mathbb{D}$. To prove the theorem, it suffices to show that the angular boundary values

$$\lim_{\lambda\to\zeta} [d(\lambda)-d(\bar{\lambda}^{-1})] \tag{7.2}$$

exist and are equal to zero for a.e. ζ, $\zeta\in\Sigma$. Indeed, if this is the case, by a Lusin-Privalov theorem, $d\equiv 0$ on the whole of $\mathbb{C}$. This contradicts the obvious fact that $d(\infty)=1$.

Now let us prove that the limit in (7.2) is zero. Let $K=K'K''$ be any factorization of K into the product of two Hilbert-Schmidt operators. Then

$$d(\mu)=\det_2(I+K''(L-\mu I)^{-1}K').$$

Since $\det_2$ is continuous with respect to the γ_2-norm of its argument, the theorem will follow if we prove that

$$\lim_{\lambda\to\zeta} \|K''((L-\lambda I)^{-1}-(L-\bar{\lambda}^{-1}I)^{-1}K'\|_{\gamma_2}=0$$

for a.e. ζ, $\zeta\in\Sigma$. To do this, it suffices to verify two following conditions (see Lemma 2.1 in [17]):

(i) the function $\|K''((L-\lambda I)^{-1}-(L-\bar{\lambda}^{-1}I)^{-1}K'\|_{\gamma_2}$ is bounded in an angle with the vertex at ζ for a.e. ζ, $\zeta\in\Sigma$;

(ii) for a.e. ζ, $\zeta\in\Sigma$, the operator-function of λ $K''((L-\lambda I)^{-1}-(L-\bar{\lambda}^{-1}I)^{-1})K'$ has a zero angular limit in the weak operator topology, λ tending to ζ.

The first assertion was proved in the preceding lemma. The second assertion can be deduced from the formula (6.1). Indeed, given two arbitrary $x,y\in H$, there exists an exceptional set Γ, $\Gamma\subset\Sigma$, of measure zero, such that for every ζ, $\zeta\in\Sigma\setminus\Gamma$, the angular limit

$$\lim_{\lambda\to\zeta} ([(L-\lambda I)^{-1}-(L-\bar{\lambda}^{-1}I)^{-1}]K'x, K''^{*}y)$$

exists and is zero. Since the space H is separable, we can choose a dense countable set $\{x_K\oplus y_K\}$ in $H\oplus H$. To each pair $x_K\oplus y_K$ corresponds the exeptional set Γ_K, $\Gamma_K\subset\Sigma$. For all ζ, $\zeta\in\Sigma\setminus\bigcup\Gamma_K$, such that the function $\lambda\mapsto\|K''((L-\lambda I)^{-1}-(L-\bar{\lambda}^{-1}I)^{-1})K'\|$ is bounded in an angle with the vertex in ζ, the condition (ii) is satisfied (by the Banach-Steinhaus theorem). It is rest to note that such points ζ form a subset of full measure in Σ. ●

In the proof of the converse assertion we shall use the concept of simply-invariant subspaces. Let M be an i n v e r -

t i b l e operator, and let G be an invariant subspace of M. Then G is said to be s i m p l y - i n v a r i a n t, if $G \notin \operatorname{Lat} M^{-1}$. The following simple lemma shows that the preimage of a simply-invariant subspace by an intertwining transformation is simply-invariant.

LEMMA 7.4. Let M_1 and M_2 be invertible operators in the spaces H_1 and H_2. Let Φ be a transformation of H_1 into H_2 such that $\Phi M_1 = M_2 \Phi$, and let x_1 be an element of H_1, $x_2 \overset{def}{=} \Phi x_1$. Then if the subspace $span\{M_2^n x_2 : n \geqslant 0\}$ is simply invariant with respect to M_2, then $span\{M_1^n x_1 : n \geqslant 0\}$ is also simply-invariant.

PROOF. Observe that the subspace $span\{M^n x : n \geqslant 0\}$ is not simply-invariant (M is an invertible operator on H, $x \in H$) iff there exists a sequence of polynomials $\{p_n\}$ such that $p_n(0) = 0$, and $\lim p_n(M)x = x$. Assume now that $span\{M_1^n x_1 : n \geqslant 0\}$ is not simply-invariant, and $\{p_n\}$ is the corresponding sequence. Then $\lim \Phi p_n(M_1) x_1 = \Phi x_1$ and $\lim p_n(M_2) x_2 = x_2$. Hence $span\{M_2^n x_2 : n \geqslant 0\}$ is not simply-invariant. ●

Now we are able to prove the following theorem, in which, as in the previous one, L is a model operator, and θ_i admit scalar multiples.

THEOREM 7.5. If the operators $\theta(\zeta)$ are non-isometries for a.e. ζ, $\zeta \in \mathbb{T}$, then $\operatorname{Lat} L \neq \operatorname{Lat} R_\lambda$, for any λ, $\lambda \in \mathbb{D} \setminus \sigma(\lambda)$.

PROOF. The formula (5.3) provides an expression for the unbounded operator Q^{-1} (see Proposition 5.5):

$$Q^{-1} = I + \pi_* A \theta_3^{*-1} \pi^* - \pi (I + \theta_0^* A) \theta_1^{-1} \pi_*^* .$$

Hence, the operator $\Phi \overset{def}{=} \delta_1 \tau^* \cdot Q^{-1}$ can be boundedly extended to all of N_a,

$$\Phi = \delta_1 \tau^* - \Delta (I + \theta_0^* A) \Upsilon_1 \pi_*^* : N_a \to clos\, \Delta L^2(E),$$

and Proposition 5.5 yields

$$\Phi R_\lambda | N_a = (z - \lambda)^{-1} \Phi .$$

A function g, $g \in L^2(E)$, generates a simply-invariant subspace of the operator $(z-\lambda)^{-1}$ if

$$\int_0^{2\pi} \log \|g(e^{it})\|_E \, dt > -\infty$$

(this is an easy consequence of the Szegö theorem and of the fact that $\mathrm{Lat}\, z^{-1} = \mathrm{Lat}\,(z-\lambda)^{-1}$). Since $\mathrm{Lat}\, L = \mathrm{Lat}(L-\lambda)$, we have $\mathrm{Lat}\, R_\lambda \not\subset \mathrm{Lat}\, L$ iff R_λ has a simply-invariant subspace.

Note that $\delta_1 \tau^* \mathcal{N} = \delta_1 \tau^* Q^{-1} Q \mathcal{N} \subset \Phi N_a$, thus, according to Lemma 7.4, it suffices to prove that there exists an element x , $x \in \mathcal{N}$, such that

$$\int_0^{2\pi} \log \|(\tau^* x)(e^{it})\|_E^2 \, dt > -\infty . \tag{7.3}$$

By the assumtion, $\Delta(\zeta) E \neq \{0\}$ for a.e. ζ , $\zeta \in \mathbb{T}$. Hence, we can take a vector f , $f \in \mathrm{clos}\, \Delta L^2(E)$, so that $\|f(\zeta)\|_E = 1$ a.e. on $\mathbb{T}$. Let us verify that $x = (\bar{\delta}_4 \tau + \pi_* \Upsilon_4^* A \Delta) f$ belongs to $\mathcal{N}$ and satisfies the condition (7.3).

$$A\pi^* x = A(\bar{\delta}_4 \Delta + \theta^* \Upsilon_4 A \Delta) f = (\theta_4^* + A\theta^*) \Upsilon_4^* A \Delta f =$$
$$= (I + A\theta_0^*) \Upsilon_4^* A \Delta f = (I + A\theta_0^*) \pi_*^* x ,$$

i.e. $x \in \mathcal{N}$. Next, $\tau^* x = \bar{\delta}_4 f$, $\|(\tau^* x)(\zeta)\|_E = |\delta_4(\zeta)|$, and (7.3) is valid. ●

The obtained results can be summarized in

THEOREM 7.6. Let an operator L admit a model representation in which the functions θ_i have scalar multiples. The following assertions are equivalent

1) $\sigma_c(L)$ is stable **under rank one perturbations.**
2) $\sigma_c(L)$ is stable **under trace class perturbations.**
3) $\mathrm{Lat}\, L \supset \mathrm{Lat}\, R_\lambda$ **for any (some) point** λ , $\lambda \in \mathbb{D} \setminus \sigma(L)$.
4) $L \in \mathcal{R}(R_\lambda)$ **for any (some) point** λ , $\lambda \in \mathbb{D} \setminus \sigma(L)$.
5) **There exists a set** Σ , $|\Sigma| > 0$, **such that the operators** $\theta(\zeta)$ **are unitary (isometrical) for every** ζ , $\zeta \in \Sigma$.
6) **The Lebesgue measure is not absolutely continuous with respect to the spectral measure of** $z \,|\, \mathcal{N}$.

PROOF. We have already proved that $1 \Rightarrow 3 \Rightarrow 5 \Rightarrow 2 \Rightarrow 1$: the first implication follows from Lemma 7.1, the second - from Theorem 7.5, the third - from Theorem 7.2, the last is evident. We apply the obtained equivalence $3 \Leftrightarrow 5$ to the operator $L(n) \overset{\mathrm{def}}{=} L \oplus \ldots \oplus L$ (n times). Obviously, the operator $L(n)$ satisfies the condition of the theorem, and the assertion 5) holds (or does not hold) for it simultaneously with the operator L. Hence to prove $3 \Leftrightarrow 4$ it suffices to refer to Sarason's lemma [18] :

$$L \in \mathcal{R}(R_\lambda) \Leftrightarrow (\mathrm{Lat}\, L(n) \supset \mathrm{Lat}\, R_\lambda(n), \forall n).$$

Finally, we note that the equivalence $5 \Leftrightarrow 6$ was established in the previous section. ●

REMARK 7.7. For an operator L with $I-L^*L \in \gamma_1$ and $\sigma(L) \not\supset \mathbb{D}$, the established conditions are equivalent to the following: there exists a set Σ, $|\Sigma|>0$, such that the operators $Q_L(\zeta)$ are J-unitary for every ζ, $\zeta \in \Sigma$, and $\mathcal{E}(\Sigma)=0$, where $\mathcal{E}$ is the spectral measure of the unitary part of the operator L. (See Corollary 6.11). In the case of unitary operators this provides the theorem of N.K.Nikol'-skii [10] .

In the conclusion, we mention that the spectrum $\sigma_c(L)$ is stable under γ-perturbations, where γ is an arbitrary cross-normed ideal, $\gamma \supsetneqq \gamma_1$, iff $\sigma_c(L) \neq \mathbb{T}$. This is true not only for the operators considered in this section but in a much more general setting (cf. [11]).

References

I. Davis C., Foiaş C. Operators with bounded characteristic functions and their J-unitary dilations. Acta Sci. Math. (Szeged), 1971, 32, 127-140.

2. Sz.-Nagy B., Foiaş C. Analyse harmonique des opérateurs de l'espace de Hilbert, Masson et Cie., Akademiai Kiadó, 1967.

3. Набоко С.Н. Абсолютно непрерывный спектр недиссипативного оператора и функциональная модель I, II. Зап.научн. семин.ЛОМИ, 1976, 65, 90-102; 1977, 73, 118-135.

4. Ball J., Lubin A. On a class of contraction perturbations of restricted shifts. Pacific J.Math., 1976, 63, N 2.

5. Clark D. One dimensional perturbations of restricted shifts. J.Analyse Math., 1972, 25, 169-191.

6. Fuhrmann P. On a class of finite dimensional contrac-

tive perturbations of restricted shifts of finite multiplicity. Isr. J. Math., 1973, 16, 162-175.

7. W e y l H. Über beschränkte quadratische Formen deren Differenz vollstetig ist. Rend.Circ.Math.Palermo, 1909, 27, 373-392.

8. Г о х б е р г И.Ц., К р е й н М.Г. Введение в теорию линейных несамосопряженных операторов в гильбертовом пространстве. "Наука", М., 1965.

9. H a l m o s P. A Hilbert space problem book, Van Nostrand, 1967.

10. Н и к о л ь с к и й Н.К. О возмущениях спектра унитарных операторов. Матем.заметки, 1969, 5, 341-349.

11. A p o s t o l C., P e a r c y C., S a l i n a s N. Spectra of compact perturbations of operators. Indiana Univ. Math.J., 1977, 26, 345-350.

12. D o u g l a s R. Canonical models, Topics in Operator Theory (ed. by C.Pearcy). Amer.Math.Soc.Surveys, 1974, 13, 161-218.

13. П а в л о в Б.С. Об условиях отделимости спектральных компонент диссипативного оператора. Изв. АН СССР, сер.матем., 1975, 39, 123-148.

14. H e l s o n H. Lectures on invariant subspaces. N.Y.-London, 1964.

15. В а с ю н и н В.И. Построение функциональной модели Б.Секефальви-Надя - Ч.Фойаша. Зап.научн.семин. ЛОМИ, 1977, 73, 16-23.

16. R a d j a v i H., R o s e n t h a l P. Invariant subspaces, Springer-Verlag, 1973.

17. Б и р м а н М.Ш., Э н т и н а С.Б. Стационарный подход в абстрактной теории рассеяния. Изв. АН СССР, сер. матем., 1967, 31, 401-430.

18. S a r a s o n D. Invariant subspaces and unstarred operator algebras. Pacific J.Math., 1966, 17, 511-517.

19. W e i n s t e i n A. Études des spectres des équations aux derivées partieless. Mem.Sci.Math. 1937, 88.

20. A r o n s z a j n N., B r o w n R.D. Finite-dimensional perturbations of spectral problems and variational approximation methods for eigenvalue problems. Studia Math.,1970, 36, 1-76.

21. K a t o T. Perturbation theory for linear operators. Berlin, Springer-Verlag, 1966.

N.A.Shirokov

DIVISION AND MULTIPLICATION BY INNER FUNCTIONS IN SPACES OF ANALYTIC FUNCTIONS SMOOTH UP TO THE BOUNDARY

0. Introduction.

Let N be the Nevanlinna class of functions analytic in the unit disc $\mathbb{D}$ [1] . Every function f , $f\in N$,admits the factorization in the form $f = F_f I_f$, where F_f is an outer function and I_f is a ratio of two inner functions [1] , [2] , namely,

$$F_f(z) = \exp\left(\frac{1}{2\pi}\int_{\partial\mathbb{D}} \frac{\zeta+z}{\zeta-z} \log|f(\zeta)|\,|d\zeta|\right),$$

$$I_f = B_f S_f ,$$

B_f being a Blaschke-product,

$$B_f(z) = cz^N \prod_{\alpha\in(f^{-1}(0)\cap\mathbb{D})\setminus\{0\}} \frac{\bar\alpha}{|\alpha|}\,\frac{\alpha-z}{1-\bar\alpha z}, \qquad |c| = 1,$$

and

$$S_f(z) = \exp\left(-\int_{\partial\mathbb{D}} \frac{\zeta+z}{\zeta-z}\, d\mu(\zeta)\right),$$

where μ is a real measure singular with respect to the Lebesgue measure on $\partial\mathbb{D}$. A function $f\in N$ is said to belong to the S m i r n o v c l a s s $\mathcal{D}$, if $|I_f(z)|\leq 1$, $z\in\mathbb{D}$ or in other terms if and only if the measure μ is positive. Let X be a subset of $\mathcal{D}$. We say (following V.P.Havin [6]) that the class X possesses the (F)-p r o p e r t y if for any function f , $f\in X$, and for any inner function I $fI^{-1}\in\mathcal{D}$ implies $fI^{-1}\in X$. It is sometimes important to know whether a given class X has the (F)-property (for example in connection with uniqueness theorems, with the description of ideals in algebras of analytic functions and even in the theory of Gaussian processes). Classical Hardy spaces H^p , $0<p\leq\infty$, [1] possess the (F)-property - it is the theorem of V.I.Smirnov [1] . It is not very difficult to prove [2] that the space C_A of functions analytic in the disc $\mathbb{D}$ and continuous in its closure possesses the (F)-property. The first essential result applicable to a space of smooth functions (in a sense) was obtained by L.Carleson [3] . He has found a formula for the Dirichlet integral $\int_{\mathbb{D}}|f'|^2 d\sigma$ from which the (F)-property of the class of analytic functions with bounded Dirichlet integral can be easily deduced. Further progress in the problem was connected with the notion of a Toeplitz operator. This method was proposed at first for some Hilbert spaces by B.I.Korenblum [9] and then in the case of "sufficiently regular " Banach spaces by V.P.Havin [6] . The clue of the method is as follows. Let $X\subset H^1$ be a space of functions analytic in $\mathbb{D}$ and let $a\in H^\infty$. Define the operator $T_{\bar a}$ (the T o e p l i t z o p e r a t o r with the symbol a):

$$T_{\bar a} f(z) \stackrel{def}{=\!=} \frac{1}{2\pi i}\int_{\partial\mathbb{D}} \frac{f(\zeta)\,\bar a(\zeta)}{\zeta - z}\,d\zeta, \quad f\in X,\ z\in\mathbb{D}.$$

In some situations the space X is dual or predual of a space Y of functions analytic in $\mathbb{C}\setminus\mathbb{D}$ with the pairing

$$\langle f,g\rangle = \frac{1}{2\pi}\int_{\partial\mathbb{D}} fg\,|d\zeta|, \quad f\in X,\ g\in Y$$

where the integral is somehow defined. Then the operator $T_{\bar a}$ is dual or predual of the operator of multiplication by $\bar a$ in the space Y and provided the space Y is invariant under such multiplication (this property is the base of the method), the operator $T_{\bar a}$ is bounded, $\|T_{\bar a} f\|_X \leq C_a \|f\|_X$. But if I is an inner function and $f/I \in \mathcal{D}$ then $T_{\bar{I}}f = f/I$ and the (F)-property of the space X is established.

The method of Toeplitz operator can be applied to the spaces H^p_n, $n \geq 0$, $1 \leq p < \infty$, $f \in H^p_n \overset{def}{\Longleftrightarrow} f^{(n)} \in H^p$ [6], [9], [16], Λ^α_n, λ^α_n, $n \geq 0$, $0 < \alpha < 1$, $f \in \Lambda^\alpha_n (\lambda^\alpha_n) \overset{def}{\Longleftrightarrow} |f^{(n)}(z) - f^{(n)}(\zeta)| = O(o)(|z-\zeta|^\alpha)$ [6] and to some other spaces [6], [11], [12], [14], [16]. The method of Toeplitz operator is however useless for the spaces C^n_A, $n \geq 1$, $f \in C^n_A \overset{def}{\Longleftrightarrow} f^{(n)} \in C_A$ or for the spaces H^∞_n, $n \geq 1$, $f \in H^\infty_n \overset{def}{\Longleftrightarrow} f^{(n)} \in H^\infty$, which nevertheless possess the (F)-property*). So, for instance, $a(\zeta) \overset{def}{=} (\zeta-1)^i \in H^\infty$, $f(\zeta) \overset{def}{=} (\zeta-1)^{i+n} \in H^\infty_n$, but $T_{\bar{a}}f \notin H^\infty_n$. The operator $T_{\bar{B}}$, $B(z) = \prod_{n=0}^{\infty} \frac{1-2^{-n}-z}{1-(1-2^{-n})z}$, is unbounded in the space C^n_A, $n \geq 1$.

The results of this paper complete the list of basic classes of "smooth analytic functions" with the (F)-property. We shall prove, for example, that the spaces C^n_A and H^∞_n possess the (F)-property. This fact was known earlier for $n=0$ [2] and $n=1$ [15]. We shall prove also the validity of the (F)-property for spaces

$$\Lambda^n_\omega \overset{def}{=} \left\{ f \in C^n_A : |f^{(n)}(z) - f^{(n)}(\zeta)| = O(\omega(|z-\zeta|)) \right\},$$

$$\lambda^n_\omega \overset{def}{=} \left\{ f \in C^n_A : |f^{(n)}(z) - f^{(n)}(\zeta)| = o(\omega(|z-\zeta|)) \right\},$$

where $n = 0, 1, \ldots$ and ω is an arbitrary continuity modulus. These statements are contained in the theorem 1 (a more general assertion is announced in section 8 below)

THEOREM 1. The spaces C^n_A, H^∞_n, Λ^n_ω, λ^n_ω possess the (F)-property for $n = 0, 1, \ldots$ and for an arbitrary continuity modulus ω.

The method of proof is, like in [15], founded on the analysis of restrictions imposed on a function f by the presence of an inner factor I in its Nevanlinna factorization

---

*) Let us mention examples of spaces X, $X \subset H^1$, without (F)-property. V.P.Gurarii has proved that the space ℓ^1_A of all power series $\sum c_n z^n$ with $\sum |c_n| < \infty$ does not possess this property, see also [20] - [22].

(i.e. by the fact that $I_f I^{-1} \in H^{\infty}$). The passage from C_A^1 to C_A^n by $n \geq 2$ is connected with great difficulties and with some new effects. So, for $X = \Lambda_\omega^0$, λ_ω^0 , H_1^∞ or C_A^1 the inclusion $fI^{-1} \in X$ implies the inclusion $fI \in X$ for every function $f \in X$ and every inner function I . The next theorem shows that respective classes "of higher smoothness" behave differently.

THEOREM 2. If $X = \Lambda_\omega^n$, λ_ω^n , H_{n+1}^∞ or C_A^{n+1} , $n \geq 1$, then there exist a function $f \in X$ and a Blaschke product B such that $f/B \in X$, but $fB \notin X$.

For the validity of this theorem it is essential that we are dealing with a Blaschke product (not with a general inner function) and that the multiplicity of its zeros is less than $n+1$.

The case of a singular function or that of a Blaschke product with zeros of higher multiplicity differs from the situation of theorem 2, as the next theorem shows.

THEOREM 3. Let $X = \Lambda_\omega^n$, λ_ω^n , H_{n+1}^∞ or C_A^{n+1}, $n \geq 1$, and let $f \in X$.

a) If s , $s \in H^\infty$, is a singular function, $f/s \in H^1$, then $fs \in X$;

b) if B is a Blaschke product, $f/B \in H^1$, and the multiplicity of the zero of function f at any point $a \in B^{-1}(0)$ is $\geq n+1$, then $fB \in X$.

REMARKS. a) We note that theorems 1 and 2 prove the existence of a function f , $f \in X$ (X being one of spaces under consideration, $n \geq 1$), such that $fI_f \notin X$ (I_f is "the inner part" of the Nevanlinna factorization of f).

b) It is sufficient to prove theorems 1 and 3 for spaces Λ_ω^n only ($n \geq 1$, ω an arbitrary continuity modulus).

c) The proof of theorem 1 will give more than is promised in its statement, namely the existence of a constant C_n depending only on n such that for any function $f \in \Lambda_\omega^n$ with $|f^{(n)}(z) - f^{(n)}(\zeta)| \leq \omega(|z-\zeta|)$, $z, \zeta \in \mathbb{D}$, and for any inner function I , $f/I \in H^1$ the following inequality holds:

$$\left|(f/I)^{(n)}(z) - (f/I)^{(n)}(\zeta)\right| \leq C_n \omega(|z-\zeta|), \quad z, \zeta \in \mathbb{D}.$$

Theorem 3 admits an analogous improvement.

1. Two essential lemmas.

Here we shall state and prove two important facts necessary for the proof of theorem 1.

DEFINITION. a) Let us write $f\in B_{r}\Lambda^{n}_{\omega}(\Omega)$ ($r>0$, $n\geq 0$ an integer, ω a continuity modulus, $\Omega\subset\mathbb{C}$ a convex domain), if the function f is analytic in Ω , $f^{(n)}$ is continuous in $\overline{\Omega}$ and

$$|f^{(n)}(z)-f^{(n)}(\zeta)|\leq r\omega(|z-\zeta|),\ z,\ \zeta\in\overline{\Omega}\ .$$

We shall write $B\Lambda^{n}_{\omega}(\Omega)$ instead of $B_{1}\Lambda^{n}_{\omega}(\Omega)$

b) From now on C_{nK} will denote various constants depending only on n .

LEMMA 1*). Let Ω be a convex domain with the diameter ρ ; let $f\in B\Lambda^{n}_{\omega}(\Omega)$, $n\geq 0$, and suppose the domain $\overline{\Omega}$ contains $n+1$ zeros of f (with the account of their multiplicities). Then

$$|f^{(\nu)}(z)|\leq C_{n1}\rho^{n-\nu}\omega(\rho),\quad \nu=0,1,\dots,n\ . \tag{I}$$

PROOF. We shall use the induction. If $n=0$, $f\in B\Lambda^{0}_{\omega}(\Omega)$, $f(a)=0$, then

$$|f(z)|=|f(z)-f(a)|\leq\omega(\rho)\ .$$

Let us suppose the statement of lemma is true for all m , $0\leq m\leq n-1$, and take a function $f\in B\Lambda^{n}_{\omega}(\Omega)$. Put $a=a_{1}$, $f(a)=0$, $\varphi(z)=\frac{1}{z-a}f(z)$. The function φ has n zeros in $\overline{\Omega}$ and

$$\varphi(z)=\sum_{\nu=1}^{n}\frac{1}{\nu!}f^{(\nu)}(a)(z-a)^{\nu-1}+$$

$$+\frac{1}{(n-1)!}\,\frac{1}{z-a}\int_{a}^{z}(z-t)^{n-1}(f^{(n)}(t)-f^{(n)}(a))\,dt\ ,$$

*) I am indebted to V.P.Havin for this version of Lemma 1.

$$\varphi^{(n-1)}(z_1)-\varphi^{(n-1)}(z_2)=\sum_{\nu=0}^{n-1}\frac{(-1)^\nu C_{n-1}^\nu}{(z_1-a)^{\nu+1}}\int_a^{z_1}(z_1-t)^\nu(f^{(n)}(t)-f^{(n)}(a))\,dt-$$

$$-\sum_{\nu=0}^{n-1}\frac{(-1)^\nu C_{n-1}^\nu}{(z_2-a)^{\nu+1}}\int_a^{z_2}(z_2-t)^\nu(f^{(n)}(t)-f^{(n)}(a))\,dt\ , \tag{2}$$

$$\varphi^{(n)}(z)=\sum_{\nu=1}^{n}\nu C_n^\nu\frac{(-1)^\nu}{(z-a)^{\nu+1}}\int_a^{z}(z-t)^{\nu-1}(f^{(n)}(t)-f^{(n)}(a))\,dt+\frac{f^{(n)}(z)-f^{(n)}(a)}{z-a}. \tag{3}$$

We shall now estimate every term in **the right hand side of the formula** (2) of the form

$$\frac{1}{(z_1-a)^{\nu+1}}\int_a^{z_1}(z_1-t)^\nu(f^{(n)}(t)-f^{(n)}(a))\,dt-$$

$$-\frac{1}{(z_2-a)^{\nu+1}}\int_a^{z_2}(z_2-t)^\nu(f^{(n)}(t)-f^{(n)}(a))\,dt \stackrel{def}{=\!=} S_\nu^{(1)}-S_\nu^{(2)}\ .$$

We assume further that $|z_2-a|\leqslant|z_1-a| \stackrel{def}{=\!=} \Delta$. We will examine all possible situations.

Case 1: $|z_2-a|\leqslant\frac{1}{2}\Delta$. Then $|z_2-z_1|\geqslant\frac{1}{2}\Delta$ and

$$|S_\nu^{(1),(2)}|\leqslant\frac{1}{|z_{1,2}-a|^{\nu+1}}|z_{1,2}-a|^{\nu+1}\omega(|z_{1,2}-a|)\leqslant\omega(\Delta),$$

$$|S_\nu^{(1)}-S_\nu^{(2)}|\leqslant|S_\nu^{(1)}|+|S_\nu^{(2)}|\leqslant 5\omega\left(\frac{\Delta}{2}\right)\leqslant 5\omega(|z_1-z_2|)^{*)}\ . \tag{4}$$

---

*) In the inequality (4) (and further) we use the well-known property of continuity moduli: if $0<x\leqslant y$, then

$$\omega(y)\leqslant\left(1+\frac{y}{x}\right)\omega(x)$$

Case 2: $\frac{1}{2}\Delta < |z_2 - a| \leq \Delta$. Let

$$\delta \overset{def}{=\!=} |z_2 - z_1| \, , \quad \delta \leq 2\Delta \, .$$

Then

$$S_\nu^{(1)} - S_\nu^{(2)} = \left(\frac{1}{(z_1-a)^{\nu+1}} - \frac{1}{(z_2-a)^{\nu+1}}\right)\int_a^{z_1}(z_1-t)^\nu\left(f^{(n)}(t)-f^{(n)}(a)\right)dt +$$

$$+\frac{1}{(z_2-a)^{\nu+1}}\left[\int_a^{z_1}(z_1-t)^\nu\left(f^{(n)}(t)-f^{(n)}(a)\right)dt - \int_a^{z_2}(z_2-t)^\nu\left(f^{(n)}(t)-f^{(n)}(a)\right)dt\right] =$$

$$= \quad \ldots + \frac{1}{(z_2-a)^{\nu+1}}\left[\int_a^{z_1}\left((z_1-t)^\nu-(z_2-t)^\nu\right)\left(f^{(n)}(t)-f^{(n)}(a)\right)dt +\right.$$

$$\left.+\int_{z_2}^{z_1}(z_2-t)^\nu\left(f^{(\nu)}(t)-f^{(\nu)}(a)\right)dt\right] \overset{def}{=\!=} r_1 + r_2 + r_3 \, .$$

We have now

$$|r_1| \leq \frac{(\nu+1)\,\delta}{\left(\frac{\Delta}{2}\right)^{\nu+2}}\Delta^{\nu+1}\omega(\Delta) \leq 2^{\nu+2}(\nu+1)\frac{\delta}{\Delta}\left(1+\frac{\Delta}{\delta}\right)\omega(\delta) \leq 3\cdot 2^{\nu+2}(\nu+1)\,\omega(\delta)\,; \tag{5}$$

$$|r_2| \leq \frac{1}{\left(\frac{\Delta}{2}\right)^{\nu+1}}\cdot\nu(4\Delta)^\nu\delta\,\omega(\Delta) \leq 3\cdot 2^{3\nu+1}\nu\,\omega(\delta) \tag{6}$$

$$|r_3| \leq \frac{\delta^{\nu+1}}{\left(\frac{\Delta}{2}\right)^{\nu+1}}\omega(\Delta) \leq 2^{\nu+1}\left(\frac{\delta}{\Delta}\right)^{\nu+1}\left(1+\frac{\Delta}{\delta}\right)\omega(\delta) \, . \tag{7}$$

The inequalities (4)-(7) give

$$\left|\varphi^{(n-1)}(z_1) - \varphi^{(n-1)}(z_2)\right| \leq C_{n2}\,\omega(|z_1-z_2|)\,, \; z_1, z_2 \in \overline{\Omega}\,. \tag{8}$$

If we utilize the estimates like (4)-(7) in the formula (3), we obtain also

$$\left|\varphi^{(n)}(z)\right| \leq C_{n2}\frac{\omega(|z-a|)}{|z-a|} \, . \tag{9}$$

The inaqualitY (8) implies $\frac{1}{C_{n2}}\varphi \in B\Lambda_\omega^{n-1}(\Omega)$, and therefore the induction conjecture gives

$$\frac{1}{C_{n2}}|\varphi^{(\nu)}(z)| \leq C_{n3}\,\rho^{n-\nu-1}\omega(\rho)\ ,\quad \nu = 0, 1, \ldots, n-1\,. \tag{10}$$

At last,

$$f(z) = (z-a)\varphi(z),\quad f^{(\nu)}(z) = (z-a)\varphi^{(\nu)}(z) + \nu\varphi^{(\nu-1)}(z)$$

and then we get for $\nu = 0$

$$|f(z)| \leq \rho \cdot C_{n2} C_{n3}\,\rho^{n-1}\omega(\rho) = C_{n4}\,\rho^{n}\omega(\rho),$$

and for $1 \leq \nu \leq n-1$ (10) gives

$$|f^{(\nu)}(z)| \leq \rho \cdot C_{n5}\,\rho^{n-\nu-1}\omega(\rho) = C_{n5}\,\rho^{n-\nu}\omega(\rho),$$

and for $\nu = n$ (8) and (9) give

$$|f^{(n)}(z)| \leq |z-a|\,C_{n2} C_{n3}\frac{\omega(|z-a|)}{|z-a|} + nC_{n2}C_{n3}\omega(\rho) \leq (n+2)C_{n2}C_{n3}\,\omega(\rho),$$

and our induction goes. ●

LEMMA 2. a) Let $f \in B\Lambda^{n}_{\omega}(\mathbb{D})$, $n \geq 1$, $f(a) = 0$, $a \in \mathbb{D}$. Then

$$f_0(z) \stackrel{def}{=\!=} \frac{1-\bar{a}z}{z-a}\,f(z) \in B_{C_{n6}}\Lambda^{n}_{\omega}(\mathbb{D})\,.$$

b) Let Ω be the domain $\{\zeta \in \mathbb{D} : |\zeta - z_0| < h\}$, where $z_0 \in \partial\mathbb{D}$, $0 < h \leq 1$, let $f \in B\Lambda^{n}_{\omega}(\Omega)$, $n \geq 1$, and $f(a) = 0$, where $a \in \Omega$, $|a - z_0| < q_n h$, $q_n < 1$, the number q_n depending only on n.

Then

$$f_0(z) \stackrel{def}{=\!=} \frac{1-\bar{a}z}{z-a}\,f(z) \in B_{C_{n6}}\Lambda^{n}_{\omega}(\Omega).$$

PROOF. The arguments in the proof of assertions a) and b) are similar and are like the proof of lemma 1, therefore we shall describe briefly the main steps of the proof of the part b) only. It is sufficient to check the statement of lemma for any points

---

*) The sign ● here and below denotes the end of the proof.

$z_1, z_2 \in \partial\Omega$ and then make use of theorem of P.M.Tamrazov [17] about connection between the continuity moduli on the curve and in the area. The Taylor formula gives

$$f_0^{(n)}(z_1)-f_0^{(n)}(z_2)=\frac{1}{(n-1)!}\left\{\left[\frac{1-\bar{a}z_1}{z_1-a}\int_a^{z_1}(z_1-t)^{n-1}(f^{(n)}(t)-f^{(n)}(a))dt\right]^{(n)}-\right.$$

$$\left.-\left[\frac{1-\bar{a}z_2}{z_2-a}\int_a^{z_2}(z_2-t)^{n-1}(f^{(n)}(t)-f^{(n)}(a))dt\right]^{(n)}\right\}, z_1, z_2 \in \partial\Omega . \tag{11}$$

Then, like in the proof of lemma 1, we apply to (11) the formula of Newton-Leibnitz and consider the same situations as in (4) - (7) taking **into account the inequality**

$$\left|\frac{1-\bar{a}\zeta}{\zeta-a}\right| \leq c_{nz} , \qquad \zeta \in \partial\Omega$$

which holds because of the condition $|a-z_0| \leq q_n h$, and at last we obtain the necessary estimate for $z_1, z_2 \in \partial\Omega$. ●

2. Some technical preparations.

Here we shall establish facts, which will be used in the proof of lemmas 6-8, essential for the proof of theorems 1 and 3.

DEFINITION. Let $I \in H^\infty$ be an inner function, $I = BS$,

$B(z) = \prod_\alpha \frac{\bar{\alpha}}{|\alpha|}\frac{\alpha - z}{1-\bar{\alpha}z}$ a Blaschke product,

$S(z) = \exp\left(-\int_{\partial\mathbb{D}} \frac{\zeta+z}{\zeta-z}d\mu(\zeta)\right)$ a singular function. We define

$$\operatorname{spec} I \overset{def}{=\!=} \operatorname{clos}(B^{-1}(0)) \cup (\operatorname{supp}\mu) .$$

LEMMA 3. Let $z \in \partial\mathbb{D}$, $I \in H^\infty$ be an inner function, $d = \operatorname{dist}(z, \operatorname{spec} I) > 0$, $a = |I'(z)|$. Then for $|\zeta - z| \leq \frac{1}{2}d$

$$e^{-c_1(1-|\zeta|)a} \leq |I(\zeta)| \leq e^{-c_2(1-|\zeta|)a} \qquad , |\zeta| \leq 1 ,$$

$$e^{c_2(|\zeta|-1)a} \le |I(\zeta)| \le e^{c_1(|\zeta|-1)a} \qquad , \ |\zeta|\ge 1 \ ,$$

where the constants $0 < c_2 < c_1 < \infty$ are absolute.

PROOF. See [15]. ●

LEMMA 4. Let $z \in \partial\mathbb{D}$, $I \in H^\infty$ be an inner function, $d = \operatorname{dist}(z, \operatorname{spec} I) > 0$, $\sigma = \frac{1}{4}\min(d, 1/|I'(z)|)$. Then for $|\zeta - z| \le \sigma$ and for $\nu = 1, 2, \ldots$

$$|I^{(\nu)}(\zeta)| \le C_{\nu 8}\, \sigma^{-\nu} \ , \tag{12}$$

$$\left|\left(\frac{1}{I}\right)^{(\nu)}(\zeta)\right| \le C_{\nu 8}\, \sigma^{-\nu} \ . \tag{13}$$

PROOF. Lemma 3 for $|\zeta - z| \le 2\sigma$ gives

$$e^{-2c_1} \le |I(\zeta)| \ , \quad \frac{1}{|I(\zeta)|} \le e^{2c_1} \ . \tag{14}$$

The Cauchy formula for the circle $\{\zeta : |\zeta - z| = 2\sigma\}$ together with (14) implies (12) and (13). ●

LEMMA 5. Let $a > 0$, $b > 0$, $n \le \alpha \le n+1$, $n \ge 1$. Then

$$\mathcal{J}(r,a,b,\alpha) \overset{\text{def}}{=\!=} \frac{1}{2\pi}\int_{-\pi}^{\pi} \frac{1-r^2}{1-2r\cos\theta + r^2}\log(a + b|e^{i\theta}-1|^\alpha)\,d\theta \le \log\left[C_{n9}(a + b(1-r)^\alpha)\right]$$

PROOF. It is sufficient to consider the case $b = 1$ only. Inequalities

$$x^\alpha + y^\alpha \le (x+y)^\alpha \le 2^\alpha(x^\alpha + y^\alpha), \quad x + y \le \sqrt{2(x^2+y^2)} \ , \ x \ge 0, y \ge 0, \alpha \ge 1 ,$$

imply

$$\mathcal{J}(r,a,1,\alpha) \le \alpha \mathcal{J}(r, a^{1/\alpha}, 1, 1) \le \frac{\alpha}{2}\mathcal{J}(r, 2a^{2/\alpha}, 2, 2) \ .$$

We utilize now the estimate

$$a^{2/\alpha} + |e^{i\theta} - 1|^2 \le |e^{i\theta} - 1 - a^{1/\alpha}|^2 , \qquad -\pi \le \theta \le \pi ,$$

and the harmonicity in $\mathbb{D}$ of the function $\log|\zeta - 1 - a^{1/\alpha}|^2$. We get then

$$\mathcal{J}(r,a,1,\alpha) \le \frac{\alpha}{2}\log\left|2(1 - r + a^{1/\alpha})^2\right| \le \log\left[C_{n9}(a + (1-r)^\alpha)\right] . \ ●$$

3. The influence of the inner factor I on the rate of decrease of $|f|$.

In this section we find the rate of decrease of $|f|$ near

spec I caused by the smoothness of f and to the presence of the inner factor I in the Nevanlinna factorization of f. The idea to use in this connection the Taylor expansion of f at a certain point of $\mathbb{D}$ has been proposed for the first time by B.I.Korenblum [18].

LEMMA 6. Suppose $f\in B\Lambda^n_\omega(\mathbb{D})$, $n\geqslant 1$, $0<\sigma<\frac{1}{200n}$, $\rho=A\sigma$, $\frac{1}{2\sigma}\geqslant A\geqslant 100n$, $z\in\partial\mathbb{D}$ and $z_0\overset{def}{=\!=}(1-\rho)z$, $\gamma\overset{def}{=\!=}\{\zeta:|\zeta-z_0|=\sigma\}$, $\tau\overset{def}{=\!=}\{\zeta\in\partial\mathbb{D}:|\zeta-z|\leqslant 2\sigma\}$. Let us denote (this notation will be used also in the next lemmas):

$$x=x_\tau\overset{def}{=\!=}\max_{\zeta\in\tau}|f(\zeta)|$$

$$y=y_\gamma\overset{def}{=\!=}\max_{\zeta\in\gamma}|f(\zeta)|.$$

Then

$$x\leqslant 2A^n y+2\rho^n\omega(\rho) \tag{15}$$

$$|f(\zeta)|\leqslant x+12n!A^n y\left(\frac{|\zeta-z|+\sigma}{\sigma}\right)^n+4\left(|\zeta-z|+\rho\right)^{n+1}\frac{\omega(q)}{\rho},\ \zeta\in\partial\mathbb{D}. \tag{16}$$

PROOF. The Cauchy inequalities

$$|f^{(\nu)}(z_0)|\leqslant\frac{\nu!}{\sigma^\nu}y,\quad \nu=1,\dots,n$$

and the Taylor formula

$$f(\zeta_0)=f(z_0)+\sum_{\nu=1}^{n}\frac{f^{(\nu)}(z_0)}{\nu!}(\zeta_0-z_0)^\nu+\frac{1}{(n-1)!}\int_{z_0}^{\zeta_0}(\zeta_0-t)^{n-1}\left(f^{(n)}(t)-f^{(n)}(z_0)\right)dt$$

imply the following estimate (we suppose $|f(\zeta_0)|=x$)

$$x\leqslant y+\sum_{\nu=1}^{n}(A+2)^\nu y+\frac{1}{(n-1)!}(A+2)^n\sigma^n\omega((A+2)\sigma)\leqslant$$

$$\leqslant 2A^n y+2\rho^n\omega(\rho),$$

because $A\geqslant 100n$.

Applying the Taylor formula to the points z and z_0 and to the functions $f^{(\nu)}$, $\nu=1,\dots,n$, we obtain the inequalities

$$|f^{(\nu)}(z)|\leqslant 4\frac{A^{n-\nu}}{\sigma^\nu}n!\,y+\frac{1}{(n-\nu)!}\rho^{n-\nu}\omega(\rho),\quad \nu=0,1,\dots,n. \tag{17}$$

Again the Taylor formula at the points ζ and z and (17) imply (16). ●

LEMMA 7. Let the number C_2 be taken from lemma 3, the number C_{n9} from lemma 5, and

$$C_n \overset{def}{=\!=} C_{n9}\cdot 12n!\,2^{2n+20}.$$

Let the number $A = A_n$ (it is the most important number for us) be defined from the equation

$$C_n e^{-1/2\,C_2 A}\cdot A^{2n+1} = \frac{1}{8}.$$

Let $f \in B\Lambda^n_\omega(\mathbb{D})$, $n \geq 1$, $z \in \partial\mathbb{D}$, $I \in H^\infty$ be an inner function, $\ell \overset{def}{=} dist(z, spec\, I) > 0$, $|I'(z)| \overset{def}{=\!=} a > \frac{4A}{\ell}$.

Then for $\zeta \in \overline{\mathbb{D}}$, $|\zeta - z| \leq \sigma \overset{def}{=\!=} \frac{1}{a}$ the following inequalities hold:

$$|f^{(\nu)}(\zeta)| \leq C_{n10}\frac{1}{a^{n-\nu}}\,\omega\!\left(\frac{1}{a}\right), \quad \nu = 0, 1, \ldots, n. \qquad (18)$$

PROOF. We write $\rho = A\sigma$, $z_0 = (1-\rho)z$ and with the points z, z_0, with the numbers σ, ρ, A (we remark that $A\sigma < 1/4\,\ell$) and with the function f we associate the arc τ, the circle γ, the numbers $x = x_\tau$, $y = y_\gamma$ as in lemma 6. Let the point $z^0 \in \gamma$ be such that $|f(z^0)| = y$. We denote by F the outer factor in the Nevanlinna factorization of f. Using lemma 3, we get

$$y \leq |F(z^0)|\,|I(z^0)| \leq |F(z^0)|\,e^{-C_2 a(1-|z^0|)} \leq e^{-C_2/2A}|F(z^0)|.$$

We write further $z_0/|z_0| = \zeta_1 (\in \tau)$, $1 - |z^0| = \rho_1$.

Lemmas 5 and 6 imply

$$\log|F(z^0)| = \frac{1}{2\pi}\int_{\partial\mathbb{D}} \frac{1-\rho_1^2}{|\zeta - z^0|^2}\log|f(\zeta)|\,|d\zeta| \leq$$

$$\leq \frac{1}{2\pi}\int_{\partial\mathbb{D}} \frac{1-\rho_1^2}{|\zeta - z^0|^2}\log\left[x + 12n!\,A^n y\left(\frac{|\zeta - z| + \sigma}{\sigma}\right)^n + 4(|\zeta - z| + \rho)^{n+1}\frac{\omega(\rho)}{\rho}\right]|d\zeta| \leq \qquad (19)$$

$$\leq \frac{1}{2\pi}\int_{\partial\mathbb{D}} \frac{1-\rho_1^2}{|\zeta - z^0|^2}\log\left[x + 12n!\,2^{n+1}A^{2n+1} y\left(\frac{|\zeta - \zeta_1| + \rho_1}{\rho_1}\right)^{n+1} + 2^{n+10}(|\zeta - \zeta_1| + \rho_1)^{n+1}\frac{\omega(\rho_1)}{\rho_1}\right]|d\zeta| \leq$$

$$\leq \log\left[C_{n9}\cdot 12n!\,2^{2n+20}(x + A^{2n+1}y + \rho^n\omega(\rho))\right].$$

The equality $C_{n9}\cdot 12n!\cdot 2^{2n+20} = C_n$ allows us to give (19) the form

$$y \leq C_n e^{-\frac{c_2}{2}A}(x + A^{2n+1} y + \omega(\rho)\rho^n) , \tag{20}$$

and using the definition of the number A, we obtain

$$y \leq \frac{1}{8A^{2n+1}} x + \frac{1}{8} y + C_n \rho^n \omega(\rho)$$

and

$$y \leq \frac{1}{7A^{2n+1}} x + 2C_n \rho^n \omega(\rho) . \tag{21}$$

The number $A \geq 1$ depends only on n. Lemma 6 and (21) give then

$$x \leq \frac{2A^n}{7A^{2n+1}} x + C_{n12} \rho^n \omega(\rho) \leq \frac{2}{7} x + C_{n12} \rho^n \omega(\rho),$$

$$x \leq C_{n13} \rho^n \omega(\rho) \leq C_{n14} \sigma^n \omega(\sigma). \tag{22}$$

The estimate (22) is the required inequality (18) in the case $\nu = 0$. Applying the Taylor formula with the center at z_0 to the point z together with the Cauchy inequalities, lemma 6 and (21) we obtain remaining estimates in (18)$(\nu = 1, \ldots, n)$. ●

DEFINITION. Let $I \in H^\infty$ be an inner function, $I = BS$, where

$$B(z) = \prod_\alpha \frac{\bar{\alpha}}{|\alpha|} \frac{\alpha - z}{1 - \bar{\alpha} z}$$

is a Blaschke product,

$$S(z) = exp\left(-\int_{\partial\mathbb{D}} \frac{\zeta + z}{\zeta - z} d\mu(\zeta)\right)$$

is a singular function.

a) We denote for $\alpha \in spec\, I$

$$\nu_{I,n}(\alpha) = \begin{cases} n+1 , & \text{if } |\alpha| = 1 \\ \text{the multiplicity of the zero } \alpha , & \text{if } |\alpha| < 1 \end{cases}$$

b) We define also

$$d_n(z) = d_{n,I}(z) \overset{def}{=} \begin{cases} 2 , \text{ if } I \text{ is a Blaschke product with at most } n \text{ zeros;} \\ \inf\{\sigma > 0 : \sum\limits_{\substack{\alpha \in spec\, I \\ |\alpha - z| \leq \sigma}} \nu_{I,n}(\alpha) \geq n+1\} , \text{ otherwise} \end{cases}$$

c) Let $E \subset \overline{\mathbb{D}}$ be a closed set; we write

$$B|_E(z) \stackrel{def}{=} \prod_{\alpha\in \operatorname{spec} B\cap E} \frac{\bar\alpha}{|\alpha|}\frac{\alpha-z}{1-\bar\alpha z},\quad S|_E(z) \stackrel{def}{=} \exp\Big(-\int_{\partial\mathbb{D}} \frac{\zeta+z}{\zeta-z}\,d\nu(\zeta)\Big),$$

where $\nu=\mu|_E$. We write further $I|_E \stackrel{def}{=} B|_E\, S|_E$;

d) $I_{n,z}(\zeta) \stackrel{def}{=} I|_{\mathbb{D}\setminus\{\xi:|\xi-z|<d_n(z)\}}(\zeta)$;

if $z\in\partial\mathbb{D}$, we denote

$$a_n(z) \stackrel{def}{=} \left|I'_{n,z}(\zeta)\right|_{\zeta=z} = \sum_{\substack{\alpha\in \operatorname{spec} I\\ |\alpha-z|\geq d_n(z)}} \frac{1-|\alpha|^2}{|1-z\bar\alpha|^2} + 2\int_{\partial\mathbb{D}\setminus\{\zeta:|\zeta-z|<d_n(z)\}} \frac{d\mu(\zeta)}{|\zeta-z|^2}\ .$$

REMARKS. a) The expression under the infimum in the definition of d_n attains its lower bound, so instead of "inf" it is possible to write there "min".

b) If $z\in \operatorname{spec} I\cap\partial\mathbb{D}$ and $f\in\Lambda^n_\omega(\mathbb{D})$, $f/I\in H^1$, then the point z is a zero of f with the multiplicity at least $n+1$. Suppose not. Then the Taylor formula gives

$$f(\zeta)=\frac{f^{(p)}(z)}{p!}(\zeta-z)^p+o\big((\zeta-z)^p\big)\ ,\quad \zeta\to z,\quad p\leq n\ ,$$

and $f^{(p)}(z)\neq 0$, so for a certain $\rho>0$ for $0<|\zeta-z|<\rho$, $\zeta\in\mathbb{D}$ we get

$$m|\zeta-z|^p\geq|f(\zeta)|\geq c|\zeta-z|^p\ ,$$

$$m\geq\left|\frac{f(\zeta)}{(\zeta-z)^p}\right|\geq c>0\ . \tag{23}$$

The inner factors of the functions $f\in H^\infty$ and $\frac{f(\zeta)}{(\zeta-z)^p}\in H^\infty$ being equal, the point z belongs to $\operatorname{spec} I$, which is impossible because of (23) - see [2]. Therefore , we have the inequality

$$d_n(z)\leq \operatorname{dist}(z, \operatorname{supp}\mu\cup(\operatorname{spec} B\cap\partial\mathbb{D}))$$

c) If I is not a Blaschke product with not more than n zeros, the domain $\{\zeta\in\bar{\mathbb{D}}:|\zeta-z|\leq d_n(z)\}$ contains at least $n+1$ points belonging to spec I.

LEMMA 8. Let $f \in B\Lambda^n_\omega(\mathbb{D})$, $n \geq 1$, I be an inner function, $f/I \in H^1$, $z \in \partial\mathbb{D}$, $d_n = d_{n,I}(z) > 0$, $a_n(z) = a_{n,I}(z)$.

Then the following inequalities hold:

$$\left|(f/I)^{(\nu)}(z)\right| \leq C_{n15} \min\left(d_n^{n-\nu}(z)\,\omega(d_n(z)),\ \frac{1}{a_n^{n-\nu}(z)}\,\omega\left(\frac{1}{a_n(z)}\right)\right), \tag{24}$$

$$\left|f^{(\nu)}(z)\right| \leq C_{n15} \min\left(d_n^{n-\nu}(z)\,\omega(d_n(z)),\ \frac{1}{a_n^{n-\nu}(z)}\,\omega\left(\frac{1}{a_n(z)}\right)\right), \tag{25}$$

where $\nu = 0, 1, \ldots, n$.

PROOF. We begin with the estimate (25). We can suppose without loss of generality that the function I is not the Blaschke product with at most n zeros (if it is we use lemma 2), so lemma 1 holds and gives

$$\left|f^{(\nu)}(z)\right| \leq C_{n15}\, d_n^{n-\nu}(z)\,\omega(d_n(z)), \quad \nu = 0, \ldots, n. \tag{26}$$

But if $a_n(z) \leq \frac{4A_n}{d_n(z)}$, where the number A_n is taken from the lemma 7, the inequalities (24) follow from (26), and if $a_n(z) > \frac{4A_n}{d_n(z)}$, we can use (18) from lemma 7 and (25) is established.

We pass now to the proof of (24). The number $A = A_n$ is again the number from lemma 7. We shall consider two cases.

1. $a_n(z) > \frac{4A}{d_n(z)}$. We write $\sigma = \frac{1}{a_n(z)}$.

The domain $\widetilde{\Omega} = \{\zeta \in \overline{\mathbb{D}} : |\zeta - z| \leq \sigma\}$ by the definition of the number $d_n(z)$ contains at most n points of $\operatorname{spec} I$. Therefore, there exists k, $1 \leq k \leq n+1$, such that the domain $\Omega_k = \left\{\zeta \in \overline{\mathbb{D}} : \frac{k\sigma}{n+2} < |\zeta - z| < \frac{(k+1)\sigma}{n+2}\right\}$ has no point of $\operatorname{spec} I$. For this k we denote $\Omega = \left\{\zeta \in \overline{\mathbb{D}} : |\zeta - z| \leq \frac{k\sigma}{n+2} + \frac{\sigma}{2(n+2)}\right\}$. Let $J(\zeta) = I\big|_{\operatorname{spec} I \setminus \Omega}(\zeta)$. Then

$$I(\zeta) = J(\zeta)\, b(\zeta),$$

where $b(\zeta)$ is a Blaschke product with not more than n zeros. Let $f_0 = f/J$. Using lemmas 4 and 7, the equality

$$f_0^{(\nu)} = \sum_{\ell=0}^{\nu} C_\nu^\ell\, f^{(\ell)} \left(\frac{1}{J}\right)^{(\nu-\ell)}$$

and taking into account the case under consideration, we obtain

$$\left|f_0^{(\nu)}(\zeta)\right| \leq C_{n16}\, \sigma^{n-\nu}\omega(\sigma), \quad \zeta \in \Omega, \quad \nu = 0, \dots, n. \tag{27}$$

We shall show also that $f_0 \in B_{C_{n17}} \Lambda_\omega^n(\Omega)$. Let us take any points $\zeta_1, \zeta_2 \in \Omega$. Then

$$\begin{aligned} \left|f_0^{(n)}(\zeta_1) - f_0^{(n)}(\zeta_2)\right| &\leq \sum_{\nu=0}^{n-1} C_n^\nu \Big[\Big| (f^{(\nu)}(\zeta_1) - f^{(\nu)}(\zeta_2)) \left(\frac{1}{j}\right)^{(n-\nu)}(\zeta_1) \Big| + \\ &+ \Big| f^{(\nu)}(\zeta_2) \Big(\left(\frac{1}{j}\right)^{(n-\nu)}(\zeta_1) - \left(\frac{1}{j}\right)^{(n-\nu)}(\zeta_2) \Big) \Big| \Big] + \\ &+ \left| f^{(n)}(\zeta_1) - f^{(n)}(\zeta_2) \right| \frac{1}{|j(\zeta_1)|} + \left| f^{(n)}(\zeta_2) \right| \left| \frac{1}{j(\zeta_1)} - \frac{1}{j(\zeta_2)} \right| . \end{aligned} \tag{28}$$

The mean-value theorem, lemma 4 and (27) give

$$\left| f^{(\nu)}(\zeta_1) - f^{(\nu)}(\zeta_2) \right| \leq C_{n18}\, \sigma^{n-\nu-1} \omega(\sigma) \left| \zeta_1 - \zeta_2 \right|$$

$$\left| \left(\frac{1}{j}\right)^{(n-\nu)}(\zeta_1) - \left(\frac{1}{j}\right)^{(n-\nu)}(\zeta_2) \right| \leq C_{n8}\, \sigma^{\nu-n-1} \left| \zeta_1 - \zeta_2 \right|$$

$$\left| \left(\frac{1}{j}\right)^{(n-\nu)}(\zeta_1) \right| \leq C_{n8}\, \sigma^{\nu-n} \quad ,$$

and therefore (28) gives (we remind that $|\zeta_1 - \zeta_2| \leq 2\sigma$)

$$\sum_{\nu=0}^{n-1} \leq C_{n19} \frac{|\zeta_1 - \zeta_2|\, \omega(\sigma)}{\sigma} \leq C_{n20}\, \omega(|\zeta_1 - \zeta_2|) .$$

If $\nu = n$,

$$\left| f^{(n)}(\zeta_1) - f^{(n)}(\zeta_2) \right| \leq \omega(|\zeta_1 - \zeta_2|), \quad \left| f^{(n)}(\zeta_2) \right| \leq C_{n16}\, \omega(\sigma)$$

and we get at last

$$\left| f_0^{(n)}(\zeta_1) - f_0^{(n)}(\zeta_2) \right| \leq C_{n17}\, \omega(|\zeta_1 - \zeta_2|), \quad \zeta_1, \zeta_2 \in \Omega . \tag{29}$$

The choice of the domain Ω and (29) make it possible to apply lemma 2 to the function f_0 and to the finite Blaschke product b .

So $f_0/b \in B_{C_{n21}} \Lambda^n_\omega(\Omega)$. However $f_0/b = f/I \overset{def}{=\!=} F$. We apply now the Cauchy formula to the function F at the point $z_0 = \left(1 - \frac{\sigma}{n+2}\right) z$. We get

$$F^{(\nu)}(z_0) = \frac{\nu!}{2\pi i} \int_{\partial\Omega} \frac{F(\zeta)}{(\zeta - z_0)^{\nu+1}} d\zeta = \frac{\nu!}{2\pi i} \int_{\partial\Omega} \frac{f_0(\zeta)}{b(\zeta)} \frac{d\zeta}{(\zeta - z_0)^{\nu+1}} . \tag{30}$$

The domain Ω is chosen purposely so that if $\zeta \in \partial\Omega$ then $|b(\zeta)| \geqslant C_{n22} > 0$. Further, $|\zeta - z_0| \geqslant \frac{\sigma}{n+2}$, $\zeta \in \partial\Omega$, and so (27) and (30) imply

$$|F^{(\nu)}(z_0)| \leqslant C_{n23}\, \sigma^{n-\nu} \omega(\sigma), \quad \nu = 0, \ldots, n. \tag{31}$$

Now we find from (29),(31) and from the Taylor formula that

$$|F^{(n)}(z)| \leqslant |F^{(n)}(z_0)| + |F^{(n)}(z) - F^{(n)}(z_0)| \leqslant C_{n24}\, \omega(\sigma), \tag{32}$$

$$|F^{(\nu)}(z)| \leqslant \sum_{k=0}^{n-\nu} \frac{|F^{(\nu+k)}(z_0)|}{k!} |z - z_0|^k + \frac{1}{(n-\nu-1)!} \left| \int_{z_0}^{z} (z-t)^{n-\nu-1} (F^{(n)}(t) - F^{(n)}(z_0))\, dt \right| \leqslant \tag{33}$$
$$\leqslant C_{n25}\, \sigma^{n-\nu} \omega(\sigma), \quad \nu = 0, \ldots, n-1.$$

The inequalities (32) and (33) accomplish our reasoning in the case 1.

2. $a_n(z) \leqslant \frac{4A}{d_n(z)}$. Put $\sigma = \frac{1}{4} d_n(z)$ and begin the proof as in the case 1, but after the choice of the domain Ω use lemmas 2 and 4 instead of lemmas 4 and 6.

After that we can finish the proof exactly as in the case 1. ●

4. Proof of theorem I.

Let $f \in B\Lambda^n_\omega(\mathbb{D})$, let $I \in H^\infty$ be an inner function, $F \overset{def}{=\!=} f/I \in H^1$. Using the theorem of P.M.Tamrazov [17] we shall restrict ourselves only to the proof of the inequality

$$|F^{(n)}(z_1) - F^{(n)}(z_2)| \leqslant C_{n0}\, \omega(|z_1 - z_2|)$$

for $z_1, z_2 \in \partial\mathbb{D}$. We have to distinguish between two possibilities.

1. $|z_1 - z_2| \geqslant max\left(\frac{1}{4} d_n(z_1), \frac{1}{4} d_n(z_2)\right)$ *). Then owing to lem-

---

*) The numbers $d_n(z)$ and $a_n(z)$ were defined in section 3.

ma 8 we have

$$|F^{(n)}(z_1)| \le C_{n15}\,\omega(d_n(z_1)) \le C_{n0}\,\omega(|z_1-z_2|),$$

$$|F^{(n)}(z_2)| \le C_{n15}\,\omega(d_n(z_2)) \le C_{n0}\,\omega(|z_1-z_2|),$$

$$|F^{(n)}(z_1) - F^{(n)}(z_2)| \le C_{n0}\,\omega(|z_1-z_2|).$$

2. $|z_1-z_2| < max(\frac{1}{4}d_n(z_1), \frac{1}{4}d_n(z_2))$. Suppose for example, $d_n(z_1) \ge d_n(z_2)$. We consider two subcases.

2.1. $|z_1-z_2| \ge max\left(\frac{1}{100n\,a_n(z_1)}, \frac{1}{100n\,a_n(z_2)}\right)$. Again owing to lemma 8

$$|F^{(n)}(z_1)| + |F^{(n)}(z_2)| \le C_{n15}\left(\omega\left(\frac{1}{a_n(z_1)}\right) + \omega\left(\frac{1}{a_n(z_2)}\right)\right) \le$$
$$\le C_{n0}\,\omega(|z_1-z_2|)$$

2.2. $|z_1-z_2| < max\left(\frac{1}{100n\,a_n(z_1)}, \frac{1}{100n\,a_n(z_2)}\right)$. We know that $|z_1-z_2| < \frac{1}{4}d_n(z_1)$, $d_n(z_2) \le d_n(z_1)$ too. The geometrical considerations imply $d_n(z_2) \ge \frac{3}{4}d_n(z_1)$. The domains $\widetilde{\Omega}_1 = \{\zeta\in\overline{\mathbb{D}} : |\zeta-z_1| < d_n(z_1)\}$ and $\widetilde{\Omega}_2 = \{\zeta\in\overline{\mathbb{D}} : |\zeta-z_2| < d_n(z_2)\}$ contain at most n points belonging to the $spec\,I$ each and the domain $\widetilde{\Omega}_0 = \{\zeta\in\overline{\mathbb{D}} : |\zeta-z_2| < \frac{3}{4}d_n(z_1)\}$ does not intersect the domain $(\widetilde{\Omega}_1\setminus\widetilde{\Omega}_2)\cup(\widetilde{\Omega}_2\setminus\widetilde{\Omega}_1)$. Therefore reminding of our case we have

$$\frac{1}{C_{n19}}\,a_n(z_2) \le a_n(z_1) \le C_{n19}\,a_n(z_2). \tag{34}$$

We denote

$$\sigma = min\left(\frac{1}{n+2}\,\frac{d_n(z_1)}{4}, \frac{1}{n+2}\,\frac{1}{C_{n19}\cdot 100n}\,\frac{1}{a_n(z_1)}\right).$$

Then $(n+2)\sigma \le \frac{1}{4}d_n(z_1)$. The domain $\Omega = \{\zeta\in\overline{\mathbb{D}} : |\zeta-z_1| \le |z_1-z_2| + (n+2)\sigma\}$ contains at most n points of $spec\,I$. Therefore at least one of the domains $\Omega_k = \{\zeta\in\overline{\mathbb{D}} : |z_1-z_2| + k\sigma < |\zeta-z_1| \le |z_1-z_2| + (k+1)\sigma\}$, $k=0,\dots,n$ does not intersect $spec\,I$. For this k we put $\Omega = \{\zeta\in\overline{\mathbb{D}} : |\zeta-z_1| \le |z_1-z_2| + (k+\frac{1}{2})\sigma\}$.

The definition of the number σ and the condition imposed in 2.2 show that

$$\frac{\sigma}{2} \leq \operatorname{diam} \Omega \leq C_{n20}\, \sigma . \qquad (35)$$

We are able now to complete the proof of theorem 1. We denote

$$J = I\,|_{\operatorname{spec} I \setminus \Omega}\,, \quad b = I/J .$$

The function b is a finite Blaschke product with at most n zeros. Let $f_0 = f/J$. From the lemma 4, (34) and (35) we obtain

$$\left|\left(\frac{1}{J}\right)^{(\nu)}(\zeta)\right| \leq C_{n21}\, \sigma^{-\nu}, \quad \nu = 0,\dots,n+1, \quad \zeta \in \Omega . \qquad (36)$$

Expanding the function f in the Taylor series with the center at z_1 and using the inequality (24) from lemma 8 we get further

$$\left| f_0^{(\nu)}(\zeta) \right| \leq C_{n22}\, \sigma^{n-\nu}\omega(\sigma)\,, \quad \nu = 0,\dots,n. \qquad (37)$$

Utilizing (36) and (37) exactly as (28) and (29) we find that

$$\left| f_0^{(n)}(\zeta_1) - f_0^{(n)}(\zeta_2) \right| \leq C_{n23}\, \omega(|\zeta_1 - \zeta_2|), \quad \zeta_1, \zeta_2 \in \Omega .$$

But again as in lemma 8

$$F = f/I = f_0/b .$$

The function f_0 and every zero of the Blaschke product b satisfy the conditions of lemma 2. Its application gives $F = f_0/b \in B_{C_{n26}}\Lambda^n_\omega(\Omega)$ and in particular inclusions z_1, $z_2 \in \Omega$ imply

$$\left| F^{(n)}(z_1) - F^{(n)}(z_2) \right| \leq C_{n26}\, \omega(|z_1 - z_2|) .$$

Proof of theorem 1 is finished. ●

5. Proof of theorem 3.

The statements a) and b) of theorem 3 for $B\Lambda_\omega^n(\mathbb{D})$ can be embraced by the following assertion: let $I\in H^\infty$ be an inner function, and suppose $f\in B\Lambda_\omega^n(\mathbb{D})$, $f/I\in H^1$ and suppose every point $a\in \operatorname{spec} I$ is a zero of the function f of the multiplicity not less than $n+1$; then $fI\in B_{C_n}\Lambda_\omega^n(\mathbb{D})$.

The main steps of the proof of theorem 3 are similar to the corresponding steps of the proof of theorem 1, so we shall only describe it briefly. Let $\mathcal{J}$ be an inner factor in the factorization of the function f , $f=F\mathcal{J}$ then theorem 1 implies $F\in B_{C_n}\Lambda_\omega^n(\mathbb{D})$ and the inclusion $f/I\in H^1$ implies $\mathcal{J}/I\in H^\infty$. Then, if $a\in \operatorname{spec} I\cap \mathbb{D}$, the multiplicity of the zero a for I does not exceed the multiplicity of the zero a for $\mathcal{J}$ and if ν and μ are nonnegative singular measures corresponding respectively to I and $\mathcal{J}$, then $\nu\leq\mu$ and taking into account the conditions imposed on I we deduce **the inequality:**

$$|I'(\zeta)|\leq a_{n,\mathcal{J}}(\zeta)\overset{def}{=\!=}a_n(\zeta),\quad \zeta\in\partial\mathbb{D}\,. \tag{38}$$

Estimates (38), (28) and the Newton-Leibnitz formula give

$$|(fI)^{(n)}(\zeta)|\leq C_n\min\Big(\omega(d_{n,\mathcal{J}}(\zeta)),\omega\Big(\frac{1}{a_n(\zeta)}\Big)\Big),\ \zeta\in\partial\mathbb{D}. \tag{39}$$

Taking into account that $d(\zeta)\overset{def}{=\!=}d_{0,I}(\zeta)\geq d_{n,\mathcal{J}}(\zeta)$ we can transform (39):

$$|(fI)^{(n)}(\zeta)|\leq C_n\min\Big(\omega(d(\zeta)),\ \omega\Big(\frac{1}{|I'(\zeta)|}\Big)\Big),\ \zeta\in\partial\mathbb{D}\,. \tag{40}$$

From now on the proof of theorem 3 follows literally the proof of theorem 1 with an only change : the inequality (25) which was made use of in cases 1 and 2.1 has to be replaced by the inequality (40). The domain $\check{\Omega}$ does not contain points of $\operatorname{spec} I$, and therefore we put $\Omega=\check{\Omega}$ and repeat reasonings exactly as in (28)-(29), and afterwards the proof ends because in the sent situation $b\equiv 1$. ●

6. Proof of theorem 2.

We begin with two technical lemmas.

LEMMA 9. Let $a \in \mathbb{D}$, $h(z)=\frac{(z-a)^{n+1}}{1-\bar{a}z}$, $n \geq 1$. Then

$$\left|h^{(n)}\left(\frac{a}{|a|}\right)-h^{(n)}(a)\right|=\left|h^{(n)}\left(\frac{a}{|a|}\right)\right|=\frac{n!}{|a|}\left((1+|a|)^{n+1}-1\right)>1 \tag{41}$$

$$\left|h^{(\nu)}\left(\frac{a}{|a|}\right)\right| \leq C_n(1-|a|)^{n-\nu}, \quad 0 \leq \nu \leq n-1 .$$

PROOF. A straightforward verification. ●

LEMMA IO. Let us define the moduli $|g_\nu|$ of outer functions g_ν, $\nu=2,3,\dots$, in the following way:

$$|g_\nu(e^{i\theta})|=\begin{cases} |\theta|^{6n}(\pi-|\theta|)^{6n}, & -\pi<\theta<0 \\ \left(\theta-\frac{\pi}{2^{m+1}}\right)^{6n}\left(\frac{\pi}{2^m}-\theta\right)^{6n}, & \theta \in \left(\frac{\pi}{2^{m+1}},\frac{\pi}{2^m}\right), \\ & m \leq \nu-2 \text{ or } m \geq \nu+1 \\ \left(\theta-\frac{\pi}{2^{\nu+1}}\right)^{6n}\left(\frac{\pi}{2^{\nu-1}}-\theta\right)^{6n}, & \theta \in \left(\frac{\pi}{2^{\nu+1}},\frac{\pi}{2^{\nu-1}}\right); \ n \geq 1 . \end{cases} \tag{42}$$

Then $g_\nu \in C_A^{3n}$ and moreover

$$\|g_\nu\|_{C_A^{3n}} \leq C_n , \tag{43}$$

where C_n is independent of ν. If $E_\nu=\left(\{e^{\frac{\pi i}{2^m}}\}_{m=0}^{\infty} \cup \{1\}\right) \setminus \{e^{\frac{\pi i}{2^\nu}}\}$, $\nu=2,3,\dots$ then

$$|g_\nu^{(k)}(z)| \leq C_n \operatorname{dist}^{2n+2}(z,E_\nu), \ z \in \overline{\mathbb{D}}, \ 0 \leq k \leq n+1 . \tag{44}$$

PROOF. The assertions (43) and (44) can be checked with the help of a traditional method originating from [18] . ●

We begin the construction of the counterexample needed in theorem 2. Let $\Omega(y)$ be the inverse function of ω, $\Omega(y) \leq 1$. We denote

$$\varepsilon_\nu=\frac{1}{2^\nu}\Omega\left(2^{-12n(2\nu+1)}\right), \quad \nu=2,3,\dots ,$$

$$\alpha_\nu = (1-\varepsilon_\nu) e^{\frac{\pi i}{2^\nu}}, \quad \alpha_\nu^0 = e^{\frac{\pi i}{2^\nu}},$$

$$B_\nu(z) = \prod_{\substack{k=2 \\ k\neq\nu}}^{\infty} \frac{\bar\alpha_k}{|\alpha_k|} \frac{\alpha_k - z}{1-\bar\alpha_k z}, \quad B(z) = \prod_{k=2}^{\infty} \frac{\bar\alpha_k}{|\alpha_k|} \frac{\alpha_k - z}{1-\bar\alpha_k z}.$$

Now we can exhibit the desired function:

$$f(z) = \sum_{\nu=2}^{\infty} \frac{1}{\nu^2} g_\nu(z)\, B_\nu^{n+1}(z)(\alpha_\nu - z)^n. \tag{45}$$

Let us check that $f \in \Lambda_\omega^n(\mathbb{D})$, $f/B \in H^1$ (and, therefore, owing to theorem 1, $f/B \in \Lambda_\omega^n(\mathbb{D})$) but $fB \notin \Lambda_\omega^n$. At first, the estimates (43), (44) and the construction of B_ν give

$$\|f\|_{C_A^{n+1}} \le \sum_{\nu=2}^{\infty} \frac{1}{\nu^2} \|g_\nu B_\nu^{n+1}\|_{C_A^{n+1}} \|(\alpha_\nu - z)^n\|_{C_A^{n+1}} \le C_n$$

so $f \in \Lambda_\omega^n(\mathbb{D})$ for every continuity modulus ω. Further, every point α_ν is a zero of the function f of the multiplicity n, what one can see from the formula (45). Let us estimate the expression

$$v_\nu = (fB)^{(n)}(\alpha_\nu) - (fB)^{(n)}(\alpha_\nu^0).$$

Taking into account the definition of B_ν and g_ν for $k \neq \nu$ we get

$$g_k^{(r)}(\alpha_\nu^0) = 0, \quad 0 \le r \le n,$$

$$B_k^{(r)}(\alpha_\nu) = 0, \quad 0 \le r \le n,$$

hence

$$(fB)^{(n)}(\alpha_\nu) - (fB)^{(n)}(\alpha_\nu^0) =$$

$$= \frac{\bar\alpha_\nu}{|\alpha_\nu|} \frac{1}{\nu^2} \sum_{r=0}^{n} C_n^r (g_\nu B_\nu^{n+2})^{(r)}_{z=\alpha_\nu^0} \cdot \left[\frac{(\alpha_\nu - z)^{n+1}}{1-\bar\alpha_\nu z}\right]^{(n-r)}_{z=\alpha_\nu^0} = \tag{46}$$

$$=\frac{\bar{\alpha}_\nu}{|\alpha_\nu|}\frac{1}{\nu^2}g_\nu(\alpha_\nu^\circ)B_\nu^{n+2}(\alpha_\nu^\circ)\left[\frac{(\alpha_\nu-z)^{n+1}}{1-\bar{\alpha}_\nu z}\right]^{(n)}_{z=\alpha_\nu^\circ}+\frac{\bar{\alpha}_\nu}{|\alpha_\nu|}\frac{1}{\nu^2}\sum_{\tau=1}^{n}\dots .$$

For $1\leq \tau\leq n$ the estimates (41), (43), (44) give

$$\left|(g_\nu B_\nu^{n+2})^{(\tau)}_{z=\alpha_\nu^\circ}\left[\frac{(\alpha_\nu-z)^{n+1}}{1-\bar{\alpha}_\nu z}\right]^{(n-\tau)}_{z=\alpha_\nu^\circ}\right|<C_n\varepsilon_\nu , \tag{47}$$

but for $\tau=0$ (41) shows that

$$\left|g_\nu(\alpha_\nu^\circ)B_\nu^{n+2}(\alpha_\nu^\circ)\left[\frac{(\alpha_\nu-z)^{n+1}}{1-\bar{\alpha}_\nu z}\right]^{(n)}_{z=\alpha_\nu^\circ}\right|>|g_\nu(\alpha_\nu^\circ)|=2^{-6n(2\nu+1)} . \tag{48}$$

The assertions (46)-(48) imply the inequalities

$$|v_\nu|>\frac{1}{\nu^2}2^{-6n(2\nu+1)}-\frac{C_n\varepsilon_\nu}{\nu^2} , \quad \nu=2,3,\dots . \tag{49}$$

However if the function fB would belong to the $\Lambda^n_\omega(\mathbb{D})$, then for a sertain constant C the inequalities

$$|v_\nu|<C\omega(\varepsilon_\nu)=C\omega(\Omega(2^{-12n(2\nu+1)}))=C2^{-12n(2\nu+1)}$$

would be true contradicting (49). Theorem 2 is proved. ●

7. Remarks about spaces H^p_n.

The method used in the present paper permits to check the (F)-property for the spaces H^p_n, $1<p<\infty$, $n\geqslant 1$ and to prove the analogs of theorems 1-3. Namely, the following results are true.

THEOREM 4. There exist a function f, $f\in H^p_n$, $1<p<\infty$, $n\geqslant 2$, and $n\geqslant 3$, if $p=1$, and a Blaschke product B such that $f/B\in H^p_n$, but $fB\notin H^p_n$.

We remind that in H^p_1, $1\leq p\leq\infty$, such situation cannot occur. It is not clear whether the assertion of theorem 4 holds for the space H^1_2.

THEOREM 5. Let $f\in H^p_n$, $1<p<\infty$, $n\geqslant 2$.

a) If $S\in H^\infty$ is a singular function, $f/S\in H^1$, then

$fS^k \in H^p_n$, $k=1,2,\dots$;

b) If B is a Blascke product, $f/B \in H^1$ and the multiplicity of the zero of the function f for every $a \in B^{-1}(0)$ is not less than n , then $fB^k \in H^p_n$, $k=1,2,\dots$.

It is not clear whether the assertion of theorem 5 holds for the spaces H^1_n, $n \geqslant 3$ (for the space H^1_2 these assertions hold what one can deduce from the (F)-property of H^1_2).

8. Further generalizations.

In this section we announce results concerning (F)-property in domains with angles. These results cover (even in the case of the disc) the statement of theorem 1 but their proof seems to be too long to be published here.

DEFINITION. Let Ω be a bounded Jordan domain, a function f be analytic in Ω and bounded, $|f(z)| \leqslant C$. Let φ be a conformal mapping of the unit disc $\mathbb{D}$ onto Ω . Then $H(\zeta) \overset{def}{=\!=} f(\varphi(\zeta)) \in H^\infty$. We write the Nevanlinna factorization for H :

$$H = \Phi I ,$$

where Φ is an outer function, I is an inner function. We define the outer part and the inner part of the f by the equalities:

$$F_f(z) \overset{def}{=\!=} \Phi(\varphi^{-1}(z)), \quad \mathcal{J}_f(z) \overset{def}{=\!=} I(\varphi^{-1}(z)),$$

where φ^{-1} is the inverse mapping of φ .

THEOREM 6. Let Ω be a Jordan domain whose boundary consists of infinitely smooth arcs $\Gamma_1, \dots \Gamma_n$ and the arcs Γ_j and Γ_{j+1} (we set $\Gamma_{n+1} = \Gamma_1$) form the angle π/m_j , $j=1,\dots,n$, m_j are positive integers. Let an outer (in Ω) function F satisfies on $\partial\Omega$ the A_1-Muckenhoupt condition:

$$\frac{1}{|I|} \int_I |F(\zeta)||d\zeta| \leqslant C \operatorname*{ess\,inf}_I |F|, \quad I \subset \partial\Omega .$$

Let X be either the space of functions f analytic in Ω such that

$$|f^{(n)}(z)| < C_f |F(z)| , \qquad z \in \Omega ,$$

or the space of functions f analytic in Ω such that

$$\left| f^{(n)}(z_1) - f^{(n)}(z_2) \right| < C_f\, \omega\left(|F(\check{z})|\, |z_1 - z_2| \right),$$

where ω is an arbitrary continuity modulus, the point $\check{z} \in \left\{ \zeta : \left| \zeta - \frac{z_1+z_2}{2} \right| = \frac{|z_1-z_2|}{2} \right\} \overset{def}{=\!=} \gamma(z_1, z_2), \check{z} \in \overline{\Omega}$ and $dist\,(\check{z}, \partial\Omega) = \max_{\zeta \in \gamma(z_1,z_2) \cap \overline{\Omega}} dist\,(\zeta, \partial\Omega)$. Then the space X possesses the (F)-property which is defined for the domain Ω literally like in the case of the unit disc.

If $\Omega = \mathbb{D}$, $F \equiv 1$, we get theorem 1.

REMARKS. a) The assumption concerning the angles is essential: if their sizes are π/α_j, $\alpha_j, \alpha_j < \infty$ being not all positive integers, then the conclusion of the theorem fails even for $F \equiv 1$.

b) An example of the function F with A_1-Muckenhoupt condition: let μ be a nonnegative measure on $\partial\Omega$, $\mu(\partial\Omega) < \infty$, $\mu(\{\zeta\}) < 1$ for any point $\zeta \in \partial\Omega$. Then the function

$$F(z) \overset{def}{=\!=} \exp\left(-\int_{\partial\Omega} \log(z-\zeta)\, d\mu(\zeta) \right), \quad z \in \Omega,$$

is an outer function satisfying the A_1-Muckenhoupt condition (the domain Ω is as in theorem 6).

I am highly grateful to V.P.Havin for careful reading the present paper and its correction in mathematical and linguistic aspects.

References.

1. И.И.П р и в а л о в. Граничные свойства аналитических функций. М-Л, ГИТТЛ, 1950.

2. K.H o f f m a n. Banach spaces of analytic functions, Prentice-Hall, Engl.Cliffs, N.J., 1962.

3. L.C a r l e s o n. A representation formula for the Dirichlet integral, Math.Z, 1960, 73, N 2, 190-196.

4. Б.И.К о р е н б л ю м, В.С.К о р о л е в и ч. Об аналитических функциях, регулярных в круге и гладких на его границе, Мат.заметки, 1970, 7, № 2, 165-172.

5. Б.И.К о р е н б л ю м. Экстремальные соотношения для внешних функций, Мат.заметки, 1971, 10, № 1, 53-56.

6. В.П.Х а в и н. О факторизации аналитических функций, гладких вплоть до границы, Записки научных семин.ЛОМИ, 1971, 22, 202--205.

7. Ф.А.Ш а м о я н. Деление на внутреннюю функцию в некоторых

пространствах функций, аналитических в круге, Записки научных семин.ЛОМИ, 1971, 22, 206-208.

8. С.А.Виноградов, Н.А.Широков. О факторизации аналитических функций с производной из H^p, Записки научных семин.ЛОМИ, 1971, 22, 8-27.

9. Б.И.Коренблюм, В.М.Файвышевский. Об одном классе сжимающих операторов, связанных с делимостью аналитических функций, Украинский матем.журн., 1972, 14, № 5, 692-695.

10. В.Э.Кацнельсон. Замечание о канонической факторизации в некоторых пространствах аналитических функций, Записки научных семин.ЛОМИ, 1972, 30, 163-164.

11. M.Rabindranathan. Toeplitz operators and division by inner functions, Indiana Univ.Math.J., 1973, 22, N 10, 523-529.

12. Е.М.Дынькин. Гладкие функции на плоских множествах, Доклады АН СССР, 1973, 208, № 1, 25-27.

13. Ф.А.Шамоян. Об ограниченности одного класса операторов, связанных с делимостью аналитических функций, Известия АН Арм. ССР, Математика, 1973, 8, № 6, 474-490.

14. J.P.Kahane. Best approximation in $L^1(\mathbb{T})$, Bull.Amer. Math.Soc., 1974, 80, N 5, 788-804.

15. Н.А.Широков. Идеалы и факторизация в алгебрах аналитических функций, гладких вплоть до границы, Труды МИАН, 1978, 130, 196-222.

16. Ф.А.Шамоян. Об одном классе тёплицевых операторов, связанных с делимостью аналитических функций, Функц.анализ и его прил., 1979, 13, № 1, 83-84.

17. П.М.Тамразов. Контурные и телесные структурные свойства голоморфных функций комплексного переменного, Успехи матем.наук, 1973, 28, № 1, 131-161.

18. Б.И.Коренблюм. Замкнутые идеалы кольца A^n, Функц. анализ и его прил., 1972, 6, № 3, 38-53.

19. В.П.Гурарий. О факторизации абсолютно сходящихся рядов Тейлора и интегралов Фурье, Записки научных семин.ЛОМИ, 1972, 30, 15-33.

20. Н.А.Широков. Некоторые свойства примарных идеалов абсолютно сходящихся рядов Тейлора и интегралов Фурье, Записки научных семин.ЛОМИ, 1974, 39, 149-162.

21. Н.А.Широков. Деление на внутреннюю функцию не меняет класса гладкости, Доклады АН СССР, 1981.

22. J.M.Anderson. Algebras contained within H^∞, Записки научных семин.ЛОМИ, 1978, 81, 235-236.

A.L.Volberg

THIN AND THICK FAMILIES OF RATIONAL FRACTIONS

Preface

The problems investigated in this paper can be formulated as follows: given a finite positive Borel measure μ on the real line $\mathbb{R}$ (or on the unit circle $\mathbb{T}$) and a family of rational fractions $R_\Lambda = \left\{\frac{1}{z-\lambda}\right\}$, $\Lambda \subset \mathbb{C}_+ \overset{def}{=} \{\zeta \in \mathbb{C} : \operatorname{Im}\zeta > 0\}$ (or $\Lambda \subset \mathbb{D} \overset{def}{=} \{\zeta \in \mathbb{C} : |\zeta| < 1\}$), which properties of this measure and family will imply:

1) the completeness of the family R_Λ in the space $L^2(\mu)$

2) or the equivalence of the norms $\|\cdot\|_{L^2(\mu)}$ and $\|\cdot\|_{L^2(m)}$ on the linear span $\mathcal{L}_\Lambda$ of the family R_Λ (here m denotes Lebesgue measure on the line $\mathbb{R}$ or on the circle $\mathbb{T}$)

3) or the compactness of the inclusion map $j : \mathcal{L}_\Lambda \longrightarrow L^2(\mu)$, $jf = f$, $f \in \mathcal{L}_\Lambda$, $\mathcal{L}_\Lambda$ being endowed with the $L^2(m)$ -norm.

If the first possibility holds we call the family R_Λ thick with respect to the measure μ . If the second or third possibilities occur we call the family R_Λ thin with respect to the measure μ (we are interested mainly in the case of unbounded density $\frac{dm}{d\mu}$; therefore we use the therm "thin").

In this paper thick families are described for the measures with one "singular" point and some properties of regularity. Thin families are described for the measures μ with the bounded density with respect to Lebesgue measure m .

The paper is subdivided into two parts and six sections.

§1 contains main definitions and the statements of three

main theorems. We point out the connections with some other problems. We also discuss basic concepts of the proofs of these theorems. The proof of theorem 1 is based on the construction of a special domain in the complex plane $\mathbb{C}$. The construction of this domain consists of two steps.

In §2 the first step of the construction is made. This construction follows the scheme of the articles [1] and [2]. In this paragraph "a half" of theorem 1 is proved.

§3 contains the second step of the construction. Here we also finish the proof of theorem 1.

In §4, we prove theorem 1 in a very particular case by an essentially different method (using a technique of the quasi-analytic classes).

In §5, we consider thin families of rational fractions and prove theorem 2.

In §6, we prove theorem 3. Here we present some propositions ooncerning compactness of the inclusion map j and related questions. The results of this paper were partially announced in [3].

ACKNOWLEDGEMENT. The author wishes to express his deep gratitude to N.K.Nikol'skii for formulating the problem and for valuable advices. Thanks are also due to S.V.Hruščev for advices which shortened some proofs and to P.Koosis who kindly supplied me with the prepublication version of the paper [1]. Finally, the author is sincerely grateful to V.P.Havin for reading the manuscript.

1. Introduction

A. Thick families.

Let μ be a positive finite Borel measure on the real line $\mathbb{R}$. As we have mentioned before we consider only a special class of measures with one "singular" point.

DEFINITION. Let $d\mu = h\,dm$, $h \in L^1(\mathbb{R},m) \cap L^\infty(\mathbb{R},m)$, moreover, in a neighbourhood of the point $t=0$

$$\log\frac{1}{h(t)} = \begin{cases} \dfrac{\varepsilon(t)}{t}, & t>0 \\ \dfrac{\varepsilon_1(|t|)}{\sqrt{|t|}}, & t<0, \end{cases}$$

where $\lim\limits_{t\to 0+} \varepsilon(t) = \lim\limits_{t\to 0+} \varepsilon_1(t) = 0$ and

$$\lim_{t\to 0+}\frac{\varepsilon'(t)}{\varepsilon(t)}t=\lim_{t\to 0+}\frac{\varepsilon_1'(t)}{\varepsilon_1(t)}t=0.$$

Then we shall say that the measure μ (and the function h) is r e g u l a r.

Let R_Λ be a family of rational fractions $\left\{\frac{1}{z-\lambda}\right\}_{\lambda\in\Lambda}$, whose poles lie in the S t o l z d o m a i n $K_\gamma \overset{def}{=\!=} \{\zeta=x+iy\in\mathbb{C}_+: y>\gamma\cdot|x|\}$. **Recall** that $\mathcal{L}_\Lambda$ denotes the linear span of R_Λ . Now we are in a position to formulate our main theorems

THEOREM 1. Let

$$\int_{t>0}\frac{\varepsilon(t)}{t}\,dt=+\infty. \tag{1.1}$$

If

$$\int_{t<0}\frac{\varepsilon_1(|t|)}{|t|}\,dt<\infty \tag{1.2}$$

then

$$\operatorname{clos}\mathcal{L}_\Lambda=L^2(h\,dm)\Longleftrightarrow\sum_{\lambda\in\Lambda}(\operatorname{Im}\lambda)^{1/2}=\infty. \tag{1.3}$$

On the contrary if

$$\int_{t<0}\frac{\varepsilon_1(|t|)}{|t|}\,dt=\infty \tag{1.4}$$

then

$$\operatorname{clos}\mathcal{L}_\Lambda=L^2(h\,dm)\Longleftrightarrow\operatorname{card}\Lambda=\infty. \tag{1.5}$$

The fact that the measure μ has the support on the real line is unimportant. The analogous theorem is valid for measures μ, $\operatorname{supp}\mu\subset\mathbb{T}$, $\mu=h\,dm$. Certainly, we suppose that the function $t\longrightarrow h(e^{it})$ is regular and the family R_Λ has the poles in some Stolz domain $K_\alpha(1)$ in the unit disc $\mathbb{D}=\{\zeta\in\mathbb{C}:|\zeta|<1\}$ (let us recall that the S t o l z d o m a i n $K_\alpha(\xi)$ i n $\mathbb{D}$ is the interior of the convex hull of $\{\zeta\}$ and the circle $\{\zeta\in\mathbb{C}:|\zeta|=\sin\alpha\}$. To obtain the statement of the theorem for the circle $\mathbb{T}$ we must only substitute (1.3) by

$$\operatorname{clos}\mathcal{L}_\Lambda=L^2(h\,dm)\Longleftrightarrow\sum_{\lambda\in\Lambda}(1-|\lambda|)^{1/2}=\infty. \tag{1.3a}$$

Note that from (1.1) of the theorem we have

$$\int \log h\, dm = -\infty . \tag{1.6}$$

In the opposite case the following theorem(see [4]) gives a complete description of thick families R_Λ. In this theorem there are no restrictions on the set of poles Λ, except $\Lambda \subset \mathbb{C} \setminus \mathbb{T}$.

THEOREM (see [4]). If

$$\int \log h\, dm > -\infty \tag{1.7.}$$

then $\operatorname{clos} \mathcal{L}_\Lambda = L^2(h\,dm) \Longleftrightarrow \sum\limits_{\lambda\in\Lambda, |\lambda|<1} (1-|\lambda|) = \infty, \sum\limits_{\lambda\in\Lambda, |\lambda|>1} (|\lambda|-1) = \infty$.

From a different point of view the rational approximation for the case (1.7) was investigated by G.Tz.Tumarkin [5], [6].

The basic method of proof of our theorem is the construction of a region in $\mathbb{C}$ with some special properties. This method is similar to that of P.Koosis [1]. Now we shall **enlist some useful** remarks.

REMARK 1. It is clear that if R_Λ is thick with respect to $h\,dm$ and $h_1 \leq h$ a.e., then R_Λ is thick with respect to $h_1\,dm$.

REMARK 2. If the function h is such that

$$\log \frac{1}{h(t)} = \frac{\varepsilon(|t|)}{|t|}, \quad \lim_{t\to 0} \varepsilon(t) = 0, \quad \lim_{t\to 0} \frac{\varepsilon'(t)}{\varepsilon(t)}\, t = 0 \tag{1.8}$$

and the condition (1.1) is fulfilled, then from theorem 1 and remark 1 we conclude that

$$\operatorname{clos} \mathcal{L}_\Lambda = L^2(h\,dm) \Longleftrightarrow \operatorname{card} \Lambda = \infty .$$

But it turns out that in this case we may deduce this result with the help of more direct methods. (See §4, which treats this subject). We shall need some notation below.. If μ is a measure and E is a μ-measurable set then $\mu | E$ will denote the measure defined by $(\mu | E)(X) \overset{def}{=\!=} \mu(E \cap X)$ for every μ-measurable set X.

REMARK 3. We may consider more general systems of rational fractions. Namely $R_{\Lambda,k} \overset{def}{=\!=} \left\{ \frac{1}{(\zeta-\lambda)^m} : \lambda \in \Lambda,\ 1 \leq m \leq k(\lambda) \right\}$. Here k is an integer-valued function. Theorem 1 is still true for these families.

REMARK 4. One can suspect the restriction on the poles in theorem 1, namely $\Lambda \subset K_\gamma$, **is unimportant. That this is not the case is shown** by the following proposition.

PROPOSITION. Let μ be a positive finite Borel measure on the circle $\mathbb{T}$. If for every sequence $(\lambda_n)_{n\geq 1}$, $\lambda_n \in \mathbb{D}$, $\lim_{n\to\infty} \lambda_n = 1$ we have

$$\operatorname{clos} \mathcal{L}_\Lambda = L^2(\mu)$$

then there is a positive number ε such that

$$\mu \mid [e^{-i\varepsilon}, e^{i\varepsilon}] = \lambda \delta_{\{1\}} , \quad \lambda \geq 0 , \tag{1.9}$$

$\delta_{\{1\}}$ being the one-point measure.

PROOF. Suppose (1.9) does not hold. We construct a sequence $(\lambda_n)_{n\geq 1}$, $\lambda_n \in \mathbb{D}$, $\lim_{n\to\infty} \lambda_n = 1$ such that

$$\operatorname{clos} \mathcal{L}_\Lambda \neq L^2(\mu).$$

It is clear that without loss of generality we may suppose that $\mu(\{1\}) = 0$. Let $d(\lambda) \overset{def}{=\!=} \|(\zeta-\lambda)^{-1}\|_{L^2(\mu)}$.

We choose the points λ_n such that the vectors

$$e_n \overset{def}{=\!=} (\zeta - \lambda_n)^{-1} / d(\lambda_n)$$

form a Riesz basis in their linear span in $L^2(\mu)$. Firstly we show that we may choose λ_n such that $e_n \longrightarrow \mathbb{O}$ in the weak topology of $L^2(\mu)$. To show this we note that

$$1 \in \operatorname{supp} \mu \tag{1.10}$$

since (1.9) does not hold and note also that

$$\lim_{\lambda \to \zeta} d(\lambda) = +\infty \tag{1.11}$$

for μ - a.e. point $\zeta \in \mathbb{T}$. Indeed it is sufficient to prove that $d(e^{i\varphi}) = +\infty$ μ-a.e. Let $E \overset{def}{=\!=} \{e^{i\varphi} : d(e^{i\varphi}) < \infty\}$. For an arbitrary positive number ε and every point $e^{i\varphi}$, $e^{i\varphi} \in E$ there exists a neighbourhood I_φ of this point such that $\mu(I_\varphi) \leq \varepsilon |I_\varphi|$. The family of intervals $(I_\varphi)_{e^{i\varphi}\in E}$ covers the set E and it is well-known that there is a sequence of disjoint intervals $(I_{\varphi_j})_{j\geq 1}$ such that

$$E \subset \bigcup_{j\geq 1} J_{\varphi_j} , \qquad J_{\varphi_j} \overset{def}{=\!=} 5 I_{\varphi_j} .$$

So $\mu(E) \leq 5\varepsilon \sum_{j\geq 1} |I_{\varphi_j}| \leq 10\pi\varepsilon$ for an arbitrary ε, $\varepsilon > 0$. Thus (1.11) is proved. From (1.10) and (1.11) we deduce that there is a sequence $(\lambda_n)_{n\geq 1}$, $\lambda_n \in \mathbb{D}$ $\lim_{n\to\infty} \lambda_n = 1$ such that

$$\lim_{n\to\infty} d(\lambda_n) = \infty \, . \tag{1,12}$$

If we prove that for every interval $I \subset \mathbb{T}$

$$\lim_{n\to\infty} \int_I e_n \, d\mu = 0 \tag{1.13}$$

then we shall show that $e_n \to \mathbb{O}$ in the weak topology of $L^2(\mu)$. Fix an arbitrary positive number ε and choose an open interval $\mathcal{J}_\varepsilon$, $1 \in \mathcal{J}_\varepsilon$ such that

$$\mu(\mathcal{J}_\varepsilon) \leqslant \varepsilon^2 \tag{1.14}$$

(remember that we suppose that $\mu(\{1\}) = 0$).

Now we have the following chain of the inequalities

$$\left| \int_I e_n \, d\mu \right| \leqslant \frac{1}{d(\lambda_n)} \left| \int_{I \setminus \mathcal{J}_\varepsilon} \frac{d\mu(\zeta)}{\zeta - \lambda_n} \right| + \frac{1}{d(\lambda_n)} \left| \int_{\mathcal{J}_\varepsilon} \frac{d\mu(\zeta)}{\zeta - \lambda_n} \right| \leqslant$$

$$\leqslant \frac{1}{d(\lambda_n)} \cdot O(1) + \frac{1}{d(\lambda_n)} \cdot d(\lambda_n) \cdot \mu(\mathcal{J}_\varepsilon)^{1/2} \leqslant$$

$$\leqslant \frac{O(1)}{d(\lambda_n)} + \varepsilon \, .$$

Taking (1.12) into account we see that (1.13) holds.

It is clear that we may **choose a subsequence** $\{e_{n_k}\}_{k\geqslant 1}$ such that

$$< e_{n_k}, e_{n_m} >_{L^2(\mu)} \leqslant 2^{-\max(k,m)}$$

So $(e_{n_k})_{k\geqslant 1}$ forms a Riesz basis in **its linear span.** ●

B. Thin families

Let H^p denote the standard Hardy space in $\mathbb{D}$ and let $1<p<\infty$. If θ is an inner function, then $K_\theta \overset{def}{=} H^p \cap \theta \overline{H^p_0}$, where the bar stands for the complex conjugation and $H^p_0 \overset{def}{=} \{ f \in H^p : f(0) = 0 \}$. Let μ be a positive finite Borel measure on the unit circle $\mathbb{T}$ and let m be Lebesgue measure on $\mathbb{T}$.

The following questions are interesting from many points of view (see the papers cited below). We may ask: what measures μ have the property

$$C_1 \leqslant \left(\int_{\mathbb{T}} |f|^p d\mu \right) / \left(\int_{\mathbb{T}} |f|^p dm \right) \leqslant C_2 \tag{1.15}$$

where $f \in K_\theta$ and the constants $c_1, c_2, 0 < c_1 < c_2 < \infty$ are independent of f ? On the contrary: for what measures μ the operator $j: K_\theta \longrightarrow L^2(\mu)$, $jf = f$ is compact (i.e. the norm $\|\cdot\|_{L^2(\mu)}$ is essentially weaker than $\|\cdot\|_{L^2(m)}$)?

Note some previous well-known results. When $p=2$, $\theta_a \stackrel{def}{=} \exp\{-a\frac{1+z}{1-z}\}$, $a>0$, the first of the above questions has been considered by B.P.Panejah [7], [8], V.Ja.Lin [9], V.N.Logvinenko and Ju.F.Sereda [10], V.È.Kacnelson [11]. Here the first question is equivalent with the following: for which measures μ on the real line $\mathbb{R}$ the norms $(\int_{\mathbb{R}}|\varphi|^2 d\mu)^{1/2}$ and $(\int_{\mathbb{R}}|\varphi|^2 dm)^{1/2}$ are equivalent in the space of entire functions φ of exponential type not exceeding $a/2$ and such that $\int_{\mathbb{R}}|\varphi|^2 dm < \infty$? A complete **answer** has been given (in n-dimensional case and for every $p>0$) for measures $\mu = \chi_E\, dm$ [10], [11].

We shall answer the above questions for $\mu = w\,dm$, $w \in L^\infty(\mathbb{T}, m)$ and for an arbitrary inner function θ . The answer will be given in terms of **the harmonic continuation** $\hat{w}$ of the function w into the unit disc. Some results concerning this problem **with a measure** μ of more general form may be found in [9]. It is interesting to note that in [12] D.N.Clark showed that for every inner function θ there are many m-singular measures μ with the property (1.15).

If θ is a Blaschke product whose zeroes are $(\lambda_n)_{n\geq 1}$ then $K_\theta = span_{L^2(m)} R_\Lambda$ (here **span** R_Λ denotes **the c l o s e d linear span of the family** R_Λ, $R_\Lambda = \{\frac{1}{1-\bar{\lambda}_n \xi} : n \geq 1\}$). In this case the property (1.15) and the fact that the family R_Λ is thin with respect to the measure μ are equivalent.

Now we state our theorem about thin families (more generally, about the equivalence of norms in the space K_θ).

THEOREM 2. Let $p \in (1, \infty)$, $w \in L^\infty(\mathbb{T}, m)$, $w \geq 0$, $d\mu = w\,dm$. The following statements are equivalent:

a) the norms $\|\cdot\|_{L^2(\mu)}$ and $\|\cdot\|_{L^2(m)}$ are equivalent in the space K_θ;

b) if $\zeta_n \in \mathbb{D}$, $\lim \hat{w}(\zeta_n) = 0$ then $\lim\limits_{n\to\infty} |\theta(\zeta_n)| = 1$;

c) $\inf\{\hat{w}(\zeta) + |\theta(\zeta)|^\eta : \zeta \in \mathbb{D}\} > 0$.

A similar theorem is true if we everywhere replace the disc $\mathbb{D}$ by the upper half-plane $\mathbb{C}_+$. Then the statement c) for $w = \chi_E$, $\theta = e^{iaz}$, $a>0$, transforms to the property of the relative density of the set E and this is the criterion found in [8]-[11].

The problem of compactness of the inclusion map $j: K_\theta \to L^2(w\,dm)$

is considered in §6. First af all we remind a definition. Let P_+ be the orthogonal projection of $L^2(\mathbb{T})$ onto H^2 and let $P_- = I - P_+$, I being the identity operator. For a bounded measurable φ, the Toeplitz operator with symbol φ is the operator T_φ on H^2 defined by $T_\varphi h \overset{def}{=} P_+ \varphi h$ and the Hankel operator with the same symbol is defined by $H_\varphi h \overset{def}{=} P_- \varphi h$, $h \in H^2$.

By Fatou's theorem the algebra H^∞ of all bounded analytic functions on $\mathbb{D}$ may be considered as a closed subalgebra of $L^\infty(\mathbb{T})$, the algebra of all essentially bounded Lebesgue measurable functions on $\mathbb{T}$. For a given g in L^∞ let $H^\infty[g]$ denote the uniform subalgebra of L^∞ generated by H^∞ and g. An interesting example is the algebra $H^\infty[\bar{z}] = H^\infty + C(\mathbb{T})$, where C denotes the space of all continous functions on $\mathbb{T}$.

THEOREM 3. The following statements are equivalent:

a) the operator $j: K_\theta \longrightarrow L^2(w\,dm)$ is compact ;

b) the operator $T_w \mid K_\theta$ is compact;

c) $\lim\limits_{|\zeta| \to 1} \min(\hat{w}(\zeta), 1 - |\theta(\zeta)|) = 0$;

d) $H^\infty[w\bar{\theta}] \cap H^\infty[\bar{\theta}] \subset H^\infty + C$;

e) the operator $H^*_{w\bar{\theta}} H_{\bar{\theta}}$ is compact.

In conclusion we introduce some additional notation . If $\mathcal{H}$ is a Hilbert space, T is an operator in $\mathcal{H}$, $T: \mathcal{H} \to \mathcal{H}$ and E is a subspace of $\mathcal{H}$, then the symbol $T \mid E$ denotes the restriction if T onto E, $T \mid E : E \to \mathcal{H}$. We use the symbols C, C_1, C_2, C_3, C_4 for constants, moreover the letter C may denote different constants even in the same inequality.

PART I

2. <u>The first step of the construction</u>

We begin to prove the sufficiency in theorem 1. Remember that the function h is regular and the condition (1.1) is fulfilled. Let the family R_Λ be not complete in $L^2(h\,dm)$. Then there is a function p, $p \in L^2(h\,dm)$, $p \neq 0$, such that

$$\int_{\mathbb{R}} \frac{p(t)\, h(t)\, dm(t)}{t - \lambda} = 0, \quad \lambda \in \Lambda . \tag{2.1}$$

Let

$$\Phi_+(\lambda) \overset{def}{=\!=} \int_{\mathbb{R}} \frac{p(t)h(t)dm}{t-\lambda} \,, \quad \Phi_-(\lambda) \overset{def}{=\!=} \int_{\mathbb{R}} \frac{p(t)\,h(t)\,dm}{t-\bar{\lambda}}$$

in the upper half-plane $\mathbb{C}_+$. The functions Φ_+ and $\bar{\Phi}_-$ belong to the Hardy space H^2 in $\mathbb{C}_+$. It is well-known that $2\pi i\, p(t)\, h(t) = \Phi_+(t) - \Phi_-(t)$ a.e. in $\mathbb{R}$. Now it is almost obvious that $\Phi_+ \not\equiv 0$. Indeed, if it is not the case then we have $|\Phi_-(t)| = 2\pi |p(t)| \cdot |h(t)|$ and so $\int_{\mathbb{R}} \frac{\log|\Phi_-(t)|}{1+t^2}\, dm = -\infty$. But this implies that $\Phi_- \equiv 0$, since $\bar{\Phi}_- \in H^2$. Thus $p(t)\, h(t) = 0$ a.e. This is a contradiction and therefore Φ_+ is a non- zero function in H^2.

From (2.1) we see that the zero-set of Φ_+ contains the set Λ. Let

$$\varphi_1(\zeta) \overset{def}{=\!=} \frac{1}{2}\left[\Phi_+(\zeta) + \bar{\Phi}_-(\zeta)\right] = \int_{\mathbb{R}} \frac{\operatorname{Re} p(t)\, h(t)\, dm}{t-\zeta}$$

$$\varphi_2(\zeta) \overset{def}{=\!=} \frac{1}{2i}\left[\Phi_+(\zeta) - \bar{\Phi}_-(\zeta)\right] = \int_{\mathbb{R}} \frac{\operatorname{Im} p(t)\, h(t)\, dm}{t-\zeta} .$$

Let us now introduce two auxiliary functions h_1, β:

$$h_1(x) \overset{def}{=\!=} \left(\frac{1}{x}\int_x^{2x} \frac{1}{t}\int_t^{2t} h^{\frac{1}{8}}(\tau)\, d\tau\right)^8$$

$$\beta(x) \overset{def}{=\!=} |x|\, h_1^{1/4}(x) .$$

It is clear that

$$h(x) \leqslant h_1(x) \leqslant h(2x) \tag{2.2}$$

and that the functions $h_1^{1/8}, \beta$ are twice continuously differentiable and

$$h_1^{1/8}(0) = (h_1^{1/8})'(0) = (h_1^{1/8})''(0) = 0; \ \beta(0) = \beta'(0) = \beta''(0) = 0. \tag{2.3}$$

We consider the curve $\gamma \overset{def}{=\!=} \{\zeta = x+iy : y = \beta(x)\}$ and estimate $u_j(\zeta) \overset{def}{=\!=} \operatorname{Im} \varphi_j(\zeta)$, $j = 1, 2$ for $\zeta = x + i\beta(x) \in \gamma$ and for x sufficiently small:

$$u_j(x+i\beta(x)) \leqslant \int_{-\infty}^{\infty} \frac{\beta(x)\,|p(t)|\,h(t)}{(x-t)^2 + \beta^2(x)}\, dm = \int_{t:|t-x|\leqslant|x|} + \int_{t:|t-x|>|x|} \leqslant$$

$$\leq \frac{1}{\beta(x)}\left|\int_0^{2x}|p(t)|h(t)dm\right| + \frac{\beta(x)}{x^2}\int_{-\infty}^{\infty}|p(t)|h(t)dm \leq$$

$$\leq \frac{c_1 h^{\frac{1}{2}}(2x)}{\beta(x)} + \frac{c_2\beta(x)}{x^2} = \frac{c_1 h_1^{\frac{1}{2}}(2x)}{|x|h_1^{\frac{1}{4}}(2x)} + \frac{c_2|x|h_1^{\frac{1}{4}}(2x)}{x^2} \leq c\,\frac{h_1^{\frac{1}{4}}(2x)}{|x|} .$$

Now we estimate the gradients of functions u_j and $v_j = Re\,\varphi_j$ $(j=1,2)$ on the curve γ :

$$\frac{1}{2}\left|\frac{\partial u_j}{\partial x}(x+i\beta(x))\right| \leq \int_{-\infty}^{\infty}\frac{|x-t|}{(x-t)^2+\beta^2(x)}\cdot\frac{\beta(x)|p(t)|h(t)}{(x-t)^2+\beta^2(x)}\,dm =$$

$$= \int_{t:|t-x|\leq\beta(x)} + \int_{t:|t-x|>\beta(x)} \leq c\,\frac{1}{\beta(x)}\cdot\frac{h_1^{\frac{1}{4}}(2x)}{|x|} \leq \frac{c}{x^2} .$$

The same estimate is true for $\dfrac{\partial u_j}{\partial y}(\zeta)$, $\zeta\in\gamma$. Therefore on γ

$$|u_j(\zeta)| \leq c\,\frac{h_1^{\frac{1}{4}}(2x)}{x}\ ;\quad |\nabla u_j(\zeta)| = |\nabla v_j(\zeta)| \leq \frac{c}{x^2} . \tag{2.4}$$

LEMMA 1. Let W be the conformal homeomorphism of the upper half plane $\mathbb{C}_+$ onto $G \overset{def}{=\!=} \{\zeta = x+iy : y > \beta(x)\}$. Then W is smooth up to the boundary and is distortion-free, i.e.

$$c_1|W(J)| \leq |J| \leq c_2|W(J)| ,\qquad 0 < c_1 < c_2 < \infty$$

for every interval $\mathcal{J},\ \mathcal{J}\subset\mathbb{R}$.

PROOF. The function $\beta'(x)$ belongs to the class C^1. It is easy to see that the angle formed by the tangent of the boundary γ of G and by the line $\mathbb{R}$ considered as a function of the arc-length parameter on γ is continuously differentiable. The lemma now follows from the Kellog's theorem [13 , p.411] . ●

Let $g_j(\zeta) \overset{def}{=\!=} \varphi_j(W(\zeta))$, $\zeta\in\mathbb{C}_+$. From the previous lemma and the inequality (2.4) one can easily deduce the following inequalities:

$$g_j'(x) \leq \frac{c}{x^2}\ ;\ |g_j(x)| \leq \frac{c}{x}\ ;\ |\mathcal{I}m\,g_j(x)| \leq c\,\frac{h_1^{\frac{1}{4}}(cx)}{|x|} \tag{2.5}$$

for x small enough, $j=1,2$.

From the definition of the functions g_j it is clear that they are bounded outside of some neighbourhood of the point 0.

We denote the set $W^{-1}(\Lambda)$ by Λ again. Without loss of generality we may suppose that $i\in\Lambda$. Now we introduce two functions:

$$f_j(\zeta) \stackrel{def}{=\!=} \frac{\zeta^2}{\zeta^2+1}\, g_j(\zeta), \qquad \zeta\in\mathbb{C}_+ \;, \; j=1,2.$$

The inequalities (2.5) imply that the functions f_j are bounded and analytic in $\mathbb{C}_+$ and

$$|\operatorname{Im} f_j(x)| \leq c h_1^{\frac14}(cx) \qquad \text{for } x \text{ sufficiently small}, j=1,2.$$

From the definition of f_j it is clear that

$$f_1(\lambda)+if_2(\lambda)=0, \quad \lambda\in\Lambda, \quad \lambda\neq i.$$

The aim of the above reduction was the construction of the function $F \stackrel{def}{=\!=} (f_1+if_2)(f_1-if_2)=f_1^2+f_2^2$ which has the following properties:

i) F is a bounded analytic function in $\mathbb{C}_+$, $F\not\equiv 0$;

ii) $F(\lambda)=0$ for $\lambda\in\Lambda$, where Λ is a set in Stolz domain K_γ;

iii) $|F'(x)|\leq A$ for $x\in(-\alpha,\alpha)$ and for some constant A;

iv) $|\operatorname{Im} F(x)| \leq 2(|\operatorname{Im} f_1|\cdot|\operatorname{Re} f_1|+|\operatorname{Im} f_2|\cdot|\operatorname{Re} f_2|)\leq$
$\leq c\nu(x), \quad x\in(-\alpha,\alpha)$.

Here $\nu(x) \stackrel{def}{=\!=} h_1^{\frac14}(Cx)$ has the following properties:

a) $|(\nu^{\frac14})'(x)|\leq A, \quad x\in(-\alpha,\alpha)$.

b)
$$\nu(x)\leq h^{\frac14}(2Cx)=\begin{cases} \exp\left(-\frac14\,\frac{\varepsilon(2cx)}{2cx}\right), & x>0\\ \exp\left(-\frac14\,\frac{\varepsilon_1(2c|x|)}{\sqrt{2c|x|}}\right), & x<0.\end{cases}$$

The functions $h, \varepsilon, \varepsilon_1$ are from the statement of theorem 1. Without loss of generality we replace $\varepsilon(2ct)$, $\varepsilon_1(2ct)$ by $\varepsilon(t), \varepsilon_1(t)$.

Recall that $\int_0 \frac{\varepsilon(t)}{t}\,dt=+\infty$. To prove the sufficiency of the first part of theorem 1 it remains to prove lemma 2 below. The proof of this lemma is similar to the reasoning in [1], [2]. So we shall prove it omitting some tedious technique details.

LEMMA 2. If the conditions i)-iv) for the functions F and ν are fulfilled and

$$\int_0 \frac{\varepsilon(t)}{t}\,dt = +\infty\,, \tag{2.6}$$

then the zero-set Λ of the function F has the property

$$\sum_{\lambda\in\Lambda} (\mathcal{I}m\,\lambda)^{\frac{1}{2}} < \infty\,. \tag{2.7}$$

PROOF. Consider the set

$$\mathfrak{X}_0 = \{ 0 < x_0 < \alpha \;:\; |F(x_0)| \geq 2\cdot \nu^{\frac{1}{2}}(x_0) \}\,.$$

Then $\mathfrak{X}_0 \neq \emptyset$ since otherwise $|F(x)| \leq 2\nu^{\frac{1}{2}}(x)$ for $x\in(0,\alpha)$ and

$$\int_0 \log|F(t)|\,dt \asymp -\int_0 \frac{\varepsilon(t)}{t}\,dt = -\infty \qquad ^{*)}$$

and this contradicts i).

For every point $x_0\in\mathfrak{X}_0$ we construct the interval $\ell_{x_0} \overset{def}{=\!=} (x_0 - \frac{1}{2A}\cdot\nu^{\frac{1}{2}}(x_0),\, x_0 + \frac{1}{2A}\cdot\nu^{\frac{1}{2}}(x))$. From iii) and iva) it is clear that

$$|F(x)| \geq \nu^{\frac{1}{2}}(x)\,,\quad x\in \bigcup_{x_0\in\mathfrak{X}_0} \ell_{x_0}\,. \tag{2.8}$$

Let $\mathcal{O}_n \overset{def}{=\!=} \bigcup_{x_0\in\mathfrak{X}_0,\, x_0>2^{-n}} \ell_{x_0}$ be the union of finitely many disjoint intervals, $\mathcal{O}_n = \ell_1^{(n)} \cup \ell_2^{(n)} \cup \ldots \cup \ell_{N_n}^{(n)}$, the intervals being enumerated from right to left.

Note that $\ell_j^{(n)} = \ell_j^{(n+1)}$, $j = 1,\ldots, N_n - 1$.

Let $\Gamma_n \overset{def}{=\!=} [0,\alpha]\setminus\mathcal{O}_n$. The following inequalities are obvious:

$$\begin{aligned} &|F(x)| \leq 2\nu^{\frac{1}{2}}(x)\,, \quad x\in\Gamma_n\cap[2^{-n},\alpha] \\ &|\arg F(x)| \leq \frac{\pi}{2}\nu^{\frac{1}{2}}(x)\,,\quad x\in\mathcal{O}_n\,. \end{aligned} \tag{2.9}$$

Here $\arg$ stands for the principal value of the argument which varies from $-\pi$ to π. Let

$$\Psi_n(x) \overset{def}{=\!=} \begin{cases} \arg F(x) &, x\in\mathcal{O}_n \\ 0 &, x\in\mathbb{R}\setminus\mathcal{O}_n \end{cases}$$

---

*) We shall say that $\int_a^b \asymp \int_c^d$ if these integrals diverge simultaneously.

Thus from (2.9) we have

$$|\psi_n(x)| \leq \frac{\pi}{2}\nu^{\frac{1}{2}}(x). \tag{2.10}$$

Let $\tilde{\psi}_n$ denote the harmonic conjugate function

$$\tilde{\psi}_n(x) \stackrel{def}{=} \frac{1}{\pi}\int_{-\infty}^{\infty} \frac{\psi_n(t)}{x-t}\,dt .$$

Without loss of generality we may suppose that the function ψ_n is so small that $e^{2|\tilde{\psi}_n|} \in L^1_{loc}(\mathbb{R})$, $e^{\tilde{\psi}_n - i\psi_n} \in H^2(\frac{dt}{1+t^2})$.

We introduce the auxiliary function

$$\Phi_n \stackrel{def}{=} F \cdot e^{\tilde{\psi}_n - i\psi_n} .$$

The properties of Φ_n are the following :

1) $\Phi_n \in H^2(\frac{dt}{1+t^2})$; 2) the boundary values $\Phi_n(x), x \in \mathcal{O}_n$ are real. It means that there is an analytic continuation of the function Φ_n **across** $\mathcal{O}_n$ from $\mathbb{C}_+$ into the lower half plane $\mathbb{C}_-$. This new function will be called Φ_n also. We have

$$\Phi_n(\bar{\xi}) = \overline{\Phi_n(\xi)},$$

Φ_n is analytic in the **domain** $\mathcal{D}_n = \mathbb{C}_+ \cup \mathbb{C}_- \cup \mathcal{O}_n$. Let γ_n denote the boundary of $\mathcal{D}_n$, $\gamma_n \stackrel{def}{=} \partial\mathcal{D}_n$, $\gamma_n = \mathbb{R} \setminus \mathcal{O}_n$. By the reasoning similar to that of [2] we can choose a point b in the first (from the right) interval so that $|\Phi_n(b)| \geq \delta > 0$, where the number δ is independent of n. We should like to apply Jensen's inequality to the function Φ_n. That it can be done was shown in [1] and [2]. We write this inequality in the following form:

$$\sum_{\lambda\in\Lambda} G_{\mathcal{D}_n}(\lambda, b) + \int_{\gamma_n} \log^-|F(x)|\,d\omega_n(x,b) \leq \int_{\gamma_n} \log^+|F(x)|\,d\omega_n(x,b) +$$

$$+ \int_{\gamma_n} \tilde{\psi}_n(x)\,d\omega_n(x,b) + \log\frac{1}{|\Phi_n(b)|} .$$

But the first integral in the right hand part does not exceed the constant C (since F is bounded) and $\log\frac{1}{|\Phi_n(b)|} \leq \log\frac{1}{\delta}$.

So we have

$$\sum_{\lambda\in\Lambda} G_{\mathcal{D}_n}(\lambda,b)+\int_{\gamma_n}\log^-|F(x)|\,d\omega_n(x,b)\le C+\int_{\gamma_n}\tilde{\Psi}_n(x)\,d\omega_n(x,b). \tag{2.11}$$

Here $d\omega_n(\cdot,b)$ denotes the harmonic measure of the **domain** $\mathcal{D}_n$ corresponding to the point $b\in\mathcal{D}_n$ and $G_{\mathcal{D}_n}(\cdot,b)$ denotes the Green function of this **domain**

Now our aim is to prove that $J_n \overset{\text{def}}{=\!=} \int_{\gamma_n}\tilde{\Psi}_n\,d\omega_n(\cdot,b)$ is bounded independently of n :

$$\pi J_n=\int_{\gamma_n}\Big(\int_{O_n}\frac{\Psi_n(t)}{t-x}dt\Big)d\omega_n(x,b)=\int_{O_n}\Psi_n(t)\Big(\int_{\gamma_n}\frac{d\omega_n(x,b)}{x-t}\Big)dt=$$

$$=\int_{\ell_1}+\int_{O_n'}.$$

Here $O_n'\overset{\text{def}}{=\!=}O_n\setminus\ell_1$ and ℓ_1 is the first (if counted from the right) interval of the set O_n. Without loss of generality we may suppose that the first two intervals $\ell_1=(c_1,d_0)$ and $\ell_2=(c_2,d_1)$ of the set O_n become already stable, independent of n. We may suppose also that $|\Psi_n|\le 1$. Then

$$\int_{\ell_1}|\Psi_n(t)|\Big(\int_{\gamma_n}\frac{d\omega_n(x,b)}{|x-t|}\Big)dt\le\int_{c_1}^{d_0}\Big(\int_{\gamma_n\cap(-\infty,c_2]}\frac{d\omega_n(x,b)}{|x-t|}\Big)dt+\int_{c_1}^{d_0}\Big(\int_{d_1}^{c_1}\dots\Big)dt+$$

$$\int_{c_1}^{d_0}\Big(\int_{d_0}^{+\infty}\dots\Big)dt\le\frac{|d_0-c_1|}{|c_1-c_2|}+\int_{d_1}^{c_1}d\omega(x,b)\int_{c_1}^{d_0}\frac{dt}{|x-t|}+\int_{d_0}^{\infty}d\omega(x,b)\,\cdot$$

$$\cdot\int_{c_1}^{d_0}\frac{dt}{|x-t|}\le \text{const}+c\int_{d_1}^{c_1}\log\frac{1}{|x-c_1|}\,d\omega(x,b)+c\int_{d_0}^{\infty}\log\frac{c|x|}{|d_0-x|}\,d\omega(x,b)\le$$

$$\le C<\infty.$$

Here $d\omega(x,b)$ denotes the harmonic measure of the **domain** $\mathbb{C}\setminus([d_1,c_1]\cup[d_0,+\infty))$.

Thus it remains to estimate $J_n'\overset{\text{def}}{=\!=}\int_{O_n'}\Psi_n(t)\Big(\int_{\gamma_n}\frac{d\omega_n(x,b)}{x-t}\Big)dt.$

Now introduce the notation

$$G_{c,n}(t,b)\overset{\text{def}}{=\!=}\int_{\gamma_n}\log|t-x|\,d\omega_n(x,b).$$

It is clear that $G_{c,n}(\cdot,b)\in C^\infty(O_n')$, $G_{c,n}'(t,b)=-\int_{\gamma_n}\frac{d\omega_n(x,b)}{x-t}$;

$$G''_{c,n}(t,b) = -\int_{\gamma_n} \frac{d\omega_n(x,b)}{(x-t)^2} < 0 \tag{2.12}$$

for $t \in O'_n$.

Taking it into account we see that

$$J'_n = -\int_{O'_n} \Psi_n(t)\, G'_{c,n}(t,b)\, dt .$$

Now let us recall that $O'_n = \cup\, \ell_{x_0}$, $x_0 \in \mathcal{X}_0$ and $\ell_{x_0} = (x_0 - \frac{1}{2A}\nu^{1/2}(x_0),\, x_0 + \frac{1}{2A}\nu^{1/2}(x_0))$. We denote by L_{x_0} the interval with the centre x_0 and the length $5|\ell_{x_0}|$. It is a well-known fact that there is a sequence $(\ell_{x_j})_{j\geq 1}$, $x_j \in \mathcal{X}_0$ such that (ℓ_{x_j}) are disjoint and $O'_n \subset \bigcup_{j\geq 1} L_{x_j}$. If $L'_{x_j} \overset{def}{=\!=} L_{x_j} \cap O'_n$ then

$$|J'_n| \leq \sum_{j\geq 1} \int_{L'_{x_j}} |\Psi_n(t)| \cdot |G'_{c,n}(t,b)|\, dt \leq \sum_{j\geq 1} \sup_{L'_{x_j}} |\Psi_n(t)| \cdot \int_{L'_{x_j}} |G'_{c,n}|\, dt . \tag{2.13}$$

But the function $\nu^{1/2}$ is increasing and $|(\nu^{\frac{1}{2}})'| \leq A$. So we obtain that

$$\frac{2}{\pi} \sup_{L'_{x_j}} |\Psi_n(t)| \leq \sup_{L'_{x_j}} |\nu^{\frac{1}{2}}(t)| \leq \nu^{\frac{1}{2}}(x_j + \frac{5}{2A}\nu^{\frac{1}{2}}(x_j)) \leq$$
$$\leq \frac{7}{2}\nu^{1/2}(x_j) = \frac{7A}{2}|\ell_{x_j}| . \tag{2.14}$$

Now we take into account (2.12) and see that the function $G'_{c,n}(\cdot,b)$ decreases on every interval L'_{x_j}. Therefore it has at most one zero on this interval and so

$$\int_{L'_{x_j}} |G'_{c,n}(t,b)|\, dt \leq 4 \sup_{L'_{x_j}} |G_{c,n}(t,b)| . \tag{2.15}$$

From (2.13)-(2.15) we deduce that

$$|J'_n| \leq 7\pi A \sum_{j\geq 1} \ell_{x_j} \cdot \sup_{L'_{x_j}} |G_{c,n}(t,b)| . \tag{2.16}$$

But for $t \in O'_n$ we have

$$|G_{c,n}(t,b)| \leq |\log|t-b|| + G_{\mathcal{D}_n}(t,b) \leq c + G(t,b) \leq C . \tag{2.17}$$

Here $G(t,b)$ denotes the Green function of the domain $\mathbb{C} \setminus (\mathbb{R}_- \cup [d_0, \infty))$.

So from (2.16), (2.17) we deduce

$$|J_n'| \leq 7\pi A C \sum_{j\geq 1} l_{x_j} \leq 7\pi A C |O_n'| \leq 7\pi A C \alpha . \tag{2.18}$$

Now we rewrite the inequality (2.11) in the following form:

$$\int_{\mathbb{R}_-} \log^- |F(x)| d\omega_n(x,b) + \int_{\Gamma_n} \log^- |F(x)| d\omega_n(x,b) + \\ + \sum_{\lambda\in\Lambda} G_{\mathcal{D}_n}(\lambda, b) \leq C. \tag{2.19}$$

We note that the domains $\mathcal{D}_n$ increase and

$$\frac{d\omega_n(x,b)}{dx} \xrightarrow[n\to\infty]{} \frac{d\omega(x,b)}{dx}; \quad G_{\mathcal{D}_n}(\lambda,b) \xrightarrow[n\to\infty]{} G_{\mathcal{D}}(\lambda,b)$$

where $d\omega(\cdot,b)$ is the harmonic measure of the domain $\mathcal{D} = \bigcup_{n\geq 1} \mathcal{D}_n$ and $G_{\mathcal{D}}$ is its Green function. Therefore from (2.19) and Fatou's lemma we may deduce that

$$\int_{\mathbb{R}_-} \log^- |F(x)| d\omega(x,b) + \int_{\Gamma} \log^- |F(x)| d\omega(x,b) + \sum_{\lambda\in\Lambda} G_{\mathcal{D}}(\lambda,b) \leq C \tag{2.20}$$

where $\Gamma \overset{def}{=\!=} \partial\mathcal{D} \cap [0,\alpha]$.

It is clear from the construction that

$$|F(x)| \leq 2\nu^{1/2}(x) \leq 2\exp\left(-c\,\frac{\varepsilon(x)}{x}\right),$$

for $x\in\Gamma$ (see (2.9) and ivb)). Thus (2.20) implies

$$\int_0^\alpha \frac{\varepsilon(t)}{t}\, d\omega(t,b) \leq c < \infty . \tag{2.21}$$

LEMMA 3. Let the function ε be increasing, $\varepsilon(0)=0$,

$$\lim_{x\to 0} \frac{\varepsilon'(x)}{\varepsilon(x)}\, x = 0 \tag{A}$$

$$\int_0^\delta \frac{\varepsilon(x)}{x}\, dx = +\infty \tag{B}$$

and let an increasing on $[0,\delta]$ continuous function ω, $\omega(0)=0$, be such that

$$\int_\delta^\delta \frac{\varepsilon(x)}{x^\gamma}\, d\omega(x) < \infty \tag{C}$$

for some positive number γ. Then

$$\lim_{a\to 0} \frac{1}{a^{\gamma_1}} \int_0^a \frac{\omega(s)}{s^{\gamma_2}}\, ds = 0 \qquad (D)$$

for every pair of positive numbers γ_1, γ_2 satisfying $\gamma_1+\gamma_2 = \gamma+1$.

PROOF. Without loss of generality we may suppose that

$$\frac{\varepsilon'(x)}{\varepsilon(x)}\, x < \min\left(\frac{\gamma}{2}, \frac{\gamma_1}{2}, \frac{\gamma_2}{2}\right), \quad x \in (0,\delta).$$

So the function $\frac{\varepsilon(x)}{x^{\gamma}}$ decreases on $(0,\delta)$. Now from (C) we see that $\omega(x) = O\left(\frac{x^{\gamma}}{\varepsilon(x)}\right)$.

Applying (C) and the integration by parts combined with the facts mentioned above we see that

$$\int_0^{\delta} \frac{\varepsilon(x)\,\omega(x)}{x^{1+\gamma}}\, dx < \infty . \qquad (C')$$

It is clear that (A) implies: $\frac{1}{\varepsilon(t)} = O\left(\frac{1}{t^{\eta}}\right)$ for every $\eta > 0$, and therefore it is clear that the integral $\int_0 \frac{\omega(x)}{x^{\gamma_2}}\, dx$ converges. Let $W(x) \overset{def}{=} \int_0^x \frac{\omega(s)}{s^{\gamma_2}}\, ds$. Then

$$\int_0^t \frac{\varepsilon(x)}{x^{\gamma_1}}\, dW(x) = \int_0^t \frac{\varepsilon(x)\,\omega(x)}{x^{1+\gamma}}\, dx \le c < \infty$$

and so $W(x) = O\left(\frac{x^{\gamma_1}}{\varepsilon(x)}\right)$.

Another application of the integration by parts gives the chain of inequalities:

$$\infty > c \ge \int_t^{\delta} \frac{\omega(x)\,\varepsilon(x)}{x^{1+\gamma}}\, dx = \int_t^{\delta} \frac{\varepsilon(x)}{x^{\gamma_1}}\, dW(x) = \frac{\varepsilon(\delta)}{\delta^{\gamma_1}}\, W(\delta) -$$

$$- \frac{\varepsilon(t)}{t^{\gamma_1}}\, W(t) + \gamma_1 \int_t^{\delta} \frac{W(x)\,\varepsilon(x)}{x^{1+\gamma_1}} \left(1 - \frac{1}{\gamma_1}\, \frac{\varepsilon'(x)\,x}{\varepsilon(x)}\right) dx \ge -c +$$

$$+ \frac{\gamma_1}{2} \int_t^{\delta} \frac{\varepsilon(x)\,W(x)}{x^{1+\gamma_1}}\, dx .$$

Thus $\int_0 \frac{\varepsilon(x)\,W(x)}{x^{1+\gamma_1}}\, dx < \infty$. Now suppose that (D) is false. Then $W(s) \ge \eta\, s^{\gamma_1}$ for some positive number η, and we have

$$\eta \int_0^{\delta} \frac{\varepsilon(x)}{x}\, dx \le \int_0^{\delta} \frac{\varepsilon(x)\,W(x)}{x^{1+\gamma_1}}\, dx \le c < \infty .$$

This contradicts (B). ●

Let σ be a small positive number. Taking into account the inequality (2.21) and lemma 3 we may choose a number a so small that the following inequality holds:

$$\frac{1}{(2a)^{1/2}} \int_0^{2a} \frac{\omega_{\mathcal{D}}(x,b)}{x^{3/2}}\, dx \le \frac{\sigma}{3}$$

and therefore

$$\frac{1}{a^{1/2}} \int_0^{a} \frac{\omega_{\mathcal{D}}(x,b)}{x^{3/2}}\, dx \le \sigma \qquad \text{and} \quad \frac{\omega_{\mathcal{D}}(a,b)}{a} \le \sigma . \tag{2.22}$$

If we prove that

$$G_{\mathcal{D}}(\lambda, ia) \ge c(\operatorname{Im}\lambda)^{1/2} \tag{2.23}$$

for the points λ from the Stolz domain K_γ, then applying the inequality (2.20) we shall obtain that

$$\sum_{\lambda\in\Lambda} (\operatorname{Im}\lambda)^{1/2} < \infty$$

and so lemma 2 and the sufficiency of the first part of theorem 1 will be proved.

Now we prove (2.23). Let $S \overset{def}{=\!=} \mathcal{D} \cup (0,a)$, G_S be its Green function and let $S_1 = \frac{1}{a}\cdot S$ be the domain homothetic to S. It is clear that the following inequalities hold:

$$0 \le G_S(\lambda, ia) - G_{\mathcal{D}}(\lambda, ia) \le \int_0^a G_S(\lambda,t)\, d\omega_{\mathcal{D}}(t, ia) \le$$

$$\le \frac{c}{a}\int_0^1 G_S(\lambda,t)\, d\omega_{\mathcal{D}}(t,b) = \frac{c}{a}\int_0^1 G_1(\lambda_1, t_1)\, d\omega_{\mathcal{D}}(t,a,b) \le$$

$$\le \frac{c}{a}\int_0^1 G_2(\lambda_1, t_1)\, d\omega_{\mathcal{D}}(t,a,b) \overset{def}{=\!=} \mathcal{J}$$

where $\lambda_1 = \frac{\lambda}{a}$, the constant c not depending on a appears from the application of Harnack's inequality and the symbols G_1 and G_2 denote the Green functions of the regions S_1 and $S_2 \overset{def}{=\!=} \mathbb{C} \setminus \mathbb{R}_-$ respectively. Now if $G(\zeta, t) = \log\left|\frac{\zeta+t}{\zeta-t}\right|$ is the Green function of the right half-plane, we may rewrite the latter integral as

$$\mathcal{J} = \frac{c}{a}\int_0^1 G(\sqrt{\lambda_1}, t)\, d\omega_{\mathcal{D}}(t^2 a, b).$$

Integrating by parts we obtain ($\zeta \overset{def}{=} \sqrt{\lambda_1}$)

$$\mathcal{I} \leqslant \frac{c}{a}\int_0^1 \left|\frac{d}{dt} G(\zeta,t)\right| \cdot \omega_{\mathcal{D}}(t^2 a, b)\,dt + \frac{c}{a} G(\zeta,1)\,\omega_{\mathcal{D}}(a,b). \tag{2.24}$$

It is clear that

$$c_1 \leqslant \frac{Re\,\zeta}{\mathcal{I}m\,\zeta} \leqslant c_2 \quad , \quad 0 < c_1 < c_2 < \infty \tag{2.25}$$

since $\zeta = \sqrt{\frac{\lambda}{a}}$ and $\lambda \in K_\gamma$.

LEMMA 4. The condition (2.25) being satisfied, the following inequalities hold

$$\left|\frac{d}{dt} G(\zeta,t)\right| = O\left(\frac{|\zeta|}{t^2}\right); \qquad G(\zeta,1) = O(|\zeta|),$$

for $t \in (0,1)$.

PROOF. $\left|\frac{d}{dt} G(\zeta,t)\right| = 2|\zeta| \frac{\left||\zeta|^2 - t^2\right|}{|\zeta - t|^2 \cdot |\bar{\zeta} + t|^2} \leqslant \frac{c|\zeta|}{t^2}$.

From lemma 4 and (2.22), (2.24) we deduce that

$$\mathcal{I} \leqslant c|\zeta| \int_0^1 \frac{\omega_{\mathcal{D}}(t^2 a, b)}{t^2 a}\,dt + c|\zeta| \frac{\omega_{\mathcal{D}}(a,b)}{a} \leqslant$$

$$\leqslant \frac{1}{2} c|\zeta| \frac{1}{a^{1/2}} \int_0^a \frac{\omega_{\mathcal{D}}(s,b)}{s^{3/2}}\,ds + c|\zeta|\sigma \leqslant C\sigma|\zeta| = C\sigma\left|\frac{\lambda}{a}\right|^{1/2}.$$

Here the constant C is independent of a . Thus we have

$$0 \leqslant G_s(\lambda, ia) - G_{\mathcal{D}}(\lambda, ia) \leqslant 2C\sigma\left|\frac{\lambda}{a}\right|^{1/2}. \tag{2.26}$$

It is easy to see that $G_s(\lambda, ia) \geqslant c_1\left|\frac{\lambda}{a}\right|^{1/2}$ with an absolute constant c_1 . Indeed if $S_0 \overset{def}{=} \mathbb{C} \setminus (\mathbb{R}_- \cup [a, +\infty))$ then $G_s(\lambda, ia) \geqslant G_{s_0}(\lambda, ia) = G_{(1/a)s_0}(\frac{\lambda}{a}, i) \geqslant c_1\left|\frac{\lambda}{a}\right|^{1/2}$.
Now (2.26) implies (if we choose $\sigma < c_1/2C$)

$$G_{\mathcal{D}}(\lambda, ia) \geqslant \frac{c_1}{2}\left|\frac{\lambda}{a}\right|^{1/2}$$

and therefore (2.23) is proved. Thus the proof of lemma 2 is finished. ●

To prove the first part of theorem 1 it remains to show that if $\sum_{\lambda \in \Lambda} (\mathcal{I}m\,\lambda)^{1/2} < \infty$ then the family R_Λ is not complete in $L^2(h\,dm)$.

Recall that for some positive number α we have $h(t)= =\exp\left(-\frac{\varepsilon_1(|t|)}{\sqrt{|t|}}\right)$ for $t\in(-\alpha,0)$. Let φ be the conformal homeomorphism of the unit disc $\mathbb{D}$ onto the domain $G\overset{def}{=\!=} \mathbb{C}\setminus[-\alpha,0]$, $\psi\overset{def}{=\!=}\varphi^{-1}$. Let f be the outer function in $\mathbb{D}$ with the modulus

$$|f(\zeta)|=h(\varphi(\zeta)),\ \xi\in\mathbb{T}.$$

This function is correctly defined since

$$\int_{\mathbb{T}}\log|f(\zeta)||d\zeta|\asymp\int_{\partial G}\log|h(z)|\cdot|\psi'(z)||dz|\asymp-\int_0\frac{\varepsilon_1(t)}{t}dt>-\infty .$$

It is clear that the set $(\psi(\lambda))_{\lambda\in\Lambda}$ has the Blaschke property in $\mathbb{D}$, $\sum_{\lambda\in\Lambda}(1-|\psi(\lambda)|)<\infty$, and so there is a Blaschke product B whose zero-set is $(\psi(\lambda))_{\lambda\in\Lambda}$. Define the function $g: g=f\cdot B$. It is bounded and analytic in $\mathbb{D}$ and $g(\psi(\lambda))=0,\ \lambda\in\Lambda$. Therefore

$$0=2\pi i g(\psi(\lambda))=\int_{\mathbb{T}}\frac{g(\zeta)\,d\zeta}{\zeta-\psi(\lambda)}=\int_{\partial G}\frac{g(\psi(z))\,\psi'(z)\,dz}{\psi(z)-\psi(\lambda)}=$$

$$=\int_{\partial G}\frac{g(\psi(z))}{z-\lambda}dz+\int_{\partial G}g(\psi(z))\frac{\psi(\lambda)-\psi(z)-\psi'(z)(z-\lambda)}{(\psi(z)-\psi(\lambda))(\lambda-z)}dz .$$

It is clear that the last integral is equal to zero. Now if $q(t)\overset{def}{=\!=}\lim_{y\to0+}g(\psi(t+iy))-\lim_{y\to0-}g(\psi(t+iy))$ then

$$\int_{-\alpha}^{0}\frac{q(t)dt}{t-\lambda}=0,\quad \lambda\in\Lambda .$$

But from the definition of the functions g and q it is obvious that $|q(t)|\le 2h(t)$ and that $q\not\equiv0$. Now we may finish the proof.

Let

$$p(t)=\begin{cases}\dfrac{q(t)}{h(t)}, & t\in(-\alpha,0)\\ 0, & t\in\mathbb{R}\setminus(-\alpha,0).\end{cases}$$

Then $p\in L^2(h\,dm)$, $p\not\equiv0$, and $\int_{-\infty}^{\infty}\frac{p(t)h(t)dm}{t-\lambda}=0$ for every $\lambda \in \Lambda$. ●

3. The second step of the construction

Lemma 5 below contains the proof of the second part of theorem 1.

LEMMA 5. Let the conditions i)-iv) (see §2) for the functions F and ν be fulfilled and let

$$\int_0 \frac{\varepsilon(t)}{t}\,dt = +\infty \,, \tag{3.1}$$

$$\int_0 \frac{\varepsilon_1(t)}{t}\,dt = +\infty \,. \tag{3.2}$$

Then the zero-set Λ of the function F is finite.

PROOF. The proof of this lemma repeats very closely the scheme of the proof of lemma 2. We begin with the function

$$\varphi_n(\zeta) \overset{def}{=\!=} \Phi_n(\zeta^2) = F(\zeta^2)\, e^{\widetilde{\Psi}_n(\zeta^2) - i\Psi_n(\zeta^2)}$$

where the functions Φ_n, Ψ_n, $\widetilde{\Psi}_n$ were constructed in lemma 2. The function φ_n is analytic in the region $R_{n,+}$ obtained form the right half-plane by cutting out a finite number of closed intervals (one of them is infinite) of the positive half-axis $\mathbb{R}_+$. On the imaginary axis we have

$$\varphi_n(iy) = F(-y^2)\, e^{\widetilde{\Psi}_n(-y^2)}$$

and so it is clear that

$$c_1 |F(-y^2)| \le |\varphi_n(iy)| \le c_2 |F(-y^2)| \,, \tag{3.3}$$

where c_1 and c_2 do not depend on n. Following the scheme of lemma 2 we consider the set

$$\mathcal{Y}_0 = \{ iy_0 \in (-i\sqrt{\alpha}, i\sqrt{\alpha}) : |F(-y_0^2)| \ge 2\nu^{1/2}(-y_0^2) \} \,.$$

At first we note that $\mathcal{Y}_0 \ne \emptyset$. If it is not the case, then $F(x) \le 2\nu^{1/2}(x) \le 2\exp\left(-1/2\ \frac{\varepsilon_1(|x|)}{\sqrt{|x|}}\right)$ for $x \in (-\alpha, 0)$.
Applying the inequality (2.20) we obtain

$$\int_0 \frac{\varepsilon_1(|x|)}{\sqrt{|x|}}\, d\omega(x, b) < \infty \,,$$

where $d\omega$ is the harmonic measure of the **domain** $\mathcal{D}$ constructed in lemma 2. But from the properties of this **domain** one can deduce (see [2], p.76) that

$$\left|\frac{d\omega(x,b)}{dx}\right| \geq \frac{c}{\sqrt{|x|}}$$

for x sufficiently small. So we have

$$\int_0 \frac{\varepsilon_1(x)}{x}\, dx < \infty$$

and it contradicts (3.2). Now we proceed exactly as in the first step of the construction. Namely, we introduce the open subset Ω_n of the imaginary axis $i\mathbb{R}$,

$$\Omega_n = \bigcup_{\substack{iy_0 \in \mathcal{Y}_0 \\ |y_0| > 2^{-n}}} I_{y_0} ,$$

where $I_{y_0} \overset{def}{=} (iy_0 - \frac{i}{2A} v^{1/2}(-y_0^2),\ iy_0 + \frac{i}{2A} v^{1/2}(-y_0^2))$.
Let the function u_n be harmonic in the right half-plane with the boundary values

$$u_n(iy) = \begin{cases} \arg F(-y^2), & y \in \Omega_n \\ 0 & , y \in i\mathbb{R} \setminus \Omega_n . \end{cases}$$

Then $|u_n(iy)| \leq \frac{\pi}{2} v^{1/2}(-y^2)$. Let $\tilde{u}_n$ be the harmonic conjugate function. We introduce the auxiliary function

$$\varphi_{n,n}(\zeta) \overset{def}{=} \varphi_n(\zeta) e^{\tilde{u}_n(\zeta) - i u_n(\zeta)} .$$

It is clear that there is an analytic continuation of this function across Ω_n from the **domain** $R_{n,+}$ to the symmetric domain $R_{n,-}$. We denote this new function by $\varphi_{n,n}$ again. We note that

$$\varphi_{n,n}(\zeta) = \overline{\varphi_{n,n}(-\bar{\zeta})} .$$

The function $\varphi_{n,n}$ is analytic in the **domain** $R_n = R_{n,+} \cup R_{n,-} \cup \Omega_n$. Let ∂R_n be the boundary of R_n , then $\partial R_n = \Gamma_{n,1} \cup \Gamma_{n,2}$, where $\Gamma_{n,1} \overset{def}{=} \mathbb{R} \cap \partial R_n$, $\Gamma_{n,2} \overset{def}{=} i\mathbb{R} \cap \partial R_n$.
The application of the Jensen inequality (see [1], [2]) gives the following inequality which is similar to (2.11):

$$\int_{\Gamma_{n,2}} \log^- |F(-y^2)|\, d\omega_n(iy,b) + \int_{\Gamma_{n,1}} \log^- |F(x^2)|\, d\omega_n(x,b) +$$

$$\sum_{\lambda\in\Lambda} G_{R_n}(\lambda,b) \leq c + \int_{\Gamma_{n,2}} \tilde{\Psi}_n(-y^2)\, d\omega_n(iy,b) + \int_{\Gamma_{n,1}} \tilde{u}_n(|x|)\, d\omega_n(x,b) +$$

$$+ \int_{\Gamma_{n,1}} \tilde{\Psi}_n(x^2)\, d\omega_n(x,b) + \int_{\Gamma_{n,2}} \tilde{u}_n(iy)\, d\omega_n(iy,b) \overset{\text{def}}{=\!=} C + \tag{3.4}$$

$$+ \mathcal{J}_{n,1} + \mathcal{J}_{n,2} + \mathcal{J}_{n,3} + \mathcal{J}_{n,4} \ .$$

Here $d\omega_n(\cdot, b)$ denotes the harmonic measure of the **domain** R_n, corresponding to the point b , $b\in R_n$, b is independent of n. Our aim is to estimate the integral $\mathcal{J}_{n,i}$, $i=1,2,3,4$. The inequality (2.10) shows that

$$|\tilde{\Psi}_n(-y^2)| \leq \frac{1}{\pi}\int_0^\infty \frac{|\Psi_n(t)|}{t+y^2}\,dt \leq \frac{1}{2}\int_0^\alpha \frac{\nu^{1/2}(t)}{t}\,dt \leq c < \infty$$

since the function ν decreases so fast that $\int_0 \frac{\nu^{1/2}(t)}{t}\,dt < \infty$. In a similar way we can obtain that $|\tilde{u}_n(|x|)| \leq C$. Therefore $\mathcal{J}_{n,1} \leq C$, $\mathcal{J}_{n,2} \leq C$, where C does not depend on n.

To estimate $\mathcal{J}_{n,3}$ we note first that

$$\tilde{\Psi}_n(x^2) = \frac{1}{\pi}\int_0^\infty \frac{\Psi_n(t)}{x^2-t} = \frac{1}{\pi}\int_{-\infty}^\infty \frac{\tau_n(t)}{x-t}\,dt = \tilde{\tau}_n(x),$$

where

$$\tau_n(t) = \begin{cases} \Psi_n(t^2), & t>0 \\ -\Psi_n(t^2), & t<0 \ . \end{cases}$$

It is clear that $\operatorname{supp} \tau_n \subset \operatorname{clos} \mathcal{O}_n^* \subset [-\sqrt{\alpha}, \sqrt{\alpha}\,]$, where $\mathcal{O}_n^* \overset{\text{def}}{=\!=} \{x\in\mathbb{R} : x^2\in \mathcal{O}_n\}$.
Thus

$$\pi\cdot \mathcal{J}_{n,3} = \int_{\mathcal{O}_n^*} \tau_n(t) \Big(\int_{\Gamma_{n,1}} \frac{d\omega_n(x,b)}{x-t} \Big)\, dt \ .$$

We see that this integral is similar to the integral J_n , estimated in § 2. So we may reduce the estimate of this integral to the estimate of the expression (see (2.16)):

$$\mathcal{J}'_{n,3} = \sum_{j\geq 1} |I_{x_j}| \cdot \sup_{\mathcal{J}'_{x_j}} |G_{c,n}(t,b)| \ .$$

Here $(I_{x_j})_{j\geq 1}$ is a sequence of disjoint intervals, $I_{x_j}\subset \mathcal{O}_n^*$; $j=1,2,\dots$. J_{x_j} is the interval with the same centre x_j and $|J_{x_j}|=5|I_{x_j}|$. $J'_{x_j}\overset{def}{=\!=} J_{x_j}\cap \mathcal{O}_n^*$.

$$G_{c,n}(t,b)\overset{def}{=\!=}\int_{\Gamma_{n,1}}\log|x-t|\,d\omega_n(x,b).$$

It is necessary to note that $|I_{x_j}|<|x_j|$. Now we have the obvious estimate from above for $G_{c,n}(\cdot,b)$:

$$G_{c,n}(t,b)\leq|\log|t-b||+\int_{\Gamma_{n,2}}|\log|iy-t||\,d\omega_n(iy,b)+$$
$$+G_{R_n}(t,b)\leq c+\log\frac{1}{|t|}$$

for $t\in[-\sqrt{\alpha},\sqrt{\alpha}]$.

Now we have

$$J'_{n,3}\leq c+\sum_{j\geq 1}|I_{x_j}|\log\frac{1}{|x_j|}\leq c+c\int_{-\sqrt{\alpha}}^{\sqrt{\alpha}}\log\frac{1}{|t|}\,dt\leq c<\infty .$$

The estimate of $J_{n,4}$ is quite similar to that above and so $J_{n,3}\leq c$, $J_{n,4}\leq c$. Therefore from (3.4) we deduce

$$\int_{\Gamma_{n,2}}\log^-|F(-y^2)|\,d\omega_n(iy,b)+\int_{\Gamma_{n,1}}\log^-|F(x^2)|\,d\omega_n(x,b)+$$
$$+\sum_{\lambda\in\Lambda}G_{R_n}(\lambda,b)\leq c .$$

The application of Fatou's lemma gives

$$\int_{\Gamma_2}\log^-|F(-y^2)|\,d\omega(iy,b)+\int_{\Gamma_1}\log^-|F(x^2)|\,d\omega(x,b)+ \qquad (3.5)$$
$$+\sum_{\lambda\in\Lambda}G_R(\lambda,b)\leq c .$$

Here $R=\bigcup R_n$, $\Gamma_2=\partial R\cap i\mathbb{R}$, $\Gamma_1=\partial R\cap\mathbb{R}$, G_R denotes the Green function of the domain R and $d\omega(\cdot,b)$ is its harmonic measure. From the construction of this region it is easily seen that

$$|F(x^2)|\leq 2v^{1/2}(x^2)\leq 2\exp\left(-\frac{1}{2}\frac{\varepsilon(x^2)}{x^2}\right),\quad x\in[-\sqrt{\alpha},\sqrt{\alpha}];$$

$$|F(-y^2)|\leq 2v^{1/2}(-y^2)\leq 2\exp\left(-\frac{1}{2}\frac{\varepsilon_1(y^2)}{y}\right),\quad y\in[-\sqrt{\alpha},\sqrt{\alpha}].$$

Thus the inequality (3.5) is equivalent to the system of the inequalities ($\beta \overset{def}{=\!=} \sqrt{\alpha}$):

$$\int_{-\beta}^{\beta} \frac{\varepsilon^*(x)}{x^2}\, d\omega_R(x,b) < \infty \tag{3.6a}$$

$$\int_{-\beta}^{\beta} \frac{\varepsilon_1^*(y)}{y}\, d\omega_R(iy,b) < \infty \tag{3.6b}$$

$$\sum_{\lambda\in\Lambda} G_R(\lambda,b) < \infty\ . \tag{3.6c}$$

Here $\varepsilon^*(x) \overset{def}{=\!=} \varepsilon(x^2)$, $\varepsilon_1^*(x) \overset{def}{=\!=} \varepsilon_1(x^2)$. It is clear that the functions ε^*, ε_1^* have the properties (A) and (B) of lemma 3.

We want to prove that the set Λ is finite. It is clear that the set is symmetric with respect to the real and the imaginary axes. So it is enough to prove that the intersection $\Lambda \cap \{x>0\ ,\ y<0\}$ is finite. To do this it is enough to establish that if $\lambda\in\Lambda\cap\{x>0,\ y<0\}$ then $G(\lambda,d) \geqslant c$ for a fixed number $d\in\mathbb{R}$ and some positive number c not depending on n .

Let σ be a very small positive number. The application of lemma 3 gives a number a , $a>0$, so that

$$\frac{1}{2a}\int_{-2a}^{2a} \frac{\omega_R(t,b)}{t^2}\, dt < \frac{\sigma}{4}\ .$$

Therefore

$$\frac{1}{a}\int_{-a}^{a} \frac{\omega_R(t,b)}{t^2}\, dt < \sigma,\quad \frac{\omega_R(-a,b)}{a^2} < \sigma. \tag{3.7}$$

Let $S \overset{def}{=\!=} R\cup(-a,0)$. It is clear that the following chain of inequalities holds:

$$0 \leqslant G_S(\lambda,b) - G_R(\lambda,b) = \int_{-a}^{0} G_S(\lambda,t)\, d\omega_R(t,b) \leqslant$$

$$\leqslant \frac{c}{a^2}\int_{-a}^{0} \frac{a}{|t|}\, G_S(\lambda,b)\, d\omega_R(t,b)\ .$$

To obtain the last inequality one must apply Harnack's inequaluty to the function $G_S(\lambda,\cdot)$ positive and harmonic in $\{Re\,\zeta<0\}\setminus(-\infty,-a]$ (remind that $Re\,\lambda>0$)

Thus we have

$$0 \leqslant G_S(\lambda,b) - G_R(\lambda,b) \leqslant \frac{c}{a}\int_{-a}^{0} \frac{d\omega_R(t,b)}{t}\cdot G_S(\lambda,b) =$$

$$c\left(\frac{\omega_R(-a,b)}{a^2}+\frac{1}{a}\int_{-a}^{0}\frac{\omega_R(t,b)}{t^2}dt\right)\cdot G_S(\lambda,b)\leq 2C\sigma G_S(\lambda,b).$$

The last inequality follows from (3.7). Now the choice of σ $\sigma<\frac{1}{4C}$, gives

$$G_R(\lambda,b)\geqslant\frac{1}{2}G_S(\lambda,b). \tag{3.8}$$

The similar reasoning shows that

$$\frac{d\omega_S(x,b)}{|dx|}\asymp\frac{d\omega_R(x,b)}{|dx|}. \tag{3.9}$$

Another application of lemma 3 shows that

$$\lim_{\eta\to 0}\frac{1}{\eta}\int_0^{\eta}\frac{\omega_R^*(t,b)}{t}dt=0, \tag{3.10}$$

where $\omega_R^*(t,b)$ denotes the harmonic measure of $\partial R\cap\{\zeta:|\zeta|<t\}$. Taking (3.9), (3.10) into account we see that for an arbitrary small positive number σ there is a positive number η such that

$$\frac{1}{\eta}\int_0^{2\eta}\frac{\omega_S^*(t,b)}{t}dt<\sigma. \tag{3.11}$$

Now we estimate $G_S(\lambda,-\eta)$ from below (we may suppose that $\eta<\frac{a}{2}$ and so $-\eta\in S$):

$$\begin{aligned}G_S(\lambda,-\eta)&=\log\frac{1}{|\lambda+\eta|}+\int_{\partial S}\log|t-\lambda|\,d\omega_S(t,-\eta)=\\&=\log\frac{1}{|\lambda+\eta|}+J_0+J_1+J_2+J_3+I_0+I_1+I_2+I_3,\end{aligned} \tag{3.12}$$

where

$$J_i=\int_{\gamma_i\cap\{t:|t|>2\eta\}};\quad I_i=\int_{\gamma_i\cap\{t:|t|\leq 2\eta\}};$$

$$\gamma_0=\Gamma_2\cap i\mathbb{R}_+,\quad \gamma_2=\Gamma_2\cap i\mathbb{R}_-,\quad \gamma_1=\Gamma_1\cap\mathbb{R}_+,\quad \gamma_3=\Gamma_1\cap\mathbb{R}_-$$

It is clear that $I_3=0$.

We may suppose that $|\lambda|$ is small enough and so

$$J_i\geqslant\log\frac{3}{2}\eta\cdot\omega(\gamma_i\cap\{t:|t|>2\eta\},-\eta). \tag{3.13}$$

The integration by parts gives

$$I_i \geq \log \tfrac{3}{2}\eta \cdot \omega(\gamma_i \cap \{t : |t| < 2\eta\}, -\eta) - \\ - c \int_{\gamma_i \cap \{t:|t|<2\eta\}} \frac{\omega_s(t,-\eta)}{|t-\lambda|}\, dt \, .$$

The last integral does not exceed $c\int_0^{2\eta} \frac{\omega_s^*(t,-\eta)}{t}\, dt$ since the point λ lies in some Stolz domain. Thus, after another application of Harnack's inequality we have

$$I_i \geq \log \tfrac{3}{2}\eta \cdot \omega(\gamma_i \cap \{|t| < 2\eta\}, -\eta) - \frac{c}{\eta}\int_0^{2\eta} \frac{\omega_s^*(t,b)}{t}\, dt \, . \tag{3.14}$$

Now from (3.11)-(3.14) we deduce that

$$G_s(\lambda,-\eta) \geq \log \frac{1}{|\lambda+\eta|} + \log \tfrac{3}{2}\eta - 3C\sigma \, . \tag{3.15}$$

Now we choose σ so small that $\sigma < \frac{\log \frac{4}{3}}{6C}$ and note that for λ small enough $\log \frac{1}{|\lambda+\eta|} \geq \log \frac{8}{9\eta}$. So from (3.15) we have

$$G_s(\lambda,-\eta) \geq \tfrac{1}{2} \log \tfrac{4}{3} \, .$$

Applying Harnack's inequality once more and taking (3.8) into account we obtain that

$$G_R(\lambda, b) \geq c$$

for some constant c independent of λ. Thus lemma 5 and the second part of theorem 1 are proved. ●

4. The proof of theorem 1 in a particular case

In this paragraph we consider o n l y the w e i g h t s h with the following properties: a) $h \in L^1(\mathbb{R}) \cap L^\infty(\mathbb{R})$; b) $h(x) = \{ v(\frac{1}{|x|}) \}^{-2}$, $|x| \leq 1$, where the function v is logarithmically convex and rapidly increasing*) ; c) the function v is of the form $v(x) = exp(\frac{x}{g(x)})$, where

$$\lim_{x\to\infty} \frac{g'(x)\,x}{g(x)} = 0 \quad \text{and} \quad \frac{g(x)}{g(\frac{x}{g(x)})} \leq C \tag{*}$$

*) The properties of logarithmically convex, rapidly increasing functions may be found in [14] .

THEOREM 1'. Let the function h satisfy condition a)-c) and let

$$\int_0^1 \log h(x)\,dx = -2\int_1^{\infty} \frac{\log r(x)}{x^2}\,dx = -\infty . \tag{4.1}$$

Then the family $R_\Lambda = \{\frac{1}{z-\lambda}\}_{\lambda\in\Lambda}$ of rational fractions with the set of poles Λ in some Stolz domain is dense in $L^2(h\,dm)$ provided that the set Λ is infinite.

A sequence of moments corresponds to the function r :

$$R_n \overset{def}{=} \sup_{x>0} \frac{x^n}{r(x)} , \quad n = 1, 2, \ldots .$$

We introduce two functions constructed with the help of the sequence $(R_n)_{n\geq 1}$

$$T(x) \overset{def}{=} \sup_{n>0} \frac{x^n}{n!\,R_n} , \qquad R(x) \overset{def}{=} \sup_{n>0} \frac{x^n}{\sqrt{n!\,R_n}} .$$

In terms of the "moment sequence" the condition (4.1) may be reformulated in the following form

$$\sum_{n=1}^{\infty} \frac{R_n}{R_{n+1}} = \infty .$$

In what follows we shall need the fact that (4.1) implies

$$\int_1^{\infty} \frac{\log R(x)}{x^2}\,dx = \infty . \tag{4.2}$$

If it is not the case, then we have

$$\sum_{n\geq 1} \frac{1}{Q_n} < \infty , \quad \text{where } Q_n \overset{def}{=} \frac{\sqrt{(n+1)!\,R_{n+1}}}{\sqrt{n!\,R_n}} = \sqrt{n+1}\sqrt{\frac{R_{n+1}}{R_n}} .$$

This implies that $\frac{1}{Q_n} = O(\frac{1}{n})$, since $(\frac{1}{Q_n})_{n\geq 1}$ is a decreasing sequence. Therefore, we have

$$\sum_{n\geq 1} \frac{R_n}{R_{n+1}} = \sum_{n\geq 1} (n+1)\left(\frac{1}{Q_n}\right)^2 \leq c \sum_{n\geq 1} \frac{1}{Q_n} < \infty$$

and this contradicts (4.1). Thus (4.2) is proved.

PROOF OF THEOREM 1' Let the set Λ be infinite, lie in some Stolz domain K_γ and let the family R_Λ be not dense in $L^2(h\,dm)$. Then there is a function P, $P\not\equiv 0$, $p\in L^2(h\,dm)$ such that

$$\int_{\mathbb{R}} \frac{p(t)h(t)dm}{t-\lambda}=0,\quad \lambda\in\Lambda .$$

We have the analytic in $\mathbb{C}_+$ function $F(\zeta)\overset{def}{=\!=}\int_{\mathbb{R}}\frac{p(t)h(t)dm}{t-\zeta}$ which is not identically zero (see the beginning of §2) and whose zero-set contains the set Λ. Moreover, F is in the Hardy class H^2 in $\mathbb{C}_+$. It is clear that this function is infinitely differentiable in the closure of every Stolz domain K_γ, $0<\gamma<\infty$. We see also that

$$\lim_{\zeta\in K_\gamma,\,\zeta\to 0} F^{(n)}(\zeta)=0 \tag{4.3}$$

since the infinitely differentiable function F has infinitely many zeroes in K_γ.

Let

$$\Gamma_\alpha \overset{def}{=\!=}\{\zeta=x+iy\in\mathbb{C}_+ : y=tg\,\alpha\cdot x\},$$

$$P_\alpha\overset{def}{=\!=}\{\zeta\in\mathbb{C} : \mathcal{I}m(-\zeta e^{i\alpha})\geqslant 1\},\quad \alpha\in(0,\pi).$$

If $\zeta\in\Gamma_\alpha$, $\alpha\in(0,\frac{\pi}{4})\cup(\frac{3\pi}{4},\pi)$, we have the following estimate:

$$|F^{(n)}(\zeta)|\leqslant cn!\left[\sup_{x>0}\frac{\left(\frac{x}{\sin\alpha}\right)^n}{r(x)}+\frac{c^n}{(\sin\alpha)^n}\right]\leqslant\frac{cn!}{(\sin\alpha)^n}R_n . \tag{4.4}$$

We introduce a function Φ_α analytic and bounded by a constant independent of α in the half-plane P_α

$$\Phi_\alpha(w)\overset{def}{=\!=}\int_{\Gamma_\alpha}F(\zeta)e^{-i\zeta w}d\zeta .$$

From (4.4) we deduce that for $\alpha\in(0,\frac{\pi}{4})\cup(\frac{3\pi}{4},\pi)$ and for $w\in\partial P_\alpha$

$$\begin{aligned}|\Phi_\alpha(w)|&\leqslant\inf_n\frac{\sup_{\zeta\in\Gamma_\alpha}|F^{(n)}(\zeta)|}{|w|^n}\leqslant c\inf_n\frac{n!\,R_n}{(\sin\alpha)^n|w|^n}=\\&=c\frac{1}{\sup\frac{(|w|\sin\alpha)^n}{n!\,R_n}}=\frac{c}{T(|w|\sin\alpha)}.\end{aligned}\tag{4.5}$$

It is clear that $\Phi_{\alpha_1}(w) = \Phi_{\alpha_2}(w)$ for $w \in P_{\alpha_1} \cap P_{\alpha_2}$ and so we have a function Φ analytic and bounded in the domain $R \overset{def}{=} \bigcup_{0<\alpha<\pi} P_\alpha = \mathbb{C} \setminus (\operatorname{clos} D \cup \{w = u+iv : u \geq 0, |v| \leq 1\})$. It is easily seen that for $w = u+iv \in \partial P_\alpha$ where $u > 0$ and $|v| > 2$ we have

$$|w| \sin\alpha \geq c\,|\operatorname{Im} w|. \tag{4.6}$$

Now let K be a positive continuous increasing function defined for $x \geq -2$, $K(-2) = 0$, so that the curve $\gamma = \{\zeta = x+iy : y = K(x)\}$ lies in the domain R. Suppose also that

$$\frac{K(x)}{x}\downarrow, \ \lim_{x\to\infty} K(x) = \infty, \ \lim_{x\to\infty} \frac{K(x)}{x} = 0, \ \int^{\infty} \frac{K(x)}{x^2}\,dx < \infty. \quad (**)$$

The function Φ constructed above is analytic and bounded in the domain $\mathcal{D} \overset{def}{=} \mathbb{C} \setminus \{\zeta = x+iy : |y| \leq K(x)\}$. Moreover, from (4.5),(4.6) we deduce that

$$|\Phi(\zeta)| \leq \frac{c_1}{T(c_2|\operatorname{Im}\zeta|)} \tag{4.7}$$

for $\zeta \in \mathcal{D}$, $|\zeta| \geq |\zeta_0|$.

In order to prove that $\Phi \equiv 0$ we consider the conformal homeomorphism $\zeta(\omega)$ of the domain $\Pi \overset{def}{=} \{\omega = \sigma + it : |t| < \frac{\pi}{2}\}$ onto the domain $\Omega \overset{def}{=} \{\zeta = \xi + i\eta : |\eta| < \pi - \frac{2K(e^\xi)}{e^\xi}\}$ and we prove that the function $g(\omega) \overset{def}{=} \Phi(e^{\zeta(\omega)})$ is identically zero. Let $\theta(\xi) \overset{def}{=} 2\pi - 4\frac{K(e^\xi)}{e^\xi}$.

By the well-known inequalities due to S.Warshawski [15] we have

$$\sigma_2 - \sigma_1 \leq \pi \int_{\xi_1}^{\xi_2} \frac{d\xi}{\theta(\xi)} + \frac{\pi}{12}\int_{\xi_1}^{\xi_2} \frac{\theta'^2(\xi)}{\theta(\xi)}\,d\xi + O(1). \tag{4.8}$$

The function $\xi \mapsto K(e^\xi)$ increases and $\xi \mapsto \frac{K(e^\xi)}{e^\xi}$ decreases, and so $0 \leq \frac{K'(e^\xi)e^\xi}{K(e^\xi)} \leq 1$. Taking it into account we see that

$$\theta'(\xi) \leq \frac{K(e^\xi)}{e^\xi} \leq C. \tag{4.9}$$

Then for ξ_1 large enough we have

$$\int_{\xi_1}^{\xi_2} \frac{\theta'^2(\xi)}{\theta(\xi)}\,d\xi \leq c \cdot \sup_{[\xi_1,\xi_2]} \theta'(\xi) \cdot \int_{\xi_1}^{\xi_2} \frac{K(e^\xi)}{e^\xi}\,d\xi \leq c$$

since (4.9) and $(**)$ hold. From the last inequality and from (4.8) we deduce that

$$\sigma \leq \pi \int_{\xi_1}^{\xi} \frac{d\xi}{\theta(\xi)} + C \leq \frac{\xi}{2} + O\Big(\int_{\xi_1}^{\xi} \frac{K(e^{\xi})}{e^{\xi}} d\xi\Big) \leq \frac{\xi}{2} + C. \tag{4.10}$$

Thus we obtain taking into account (4.7) and (4.10) that for $\omega \in \left[n \pm \frac{i\pi}{2}, n+1 \pm \frac{i\pi}{2}\right]$ we have the following estimate for the function $g(\omega)$:

$$\log \frac{1}{|g(\omega)|} \geq \log T(c_2 K(c_3 e^{2n})) + c_4 .$$

Thus we conclude that

$$\int_1^{\infty} \log \frac{1}{|g(\sigma + \frac{i\pi}{2})|} \frac{d\sigma}{ch\, \sigma} \geq C_1 + C_2 \int_1^{\infty} \frac{\log T(c_3 K(y))}{y^{3/2}} dy. \tag{4.11}$$

Suppose now that we know how to construct the function φ behaving so that: 1) φ is a smooth and increasing function; 2) $\frac{\varphi(x)}{x}$ decreases, $\lim_{x\to\infty} \frac{\varphi(x)}{x} = 0$; 3) $\int_1^{\infty} \frac{\varphi^2(x)}{x^3} dx < \infty$; 4) $\int_1^{\infty} \frac{\log R|\varphi(x)|}{x^2} dx = \infty$. Granted this our theorem follows easily. Indeed if we choose $K(x) \overset{def}{=} \frac{\varphi^2(\sqrt{x})}{c_3}$, then $K(x)$ satisfies $(**)$ and

$$\int_1^{\infty} \frac{\log T(c_3 K(x))}{x^{3/2}} dx = 2 \int_1^{\infty} \frac{\log R[\varphi(x)]}{x^2} dx = \infty .$$

So from (4.11) and Jensen's inequality we deduce that $g \equiv 0$. Then the function Φ is identically zero also and therefore $F \equiv 0$. This is a contradiction. But the function $\varphi(x) \overset{def}{=} \frac{x}{\sqrt{g(x)} \int_1^x \frac{dt}{tg(t)}}$ obviously satisfies 1) and 2). Indeed, using $(*)$ it is easy to deduce that $R(x) \geq exp\Big(\frac{c_1 x}{\sqrt{g_2(c_2 x)}}\Big)$. Now set $f(x) = \int_1^x \frac{dt}{tg(t)}$. Then (4.1) shows that $\lim_{x\to\infty} f(x) = \infty$. Now it is clear that $\int^{\infty} \frac{\varphi^2(x)}{x^3} dx \asymp \int^{\infty} \frac{df}{f^2} < \infty$ and, taking into account that $\varphi(x) \leq x$ we obtain

$$\int_1^{\infty} \frac{\log R[\varphi(x)]}{x^2} dx \geq c_1 \int_1^{\infty} \frac{\varphi(x)}{\sqrt{g_2(c_2 \varphi(x))}} \frac{dx}{x^2} \geq$$

$$c_1\int_1^\infty \frac{dx}{x\sqrt{g(x)}\, f(x)\sqrt{g(c_2x)}} \geqslant c\int_1^\infty \frac{df(x)}{f(x)} = \infty . \qquad \bullet$$

PART II

5. The proof of theorem 2

The proof of theorem 2 is based on two lemmas due to L.Carleson [16] and S.-Y.Chang, J.Garnett [17].

LEMMA I (L.Carleson [16]). Let h be a function of norm 1 in H^∞. For ε positive and sufficiently small there is a number $\sigma(\varepsilon)$, $\lim\limits_{\varepsilon\to 0}\sigma(\varepsilon)=0$, and a system Γ of closed rectifiable curves $(\Gamma_i)_{i\geq 1}$ in $\operatorname{clos}\mathbb{D}$, with disjoint interiors, with the following properties:

i) $\{\zeta\in\mathbb{D}: |h(\zeta)|<\varepsilon\}\subset \bigcup\limits_{i=1}^{\infty} \operatorname{int}(\Gamma_i)$;

ii) for $\zeta\in\Gamma\cap\mathbb{D}$, $\varepsilon\leq|h(\zeta)|\leq\sigma(\varepsilon)$

iii) arc length measure on $\Gamma\cap\mathbb{D}$ is a Carleson measure with a Carleson constant $C(\varepsilon)$ which depends only on ε *).

LEMMA II (S.-Y.Chang, J.Garnett [17]). Let Γ be a contour constructed in the previous lemma for an inner function θ and a positive number ε. Then for every function f, $f\in H^1$,

$$\int_{\mathbb{T}} f\,\bar\theta\, d\zeta = \int_{\Gamma}\frac{f}{\theta}\,d\zeta$$

a) $\Longrightarrow$ b). We suppose that the norms $\|\cdot\|_{L^p(m)}$ and $\|\cdot\|_{L^p(w\,dm)}$ are equivalent on the space K_θ. Therefore, taking into account that $k_\theta(a,\xi)\overset{\text{def}}{=\!=}\dfrac{1-\bar\theta(a)\theta(\zeta)}{1-\bar a\zeta}\in K_\theta$ for every $a, |a|<1$ we have

$$c\int_{\mathbb{T}}\left|\frac{1-\bar\theta(a)\theta(\zeta)}{1-\bar a\zeta}\right|^p dm(\zeta)\leq\int_{\mathbb{T}}\left|\frac{1-\bar\theta(a)\theta(\zeta)}{1-\bar a\zeta}\right|^p w(\zeta)\,dm . \qquad (5.1)$$

Now we estimate both parts of this inequality.

---

*) A Carleson measure μ is a measure on $\mathbb{D}$ such that $\|f\|_{L^1(\mu)}\leq c\|f\|_{H^1}$ for every function f in the Hardy class H^1. The Carleson constant of this measure is the best constant in this inequality.

$$\int_{\mathbb{T}} |k_\theta(a,\zeta)|^p dm(\zeta) = \frac{1}{1-|a|^2} \int_{\mathbb{T}} |1-\bar{\theta}(a)\theta(\zeta)|^p |1-\bar{a}\xi|^{2-p} \frac{1-|a|^2}{|1-\bar{a}\xi|^2} dm \geqslant$$

$$\geqslant \frac{1}{1-|a|^2} \left| \int_{\mathbb{T}} (1-\bar{\theta}(a)\theta(\zeta))^p (1-\bar{a}\xi)^{2-p} \frac{1-|a|^2}{|1-\bar{a}\xi|^2} dm \right| =$$

$$= \frac{1}{1-|a|^2} \; \frac{(1-|\theta(a)|^2)^p}{(1-|a|^2)^{p-2}} = \frac{(1-|\theta(a)|^2)^p}{(1-|a|^2)^{p-1}} \; .$$

To estimate the right hand part we choose a number ε, $0<\varepsilon<\min(2,2p-2)$:

$$\int_{\mathbb{T}} |k_\theta(a,\zeta)|^p w(\zeta) dm(\zeta) \leqslant 2^p \int_{\mathbb{T}} w(\xi) \frac{1}{|1-\bar{a}\zeta|^\varepsilon} \cdot \frac{1}{|1-\bar{a}\zeta|^{p-\varepsilon}} dm \leqslant$$

$$2^p \left(\int_{\mathbb{T}} w^{2/\varepsilon}(\xi) \frac{1}{|1-\bar{a}\xi|^2} dm \right)^{\varepsilon/2} \cdot \left(\int_{\mathbb{T}} \frac{dm}{|1-\bar{a}\zeta|^{(p-\varepsilon)\cdot\frac{2}{2-\varepsilon}}} \right)^{\frac{2-\varepsilon}{2}} \leqslant$$

$$c(p) \cdot \frac{1}{(1-|a|^2)^{\varepsilon/2}} \cdot \left(\int_{\mathbb{T}} w(\zeta) \frac{1-|a|^2}{|1-\bar{a}\zeta|^2} dm \right)^{\varepsilon/2} \cdot \frac{1}{(1-|a|^2)^{p-1-\varepsilon/2}} =$$

$$= c(\hat{w}(a))^{\varepsilon/2} \cdot \frac{1}{(1-|a|^2)^{p-1}} \; .$$

Thus from (5.1) we deduce that

$c\hat{w}(a) \geqslant (1-|\theta(a)|^2)^{\frac{2p}{\varepsilon}}$ and so b) holds

b)$\Longrightarrow$c). It is obvious.

c)$\Longrightarrow$a). Let $\hat{w}(\zeta)+|\theta(\xi)| \geqslant \eta > 0$, for every $\zeta\in\mathbb{D}$. Let $E \overset{def}{=\!=} \{e^{it} : w(e^{it}) > \eta/2\}$. If ω_E denotes the harmonic measure of the set E then it is clear that

$$\omega_E(\zeta) + |\theta(\zeta)| \geqslant \sigma > 0 \,, \quad \zeta\in\mathbb{D} \,. \tag{5.2}$$

We shall prove, that for every $f\in K_\theta$

$$\|f\|_{L^2(m)} \leqslant c\|f\|_{L^2(E)} \overset{def}{=\!=} c\left(\int_E |f|^2 dm\right)^{1/2} . \tag{5.3}$$

It is obvious that this implies a).

Suppose first that we know how to construct the function u with the following properties: 1) $u-\mathbb{1}\in\theta H^\infty$; 2) $\|u\|_{L^\infty(E')} \leqslant \sigma$ where $E' \overset{def}{=\!=} \mathbb{T}\setminus E$ and σ is a small positive number. Granted this it is easy to prove (5.3). Indeed 1) implies that $T_{\bar{u}}|K_\theta = I|K_\theta$

where I stands for the identity operator. Thus for every $f \in K_\theta$ we have

$$\|f\|_p^p = \|T_{\bar{u}} f\|_p^p \le C_p \|\bar{u} f\|_p^p = c_p \int_E |u|^p |f|^p + c_p \sigma^{-p} \int_{E'} |f|^p dm .$$

Now if we choose σ so small that $\sigma^p c_p < \frac{1}{2}$ then

$$\|f\|_p^p \le 2 C_p \|u\|_\infty^p \int_E |f|^p dm$$

and (5.3) holds.
Our aim is to prove that

$$dist_{L^\infty(E')} (\bar{\theta}, H^\infty) = 0 . \tag{5.4}$$

Granted this we may find for an arbitrary σ, $\sigma > 0$, a function h, $h \in H^\infty$ so that $\|\bar{\theta} + h\|_{L^\infty(E')} < \sigma$. Then for $u = 1 + \theta h$ the properties 1) and 2) are obviously fulfilled.
It remains to prove (5.4). Fix a number ε so small that $\sigma(\varepsilon) < \delta/2$ (σ is the function from lemma I and δ is the same as in the inequality (5.2)). An application of lemma I gives the contour Γ on which

$$\varepsilon \le |\theta| \le \sigma(\varepsilon) < \delta/2 .$$

Now from (5.2) we have $\omega_E(\zeta) \ge \delta/2$ for $\zeta \in \Gamma$. Let F be an outer function with the modulus

$$|F(e^{it})| = e^{-\omega_E(e^{it})} .$$

Taking into account that $|F| = 1$ on E' and that $F^{-1} \in H^\infty$ we have (for an arbitrary positive integer n)

$$dist_{L^\infty(E')}(\bar{\theta}, H^\infty) = \inf_{h \in H^\infty} \|\bar{\theta} + h\|_{L^\infty(E')} = \inf_{h \in H^\infty} \|F^n \bar{\theta} + F^n h\|_{L^\infty(E')}$$

$$= \inf_{h \in H^\infty} \|F^n \bar{\theta} + h\|_{L^\infty(E')} \le \inf_{h \in H^\infty} \|F^n \bar{\theta} + h\|_{L^\infty(\mathbb{T})} .$$

Now by the standard duality argument

$$\inf_{h \in H^\infty} \|F^n \bar{\theta} + h\|_\infty = \frac{1}{2\pi} \sup_{f \in H^1, \|f\| \le 1} \left(\left| \int_{\mathbb{T}} F^n \bar{\theta} f \, d\zeta \right| \right) .$$

An application of lemma II gives

$$\left| \int_{\mathbb{T}} F^n \bar{\theta} f \, d\zeta \right| = \left| \int_\Gamma \frac{F^n f}{\theta} d\zeta \right| .$$

But the last expression does not exceed

$$\int_\Gamma \frac{|F|^n |f|}{|\theta|} |d\zeta| \le \frac{1}{\varepsilon} \int_\Gamma e^{-n\omega_E(\zeta)} |f| |d\zeta| \le \frac{e^{-\frac{n\delta}{2}}}{\varepsilon} c(\varepsilon) \|f\|_1 .$$

By the choice of n the last expression can be made arbitrarily small, and (5.4) is proved. ●

6. Compactness of the inclusion map j and related questions. Proof of theorem 3.

In this section we shall be interested in the following question: for what symbols φ the operator $T_\varphi | K_\theta$ is compact? This question is closely related to the problem concerning the criterion of compactness for the operators $H_v^* H_u$ (see [18]). Indeed, the compactness of the operator $H^*_{\bar\varphi\bar\theta} H_{\bar\theta}$ is equivalent to the compactness of the operator $T_\varphi | K_\theta$. We know the answer only for the "almost" real symbols φ (see proposition 7 below). But for the proof of theorem 3 we need only positive symbols φ . Lemma 6 treats this subject.

Let $\mathbb{N}$ stand for the set of positive integers. Let $P_a(\zeta) \overset{def}{=} \frac{1-|a|^2}{|1-\bar a \zeta|^2}$, $a \in \mathbb{D}$, $\zeta \in \mathbb{T}$, denote the Poisson kernel and for $f \in L^1$, $\hat f(a) \overset{def}{=} \int_{\mathbb{T}} f P_a \, dm$.

LEMMA 6. If the function φ satisfies the inequality

$$dist(sign\, \varphi, H^\infty + C) < 1$$

then the following statements are equivalent:

a) $H^*_{\bar\varphi\bar\theta} H_{\bar\theta}$ is compact;

b) $T_\varphi | K_\theta$ is compact;

c) $\lim_{|\zeta| \to 1} \min(|\hat\varphi|(\zeta), 1 - |\theta(\zeta)|) = 0$;

d) $H^\infty[\bar\varphi\bar\theta] \cap H^\infty[\bar\theta] \subset H^\infty + C$.

PROOF. a)$\Leftrightarrow$b). It is obvious.

b)$\Rightarrow$c). Let $g \in H^\infty + C$ and suppose that $\|sign\varphi - g\|_\infty \le q < 1$. We denote $g_1 = \varphi + g$, $g_2 = \varphi - g$. Now take a sequence $(\zeta_n)_{n \ge 1}, \zeta_n \in \mathbb{D}$ such that $|\theta(\zeta_n)| \le \gamma < 1, |\zeta_n| \to 1$. Then it is clear that the functions $r_n \overset{def}{=} (1 - |\zeta_n|^2)^{1/2} k_\theta(\zeta_n, \zeta)$ converge weakly to zero and that $r_n \in K_\theta$. Therefore $\| P_+(g_1 + g_2) r_n \|_2 \to 0$, and so

$$\left| \|P_+ g_1 r_n\|_2 - \|P_+ g_2 r_n\|_2 \right| \to 0 .$$

Taking into account that $H_{g_1-g_2} = 2H_g$ is compact, we obtain that $\| P_-(g_1-g_2)r_n\|_2 \longrightarrow 0$ and so

$$\big|\, \| P_- g_1 r_n\|_2 - \| P_- g_2 r_n \|_2 \,\big| \longrightarrow 0 .$$

Thus, we have

$$\big|\, \|(\varphi+g) r_n\|_2^2 - \|(\varphi-g) r_n\|_2^2 \,\big| \longrightarrow 0 ;$$

after computing this expression we see that

$$\Big| \operatorname{Re} \int_{\mathbb{T}} \bar{\varphi} g P_{\zeta_n} \, dm \Big| \longrightarrow 0 .$$

Now

$$\Big| \operatorname{Re} \int_{\mathbb{T}} \bar{\varphi} g P_{\zeta_n} dm \Big| \geqslant \Big| \int_{\mathbb{T}} \bar{\varphi} \operatorname{sign} \varphi \, P_{\zeta_n} dm \Big| - \int_{\mathbb{T}} |\varphi| \cdot | g - \operatorname{sign} \varphi | P_{\zeta_n} dm \geqslant$$

$$\geqslant (1-q) \, |\widehat{\varphi}| (\zeta_n) .$$

Thus $|\widehat{\varphi}|(\zeta_n) \longrightarrow 0$.

c) $\Longrightarrow$ d). Here we shall apply the reasoning of [18]. Using the reasoning absolutely similar to that of lemma 5 [18] we may prove that for every pair of numbers m and n, $m, n \in \mathbb{N}$ the operator $H^*_{\bar{\varphi}^m \bar{\theta}^m} H_{\bar{\theta}^n}$ is compact. Thus for every h_1 and h_2 in H^∞ the operator $H^*_{\bar{\varphi}^m \bar{\theta}^m \bar{h}_1} H_{\bar{\theta}^n h_2} = T_{\bar{h}_1} H^*_{\bar{\varphi}^m \bar{\theta}^m} H_{\bar{\theta}^n} T_{h_2}$ is compact and therefore for every function $g \in H^\infty[\bar{\varphi}\bar{\theta}] \cap H^\infty[\bar{\theta}]$ the operator $H^*_g H_g$ is compact. It means that $g \in H^\infty + C$.

d) $\Longrightarrow$ a). It was proved in [18]. ●

To prove theorem 3 it remains to show, that the conditions a)-c) of the theorem imply that the inclusion map j : $K_\theta \to L^2(w\,dm)$ is compact. But the compactness of j is equivalent to the compactness of two operators:

$$P_+ w^{1/2} \theta P_- \bar{\theta} | H^2, \quad P_- w^{1/2} \theta P_- \bar{\theta} | H^2 .$$

The compactness of the first operator is equivalent to the compactness of $H^*_{w^{1/2}\bar{\theta}} H_{\bar{\theta}}$ and the compactness of the second to the compactness of $H^*_{w^{1/2}} H_{\bar{\theta}}$. But if we take into account the condition c) of the theorem and use the reasoning of lemma 5 [18] we shall see that the operators $H^*_{w^{1/2}\bar{\theta}} H_{\bar{\theta}}$ and $H^*_{w^{1/2}} H_{\bar{\theta}}$ are compact.

Lemma 6 does not give any criterion of compactness of the

operators $T_\varphi | K_\theta$ even for a real function φ.

PROPOSITION 7. If there are numbers, δ, $\delta>0$, and k, $k\in\mathbb{N}$ satisfying

$$|\arg\varphi^k|\le\frac{\pi}{2}-\delta,$$

then the statements a)-d) of lemma 6 are equivalent.

PROOF. It is clear that the only implication that we need to prove is b)⇒c).

Let the sequence $(\zeta_n)_{n\ge1}$, $\xi_n\in\mathbb{D}$ satisfy $|\theta(\zeta_n)|\le\gamma<1$, $|\zeta_n|\to1$. For the functions $r_n \overset{def}{=\!=} (1-|\zeta_n|^2)^{1/2} R_\theta(\xi_n,\xi)$ we have

$$\|T_\varphi r_n\|_2 \xrightarrow[n\to\infty]{} 0. \tag{6.1}$$

Let $f_n \overset{def}{=\!=} \varphi(\zeta)(1-\overline{\theta}(\zeta_n)\theta(\zeta))$. Now we compute $T_\varphi r_n$:

$$(T_\varphi r_n)(\lambda)=(1-|\zeta_n|^2)^{1/2}\Big(P_+\frac{f_n(\zeta)}{1-\overline{\zeta}_n\zeta}\Big)(\lambda)=$$

$$=\frac{(1-|\zeta_n|^2)^{1/2}}{2(1-\overline{\zeta}_n\lambda)}\int_{\mathbb{T}} f_n(\zeta)\big(S_\lambda(\zeta)+\overline{S}_{\zeta_n}(\zeta)\big)\,dm(\zeta), \tag{6.2}$$

where S_λ is the Schwartz kernel, $S_\lambda(\zeta)=\frac{\zeta+\lambda}{\zeta-\lambda}$, $\zeta\in\mathbb{T}$, $\lambda\in\mathbb{D}$.
Let $F_n(\lambda) \overset{def}{=\!=} \int_{\mathbb{T}} f_n(\zeta) S_\lambda(\zeta)\,dm$,

$$F_n^*(\lambda) \overset{def}{=\!=} \int_{\mathbb{T}} \overline{f}_n(\zeta) S_\lambda(\zeta)\,dm.$$

Taking into account (6.1) and (6.2), we see that

$$\mathcal{J}_n \overset{def}{=\!=} \int_{\mathbb{T}} |F_n(\lambda)+\overline{F}_n^*(\zeta_n)|^2 P_{\zeta_n}(\lambda)\,dm(\lambda) \xrightarrow[n\to\infty]{} 0.$$

By the Hölder inequality it is obvious that

$$\lim_{n\to\infty}|F_n(\zeta_n)+\overline{F}_n^*(\zeta_n)|=0.$$

Thus we have

$$\lim_{n\to\infty}\int_{\mathbb{T}}\mathrm{Re}\, f_n\cdot P_{\zeta_n}\,dm=\lim_{n\to\infty}\mathrm{Re}\int_{\mathbb{T}}\mathrm{Re}\, f_n S_{\zeta_n}\,dm=\frac{1}{2}\lim_{n\to\infty}\mathrm{Re}\big(F_n(\zeta_n)+F_n^*(\zeta_n)\big)=0,$$

$$\lim_{n\to\infty}\int_{\mathbb{T}}\mathrm{Im}\, f_n\cdot P_{\zeta_n}\,dm=\lim_{n\to\infty}\mathrm{Re}\int_{\mathbb{T}}\mathrm{Im}\, f_n\cdot S_{\zeta_n}\,dm=\frac{1}{2}\lim_{n\to\infty}\mathrm{Im}\big(F_n(\zeta_n)-F_n^*(\zeta_n)\big)=0$$

and therefore

$$\lim_{n\to\infty}\int_{\mathbb{T}} f_n P_{\zeta_n}\,dm=0. \tag{6.4}$$

Now introduce the following notation : $a_n(\zeta)=\mathrm{Re}\, f_n(\zeta)$,

$b_n(\zeta) = \operatorname{Im} f_n(\zeta)$. If u is a function harmonic in $\mathbb{D}$ then $\tilde{u}$ denotes its harmonic conjugate, $\tilde{u}(0) = 0$. Using this notation we have

$$\mathcal{I}_n = \int_{\mathbb{T}} | a_n(\lambda) + i b_n(\lambda) + i(\tilde{a}_n(\lambda) + i\tilde{b}_n(\lambda)) +$$

$$+ a_n(\zeta_n) + i b_n(\zeta_n) - i(\tilde{a}_n(\zeta_n) + i\tilde{b}_n(\zeta_n)) |^2 P_{\zeta_n}(\lambda) dm(\lambda) .$$

Taking into account (6.4) and the fact that $\lim\limits_{n\to\infty} \mathcal{I}_n = 0$ we have

$$I_n \overset{def}{=} \int_{\mathbb{T}} | b_n(\lambda) + (\tilde{a}_n(\lambda) - \tilde{a}_n(\zeta_n)) |^2 P_{\zeta_n}(\lambda) dm(\lambda) \xrightarrow[n\to\infty]{} 0 .$$

Now we introduce an auxiliary function,

$$p_n(\lambda) \overset{def}{=} [a_n(\lambda) + i(\tilde{a}_n(\lambda) - \tilde{a}_n(\zeta_n))](1 - \bar{\theta}(\zeta_n)\theta(\lambda)) .$$

This function p_n belongs to every Hardy class H^p , $p<\infty$ in $\mathbb{D}$. Moreover, it is clear that for every K , $K>0$

$$\int_{\mathbb{T}} | p_n(\lambda) |^K P_{\zeta_n}(\lambda) dm(\lambda) \le C(K) \qquad (6.5)$$

where $C(K)$ does not depend on n.

Now the following chain of inequalities is obvious:

$$\int_{\mathbb{T}} | \bar{f}_n(\lambda)(1-\bar{\theta}(\zeta_n)\theta(\lambda)) - p_n(\lambda) |^2 P_{\zeta_n}(\lambda) dm(\lambda) \le$$

$$\| 1 - \bar{\theta}(\zeta_n)\theta(\lambda) \|_\infty^2 \cdot \int_{\mathbb{T}} | (a_n(\lambda) - i b_n(\lambda)) - (a_n(\lambda) + i(\tilde{a}_n(\lambda) - \tilde{a}_n(\zeta_n))) |^2 \cdot$$

$$\cdot P_{\zeta_n}(\lambda) dm(\lambda) \le 4 \cdot | I_n | \longrightarrow 0 .$$

But $\bar{f}_n(\lambda)(1-\bar{\theta}(\zeta_n)\theta(\lambda)) = \varphi(\lambda) | 1 - \bar{\theta}(\zeta_n)\theta(\lambda) |^2$. Now let the number K be the same as in the statement of proposition 7.

$$\left| \int_{\mathbb{T}} \varphi^k(\lambda) | 1-\bar{\theta}(\zeta_n)\theta(\lambda) |^{2k} P_{\zeta_n}(\lambda) dm \right| \le \left| \int_{\mathbb{T}} p_n^k(\lambda) P_{\zeta_n}(\lambda) dm \right| +$$

$$+ \int_{\mathbb{T}} | \varphi^k(\lambda) | 1 - \bar{\theta}(\zeta_n)\theta(\lambda) |^{2k} - p_n^k(\lambda) | P_{\zeta_n}(\lambda) dm = I_1 + I_2$$

$$I_1 = | p_n^k(\zeta_n) | = | a(\zeta_n)(1 - |\theta(\zeta_n)|^2) |^k \xrightarrow[n\to\infty]{} 0$$

, since $\lim\limits_{n\to\infty} a_n(\zeta_n) = 0$ (see (6.4)).

$$I_2 \leq \left(\int_{\mathbb{T}} \left| \bar{f}_n(\lambda)(1-\bar{\theta}(\zeta_n)\theta(\lambda)) - p_n(\lambda) \right|^2 P_{\zeta_n}(\lambda)\, dm \right)^{1/2} \cdot$$

$$\cdot \left(\int_{\mathbb{T}} \left| \sum_{j=0}^{k-1} \varphi^{k-1-j}(\lambda) \cdot \left|1-\theta(\zeta_n)\theta(\lambda)\right|^{2(k-1-j)} p_n^j(\lambda) \right| P_{\zeta_n}(\lambda)\, dm \right)^{1/2} .$$

Taking into account (6.5) and (6.6) we see that $I_2 = o(1)$. But $|\arg \varphi^k| \leq \frac{\pi}{2} - \delta$ and therefore $\operatorname{Re} \varphi^k \geqslant c|\varphi|^k$ for some constant c, $c > 0$.

Thus we conclude that

$$\int_{\mathbb{T}} |\varphi|^k P_{\zeta_n}\, dm \leq \frac{1}{c\,(1-\gamma)^{2k}} \left| \int_{\mathbb{T}} \varphi^k(\lambda) \cdot |1-\bar{\theta}(\zeta_n)\theta(\lambda)|^{2k} P_{\zeta_n}(\lambda)\, dm \right| \xrightarrow[n\to\infty]{} 0 .$$

So $\lim_{n\to\infty} |\hat{\varphi}|(\zeta_n) = 0$. ●

COROLLARY 1. If the function $\operatorname{sign}\varphi$ has a finite range, then the conditions a)-d) of lemma 6 are equivalent.

COROLLARY 2. If the function φ takes the values ± 1 only, then the operator $T_\varphi | K_\theta$ is not compact unless θ is a finite Blaschke product.

References

1. P.K o o s i s. Harmonic estimation in certain slit regions and a theorem of Beurling and Malliavin. Acta Math. 1979, 142, 275-304.
2. А.Л.В о л ь б е р г. Одновременная аппроксимация полиномами на окружности и внутри круга. Записки научн.сем.ЛОМИ, 1979, 92, 60-84.
3. А.Л.В о л ь б е р г. Полнота рациональных дробей в весовых L^p-пространствах на окружности. Функц.анализ и его приложения, 1980.
4. Н.И.А х и е з е р. Лекции по теории аппроксимации, ОГИЗ, Гостехиздат, 1947.
5. Г.Ц.Т у м а р к и н. Необходимые и достаточные условия для возможности приближения функции на окружности рациональными дробями, выраженные в терминах, непосредственно связанных с распределением полюсов аппроксимирующих дробей. Изв.АН СССР. Серия матем. 1966, 30, № 5, 969-980.
6. Г.Ц.Т у м а р к и н. Приближение функций рациональными дробями с заранее заданными полюсами. Доклады АН СССР, 1954, 98, № 6, 909-912.

7. Б.П.П а н е я х. О некоторых задачах гармонического анализа. Доклады АН СССР, 1962, 142, № 5, 1026-1029.

8. Б.П.П а н е я х. Некоторые неравенства для функций экспоненциального типа и априорные оценки для общих дифференциальных операторов, Успехи матем.наук, 1966, 21, № 3, 75-114.

9. В.Я.Л и н. Об эквивалентных нормах в пространстве суммируемых с квадратом целых функций экспоненциального типа, Матем.сб. 1965, 67(109), № 4, 586-608.

10. В.Н. Л о г в и н е н к о, Ю.Ф. С е р е д а. Эквивалентные нормы в пространстве целых функций экспоненциального типа. Теория функций, функциональный анализ и их приложения. Вып.19. Республиканский научный сборник, Харьков. 1973.

11. В.Э. К а ц н е л ь с о н. Эквивалентные нормы в пространстве функций экспоненциального типа. Матем.сб. 1973, 92(134), № 1.

12. D.N.C l a r k. One dimensional perturbations of restricted shift. J.anal.math., 1972, 25, 169-191.

13. Г.М.Г о л у з и н. Геометрическая теория функций комплексного переменного. Москва, Наука, 1966.

14. S.M a n d e l b r o j t. Séries adhérentes. Régularisation des suites. Applications.Paris. 1952.

15. S.E.W a r s h a w s k i. On conformal mapping of infinite strips. Trans.Amer.Math.Soc. 1942, 51.

16. L.C a r l e s o n. The corona problem. Lect.Notes in Math.118, Springer-Verlag, 1972.

17. S.-Y. C h a n g, J. G a r n e t t. Analyticity of functions and subalgebras of L^∞ containing H^∞. Proc.Amer.Math.Soc. 1978, 72, N 1, 41-46.

18. S. A x l e r, S.-Y. C h a n g, D. S a r a s o n. Product of Toeplitz operators. Integr.equat.and operator theory. 1978, 1, N 3, 285-309.

ɔl. 700: Module Theory, Proceedings, 1977. Edited by C. Faith and Wiegand. X, 239 pages. 1979.

ɔl. 701: Functional Analysis Methods in Numerical Analysis, Pro- ɘedings, 1977. Edited by M. Zuhair Nashed. VII, 333 pages. 1979.

ɔl. 702: Yuri N. Bibikov, Local Theory of Nonlinear Analytic Ordinary ifferential Equations. IX, 147 pages. 1979.

ɔl. 703: Equadiff IV, Proceedings, 1977. Edited by J. Fábera. XIX, 41 pages. 1979.

ɔl. 704: Computing Methods in Applied Sciences and Engineering, 77, I. Proceedings, 1977. Edited by R. Glowinski and J. L. Lions. , 391 pages. 1979.

ɔl. 705: O. Forster und K. Knorr, Konstruktion verseller Familien ɔmpakter komplexer Räume. VII, 141 Seiten. 1979.

ɔl. 706: Probability Measures on Groups, Proceedings, 1978. dited by H. Heyer. XIII, 348 pages. 1979.

ɔl. 707: R. Zielke, Discontinuous Čebyšev Systems. VI, 111 pages. 79.

ɔl. 708: J. P. Jouanolou, Equations de Pfaff algébriques. V, 255 ges. 1979.

ɔl. 709: Probability in Banach Spaces II. Proceedings, 1978. Edited A. Beck. V, 205 pages. 1979.

l. 710: Séminaire Bourbaki vol. 1977/78, Exposés 507–524. IV, 8 pages. 1979.

. 711: Asymptotic Analysis. Edited by F. Verhulst. V, 240 pages. 79.

. 712: Equations Différentielles et Systèmes de Pfaff dans le amp Complexe. Edité par R. Gérard et J.-P. Ramis. V, 364 pages. 79.

. 713: Séminaire de Théorie du Potentiel, Paris No. 4. Edité par lirsch et G. Mokobodzki. VII, 281 pages. 1979.

. 714: J. Jacod, Calcul Stochastique et Problèmes de Martingales. 539 pages. 1979.

. 715: Inder Bir S. Passi, Group Rings and Their Augmentation als. VI, 137 pages. 1979.

. 716: M. A. Scheunert, The Theory of Lie Superalgebras. X, 271 es. 1979.

. 717: Grosser, Bidualräume und Vervollständigungen von Ba- hmoduln. III, 209 pages. 1979.

718: J. Ferrante and C. W. Rackoff, The Computational Com- xity of Logical Theories. X, 243 pages. 1979.

719: Categorial Topology, Proceedings, 1978. Edited by H. rlich and G. Preuß. XII, 420 pages. 1979.

720: E. Dubinsky, The Structure of Nuclear Fréchet Spaces. 97 pages. 1979.

721: Séminaire de Probabilités XIII. Proceedings, Strasbourg, 7/78. Edité par C. Dellacherie, P. A. Meyer et M. Weil. VII, 647 es. 1979.

722: Topology of Low-Dimensional Manifolds. Proceedings, 7. Edited by R. Fenn. VI, 154 pages. 1979.

723: W. Brandal, Commutative Rings whose Finitely Generated dules Decompose. II, 116 pages. 1979.

724: D. Griffeath, Additive and Cancellative Interacting Particle tems. V, 108 pages. 1979.

725: Algèbres d'Opérateurs. Proceedings, 1978. Edité par e la Harpe. VII, 309 pages. 1979.

726: Y.-C. Wong, Schwartz Spaces, Nuclear Spaces and Tensor lucts. VI, 418 pages. 1979.

727: Y. Saito, Spectral Representations for Schrödinger Opera- With Long-Range Potentials. V, 149 pages. 1979.

728: Non-Commutative Harmonic Analysis. Proceedings, 1978. ed by J. Carmona and M. Vergne. V, 244 pages. 1979.

Vol. 729: Ergodic Theory. Proceedings, 1978. Edited by M. Denker and K. Jacobs. XII, 209 pages. 1979.

Vol. 730: Functional Differential Equations and Approximation of Fixed Points. Proceedings, 1978. Edited by H.-O. Peitgen and H.-O. Walther. XV, 503 pages. 1979.

Vol. 731: Y. Nakagami and M. Takesaki, Duality for Crossed Products of von Neumann Algebras. IX, 139 pages. 1979.

Vol. 732: Algebraic Geometry. Proceedings, 1978. Edited by K. Lønsted. IV, 658 pages. 1979.

Vol. 733: F. Bloom, Modern Differential Geometric Techniques in the Theory of Continuous Distributions of Dislocations. XII, 206 pages. 1979.

Vol. 734: Ring Theory, Waterloo, 1978. Proceedings, 1978. Edited by D. Handelman and J. Lawrence. XI, 352 pages. 1979.

Vol. 735: B. Aupetit, Propriétés Spectrales des Algèbres de Banach. XII, 192 pages. 1979.

Vol. 736: E. Behrends, M-Structure and the Banach-Stone Theorem. X, 217 pages. 1979.

Vol. 737: Volterra Equations. Proceedings 1978. Edited by S.-O. Londen and O. J. Staffans. VIII, 314 pages. 1979.

Vol. 738: P. E. Conner, Differentiable Periodic Maps. 2nd edition, IV, 181 pages. 1979.

Vol. 739: Analyse Harmonique sur les Groupes de Lie II. Proceedings, 1976–78. Edited by P. Eymard et al. VI, 646 pages. 1979.

Vol. 740: Séminaire d'Algèbre Paul Dubreil. Proceedings, 1977–78. Edited by M.-P. Malliavin. V, 456 pages. 1979.

Vol. 741: Algebraic Topology, Waterloo 1978. Proceedings. Edited by P. Hoffman and V. Snaith. XI, 655 pages. 1979.

Vol. 742: K. Clancey, Seminormal Operators. VII, 125 pages. 1979.

Vol. 743: Romanian-Finnish Seminar on Complex Analysis. Proceedings, 1976. Edited by C. Andreian Cazacu et al. XVI, 713 pages. 1979.

Vol. 744: I. Reiner and K. W. Roggenkamp, Integral Representations. VIII, 275 pages. 1979.

Vol. 745: D. K. Haley, Equational Compactness in Rings. III, 167 pages. 1979.

Vol. 746: P. Hoffman, τ-Rings and Wreath Product Representations. V, 148 pages. 1979.

Vol. 747: Complex Analysis, Joensuu 1978. Proceedings, 1978. Edited by I. Laine, O. Lehto and T. Sorvali. XV, 450 pages. 1979.

Vol. 748: Combinatorial Mathematics VI. Proceedings, 1978. Edited by A. F. Horadam and W. D. Wallis. IX, 206 pages. 1979.

Vol. 749: V. Girault and P.-A. Raviart, Finite Element Approximation of the Navier-Stokes Equations. VII, 200 pages. 1979.

Vol. 750: J. C. Jantzen, Moduln mit einem höchsten Gewicht. III, 195 Seiten. 1979.

Vol. 751: Number Theory, Carbondale 1979. Proceedings. Edited by M. B. Nathanson. V, 342 pages. 1979.

Vol. 752: M. Barr, *-Autonomous Categories. VI, 140 pages. 1979.

Vol. 753: Applications of Sheaves. Proceedings, 1977. Edited by M. Fourman, C. Mulvey and D. Scott. XIV, 779 pages. 1979.

Vol. 754: O. A. Laudal, Formal Moduli of Algebraic Structures. III, 161 pages. 1979.

Vol. 755: Global Analysis. Proceedings, 1978. Edited by M. Grmela and J. E. Marsden. VII, 377 pages. 1979.

Vol. 756: H. O. Cordes, Elliptic Pseudo-Differential Operators – An Abstract Theory. IX, 331 pages. 1979.

Vol. 757: Smoothing Techniques for Curve Estimation. Proceedings, 1979. Edited by Th. Gasser and M. Rosenblatt. V, 245 pages. 1979.

Vol. 758: C. Năstăsescu and F. Van Oystaeyen; Graded and Filtered Rings and Modules. X, 148 pages. 1979.

Vol. 759: R. L. Epstein, Degrees of Unsolvability: Structure and Theory. XIV, 216 pages. 1979.

Vol. 760: H.-O. Georgii, Canonical Gibbs Measures. VIII, 190 pages. 1979.

Vol. 761: K. Johannson, Homotopy Equivalences of 3-Manifolds with Boundaries. 2, 303 pages. 1979.

Vol. 762: D. H. Sattinger, Group Theoretic Methods in Bifurcation Theory. V, 241 pages. 1979.

Vol. 763: Algebraic Topology, Aarhus 1978. Proceedings, 1978. Edited by J. L. Dupont and H. Madsen. VI, 695 pages. 1979.

Vol. 764: B. Srinivasan, Representations of Finite Chevalley Groups. XI, 177 pages. 1979.

Vol. 765: Padé Approximation and its Applications. Proceedings, 1979. Edited by L. Wuytack. VI, 392 pages. 1979.

Vol. 766: T. tom Dieck, Transformation Groups and Representation Theory. VIII, 309 pages. 1979.

Vol. 767: M. Namba, Families of Meromorphic Functions on Compact Riemann Surfaces. XII, 284 pages. 1979.

Vol. 768: R. S. Doran and J. Wichmann, Approximate Identities and Factorization in Banach Modules. X, 305 pages. 1979.

Vol. 769: J. Flum, M. Ziegler, Topological Model Theory. X, 151 pages. 1980.

Vol. 770: Séminaire Bourbaki vol. 1978/79 Exposés 525–542. IV, 341 pages. 1980.

Vol. 771: Approximation Methods for Navier-Stokes Problems. Proceedings, 1979. Edited by R. Rautmann. XVI, 581 pages. 1980.

Vol. 772: J. P. Levine, Algebraic Structure of Knot Modules. XI, 104 pages. 1980.

Vol. 773: Numerical Analysis. Proceedings, 1979. Edited by G. A. Watson. X, 184 pages. 1980.

Vol. 774: R. Azencott, Y. Guivarc'h, R. F. Gundy, Ecole d'Eté de Probabilités de Saint-Flour VIII-1978. Edited by P. L. Hennequin. XIII, 334 pages. 1980.

Vol. 775: Geometric Methods in Mathematical Physics. Proceedings, 1979. Edited by G. Kaiser and J. E. Marsden. VII, 257 pages. 1980.

Vol. 776: B. Gross, Arithmetic on Elliptic Curves with Complex Multiplication. V, 95 pages. 1980.

Vol. 777: Séminaire sur les Singularités des Surfaces. Proceedings, 1976-1977. Edited by M. Demazure, H. Pinkham and B. Teissier. IX, 339 pages. 1980.

Vol. 778: SK1 von Schiefkörpern. Proceedings, 1976. Edited by P. Draxl and M. Kneser. II, 124 pages. 1980.

Vol. 779: Euclidean Harmonic Analysis. Proceedings, 1979. Edited by J. J. Benedetto. III, 177 pages. 1980.

Vol. 780: L. Schwartz, Semi-Martingales sur des Variétés, et Martingales Conformes sur des Variétés Analytiques Complexes. XV, 132 pages. 1980.

Vol. 781: Harmonic Analysis Iraklion 1978. Proceedings 1978. Edited by N. Petridis, S. K. Pichorides and N. Varopoulos. V, 213 pages. 1980.

Vol. 782: Bifurcation and Nonlinear Eigenvalue Problems. Proceedings, 1978. Edited by C. Bardos, J. M. Lasry and M. Schatzman. VIII, 296 pages. 1980.

Vol. 783: A. Dinghas, Wertverteilung meromorpher Funktionen in ein- und mehrfach zusammenhängenden Gebieten. Edited by R. Nevanlinna and C. Andreian Cazacu. XIII, 145 pages. 1980.

Vol. 784: Séminaire de Probabilités XIV. Proceedings, 1978/79. Edited by J. Azéma and M. Yor. VIII, 546 pages. 1980.

Vol. 785: W. M. Schmidt, Diophantine Approximation. X, 299 pages. 1980.

Vol. 786: I. J. Maddox, Infinite Matrices of Operators. V, 122 pages. 1980.

Vol. 787: Potential Theory, Copenhagen 1979. Proceedings, 1979. Edited by C. Berg, G. Forst and B. Fuglede. VIII, 319 pages. 1980.

Vol. 788: Topology Symposium, Siegen 1979. Proceedings, 1979. Edited by U. Koschorke and W. D. Neumann. VIII, 495 pages. 1980.

Vol. 789: J. E. Humphreys, Arithmetic Groups. VII, 158 pages. 1980.

Vol. 790: W. Dicks, Groups, Trees and Projective Modules. IX, 12[illegible] pages. 1980.

Vol. 791: K. W. Bauer and S. Ruscheweyh, Differential Operators for Partial Differential Equations and Function Theoretic Applications. V, 258 pages. 1980.

Vol. 792: Geometry and Differential Geometry. Proceedings, 1979. Edited by R. Artzy and I. Vaisman. VI, 443 pages. 1980.

Vol. 793: J. Renault, A Groupoid Approach to C*-Algebras. III, 16[illegible] pages. 1980.

Vol. 794: Measure Theory, Oberwolfach 1979. Proceedings 197[illegible] Edited by D. Kölzow. XV, 573 pages. 1980.

Vol. 795: Séminaire d'Algèbre Paul Dubreil et Marie-Paule Malliavi[illegible] Proceedings 1979. Edited by M. P. Malliavin. V, 433 pages. 198[illegible]

Vol. 796: C. Constantinescu, Duality in Measure Theory. IV, 19[illegible] pages. 1980.

Vol. 797: S. Mäki, The Determination of Units in Real Cyclic Sext[illegible] Fields. III, 198 pages. 1980.

Vol. 798: Analytic Functions, Kozubnik 1979. Proceedings. Edite[illegible] by J. Ławrynowicz. X, 476 pages. 1980.

Vol. 799: Functional Differential Equations and Bifurcation. Procee[illegible]ings 1979. Edited by A. F. Izé. XXII, 409 pages. 1980.

Vol. 800: M.-F. Vignéras, Arithmétique des Algèbres de Quaternion[illegible] VII, 169 pages. 1980.

Vol. 801: K. Floret, Weakly Compact Sets. VII, 123 pages. 198[illegible]

Vol. 802: J. Bair, R. Fourneau, Etude Géometrique des Espac[illegible] Vectoriels II. VII, 283 pages. 1980.

Vol. 803: F.-Y. Maeda, Dirichlet Integrals on Harmonic Spaces. 180 pages. 1980.

Vol. 804: M. Matsuda, First Order Algebraic Differential Equatio[illegible] VII, 111 pages. 1980.

Vol. 805: O. Kowalski, Generalized Symmetric Spaces. XII, 1[illegible] pages. 1980.

Vol. 806: Burnside Groups. Proceedings, 1977. Edited by J. L. Me[illegible]nicke. V, 274 pages. 1980.

Vol. 807: Fonctions de Plusieurs Variables Complexes IV. P[illegible]ceedings, 1979. Edited by F. Norguet. IX, 198 pages. 1980.

Vol. 808: G. Maury et J. Raynaud, Ordres Maximaux au Sens [illegible] K. Asano. VIII, 192 pages. 1980.

Vol. 809: I. Gumowski and Ch. Mira, Recurences and Discr[illegible] Dynamic Systems. VI, 272 pages. 1980.

Vol. 810: Geometrical Approaches to Differential Equations. P[illegible]ceedings 1979. Edited by R. Martini. VII, 339 pages. 1980.

Vol. 811: D. Normann, Recursion on the Countable Functiona[illegible] VIII, 191 pages. 1980.

Vol. 812: Y. Namikawa, Toroidal Compactification of Siegel Spac[illegible] VIII, 162 pages. 1980.

Vol. 813: A. Campillo, Algebroid Curves in Positive Characteris[illegible] V, 168 pages. 1980.

Vol. 814: Séminaire de Théorie du Potentiel, Paris, No. 5. Proce[illegible]ings. Edited by F. Hirsch et G. Mokobodzki. IV, 239 pages. 19[illegible]

Vol. 815: P. J. Slodowy, Simple Singularities and Simple Algebr[illegible] Groups. XI, 175 pages. 1980.

Vol. 816: L. Stoica, Local Operators and Markov Processes. [illegible] 104 pages. 1980.